Inelastic Behaviour of Plates and Shells

International Union of Theoretical and Applied Mechanics

Inelastic Behaviour of Plates and Shells

IUTAM Symposium, Rio de Janeiro, Brazil
August 5-9, 1985

Editors
L. Bevilacqua, R. Feijóo and R. Valid

Springer-Verlag
Berlin Heidelberg NewYork
London Paris Tokyo

Dr.Luiz Bevilacqua
Dr. Raúl Feijóo
Dr. Roger Valid

Pontifícia Universidade Católica
do Rio de Janeiro PUC-RJ
Rua Marquès de Sao Vicente 225
Gávea
Rio de Janeiro – Brazil

ISBN 3-540-16366-2 Springer-Verlag Berlin Heidelberg New York
ISBN 0-387-16366-2 Springer-Verlag New York Berlin Heidelberg

Library of Congress Cataloging in Publication Data.

Library of Congress Cataloging-in-Publication Data
Inelastic behaviour of plates and shells.
(IUTAM-Symposien)
Festschrift in memory of professor Antoni Sawczuk.
Cosponsored by the International Union of Theoretical and Applied Mechanics
and the International Council of Scientific Unions.
1. Plates (Engineering)--Congresses. 2. Shells (Engineering)--Congresses. 3. Sawczuk, A. (Antoni)
I. Bevilacqua, L. (Luiz), 1937 –
II. Feijóo, R. (Raul)
III. Valid, Roger.
IV. Sawczuk, A. (Antoni)
V. International Union of Theoretical and Applied Mechanics.
VI. International Council of Scientific Unions.
VII. Series.
TA660.P6I44 1986 624.1'776 86-17760
ISBN 0-387-16366-2 (U.S.)

Offsetprinting: Mercedes-Druck, Berlin
Bookbinding: Lüderitz & Bauer, Berlin
2161/3020-543210

In memory of
Professor Antoni Sawczuk

Professor Antoni Sawczuk
(1927–1984)

Preface

During the last ten years a considerable volume of information has been accumulated regarding the inelastic behaviour of materials. The increasing number of communications published in specialised journals and also the frequency of meetings in these fields, indicates a considerable research effort aimed at such topics as plasticity, creep, fatigue, visco-plasticity and the like.

This fact encouraged a group of Brazilian researchers, stimulated enthusiastically by Professor P. Germain, to submit a proposal for a Symposium on the "Inelastic Behaviour of Plates and Shells" to the General Assembly of IUTAM. Brazil had recently joined IUTAM and the Brazilian Association of Mechanical Sciences was eager to host an IUTAM meeting.

In the selection of the subject, it was taken into account, besides a promising number of original contributions, the interest to be raised amongst the Brazilian researchers and engineers, in order to maximise the participation of the host country.

The recent steps taken in this country towards the development of the aero-space industry, the construction of nuclear power plants and the off-shore exploration of petroleum have required an intensification of research activities in several fields, structural behaviour of plates and shells being one of the most important. Therefore, the suggested theme would attract the interest of a significant group of Brazilian researchers and engineers and match the necessity for exchanging experience among leading scientists working in those fields.

The organization of the scientific programme owes a great deal of its success to Professor A. Sawczuk, who was not spared the best of his enthusiasm and efforts to prepare a list of topics and potential contributors in cooperation with the other members of the Scientific Committee. Unfortunately, he did not live to see the success of the meeting which he so carefully helped to organise. Then Professor R. Valid accepted to replace him as Co-chairman.

ation. In Grenoble he received a doctorate honoris causa at the Polytechnic Institute. In 1983 he was elected Resident Rector of the International Centre of Mechanical Sciences (CISM) in Udine, Italy, where he worked until his sudden death in 1984. Professor Sawczuk organized numerous scientific meetings, international symposia and conferences. Amongst others he was one of the co-organizers of the SMiRT- conferences devoted to structural mechanics in reactor technology. Until his death he was a member of the General Assembly of IUTAM and a member of IASS and RILEM. Professor Sawczuk was, moreover, a member of editorial Committees or Advisory Boards of a number of esteemed scientific journals, the International Journal of Mechanical Sciences, Archives of Mechanics, Journal de Mécanique Théorique et Appliquée and the Journal of Structural Mechanics and Engineering Transactions.

Professor Sawczuk's contribution to mechanical sciences is outstanding. He published over one hundred scientific papers and several monographs, from which the first published in 1963 'Tragfähigkeits-Theorie der Platten' written with the late Professor Th. Jaeger still remains the fundamental book in the theory of the load-carrying capacity of plates.

One of his main fields of scientific activity were problems of structural plasticity. Over half his publications were devoted to these problems. In this field he educated numerous young scientists who developed his ideas and his work. His scientific results have been widely referred to in the world literature of the subject.

His other field of scientific activity concerned problems related to the fundamentals of the mechanics of continuous media, especially problems dealing with the mechanics of anisotropic continua. Also in this field he achieved significant results among others concerning new concepts in the description of the strain hardening of materials in which he again educated numerous young people.

The decease of Professor Antoni Sawczuk is a great loss to the entire scientific community in the field of mechanics.

Professor W. Szczepiński
Polish Academy of Sciences

Short Biography

ANTONI SAWCZUK
1927-1984

Professor Antoni Sawczuk, a long-time member of the Editorial Board of the International Journal of Mechanical Sciences, died at the age of 57 on 27 May 1984 in Grenoble, France. Born in 1927 at Komarno in Poland, he graduated from the Warsaw Technical University in 1951 and started his work as a designer in industry, which he continued until 1954. At the same time he started his scientific activities in the Department of Civil Engineering at the Warsaw Technical University teaching the Strength of Materials. Under the guidance of Professor Waclaw Olszak he received his Ph.D. degree there in 1953 and habilitation in 1960.

In 1961 he began to work at the Institute of Fundamental Technological Research. In 1969 after the retirement of Professor Olszak he became the Head of the Department of Continuous Media at the Institute. Professor Sawczuk held this post until 1981. In 1969 he was elected a corresponding member of the Polish Academy of Sciences and in 1983 he received the full membership of the Academy. For a long time until 1984 Professor Sawczuk served as the President of the Committee of Mechanics of the Polish Academy of Sciences.

Professor Sawczuk had close ties with the scientific community across the world, which started with his fellowship in the academic year 1958-1959 at the Illinois Institute of Technology and at Brown University, Providence, where he worked with Professors P.G. Hodge and W. Prager. His cooperation with the Illinois Institute of Technology continued when he was a visiting professor in the academic year 1964-1965. He was appointed a visiting professor also at Grenoble University, France, where he organized a group working with him in the mechanics of plastic deform-

We would like to thank the members of the Scientific Committee for their advice and help. We are especially indebted to Professors P. Germain and W. Koiter, who encouraged this country to join IUTAM and afterwards to organise the Symposium.

We would like to express our gratitude to IUTAM and ICSU for co-sponsoring the event and also to several Brazilian agencies and industries, FINEP, CNPq and Companhia Vale do Rio Doce, amongst others, for their support. The cooperation of the cultural attachés representing the different countries whose delegations attended the Symposium, were also of great help. To the Catholic University of Rio de Janeiro(PUC/RJ) who kindly hosted the event, to the members of the Local Organising Committee and to Ms. Maggie Corson and Anna Ottino, who performed all the secretarial tasks, we acknowledge our deepest gratitude.

A number of the papers have been retyped in order to give a more uniform presentation. This required additional support which was kindly given by Xerox do Brasil and PROMON Engenharia.

Finally, we would like to thank the editorial staff of Springer-Verlag for their advice and effort in publishing these proceedings and Pergamon Press for authorising the publication of Professor Sawczuk's biography.

L. Bevilacqua, R. Feijóo and R. Valid
December 1985

Scientific Committee

L. BEVILACQUA (Chairman)	- Rio de Janeiro, Brazil
G.D. GALLETLY	- Liverpool, UK
P. GERMAIN	- Paris, France
S. KALITSZKY	- Budapest, Hungary
W.T. KOITER	- Delft, The Netherlands
G. MAIER	- Milan, Italy
P.M. NAGHDI	- Berkeley, USA
A. SAWCZUK*	- Udine, Italy
R. VALID (Co-Chairman)	- Chatillon-sous-Bagneux, France
Y. YAMOMOTO	- Tokyo, Japan

*Professor Sawczuk gave his best cooperation to the organisation of this Symposium until May 1984 when, unfortunately, after a brief stay in hospital, he passed away.

Local Organizing Committee

L. BEVILACQUA	- Pontifícia Universidade Católica, Rio de Janeiro
A. BLASS	- Universidade Federal de Santa Catarina, Florianopolis
R. FEIJÓO	- Laboratório Nacional de Computação, Rio de Janeiro
L.C. MARTINS	- Universidade Federal do Rio de Janeiro
R. SAMPAIO	- Pontifícia Universidade Católica, Rio de Janeiro
E. TAROCO	- Laboratório Nacional de Computação, Rio de Janeiro

List of Participants

AKIYAMA, H.	University of Tokyo, Japan.
AL QURESHI, H.	ITA-IEMT-CTA, São José dos Campos Brazil.
DE ALMEIDA, M.C.	Universidade Federal de Paraná, Brazil.
ALTMAN, W.	ITA-CTA, São José dos Campos, Brazil.
ALVES, D.B.	Universidade Federal de Santa Catarina, Brazil.
DE ANDRADE, S.A.L.	Pontifícia Universidade Católica do Rio de Janeiro, Brazil.
BALTOV, A.	Bulgarian Academy of Sciences, Bulgaria.
BANICHUK, N.	USSR Academy of Sciences, USSR.
BARBOSA, H.J.C.	Laboratório Nacional de Computação Científica, Rio de Janeiro, Brazil.
BARCELLOS, C.S.	Universidade Federal de Santa Catarina, Brazil.
BASOMBRIO, F.G.	Centro Atomico de Bariloche, Argentina.
BATISTA, R.	COPPE/Universidade Federal do Rio de Janeiro, Brazil.
BEVILACQUA, L.	Pontífica Universidade Católica do Rio de Janeiro, Brazil.
BOLOTIN, V.V.	USSR Academy of Sciences, USSR.
BRILLA, J.	Comenius University, Mlynská Dolina Czechoslovakia.
CALLADINE, C.R.	University of Cambridge, UK.
DE CASTRO, J.T.P.	Pontifícia Universidade Católica do Rio de Janeiro, Brazil.
COMBESCURE, A.	Centre d'Etudes Mecaniques et Thermiques, Gif-sur-Yvette, France.
CONCI, A.	Pontifícia Universidade Católica do Rio de Janeiro, Brazil.
COSTA MATTOS, H.	COPPE/Universidade Federal do Rio de Janeiro, Brazil.
CREUS, G.J.	Universidade Federal do Rio Grande do Sul, Porto Alegre, Brazil.

CYRAS, A.	Vilnius University, USSR.
DAVET, J.	Laboratoire de Mecanique des Structures, Grande Voie des Vignes, France.
DESTUYNDER, P.	Ecole Centrale des Arts e Manufactures, Grande Voi des Vignes,France.
DULACKSA, E.	Institute for Architectural Development, Budapest, Hungary.
DUMONT, N.A.	Pontifícia Universidade Católica do Rio de Janeiro, Brazil.
ESTEFEN, S.F.	COPPE/Universidade Federal do Rio de Janciro, Brazil.
FEIJÓO, R.A.	Laboratório Nacional de Computação Científico, Rio de Janeiro, Brazil.
FREIRE, J.	Pontifícia Universidade Católica do Rio de Janeiro, Brazil.
GALEÃO, A.C.	Laboratório Nacional de Computação Científica, Rio de Janeiro, Brazil.
DE GAMA, R.M.S.	COPPE/Universidade Federal do Rio de Janeiro, Brazil.
GATTASS, M.	Pontifícia Universidade Católica do Rio de Janeiro, Brazil.
GERMAIN, P.	Acadêmie des Sciences, Paris, France.
GHAVAMI, K.	Pontifícia Universidade Católica do Rio de Janeiro, Brazil.
GLOCKNER, P.G.	University of Calgary, Canada.
GODOY, L.A.	Universidad Nacional de Córdoba Argentina.
GONÇALVES, P.B.	Pontifícia Universidade Católica do Rio de Janeiro, Brazil.
GONÇALVES FILHO, O.J.A.	CNEN, Rio de Janeiro, Brazil.
GROEHS, A.G.	Universidade Federal do Rio Grande do Sul, Porto Alegre, Brazil.
GUERREIRO, J.N.C.	Laboratório Nacional de Computação Científica, Rio de Janeiro, Brazil.
HALBRITTER, A.	NUCLEN, NUCLEBRAS, Rio de Janeiro, Brazil.
HORVAY, G.	North Carolina State University, U.S.A.

IDELSHON, S.	INTEC, Santa Fe, Argentina.
IGUTI, F.	UNICAMP, Campinas, Brazil.
KOITER, W.T.	University of Delft, The Netherlands.
KUZNETSOV, E.	University of Illinois, Urbana, U.S.A.
LAMPI, L.H.	EMBRAER, São José dos Campos, Brazil.
LECKIE, F.A.	University of Illinois, Urbana, U.S.A.
LEMAITRE; J.	ENSET, Crachen, France.
LEON, J.	Universidad Simon Bolivar, Caracas Venezuela.
DE LIMA, F.S.	Instituto de Estudos Avançados, São José dos Campos, Brazil.
LITTLE, G.H.	University of Birmingham, UK.
LOPES, R.C.	Universidade do Amazonas, Manaus Brazil.
LUKASIEWICZ, S.	University of Calgary, Canada.
MAMIYA, E.M.	COPPE/Universidade Federal do Rio de Janeiro, Brazil.
MANG, H.A.	Technical University of Vienna, Austria.
MARTIN, J.B.	University of Cape Town, South Africa.
MARTINS, L.C.	COPPE/Universidade Federal do Rio de Janeiro, Brazil.
NITZCHE, F.	D.F. Vasconcellos, São José dos Campos, Brazil.
DE OLIVEIRA, A.M.	ITA-CTA, São José dos Campos, Brazil.
ONAT, E.T.	Yale University, U.S.A.
PAMPLONA, D.	Pontifícia Universidade Católica do Rio de Janeiro, Brazil.
PEREIRA, N.Z.	Pontifícia Universidade Católica do Rio de Janeiro, Brazil.
PFEIL, M.	Pontifícia Universidade Católica do Rio de Janeiro, Brazil.
PONTER, A.	University of Leicester, UK.
RANGEL, A.G.P.	IEAv, São José dos Campos, Brazil.

REID, S.R.	University of Manchester, UK.
RIERA, J.	Universidade Federal do Rio Grande do Sul, Porto Alegre, Brazil.
RIMROTT, F.P.J.	University of Toronto, Canada.
RIZZI, P.	ITA-IEA, São José dos Campos, Brazil.
ROCHINHA, F.A.	COPPE/Universidade Federal do Rio de Janeiro, Brazil.
ROSAS E SILVA, R.	Pontifícia Universidade Católica do Rio de Janeiro, Brazil.
ROZVANY, G.I.N.	Monash University, Australia.
RUAS DE BARROS, V.	Pontifícia Universidade Católica do Rio de Janeiro, Brazil.
SAMPAIO, R.	Pontifícia Universidade Católica do Rio de Janeiro, Brazil.
SAVE, M.	Faculté Polytechnique de Mons, Belgium.
SAVIO, H.R.C.	EMBRAER, São José dos Campos, Brazil.
SEGENREICH, S.A.	Pontifícia Universidade Católica do Rio de Janeiro, Brazil.
SILVA, M.A.G.	Universidade Tecnica de Lisboa, Portugal.
SOUZA, M.A.	Pontifícia Universidade Católica do Rio de Janeiro, Brazil.
STUCKENBRUCK, S.	Pontifícia Universidade Católica do do Rio de Janeiro, Brazil.
TAROCO, E.	Laboratório Nacional de Computação Científica, Rio de Janeiro, Brazil
TUPINAMBA, P.M.	CONSUB Consultoria e Projeto/SA, Rio de Janeiro, Brazil.
VALID, R.	ONERA, Chatillon-sous-Bagneux, France.
VANDEPITTE, D.	Rijksuniversiteit-Gent, Belgium.
VODENICHAROV, S.B.	Bulgarian Academy of Sciences, Sofia Bulgaria.
WUNDERLICH, W.	Ruhr Universitat Bochum, Federal Republic of Germany.

YUHARA, T. — Nagasaki Technical Institute, Japan.

ZAK, M. — Jet Propulsion Laboratory, California Institute of Technology, U.S.A.

ZYCZKOWSKI, M. — Technical University of Cracow, Poland.

Programme

Monday, August 5

Opening address: Luiz Bevilacqua, Chairman and José Paulo de Almeida e Albuquerque, Vice President for Academic Affairs, PUC/RJ.

Morning

(Chairman: F.A. Leckie)

A.R.S. Ponter

F.P.J. Rimrott

P. Destuynder

Afternoon

(Chairman: G.I. Rozvany)

C.R. Calladine

M.A. Souza

P.G. Glockner

Tuesday, August 6

Morning

(Chairman: P. Germain)

V.V. Bolotin

F.A. Leckie

M. Zak

E. Dulácska

Afternoon

(Chairman: M. Save)

G.Horvay

R. Valid

T. Yuhara

D. Vandepitte

J. Brilla

Wednesday, August 7

Morning

(Chairman: W.T. Koiter)

N. Banichuk

A. Cyras

G.I.N. Rozvany

M. Zyczkowski

K. Ghavami

Thursday, August 8

Morning

(Chairman N. Banichuk)

M. Save

A. Baltov

H.A. Mang

A. Combescure

Afternoon

(Chairman: F.P.J. Rimrott)

W. Wunderlich

N.Z. Pereira

Friday, August 9

Morning

(Chairman: J. Lemaitre)

S. Lukasiewicz

S.R.Reid

E.T. Onat

J.B. Martin

G.H. Little

Closing Address: Professores P. Germain, W.T. Koiter and L. Bevilacqua.

Contents

The Effects of Thermal Loading on the Deformation of Shell Structures

A.R.S. Ponter*, K.F. Carter* and S. Karadeniz+

*Department of Engineering, University of Leicester, U.K.
+Faculty of Engineering, University of Enciyes, Turkey.

Summary

The effect of severe cyclic thermal loading on the plastic and creep deformation of shell structures may be understood through the use of shakedown theory and its extension to time dependent material properties. The paper discusses methods of estimating shakedown limits, deformation in excess of shakedown, and steady state creep strain rates using solution technique based on this theory with the objective of forming a broad understanding of the material and structural phenomena involved. The incentive for this work arises from problems associated with the structural design of the U.K. Commercial Sodium Cooled Fast-Breeder Reactors.

1) Introduction

Sodium Cooled Fast-Breeder Reactors [1] retain high neutron flux densities within the core and efficient heat transfer from the core by using liquid sodium as a coolant. These advantages are offset by severe thermal loading conditions, particularly for components within and adjacent to the reactor core. A typical component is subjected to quite moderate dead loading (never greater than 0.25 of the plastic limit load) but is likely to be subjected to periodic severe temperature fluctuation when the reactor is shut-down. These transients can produce elastic stresses well in excess of the yield stress and the temperatures can locally reach 600°C, well into the creep range of the designated structural alloy, ASME type 316 Austenitic stainless steel, as creep occurs at temperatures in excess of 480°C. The principal design concerns are that components, most of which are thin shells, should not suffer unacceptable incremental strain growth or long term creep or fatigue failure. This paper describes a sequence of theoretical studies, with correlations with some experimental data, which were developed to provide an understanding of the relationship between material properties and structural deformations under these conditions. The basic theoretical tools are those of classical plasticity, the shakedown theorems [2,3] and their extension to time dependent material properties [4,5,6]. The behaviour in excess of the shakedown limit for strain hardening materials is understood by expanding the solution about the perfectly plastic shakedown limit. The rate of ratchet of a perfectly plastic structure in excess of shakedown may also be estimated in this way.

Section 2) discusses the use of the upper bound shakedown theorem as a basis for a finite element technique and the solution so obtained are

Inelastic Behaviour of Plates and Shells
IUTAM Symposium Rio de Janeiro 1985
Editors: L. Bevilacqua, R. Feijóo and R. Valid

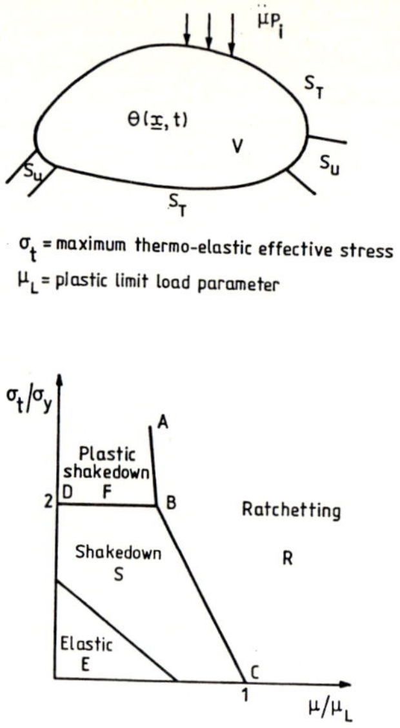

Fig.1 Schematic representation of general problem.

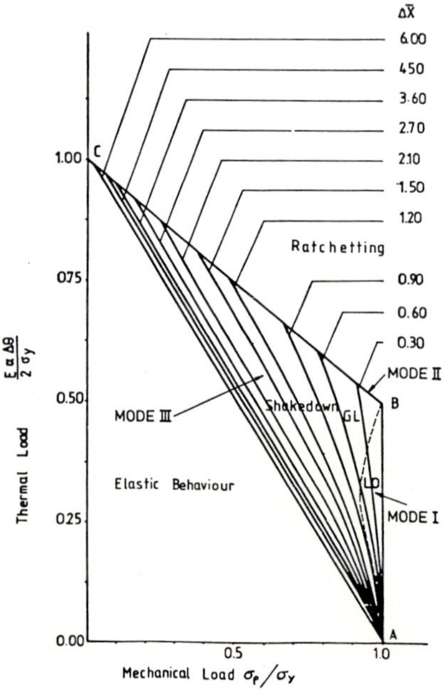

Fig.2 Interaction diagram for cylinder subject to axial stress σ_p and moving temperature discontinuity. Temperature independent yield stress.

obtained in the form of interaction diagrams as an aid to design and design code formulation. In section (3) the behaviour for loads in excess of the shakedown limit is estimated by expanding the perfectly plastic solution and a strain hardening solution about the shakedown solution. It is shown that the number of cycles required to reach an asymptotic state for a strain hardening material can be very large for loads in excess of shakedown.

In the final section (4) a method of estimating creep deformation rates is described, which is based on the idea of rapid cycle and slow cycle solutions. Comparison with experimental data suggests that the most significant feature in interpreting these solutions is an understanding of the load level in relation to the shakedown limit. The observation indicates that tests carried out in excess of the shakedown boundary are poor indicators of the validity for use in design of finite element techniques using complex constitutive relations.

2) The General Problem and Shakedown Theory

The general problem is shown schematically in Fig.1. A structure with volume V and surface S is subjected to constant loads μP_i where μ is a load parameter, and a cyclic history of temperature $\theta(x,t)$ which varies with spatial co-ordinates $\underline{x}$, and time t. For the elastic-perfectly plastic case with uniaxial yield stress σ_y the general features of the behaviour are shown in terms of a Bree diagram [7] with axis μ/μ_L where μ_L the limit value for uniform temperatures, and σ_t/σ_y, where σ_t is the maximum effective thermo-elastic stress in the body. For proportional thermal loading where

$$\theta = \theta_o + g(x)\, h(t) \Delta\theta \tag{1}$$

and

$$o \leq g(x) \leq 1 \; , \quad o \leq h(t) \leq 1 \tag{2}$$

then

$$\sigma_t = kE\alpha\Delta\theta \tag{3}$$

where E and α are the (assumed) temperature independent elastic modulus and linear coefficient of expansion, and k, which depends on Poissons ratio, lies in the range $o \leqslant k \leqslant 1$.

The diagram is subdivided into four regions: E where purely elastic behaviour occurs: S where shakedown occurs with elastic behaviour after some inelastic strain during the first few cycles; F where plastic strains occur over a confined volume of material but no incremental growth occurs, producing a reversed plasticity or plastic shakedown condition, and : R where incremental growth occurs during each cycle. Shakedown theory is concerned with the boundary CBD, whereas the ratchet boundary CBA provides the significant design limit, if the separate problem of fatigue is taken into account.

For most thin shell problems where the temperature variations through the thickness of the shell are small compared with variations along the shell surface, the region F is small or entirely absent and for small values of μ with increasing σ_t (i.e. $\Delta\theta$) the shell moves through the elastic and shakedown regions to incremental displacement growth in the R region. The question of the existence of the F region has been discussed by Karadeniz and Ponter [8] who have shown, using an extension of the shakedown theorems, that problems for which no F region exists can be characterised by the following property. For values of σ_t above the line BD there exists a volume of material where the thermo-elastic stress history cannot be contained within the yield surface by a rigid body translation in stress space. If this volume contains a mechanism of deformation on which the applied load does positive work, then no F region exists. For example, for axisymmetric shells if this condition occurs through the thickness of the shell a simple local extensional deformation mode would be activated by any axial load arising from P . For this reason, the shakedown limit is of particular significance for thin shells, and the transition from shakedown to ratchetting a significant event.

The shakedown boundary CBD can be estimated from above and below by the shakedown theorems [2,3]. The theory applies to a material where the elastic component of strain $\underline{\varepsilon}^e$ is given by

$$\underline{\varepsilon}^e = \underset{\sim}{C}\, \underline{\sigma} \tag{4}$$

where C is a temperature independent matrix of elastic constants. An increment of plastic strain $d\underline{\varepsilon}^P$ is given in terms of a yield function $f(\underline{\sigma}') - \sigma_y(\theta) = o$ by the associated flow rule

$$d\underline{\varepsilon}^P = d\lambda \frac{\partial f}{\partial \sigma'} \tag{5}$$

where σ is the deviatoric stress, $d\lambda$ a plastic multiplier and $\sigma_y(\theta)$ a temperature dependent uniaxial yield stress, and f a function which is homogeneous of degree unity in the deviatoric stress components. The total strain increment is then given by

$$d\underline{\varepsilon} = d\underline{\varepsilon}^e + d\underline{\varepsilon}^P + \alpha\underline{\delta} d\theta \tag{6}$$

where $\underline{\delta}$ is the Kroneker delta and α the (constant) linear coefficient of thermal expansion. Of these assumptions the temperature independence of C and α is the most questionable. We find that inclusion of strain hardening effects can be included to some extent by treating it as a perturbation about the perfectly plastic model.

The upper bound theorem may be expressed in the following form. A kinematically admissible history of plastic strain $\dot{\underline{\varepsilon}}^c$ satisfies the condition that the net increment of strain over the cycle

$$\Delta\underline{\varepsilon}^c = \int_o^{\Delta t} \dot{\underline{\varepsilon}}\, dt \tag{7}$$

shall be compatible with a displacement fields $\Delta\underline{u}^c$. The upper bound then becomes

$$\mu \int_S \underline{P} \Delta \underline{u}^c \, dS \leq \int_V \int_o^{\Delta t} (\underline{\sigma}^c - \hat{\underline{\sigma}}^\theta)\, \dot{\underline{\varepsilon}}^c \, dt\, dV \tag{8}$$

where $\underline{\sigma}^c$ is the stress on the yield surface corresponding to $\dot{\varepsilon}^P = \dot{\varepsilon}^c$ and $\hat{\sigma}^\theta(\underline{x},t)$ is the thermo-elastic stress history. The application of this theory to the solution of problems requires a specification of both $\Delta\underline{\varepsilon}^c$ and the entire history $\dot{\varepsilon}^c$. This difficulty may be avoided if a linearized yield surface is used (such as the Tresca yield condition in terms of principal stresses) consisting in hyerplanes;

$$\underline{N}^{i}\underline{\sigma}' = \sigma_{y}^{i} \quad , \quad i = 1...m \tag{9}$$

where σ_y^i is the yield surface corresponding to the i-th hyperplane in deviatoric stress space. If the corresponding plastic multipliers are denoted by λ_i then

$$\dot{\underline{\varepsilon}}^{c} = \underline{N}^{i}\dot{\lambda}_{i} \quad , \quad \Delta\underline{\varepsilon} = \underline{N}^{i}\lambda_{i} \tag{10}$$

where

$$\lambda_{i} = \int_{o}^{\Delta t} \dot{\lambda}_{i} \, dt \tag{11}$$

The volume integrand of the upper bound (8) may now be expressed as [9]

$$\int_{o}^{\Delta t} (\underline{\sigma}^{c} - \hat{\underline{\sigma}}^{\theta})\dot{\underline{\varepsilon}}^{c} \, dt = \Sigma C_{i}\lambda_{i} \tag{12}$$

where

$$C_{i} = \min_{o<t<\Delta t} \left\{ (\underline{\sigma}^{c}(t) - \hat{\underline{\sigma}}^{\theta}(t)) \, \underline{N}^{i} \right\} \quad \text{for } i = 1....m \tag{13}$$

The pointwise minimization process identifies the instant when the distance in stress space between $\hat{\underline{\sigma}}^{\theta}$ and the i-th hyperplane is a minimum. this minimization can be carried out at each point in the structure for each i, independently of the values of λ_i or their spatial variation. A finite element technique may then be devised by specifying finite element approximation to both the individual λ_i's and to the displacement field $\Delta\underline{u}$. The upper bound (8) becomes a linear function of nodal values by subjecting the displacement field to the constraint

$$\int_{S} \underline{P}.\Delta\underline{u} \, dS = 1. \tag{14}$$

The addition of equations which relate the nodal values of the λ_i and $\Delta\underline{u}$ yields a linear programming problem. There are two practical difficulties. It is necessary to find spatial variations of λ_i and $\Delta\varepsilon^c$ so that the relationship between them is consistantly satisfied throughout each element, and this cannot always be achieved. However, for axisymmetric shells and the Tresca yield condition such functions can be found. The second problem arises from the large number of variables involved (typically six λ_i and two components of $\Delta\underline{u}$ per node for an axisymmetric shell) which limits the size of problem which can be solved. Such a technique has been used successfully for cylindrical shells by Karadeniz and Ponter [7,10] and has been subsequently extended to axisymmetric shells, using simple linear displacement variation with curvatures concentrated on hinge circles at nodal points. By successive refinement of the element length in regions of rapid variation of $\hat{\underline{\sigma}}^{\theta}$ convergent values of μ can be calculated to three significant figures.

The results of such calculation are shown in Figs. (2), (3), (4) and (5). The first three figures refer to the problem of a long cylindrical shell of radius R and thickness h subjected to an axial stress

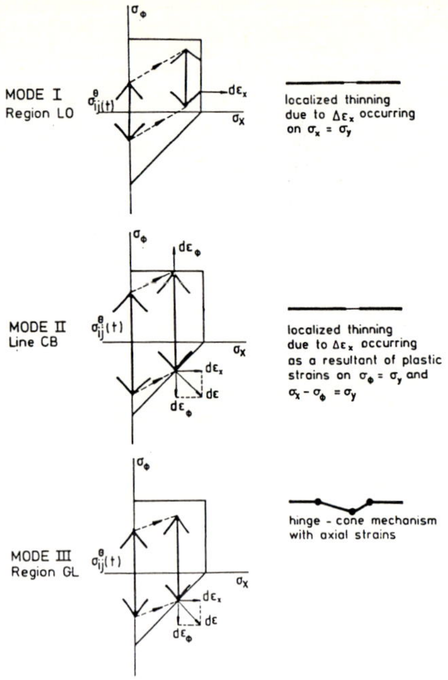

Fig.3 Schematic representation of mechanisms of deformation for problem of Fig.(2) and the corresponding stress state at shakedown in terms of σ_x and σ_ϕ the axial and circumferential stress components.

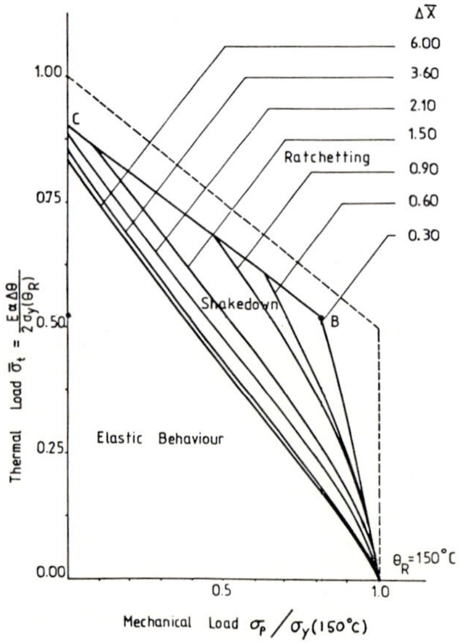

Fig.4 The same problem as that of Fig.(2) but with a temperature dependent yield stress.

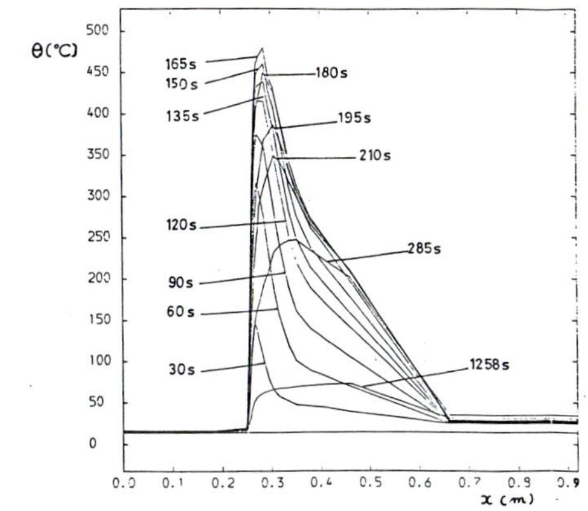

Fig.5 The history of axial temperature at a sequence of instants during the cycle of the Lyon test.

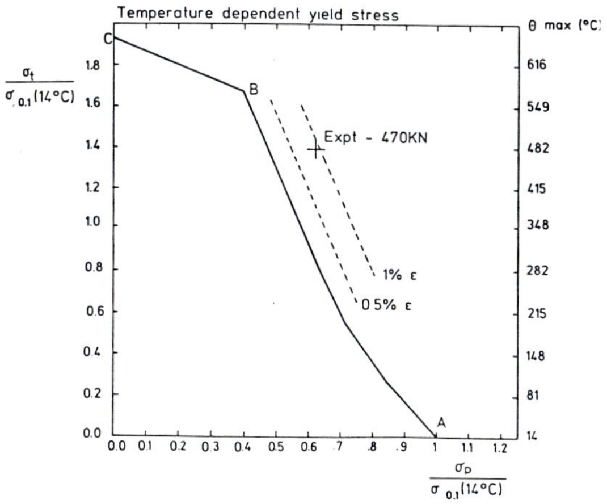

Fig.6 Interaction diagram for the Lyon test.

σ_p . A temperature discontinuity $\Delta\theta$ occurs at the mid-point of the length of the cylinder and this discontinuity repeatedly traverses a length of cylinder Δx .The maximum thermo-elastic stress, which occurs in the hoop stress, corresponds to a value of k=0.5 in equation (3) and Fig. (2) shows a sequence of shakedown boundaries corresponding to a range of values of

$$\Delta\bar{x} = \Delta X\beta/\pi \qquad \text{where} \quad \beta = \left\{\frac{3(1 - \nu^2)}{R^2\, h^2}\right\}^{0.25} \tag{15}$$

and π/β is the decay length of the thermo-elastic stresses away from the temperature discontinuity. A temperature independent yield stress is assumed. The optimal mechanisms are shown in Fig.(3) for various regions of the diagrams together with a schematic representation of the thermo-elastic histories in terms of axial stress σ_x and hoop stress σ_ϕ . For sufficiently large values of $\Delta\bar{x} > 3.6$ the shakedown boundary and the boundary to the elastic region are nearly identical. The mode of deformation (Mode III) consists of an inward movement of the cylinder. For $\sigma_p = 0$, however, a localized reversed plasticity mechanism (Mode II) operates and no incremental growth would occur. When the effect of temperature on the yield stress is included the relative position of these mechanisms in the diagrams changes as shown in Figure (4). The most significant feature of this diagram is that for $\Delta\bar{x} > 2.10$ and $\sigma_p = 0$, the shakedown limit involves a Mode III mechanism involving an inward movement, i.e. it implies that the shell would ratchet purely under the action of thermal fluctuation alone. There is some experimental support for this observation [11]. The last example shown in Figs. (5) and (6) simulate an experiment carried out at the Institute National Des Sciences Applique de Lyon where a cylinder of Type 316 stainless steel was subjected to an axial load and a complex cyclic temperature history induced by diverting the hot gases from a burner onto a narrow hoop of cylinder. The temperature history for a single cycle is shown in Fig. (5) and the interaction diagram obtained by linearly scaling the temperature history is shown in Fig. (6). The shakedown boundary consists of two parts AB and BC along which the mechanism remains constant. Along AB a localised axial extension occurs activated by the fluctuating axial bending moment (rather like the classic Bree problem [7] where the bending moment arises from a through-thickness temperature gradient) and along BC a reverse plasticity mechanism occurs activated by the fluctuation in the hoop stress. An experimental point is shown in the diagram as being outside the shakedown boundary and contours of accumulated plastic strain are shown based on a theory described in the next section. The prediction of about 1% strain after a number of cycles is a reasonable estimate of the observed values.

The method is an example of a form of analysis which provides information which may be used directly for design. A large number of diagrams of the type shown in Figs. (3)-(6) are being generated as an atlas of special cases both for direct use in design and as a basis for formulation of less conservative design rules than those currently in use.

3) Deformation Rates and Accumulated Plastic Strains for Loads in Excess of the Shakedown Limit

The one available complete solution [7] indicates that the rate of strain growth becomes significant even for small increases in μ above the shakedown limit. An understanding of this phenomena can be obtained by considering a shell at the shakedown limit, $\mu = \mu^s$, for some σ_t. The applied load is then increased by $\Delta\mu \underline{P}$ so that the stress history becomes

$$\underline{\sigma}(\underline{x},t) = (\mu\hat{\underline{\sigma}}^P + \hat{\underline{\sigma}}^\theta + \bar{\underline{\rho}}) + (\Delta\mu\hat{\underline{\sigma}}^P + \Delta\underline{\rho}(\underline{x},t)) \tag{16}$$

where the first term is the shakedown solution for $\Delta\mu = 0$, and $\hat{\underline{\sigma}}^P$ is the elastic stresses corresponding to $\underline{P}$ and $\bar{\rho}$ a constant residual stress field. The second term is the change in the stress history in the cyclic state due to the increase in $\Delta\mu$. We assume, for sufficiently small $\Delta\mu$ that the shell will deform in the deformation mode of the shakedown solution $\Delta\underline{u}$. A very general result can be derived for the relationship between $\Delta\mu$ and the magnitude of displacement $\Delta\underline{u}/\Delta N$ per cycle, but a simpler result is possible for the case of proportional thermal loading (equations (1) and (2)) where the increment $\Delta\underline{\varepsilon}^P$ is composed of two parts

$$\Delta\underline{\varepsilon}^P = \Delta\underline{\varepsilon}^1 + \Delta\underline{\varepsilon}^2 \tag{17}$$

where $\Delta\varepsilon^1$ and $\Delta\varepsilon^2$ are individually generally not compatible. If $\Delta\rho$ is the residual stress field caused by imposing the strain distribution $\Delta\underline{\varepsilon}^1$ in an initially stress free body, then the rate of ratchet per cycle is given by [12]

$$\Delta\mu\underline{P}\,\frac{\Delta\underline{u}}{\Delta N} = -\int_V \Delta\underline{\rho}\,\Delta\underline{\varepsilon}^1\, dV \tag{18}$$

For the case when $\Delta\underline{\varepsilon}^1$ and $\Delta\varepsilon^2$ occur in distinct non-overlapping parts of the volume V then it is possible to show [12] that

$$\Delta\mu\underline{P}\,\frac{\Delta u}{\Delta N} \leqslant \tfrac{1}{2}\, U(\Delta\underline{\varepsilon}) \tag{19}$$

where U is the elastic strain energy of the body corresponding to the strain field $\Delta\underline{\varepsilon}$. If we assume, in addition, that the structure is kinematically determinate, i.e. $\Delta\varepsilon$ is proportional to the elastic strains corresponding to P acting alone then

$$\frac{\Delta\underline{u}}{\Delta N} \geqslant 4\Delta\underline{u}^e, \tag{20}$$

where $\Delta\underline{u}^e$ is the elastic displacement due to $\Delta\mu\underline{P}$. This lower bound rate is very high as a 1% increase in $\mu\underline{P}$ above the shakedown limit would cause an accumulation of displacement equal to the elastic displacement due to μP in 25 cycles. For the Bree problem equality holds in equation (20) for small $\Delta\mu$. Although inequality (20) is directly applicable to comparatively few problems, the rate of ratchet is found to be generally of this order of magnitude. In Fig.(7) contours of constant ratchet rate [13] are shown for the problem of Fig.(2) for a movement of a temperature discontinuity $\Delta x = a$ which is small compared with $\sqrt{Rh}$ for both a Tresca and a Von Mises yield condition. For the Tresca yield condition

$$\frac{\Delta u}{\Delta N} = \Delta u^e \tag{21}$$

if the elastic solution assumes infinite stiffness outside the length, a, of cylinder which suffers incremental growth due to a reverse plasticity

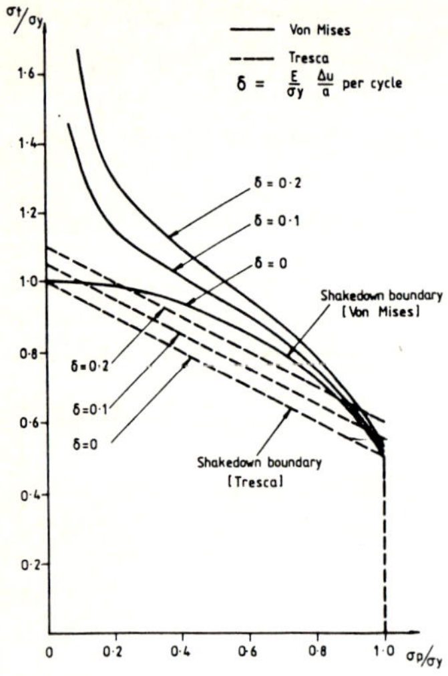

Fig.7
Shakedown boundaries and contours of constant ratchet rate for moving temperature discontinuity when Δx is small.

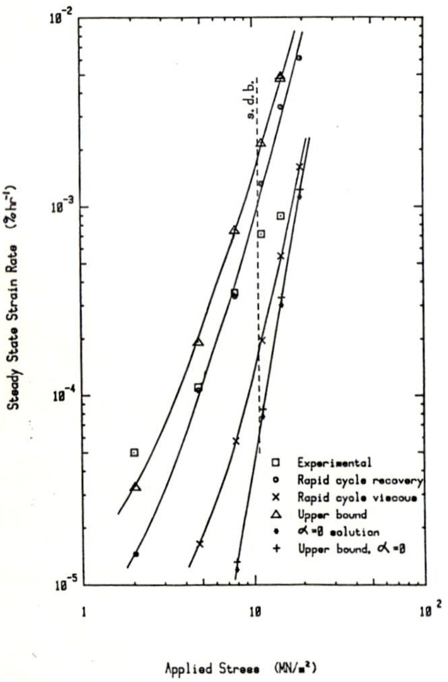

Fig.8
Steady state creep rates for tests on copper plates [16]; comparison between experimental values and rapid and slow cycle solutions.

mechanism. For the Von Mises yield condition the rate is both greater and less than this value depending upon the position on the shakedown boundary. Although the shakedown limits differ by only 15% the contours differ by a much greater amount as the ratchet rate is sensitive to the local curvature of the yield surface, particularly for small values of the axial stress σ_P.

For a work hardening materials these ratchet rates will not continue indefinitely but will decrease from close to the perfectly plastic rate until an asymptotic state is reached. An estimate of the number of cycles required to reach this state and the accumulated displacement can be found using a similar type of analysis to that described above for a linear kinematic hardening material with a uniaxial relationship between increments of stress $d\sigma$ and plastic strain $d\varepsilon^P$ of the form

$$d\varepsilon^P = \frac{K}{E} d\sigma \tag{22}$$

where K for 316 stainless steel is about 40. If we assume that the stress and strain state at the shakedown limit is the same as that of a perfectly plastic material and the load is then increased by $\Delta\mu \underline{P}$, it can be shown [14] that the asymptotic displacement (again assuming the shakedown state displacement field) $\Delta \underline{u}^a$ is bound from above by

$$\underline{P}\, \Delta\underline{u}^a \leq K\, \underline{P}\, \Delta\underline{u}^e \tag{23}$$

where Δu^e is the elastic displacement due to $\Delta\mu\underline{P}$ assuming elastic incompressibility. For kinematically determinate structures equality holds in (23).

If we assume equality in (23) and (20) and ignore the differences between the definitions of $\Delta\underline{u}^e$ in the two cases (i.e. the value of Poissons ratio) an estimate of the number of cycles N_o to reach the asymptotic value $\Delta\underline{u}^a$ at the perfectly plastic rate is given by

$$\Delta\underline{u}^a \doteq N_o \frac{d\underline{u}}{dN} \tag{24}$$

For the rates given by (20) and (21) this yields $N_o = K/4$ and K respectively (i.e. $N_o = 10$ and 40 for K=40). In practice the asymptotic state is approached within say 10% for $N \simeq 3N_o$ implying that at least 30 cycles are required. This relatively slow rate of convergence and its independants of the size of the asymptotic state has caused considerable difficulty in the generation of finite element solutions and the interpretation of experimental results, as computer solutions and laboratory tests are often terminated before sufficient number of cycles have elapsed. As the growth in strain in the initial cycles often looks rather like a creep curve experiments have been interpreted erroneously as indicating the presence of creep strain or material ratchetting. The presence of creep can only be assessed by noting whether the displacement rate eventually becomes constant and does not reduce to zero.

4) Estimates of Creep Deformation Rates

If the temperature remains sufficiently high (for steels above about 480°C) for substantial periods during the cycle, time dependent strains occur which result in an asymptotic rate of deformation per cycle which is non-zero. The development of constitutive equations which are capable of describing creep behaviour for histories of stress and temperature remain under development. It is possible, however, to estimate deformation rates on the basis of a simple theory which requires knowledge of constant stress and temperature data and the evaluation of a single residual stress field. The basis of the technique is the "rapid cycle" solution which assumes that the cycle time Δt is very small compared with characteristic material time scales [6] and has upper bounding properties. The steady state, achieved after sufficient number of cycles, is then given by the condition that the average rate of strain over an entire cycle $\dot{\tilde{\underline{\varepsilon}}}$, given by

$$\dot{\tilde{\underline{\varepsilon}}} = \lim_{\Delta t \to o} \int_o^{\Delta t} \dot{\underline{\varepsilon}}\,(\sigma^*)\,dt \tag{25}$$

is compatible with a displacement rate field $\tilde{\underline{u}}$. The history of stress $\underline{\sigma}^*$ is given by

$$\underline{\sigma}^* = \mu\,\hat{\underline{\sigma}}^P + \hat{\underline{\sigma}}^\theta + \bar{\underline{\rho}} \tag{26}$$

where $\bar{\rho}$ is a constant residual stress field which may be evaluated from the compatibility of $\dot{\tilde{\varepsilon}}$. The precise form of (25) depends upon the creep constitutive relationship.

The steady state creep behaviour of most metals can be described by

$$\dot{\underline{\varepsilon}}^{ss} = \frac{A\dot{\varepsilon}_o}{\sigma_o^n}\,\phi(\underline{\sigma}')\,\frac{\partial\phi}{\partial\sigma'}\,e^{\gamma(\theta-\theta_o)} \tag{27}$$

where $\phi(\sigma')$ is usually the Von Mises effective stress and $\dot{\varepsilon}_o$ is the uniaxial creep rate at constant stress σ_o and temperature θ_o. The quantity A ensures consistency with uniaxial behaviour and γ is a material constant. A number of constitutive equations may be constructed which are consistent with this equation but differ for variable stress and temperature. Two extreme cases seem to be a purely viscous material (i.e. equation (27) always holds) and a recovery model, the Bailey-Orowan model, which assumes the material has an internal yield stress [5]. The expression equation (25) for these two cases, takes the form,

$$\dot{\tilde{\underline{\varepsilon}}} = \frac{1}{\Delta t}\int_o^{\Delta t} \dot{\underline{\varepsilon}}^{ss}(\underline{\sigma}^*(t))\,dt \qquad \text{viscous model} \tag{28}$$

$$\dot{\tilde{\underline{\varepsilon}}} = \dot{\underline{\varepsilon}}^{ss}(\underline{\sigma}^*(t_o)) \qquad \text{recovery model} \tag{29}$$

where $\qquad \phi(\underline{\sigma}^*(t_o)) \geq \phi(\underline{\sigma}^*(t)) \quad o < t < \Delta t,\ o < t_o < \Delta t$

The expression (29) always gives the larger strain rate. In (29) if the extreme stress occurs at more than one instant in the cycle, say t_o and t_1, then the average rate is given by

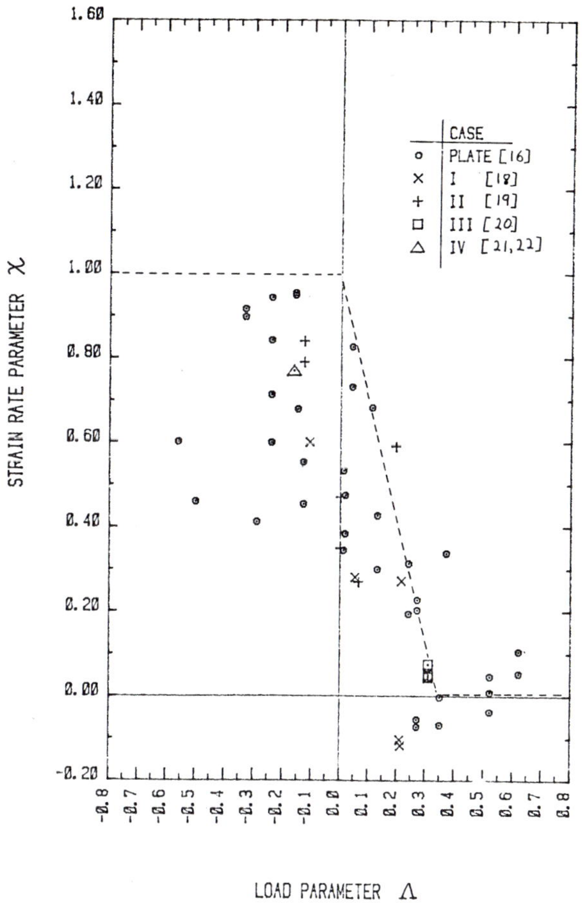

Fig.9 Non-dimensional plot of a range of structural creep data. showing change from rapid cycle solution (X=1) to slow cycle solution (X=0) as the load crosses the high temperature shakedown boundary (Λ =0).

$$\left.\begin{array}{l}\dot{\underline{\varepsilon}} = q\,\dot{\underline{\varepsilon}}^{ss}(\underline{\sigma}^{*}(t_o)) + (1-q)\,\dot{\underline{\varepsilon}}^{ss}(\underline{\sigma}^{*}(t_1)) \\ o \leq q \leq 1, \quad \left.\begin{array}{l}\phi(\underline{\sigma}^{*}(t_1)) \\ \phi(\underline{\sigma}^{*}(t_o))\end{array}\right\} \geq \phi(\underline{\sigma}^{*}(t))\end{array}\right\} \tag{30}$$

Fig.(8) shows a comparison between average experimental strain rates on copper plates thermally cycled by means of infra-red heaters and these rapid cycle solutions, generated using finite element methods[16].The plates were subjected to a constant thermal cycle (Δt = 2 hours) and a sequence of increasing in-plane average stress which remained constant until a steady state was reached. The shakedown load assuming a yield stress σ_y equal to the 0.1% off-set stress at the maximum temperature, 250°C, is shown as the dashed line s.d.b in the figure. The results of some simple bounding calculations are also included. The most important feature of this experiment, and others in the same series, is that the rates within the shakedown boundary are very close to the rapid cycle solution for the recovery model, but the experimental points move over for increasing σ_p to the solution marked "α=o" which includes the variation of material properties with temperatures but assumes a zero coefficient of thermal expansion. This later solution closely approximates a "slow cycle" solution where Δt is assumed to be large with respect to material time scales. This transition appears to be due to increased strain rates within the cycle for stresses above yield. The generality of this behaviour can be judged by plotting the results of a number of tests involving a range of cycle times [16,18-22] against two parameters,

$$X = (\dot{\varepsilon}^{exp} - \dot{\varepsilon}_o)/(\dot{\varepsilon} - \dot{\varepsilon}_o) \tag{31}$$

where $\dot{\varepsilon}$ is the recovery rapid cycle solution, and where $\dot{\varepsilon}_o$ is the value for $\alpha = o$, and

$$\Lambda = (P - P_s)/P_L \tag{32}$$

where P, P_S and P_L are the applied load, the shakedown load and limit load respectively, all using $\sigma_{0.1}$ at the maximum temperature. The resultant diagram, using either local or average strain rates [17] is shown in Fig.(9). Although there is a great deal of scatter the data points are consistantly less then X = 1 (i.e. rapid cycle) for Λ<o (within shakedown) and then decline to X=0 (i.e. slow cycle) as Λ exceeds about 0.35. This type of information may well be sufficient for estimating creep deformation rates from rapid cycle solutions. This result has some significants for the interpretation of structural feature tests which are customarily used to "validate" computer codes and the associated constitutive relationships. The slow cycle solution is identical for all constitutive equations capable of predicting the same constant stress and temperature data. The rapid cycle solutions, however, can differ depending upon the constitutive assumptions, and this can be seen in Fig.(8) where the rapid cycle solutions for both the viscous and recovery model are shown. If features tests are performed at loads in excess of shakedown, a response close to the slow cycle solution will be obtained

whereas the design situation is usually within shakedown. In other words, it is inadvisable to conduct tests at markedly greater loads than those expected in practice.

Acknowledgements.

The support of the Science and Engineering Research Council, the United Kingdom Atomic Energy Authority and the EEC Directorate for Science and Technology are gratefully acknowledged for various parts of the work reported in this paper.

References:

1) Holmes, J.A.G.: "High temperature problems associated with the design of the commercial fast reactor", Creep in Structures (Ed. Ponter, A.R.S. and Hayhurst, D.R.). Springer-Verlag, Berlin (1981) pp 279-287.

2) Koiter, W.T.: "General Theorems for elastic-plastic solids", Progress in Solid Mechanics (Ed. Hill, R. and Sneddon, I.) North Holland Press, 1960, p167.

3) Prager, W.: "Shakedown in elastic-plastic media subjected to cycles of load and temperature", Symposium Su la Plasticita' Nella Scienza Delle Construzioni, Bologna, 1957, pp 239-244.

3) Ponter, A.R.S.: "Deformation, displacement and work bounds for structures in a state of creep and subjected to variable loading", Trans ASME, Jn. Appl. Mech., Vol. 39, 1972, pp 953.

5) Ponter, A.R.S.: "Deformation bounds for the Bailey-Orowan theory of creep", Trans ASME, Jn. Appl. Mech. Vol. 42, 1975, p.619.

6) Ponter, A.R.S.: "The analysis of cyclically loaded creeping structure for short cycle times", Int. Jn. Solids and Structures, Vol. 12, (1976) p.809.

7) Bree, J.: "Elastic-plastic behaviour of thin tubes subjected to internal pressure and intermittent high heat fluxes with application to fast nuclear reactor fuel elements". J. Strain Analysis, Vol. 2., 1967, pp 226-238.

8) Ponter, A.R.S. and Karadeniz, S.: "An extended shakedown theory for structures which suffer cyclic thermal loading", Part I and II, to appear Trans ASME, Jn. Applied Mech. 1985.

9) Karadeniz, S. and Ponter, A.R.S.: "A linear programming upper bound approach to the shakedown limit of thin shells subjected to variable thermal loading", Jn. of Strain Analysis, Vol. 19, 1984, pp. 221-230.

10) Karadeniz, S.: Phd. Thesis, University of Leicester, 1984.

11) Bell, R.T.: "Ratchetting experiments on thin cylinders subjected to axially moving temperature fronts", UKAEA, Risley Nuclear Power Development Establishment, Report ND-R-835(R), Risley, U.K., Oct. 1980.

12) Ponter, A.R.S. and Cocks, A. C. F.: "The incremental strain growth of an elastic-plastic body loaded in excess of the shakedown limit". Trans ASME, Jn of Appl.Mech., Paper WA/APM-10,1984.

13) Ponter, A. R. S. and Cocks, A. C. F.: "The incremental strain growth of elastic-plastic bodies subjected to high levels of cyclic thermal loading", Trans ASME, Jn. of Appl., Mech., Paper WA/APM-11 1984.

14) Cocks, A. F. and Ponter, A. R. S. "The accumulation of plastic strains in excess of the shakedown limit for weakly strain hardening materials", to appear.

15) Odquist, F. K. G.: "Mathematical Theory of Creep and Creep Rupture" Oxford University Press,2nd edition,1974.

16) Ponter, A. R. S. Jakeman, R. R. and Morrison, C. J.: "An experimental study of simplified methods for the prediction of the deformation of structures subject to severe cyclic thermal loading", to appear Jn of Strain Analysis, 1985.

17) Jakeman, R. R., and Ponter, A. R. S.: "Comparisons of available creep structural ratchetting data and the consequences for fast reactor design", to appear.

18) Ponter, A. R. S. and Megahed, M. M.: "Creep and plastic ratchetting in cyclically thermally loaded structures", IUTAM Symposium on Physical Non-Linearities in Structural Analysis", Senlis 1980 (Ed. Hult and J. Lemaitre), Springer Verlag, Berlin, 1981, pp220-227.

19) Megahed, M. M., Ponter, A. R. S. and Morrison C. J.: "An experimental and theoretical investigation into the creep properties of a simple structure of 316 stainless steel", Int. Jn. Mech. Sci., Vol.26, No.3 1984, pp149-164.

20) Ainsworth, R. A: "An experimental study of a three bar structure subjected to variable temperature", Int. Jn. Mech. Sci.,Vol 19. 1977, pp247-256.

21) Corum, J.M., Young, H C. and Grindell, A.C: "Thermal ratchetting in pipes subjected to intermittent thermal downshock at elevated temperatures". 2nd Nat. Cong. of Press.Vess. Piping, San Francisco, Cal., ASME 1975, pp 47-58.

22) Corum, J. M: "Material property data for elastic-plastic creep analysis of benchmark problems", ibid, pp99-109.

Wrinkling in Thin Plates due to In-plane Body Forces

F.P.J. RIMROTT and MICHAL CVERCKO
University of Toronto
Toronto, Ontario, Canada M4S 1A4.

Summary

In-plane body forces, i.e. the weight, of a vertically mounted blanket, i.e. a plate of near-zero flexural stiffness and near-zero compressive strength, lead to the formation of curved wrinkles. Wrinkle lines bordering each finite size tension strip, and tension lines within each tension strip, assume shapes that are governed by the differential equation

$$\frac{d^2y}{dx^2} \pm \frac{q}{H} = 0$$

where q is the reduced distributed weight force, and H a horizontal pretension force parameter.

The case of a blanket with sinusoidal boundaries, supported at two corner points is discussed in detail.

Nomenclature

B = Bending energy stored in wrinkles, J

F = Horizontal component of tension strip tension force, N

H = Horizontal force component parameter, N

P = Potential energy change, J

Q = $\dfrac{a_{i-1}}{a_i} = \dfrac{m_{i-1}}{m_i}$

T = Tension strip tension force, N

U = Work done by horizontal force, J

V = Vertical component of tension strip tension force, N

W = Weight force of blanket, N

a = Amplitude of cosine curve, m

e = Horizontal displacement, m

h = Height of blanket, m

Inelastic Behaviour of Plates and Shells
IUTAM Symposium Rio de Janeiro 1985
Editors: L. Bevilacqua, R. Feijóo and R. Valid

l = Distance between supports, m

q = Reduced distributed load, N/m

α = Slope angle, rad

γ = Specific weight of blanket (× blanket thickness), N/m^2

σ = Stress (× blanket thickness), N/m

τ = Shear stress (× blanket thickness), N/m

Introduction

There are new spacecraft designs, such as the one for 'Olympus' (L-Sat, ↑1986), that call for large solar power collection panels, which are loaded, as a result of spacecraft manoeuvres, not only normal to the panel, but also in the plane of the panel. Spacecraft engineers refer to the structural substrate of solar panels as blankets. These blankets are typically made of a plastic material, such as kapton. What engineers are very much afraid of, is that wrinkles might form in this substrate due to the in-plane component of the load. Such concerns have led to the study of a novel class of problems, concerned with the occurrence of wrinkling due to distributed body forces.

By extending the tension field theory [1,2,3,4] for thin plates to include body forces, it can be shown [5,6,7] that tension lines, and wrinkles for that matter, are no longer straight, but curved, an effect brought about by the distributed in-plane body forces.

Fortunately, in-plane loading of thin plates can easily be simulated in the laboratory, by suspending the plate vertically and using its weight as the distributed body force, and, if the self-weight proves insufficient, by adding small uniformly distributed wafers, which are glued to the plate surface and serve to increase the load without introducing stiffness.

The case of a plate with sinusoidal boundaries and suspended at two corner points is discussed in detail. The tension line pattern is shown to consist then of sinusoidal lines, and theory and experiment are compared, showing good agreement.

The subsequent discussion is arbitrarily subdivided into Pre-Wrinkling and Post-Wrinkling. Under prewrinkling conditions there is no sideways fallout of the blanket, under postwrinkling conditions there is. The lateral fallout turns out to be of great significance, because it forces the horizontal force H_i of each tension strip into the middle of the tension strip, leading to a uniform σ_x tension stress distribution and the occurrence of a (small) σ_y compression stress, in contra-distinction to the prewrinkling tension line field, where σ_x is varying and there is no compression stress.

Pre-Wrinkling Tension Line Field

We assume the blanket material to be inextensional. If no horizontal displacement of the supports has taken place (e.g. when $e = 0$ in Figure 14), then there is no change in the shape of the blanket. There is also no lateral displacement, i.e. in z-direction (Figure 7), of the blanket.

Equilibrium

The blanket is assumed to be thin. Thus $\sigma_z = 0$. The blanket is also assumed to be incapable of supporting compressive stress. The blanket shown in Figure 1, which is subjected to its own load, is about to exhibit a wrinkle pattern. Along the wrinkles, one principal stress is acting, which is tensile, and the lines along this stress are consequently either *wrinkle lines* or *tension lines*. Accross the wrinkles, the other principal stress is acting; its magnitude is zero.

In Figure 2 a blanket element is shown, bounded by two vertical lines at x and $x + dx$, and two adjacent tension lines characterized by y and $y + dy$.

Equilibrium in horizontal direction requires that

$$dH + \frac{\partial dH}{\partial x} dx = dH \tag{1}$$

or

$$\frac{\partial dH}{\partial x} = 0 \tag{2}$$

or expressed in words, there is no change in the horizontal component dH of the tension force dT in the tension strip of width dy.

Equilibrium in vertical direction requires that

$$\frac{\partial\, dV}{\partial x}\, dx = \gamma\, dx\, dy \qquad (3)$$

where γ is the weight/blanket area.
Since

$$dV = \frac{\partial y}{\partial x}\, dH \qquad (4)$$

equation (3) may also be written

$$\frac{\partial^2 y}{\partial x^2}\, dH\, dx + \frac{\partial y}{\partial x}\,\frac{\partial\, dH}{\partial x}\, dx = \gamma\, dx\, dy \qquad (5)$$

Because of equation (2), the second term on the left hand side of equation (5) vanishes, such that the latter might be written

$$\frac{\partial^2 y}{\partial x^2} - \gamma\,\frac{dy}{dH} = 0 \qquad (6)$$

If H increases as y increases, then

$$\lim_{H\to 0} \frac{dy}{dH} = \frac{\partial y}{\partial H} \qquad (7)$$

If H decreases as y increases

$$\lim_{H\to 0} \frac{dy}{dH} = -\frac{\partial y}{\partial H} \qquad (8)$$

Incorporating (7) and (8) into (6), the equation of equilibrium becomes

$$\boxed{\frac{\partial^2 y}{\partial x^2} \mp \gamma\,\frac{\partial y}{\partial H} = 0} \qquad (9)$$

Solutions $y = y(x,H)$ of the partial differential equation (9) represent a family of tension lines. Equation (9) is the same as that given by Mansfield [5], with the exception of the $\mp$ sign, which is new and allows for cos-shaped tension lines as well as for Cosh-shaped tension lines.

The quantity H appearing in equation (9) is a parameter, representing the horizontal component of the tension force. It has the dimension N . The parameter H is zero for the zeroeth tension line of a field. For the last tension line of the field it equals F, i.e. the horizontal component of the total tension force T .

Stresses

The x-component of the stress in the blanket is

$$\sigma_x = \mp \frac{\partial H}{\partial y} \tag{10}$$

where the sign must be chosen such that σ_x is positive, and y must be an appropriately chosen vertical coordinate.

The blanket material is assumed to be *inextensional* in tension. Further, it can only carry *tensile* stresses. The three principal stresses in the material are thus characterized by

$$\left.\begin{aligned} \sigma_{I} &\geq 0 \\ \sigma_{II} &\geq 0 \\ \sigma_{III} &= 0 \end{aligned}\right\} \tag{11}$$

where σ_{III} is in the direction normal to the blanket plane. The tendency to form wrinkles (Figure 1) occurs in areas where the second principal stress also vanishes, i.e. when

$$\left.\begin{aligned} \sigma_{I} &\geq 0 \\ \sigma_{II} &= 0 \\ \sigma_{III} &= 0 \end{aligned}\right\} \tag{12}$$

Coordinate transformation and equations (2) immediately lead to

$$\sigma_x \sigma_y = \tau_{xy}^2 \qquad (13)$$

$$\left.\begin{aligned} \sigma_x &= \sigma_I \cos^2 \alpha \\ \sigma_y &= \sigma_I \sin^2 \alpha = \sigma_x \tan^2 \alpha \\ \tau_{xy} &= \sigma_I \sin \alpha \cos \alpha = \sigma_x \tan \alpha \end{aligned}\right\} \qquad (14)$$

where $\frac{dy}{dx} = \tan \alpha$ is the slope of the tension line $y = f(x,H)$ (Figure 2) passing through the point P at which the stress is being determined.

<u>A Simple Solution</u>

By inspection we conclude that a family of curves of the type

$$y = a_0 \, e^{-\frac{H \pi^2}{\gamma l^2}} \cos \frac{\pi}{l} x \qquad (15)$$

satisfies equation (9) if the minus sign is valid. Note that the bottom boundary of the blanket is given by equation (11) with $H = 0$, i.e. the bottom boundary is defined by the zeroeth tension line (Figure 3). Also note that the top boundary of the blanket is given by equation (11) with $F = H_n$, i.e. the top boundary is defined by the last (the nth) tension line, where n is an arbitrary positive integer. It can be shown that

$$F = \frac{\gamma l^2}{\pi^2} \ln \frac{a_0}{a_n} \qquad (16)$$

while the parameter

$$H = \frac{\gamma l^2}{\pi^2} \ln \frac{a_0 \cos \pi x/l}{y} \qquad (17)$$

The stress component in x-direction is

$$\sigma_x = - \frac{\partial H}{\partial y} = \frac{\gamma l^2}{\pi^2 y} \qquad (18)$$

The relatively simple solution (11) is only valid for a blanket whose boundaries a cosine-shaped (Figure 3). The lower boundary is given by

$$y = a_0 \cos \frac{\pi}{l} x$$

The upper boundary is given by

$$y = a_n \cos \frac{\pi}{l} x$$

with $a_n = a_0 \, e^{-\frac{\pi^2 H_n}{l^2 \gamma}}$.

In the limit, the upper boundary may be a straight line, which implies $a_n = 0$ and $H_n = \infty$.

In Figure 4, the stress distributions in a blanket with cosine-shaped boundaries is given, for a specific example.

Other Solutions

The simple solution (11) of the differential equation (9) requires, that the bottom boundary of the blanket is cosine-shaped. In practice, blankets of *rectangular* shape are, of course, very common. Mansfield [5] has shown that

$$y = -\frac{4}{\pi} a_0 \sum_{m=0}^{\infty} \frac{(-1)^m}{2m+1} \, e^{-(2m+1)^2 \frac{H\pi^2}{\gamma l^2}} \cos(2m+1)\frac{\pi}{l} x \qquad (19)$$

gives the tension lines for such blankets, as shown in Figure 5.

Another case of interest is represented in Figure 6, where a cos-field and a Cosh-field of tension lines occur [7]. The latter results, when the + sign is used in equation (9), which then has a solution

$$y = h - A_0 \, e^{-\frac{H c^2}{\gamma l^2}} \operatorname{Cosh} \frac{c}{l} x \qquad (20)$$

with $c = 2\,\mathrm{ArCosh}\,\frac{h}{A_0}$.

Equation (19) and (20) have been presented to indicate the variety of solutions available. For purposes of the present paper we restrict ourselves to a consideration of the simple solution (11) which implies that the blanket boundaries are cosine-shaped as in Figure 3.

Post-Wrinkling Tension Line Field

Let us assume that a cosine field of tension lines had formed in a blanket with cosine shaped boundaries, and supported at two points as shown in Figure 7. As soon as the supports are moved in by a small amount e (Figure 14), the blanket will descend, the upper boundary more so than the lower. This will cause the blanket to bulge sideways, with no wrinkles forming if the blanket had sufficient bending stiffness, with a finite number of wrinkles for limited bending stiffness, and an infinite number of wrinkles in case of zero bending stiffness.
Strictly speaking, the blanket material was assumed to be incapable of sustaining compressive stress, which would also mean that the material had zero bending stiffness. In reality there is, of course, some compressive strength and some bending stiffness present, be they ever so small.

Evidence of the presence of some small bending stiffness is the formation of a *finite* number of wrinkles, which is typically observed in experiments.

Let us put on record two consequences of the formation of a finite number of wrinkles

(1) The blanket moves into a zig-zag pattern. In the process, each finite-size tension strip assumes a (slight) out-of-plane curvature, enhancing the material's capability to sustain (small) compressive stresses.

(2) Equilibrium of each finite-size tension strip requires that the resultant horizontal force component F is located along

the centre ($y_m = m \cos \frac{\pi}{l} x$) of each tension strip, bringing about a uniform distribution of the x-component of the stress.

In order to investigate the stresses in a tension strip after wrinkles have formed, we must thus relax the stipulation of zero compressive stress made in the beginning.

Equilibrium with Lateral Fallout

The formation of wrinkles makes a lateral fallout of the blanket inevitable. The lateral fallout brings about a uniform distribution of the horizontal stress σ_x. It also leads to only a fraction γ_f of the specific weight γ acting as a load on the 'cable'. From Figure 8,

$$\frac{\gamma_f}{\gamma} = \frac{a}{m}$$

or

$$\gamma_f = \frac{a}{m} \gamma . \tag{21}$$

Using γ_f (instead of γ) in equation (9)

$$\frac{\partial^2 y}{\partial x^2} \mp \frac{a}{m} \gamma \frac{\partial y}{\partial H} = 0 \tag{22}$$

For mathematical convenience, let us now select e.g. the third tension strip. Then

$$\frac{\partial H}{\partial y} = -\sigma_x = -\frac{H}{a_2 \cos \frac{\pi}{l} x - y} \tag{23}$$

If we write for

$$\frac{a}{m_3} \gamma (a_2 \cos \frac{\pi}{l} x - y) = q \tag{24}$$

then eventually [7],

$$\boxed{\frac{d^2 y}{dx^2} \pm \frac{q}{H} = 0} \tag{25}$$

where the minus sign in equation (25) corresponds to the plus sign in equation (22).

Solution

For e.g. the third tension strip, with a bottom boundary of

$$y_2 = a_2 \cos \frac{\pi}{l} x \tag{26}$$

equation (19) would read

$$\frac{d^2y}{dx^2} + \frac{q}{H} = 0 \tag{27}$$

with a reduced distributed load

$$q = \frac{a}{m_3} \gamma (a_2 \cos \frac{\pi}{l} x - y) \tag{28}$$

as an inspection of Figure 9 shows.

Let a trial solution of equations (27) and (28) be

$$y = a \cos \frac{\pi}{l} x \tag{29}$$

for which we find

$$a = a_2 - m_3 \frac{\pi^2}{l^2} \frac{H}{\gamma} \tag{30}$$

The quantity H serves again as a parameter. When $H = 0$,

$$a = a_2$$

when $H = \frac{1}{2} F_3$,

$$a = m_3$$

when $H = F_3$

$$a = a_3$$

where $F_3 = \frac{\gamma l^2}{\pi^2 m_3}(a_2 - a_3)$ (31)

The weight force of the third tension strip is

$$W_3 = 2\gamma(a_2 - a_3)\int_0^{l/2} \cos\frac{\pi}{l}x = \frac{2\gamma l}{\pi}(a_2 - a_3) \qquad (32)$$

or

$$W_3 = \frac{2\gamma l}{\pi} a_3 (Q - 1) \qquad (33)$$

with $Q = \frac{a_2}{a_3}$,

or

$$W_3 = \frac{2\gamma l}{\pi} a_0 \frac{Q - 1}{Q^3}$$

In general, for the ith tension strip

$$W_i = \frac{2\gamma l}{\pi} a_0 \frac{Q - 1}{Q^i} \qquad (34)$$

with $Q = \sqrt[n]{\frac{a_0}{a_n}}$.

For the whole blanket

$$W = \frac{2\gamma l}{\pi}(a_0 - a_n) = \frac{2\gamma l a_0}{\pi} \quad \frac{Q^n - 1}{Q^n} \qquad (35)$$

The horizontal force F_3 can also be written

$$F_3 = \frac{2\gamma l^2}{\pi^2} \frac{Q - 1}{Q + 1} \qquad (36)$$

The slope of the tension force T_3 at the support is obtainable from

$$\tan \alpha_3 = \frac{W_3/2}{F_3} = \frac{\pi a_3}{2l}(Q + 1) \qquad (37)$$

See Figure 10.

Compression Stress

After wrinkles have formed, the blanket material can withstand a (small) compressive stress because each finite tension strip now has a (slight) out-of-plane curvature.

The Rimrott-Kingsland form (19) and the Mansfield equation (9) give the same result, if the latter is modified to read

$$\frac{\partial^2 y}{\partial x^2} - \frac{a}{m}\gamma\frac{\partial y}{\partial H} = 0 \qquad (39)$$

i.e. when the specific weight is no longer uniform but lighter in the upper half of the tension strip and heavier in the lower half.

In other words, there is superimposed on the specific weight a supplementary specific force

$$\gamma_c = \gamma(1 - \frac{a}{m}) \qquad (40)$$

This force is produced by a compression stress in y-direction amounting to

$$\sigma_y = \int_y^{y_2} \gamma_c\, dy = \gamma\left[\int_a^{a_2} (1 - \frac{a}{m_3})\, da\right] \cos\frac{\pi}{l}x \qquad (41)$$

which, upon integration, gives

$$\sigma_y = -\frac{\gamma}{2m_3}(a_2 - a)(a - a_3)\cos\frac{\pi}{l}x \qquad (42)$$

It reaches a maximum magnitude along the centre line $m_3\cos\frac{\pi}{l}x$ of the tension strip.

$$\max|\sigma_y| = \frac{\gamma}{2m_3}(a_2 - m_3)(m_3 - a_3)\cos\frac{\pi}{l}x = \frac{\gamma m_3}{2}\frac{(Q-1)^2}{(Q+1)^2}\cos\frac{\pi}{l}x \qquad (43)$$

The greatest magnitude is at $x = 0$, and amounts to

$$\frac{\gamma m_3}{2} \frac{(Q - 1)^2}{(Q + 1)^2} \tag{44}$$

We had arbitrarily chosen the third tension strip. In the first tension strip the greatest magnitude would be

$$\frac{\gamma m_1}{2} \frac{(Q - 1)^2}{(Q + 1)^2} \tag{45}$$

and since $m_1 > m_3$, expression (45) will be larger than expression (44). If there are n tension strips in a wrinkled blanket, then, with $Q = \sqrt[n]{a_0/a_n}$,

$$m_1 = \frac{a_0 - a_1}{2} = \frac{a_0}{2}(1 + \frac{1}{Q}) = \frac{a_0}{2} \frac{Q + 1}{Q} \tag{46}$$

such that the compression stress in the whole blanket will have a greatest magnitude of

$$\frac{\gamma a_0}{4} \frac{(Q - 1)^2}{Q(Q + 1)} \tag{47}$$

If the blanket material has a critical wrinkling stress of σ_{cr} as a characteristic property then

$$\frac{\gamma a_0}{4} \frac{(Q - 1)^2}{Q(Q + 1)} \leq |\sigma_{cr}| \tag{48}$$

Since σ_{cr} cannot be exceeded, the number of wrinkles forming will adjust itself such that inequality (48) is observed.

In Figures 11 and 12 experimental evidence is presented that shows that the number of wrinkles increases as the specific weight γ of the blanket increases. The number of wrinkles also increases as the depth a_0 of the blanket increases.

The Maximum Tension Force Criterion

Let us assume that the (small) compressive stress had reached a critical value in the blanket (with cosine boundaries) of Figure 13. Consequently a buckle (= wrinkle) was beginning to form, obviously at the centre, where the compressive stress reaches the critical value first. Will the wrinkle stay there, or will it locate itself at some other tension line y_1 ?

The question may also be rephrased by asking which wrinkle line arrangement can carry the maximum (horizontal component H of the) tension force.

In Figure 13 two views of the same blanket, bounded by

$$y_2 = a_2 \cos\frac{\pi}{l}x \tag{49}$$

and

$$y_0 = a_0 \cos\frac{\pi}{l}x \tag{50}$$

are shown. In Figure 13b a wrinkle has appeared, along

$$y_1 = a_1 \cos\frac{\pi}{l}x \tag{51}$$

The question is, at which amplitude a_1 does the wrinkle line locate itself. We write

$$F = F_1 = F_2 = \frac{\gamma l^2}{\pi^2}\left(\frac{1}{m_1}(a_0 - a_1) + \frac{1}{m_2}(a_1 - a_2)\right) \tag{52}$$

with

$$m_1 = \frac{a_0 - a_1}{2} \tag{53}$$

$$m_2 = \frac{a_1 - a_2}{2} \tag{54}$$

and let

$$\frac{\partial F}{\partial a_1} = 0 \tag{55}$$

which we find satisfied for $\frac{a_0}{a_1} = \frac{a_1}{a_2}$.

Thus we conclude, that the wrinkle does not locate itself

at the centreline of the blanket (Figure 13), but such that a_2, a_1 *and* a_0 *form a geometric progression.*

The same result, viz.

$$\frac{a_0}{a_1} = \frac{a_1}{a_2} = \frac{a_2}{a_3} = \ldots\ldots = Q = \sqrt[n]{\frac{a_0}{a_n}} \qquad (56)$$

is obtained for a field with $(n - 1)$ wrinkles.

By similar reasoning, it can be shown that the amplitudes of the tension strip centrelines also form a geometric progression

$$\frac{m_1}{m_2} = \frac{m_2}{m_3} = \ldots\ldots\ldots = \frac{m_{n-1}}{m_n} = Q = \sqrt[n-1]{\frac{m_1}{m_n}} \qquad (57)$$

with the same ratio as equations (56).

If $(n - 1)$ wrinkles have formed, then the horizontal component of the total tension force is

$$F = \frac{\gamma l^2}{\pi^2} 2n \frac{\sqrt[n]{\frac{a_0}{a_n}} - 1}{\sqrt[n]{\frac{a_0}{a_n}} + 1} \qquad (58)$$

Each tension strip carries

$$F_i = \frac{F}{n} = \frac{2\gamma l^2}{\pi^2} \frac{\sqrt[n]{\frac{a_0}{a_n}} - 1}{\sqrt[n]{\frac{a_0}{a_n}} + 1} \qquad (59)$$

The amplitude m_i of the centre line of the ith tension strip is

$$m_i = \frac{1}{2}(a_{i-1} + a_i) \qquad (60)$$

The stress component σ_x in the ith tension strip is

$$\sigma_{xi} = \frac{\gamma l^2}{\pi^2 m_i \cos \frac{\pi}{l} x} = \frac{F}{n(a_{i-1} - a_i)\cos \frac{\pi}{l} x} \qquad (61)$$

If the number of wrinkles is infinity, then equation (58) will become

$$F = \frac{\gamma l^2}{\pi^2} \ln \frac{a_0}{a_n} \tag{62}$$

and the sum of all (infinitesimal) tension strips between $y = a \cos \frac{\pi}{l} x$ and $y_0 = a_0 \cos \frac{\pi}{l} x$ will carry

$$H = \frac{\gamma l^2}{\pi^2} \ln \frac{a_0}{a} \tag{63}$$

The stress component σ_x is then

$$\sigma_x = -\frac{\partial H}{\partial y} = -\frac{\partial H}{\partial a}\frac{\partial a}{\partial y} = \frac{\gamma l^2}{\pi^2 y} = \frac{\gamma l^2}{\pi^2 a \cos \frac{\pi}{l} x} \tag{64}$$

Equations (62), (63), and (64) are seen to correspond to the prewrinkling equations (16), (17), and (18).

Rotation of Moment

In Figure 10, the third tension strip is shown. The horizontal forces F_3 can be looked upon as forming a couple of magnitude

$$M_3 = F_3 m_3 \tag{65}$$

If the supported end is displaced horizontally by a distance e (figure 14), the moment M_3 can be thought of as rotating through an angle θ_3. The work done by F_3 and the work done by M_3 must be the same.

$$U_3 = F_3 e = M_3 \theta_3 \tag{66}$$

giving

$$\theta_3 = \frac{1}{m_3} e \tag{67}$$

If the ratio m_{i-1}/m_i = constant, then

$$\frac{\theta_2}{\theta_3} = \frac{m_3}{m_2} = \frac{1}{Q} \quad (68)$$

or

$$\theta_3 = Q\,\theta_2 \quad (69)$$

i.e. the upper tension strips rotate downwards more than the lower tension strips.

In general $$\theta_i = \frac{2Q^i}{a_0(Q + 1)}\,e \quad (70)$$

The centre point m_3 of the centreline of the third tension trip will move down a distance

$$\Delta m_3 = \frac{l}{2}\,\theta_3 = \frac{l}{2m_3}\,e \quad (71)$$

In general

$$\Delta m_i = \frac{l}{2m_i}\,e \quad (72)$$

or

$$\Delta m_i = \frac{l}{a_0}\,\frac{Q^i}{Q + 1}\,e \quad (73)$$

Energy Considerations

The presence of a finite number of wrinkles suggests that the blanket material can be idealized and considered to have the following properties

(1) It is inextensional in tension.
(2) Its tensile strength is unlimited.
(3) It can carry a (small) compression stress, limited by a critical stress.
(4) The strain energy in tension is zero. In compression it is negligible.
(5) The material can bend and there is bending strain energy.

With these material characteristics we can now establish an energy balance, between potential energy P (due to lowering of the blanket weight), work U done by the horizontal force component F as it moves through a horizontal distance e, and the bending energy B stored in the wrinkles.

The weight W_1 of the first tension strip is

$$W_1 = \frac{2\gamma l}{\pi}(a_0 - a_1) = \frac{4\gamma l m_1}{\pi}\frac{Q-1}{Q+1} \tag{74}$$

It descends a distance $\frac{l}{\pi}\theta_1$ with

$$\theta_1 = \frac{1}{m_1}e \tag{75}$$

such that

$$P_1 = W_1\frac{l}{\pi}\theta_1 = \frac{4\gamma l^2}{\pi^2}\frac{Q-1}{Q+1}e \tag{76}$$

There are altogether n tension strips, such that the total potential energy change is

$$P = \frac{4n\gamma l^2}{\pi^2}\frac{Q-1}{Q+1}e \tag{77}$$

The work done by the horizontal force components F is

$$U = 2eF \tag{78}$$

with $$F = \frac{2n\gamma l^2}{\pi^2}\frac{Q-1}{Q+1} \tag{79}$$

Thus $U = P$ if no bending energy is stored in the wrinkles.

Bending Energy

In order to gain some insight into the amount of bending energy stored in the wrinkles, we assume the following scenario. (1) The forces H are each displaced a small distance e towards the centre without any wrinkles forming. (2) Thereafter a given blanket forms a wrinkle pattern which causes an increase in potential energy (and a change in the magnitude of H) but no work is done by the horizontal forces H because there is no horizontal displacement. The gain of potential energy is then stored as bending energy in the wrinkles. How much bending energy is stored in the wrinkles ?

In Table I the bending energy stored in the wrinkles is tabulated. The work done U, equation (78), by the horizontal force is normalized to

$$U^* = \frac{U}{\frac{4\gamma l^2}{\pi^2}} \tag{80}$$

with $U = \frac{4\gamma l^2}{\pi^2} \frac{Q-1}{Q+1} e$ and $Q = \frac{a_0}{a_n}$

and listed in column (3).

The potential energy P, equation (76), is normalized to

$$P^* = \frac{P}{\frac{4\gamma l^2}{\pi^2}} \tag{81}$$

with $P = \frac{4n\gamma l^2}{\pi^2} \frac{Q-1}{Q+1} e$ and $Q = \sqrt[n]{\frac{a_0}{a_n}}$

and listed in column (4).

The bending energy B in all wrinkles is normalized to

$$B^* = \frac{B}{\frac{4\gamma l^2}{\pi^2}} \tag{82}$$

obtained from

$$B^* = P^* - U^* \tag{83}$$

and listed in column (5).

The critical stress σ_{crit}, equation (48), is normalized to

$$\sigma^*_{crit} = \frac{\sigma_{crit}}{\gamma a_0/4} \tag{84}$$

and listed in column (6).

An inspection of Table I indicates how much energy is stored in the wrinkles. The more wrinkles, the more bending energy is stored. An infinite number of wrinkles means the highest bending energy storage, which is, however, finite. An inspection of the last column shows, that for zero wrinkles, the blanket must have a high critical wrinkling stress. The lower the critical wrinkling stress, the greater the number of wrinkles that form. A vanishing critical wrinkling stress, means an infinite number of wrinkles.

<u>Observations</u>

1. Blankets form a fixed number of wrinkles, that can neither be changed by external perturbation, nor by an increase or decrease of the blanket deformation.

2. Blankets exhibit a (very small) critical wrinkling stress that influences the number of wrinkles formed. The higher the critical stress of the blanket material, the smaller the number of wrinkles.

3. The critical stress of a blanket material is the higher, the stiffer the blanket material, the lower the specific weight, and the shorter the blanket height. (Installed on the Moon's surface a blanket will exhibit fewer wrinkles than on Earth.)

4. There is typically some bending stiffness, be it ever so small, in blankets, such that there is a finite number of wrinkles. An infinite number of wrinkles is thus a limit state of theoretical significance only.

5. The greater the number of wrinkles, the greater the amount of bending energy stored in the wrinkles.

6. A prewrinkling state (as described at the beginning of the paper) never really exists. It is also a limit state of theoretical significance only, associated with zero deformation ($e = 0$) of a truly inextensional material.

7. In reality (obviously due to strains near the support points where the stress after all approaches infinity) wrinkles form immediately (i.e. even when $e = 0$). Thus wrinkling in a blanket can only be prevented by ensuring that both in-plane principal stresses are positive throughout the whole blanket.

Conclusions

In order to keep the subject matter within manageable limits, only blankets with cosine-shaped boundaries suspended at the two corner points, and subjected to gravitational body forces, have been considered. It is shown that a wrinkle-free tension line field is a theoretical possibility. In practice, however, *blankets always exhibit a finite number* of wrinkles, which is associated with a lateral displacement of each tension strip between two wrinkle lines, which in turn affects the equilibrium of each blanket element.

Thus a differential equation differing from the Mansfield equation governs the stress distribution in wrinkled blankets.

References

1. Wagner, Herbert
"Ebene Blechwandträger mit sehr dünnem Stegblech"
Zeitschrift für Flugtechnik and Motorluftschiffahrt, 20, (1929), 8, 200-207; 9, 227-233, 10, 256-262; 11, 279-284; 12, 306-314.

2. Reissner, E.
"On Tension Field Theory"
Proceedings of the 5th International Congress on Applied Mechanics, (1938), 88-92.

3. Stein, M.; J.M. Hedgepeth
"Analysis of Partly Wrinkled Membranes"
NASA TN-D-813, (1961).

4. Mansfield, E.H.
"Tension Field Theory"
Applied Mechanics (M. Hetenyi and W.G. Vincenti, editors), Springer-Verlag, Heidelberg, (1969), 305-320.

5. Mansfield, E.H.
"Gravity-induced Wrinkle Lines in Vertical Membranes"
Proceedings of the Royal Society, London, A 375, (1981), 307-325.

6. Kingsland, Blair
"In-plane Loading of Blankets"
M.A.Sc. Thesis, University of Toronto, (1983), 1-130.

7. Rimrott, F.P.J.; Blair Kingsland
"Tension Line Fields in a Simply Supported Blanket Subjected to In-plane Body Forces"
Journal of Spacecraft and Rockets, 22, 2, (1985), 141-148.

(1) Number of Wrinkles	(2) Number of Tension Strips	(3) Work done by Horizontal Force	(4) Potential Energy	(5) Bending Energy in Wrinkles	(6) Critical Stress
		U^*	P^*	B^*	σ^*_{crit}
$n-1$	n				
0	1	0.945 946	0.945 946	0	0.919 669
1	2	0.945 946	1.428 571	0.482 625	0.595 238
2	3	0.945 946	1.605 276	0.659 330	0.373 037
3	4	0.945 946	1.680 816	0.734 870	0.200 000
4	5	0.945 946	1.718 807	0.772 861	0.175 753
.	.	.	.	.	.
.	.	.	.	.	.
.	.	.	.	.	.
.	.	.	.	.	.
∞	∞	0.945 946	1.791 759	0.845 813	0

Table I

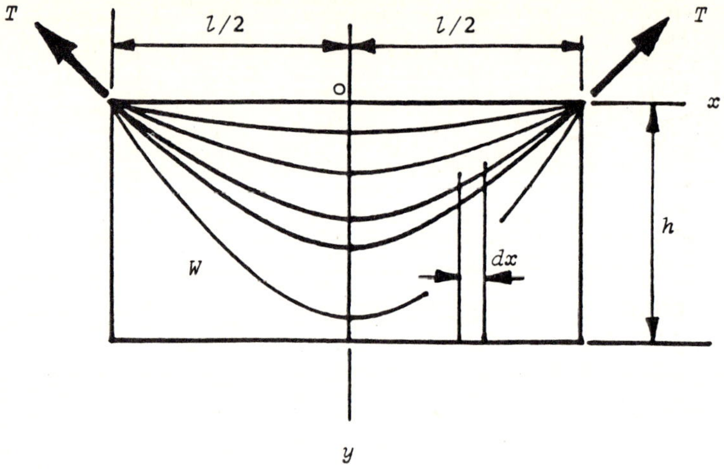

Fig. 1. Blanket suspended in Gravitational Field

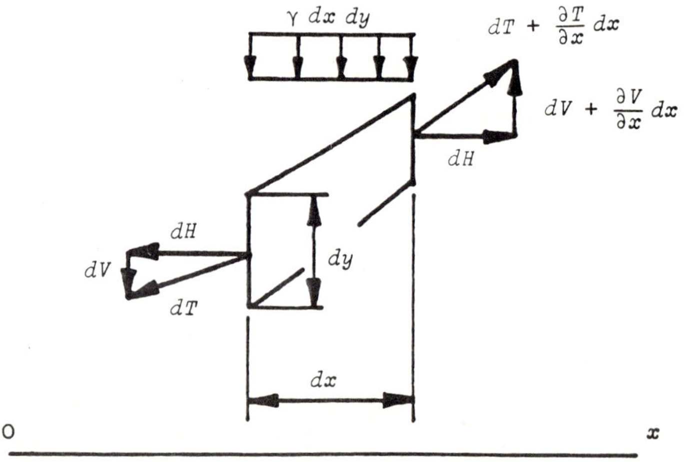

Fig. 2. Equilibrium on a Blanket Element

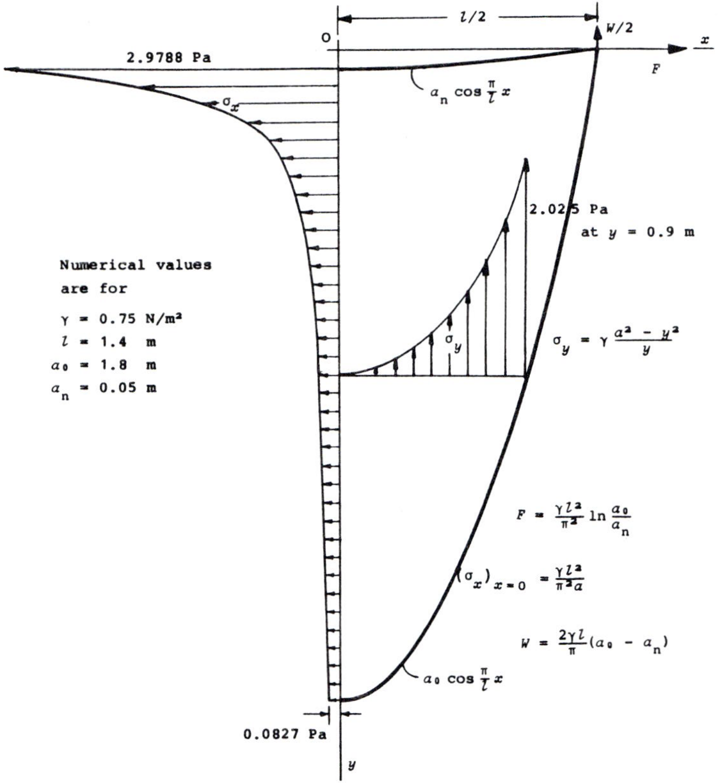

Fig. 4. Blanket prior to Formation of Wrinkles

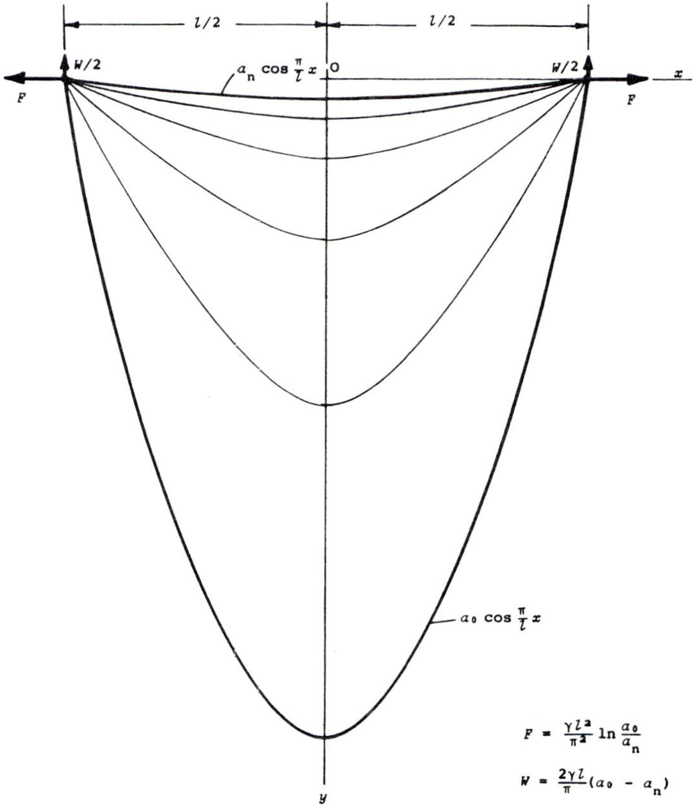

Fig. 3. Blanket with Cosine-shaped Boundaries and Cosine-shaped Tension Lines

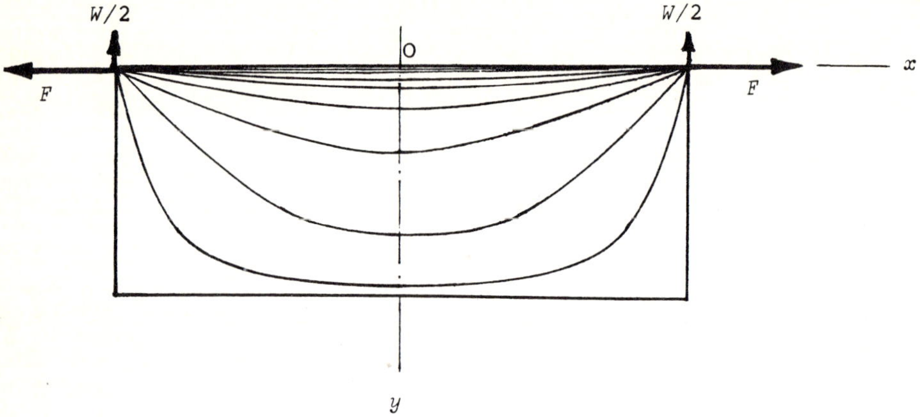

Fig. 5. Rectangular Blanket supported at Top Corners

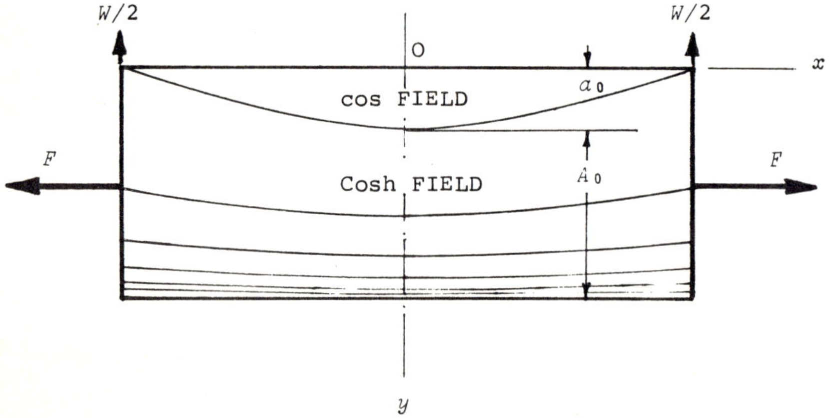

Fig. 6. Simply Supported Rectangular Blanket

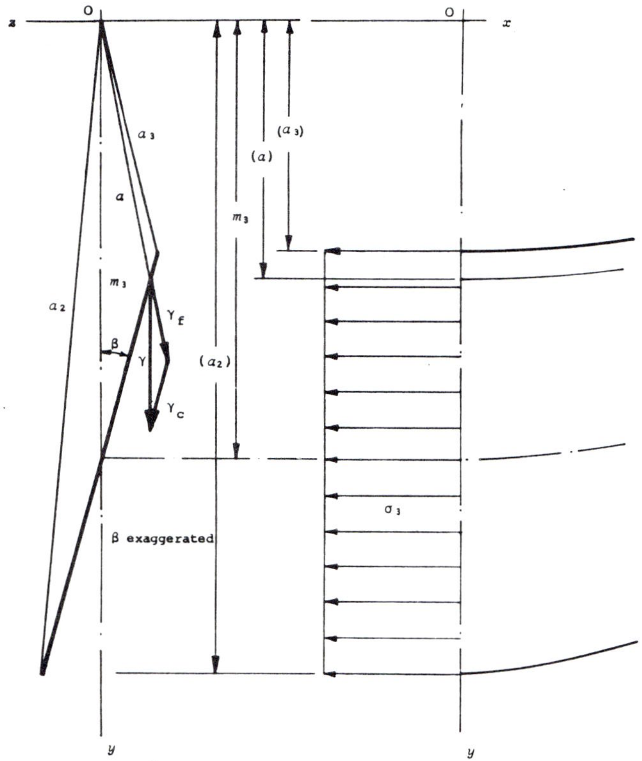

Fig. 8. Section through Tension Strip with Lateral Fallout

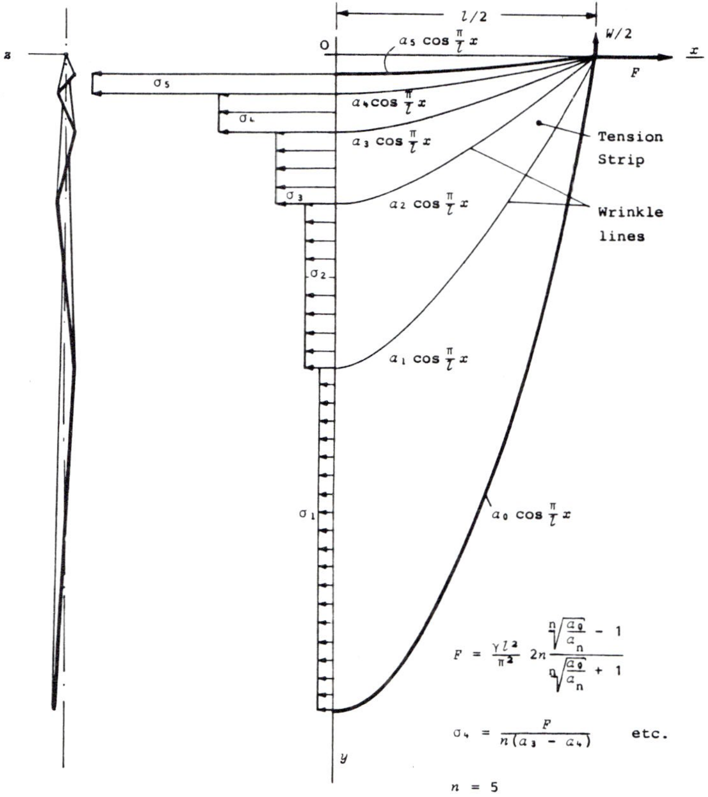

Fig. 7. Blanket with Four Wrinkles

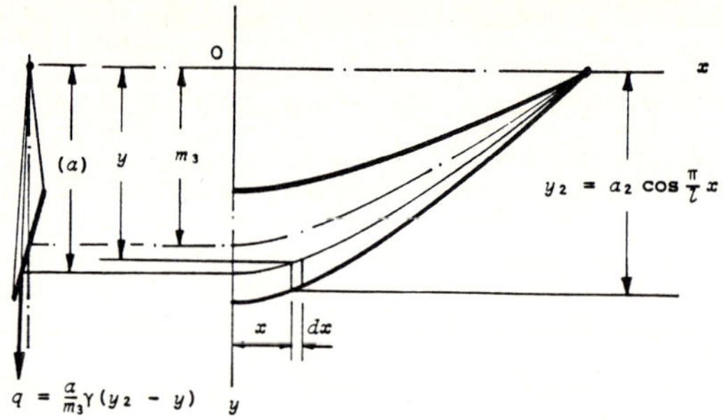

Fig. 9. Tension Lines $y = a\cos\frac{\pi}{l}x$ within the Third Tension Strip

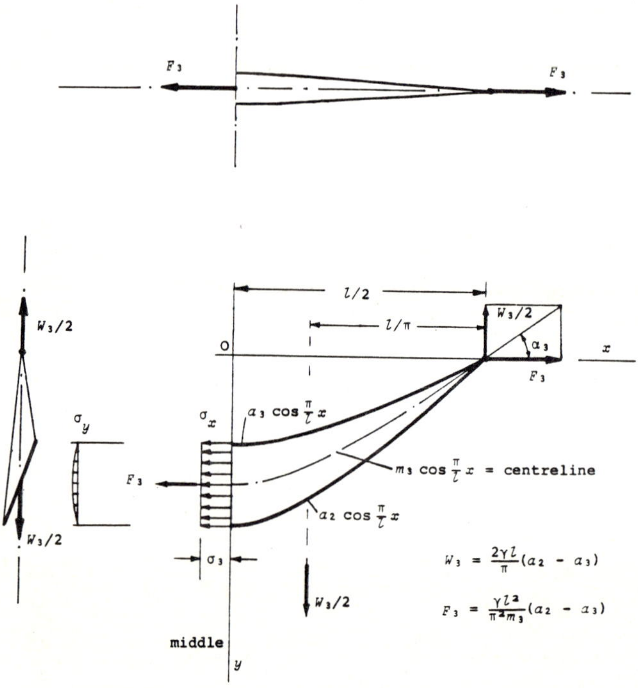

Fig. 10. The Third Tension Strip between Wrinkle Line 2 and Wrinkle Line 3

Fig. 11. A Blanket with 3 Wrinkles

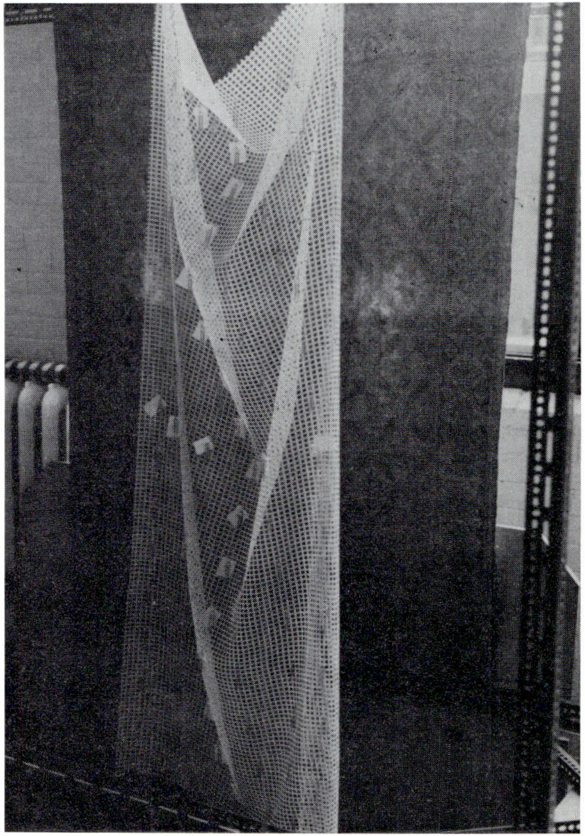

Fig. 12. The Same Blanket with Weight Wafers and 4 Wrinkles

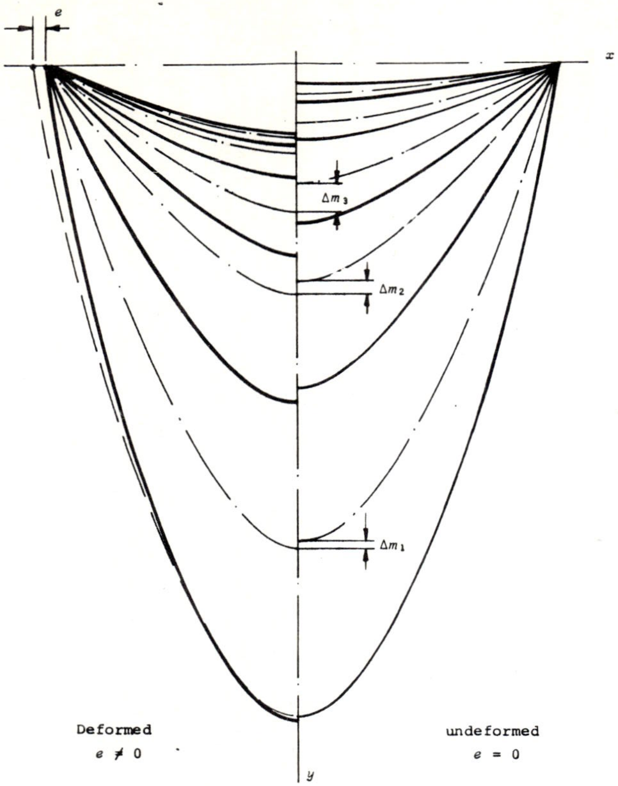

Fig. 14. The Wrinkle Pattern Remain the Same during Blanket Deformation

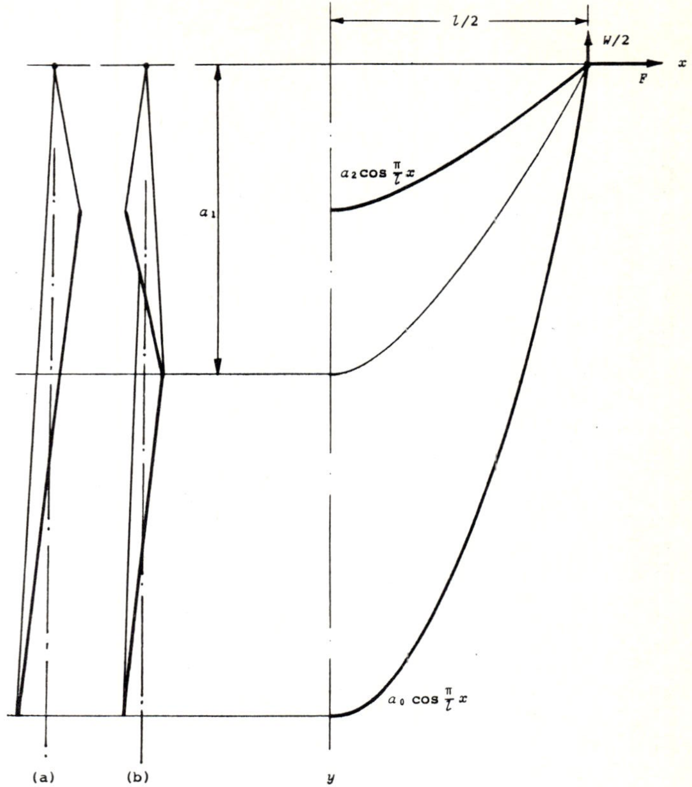

Fig. 13. The Formation of a Wrinkle

Derivation of the Inelastic Behaviour of Plates and Shells From the Three Dimensional Models and Extensions

Philippe DESTUYNDER and Quoc. Son NGUYEN*

Laboratoire de Mécanique (UA-850)
Ecole Centrale de Paris
92290 - Chatenay-Malabry (FRANCE)

*Laboratoire de Mécanique des Solides (UA-317)
Ecole Polytechnique
91140 - Palaiseau (FRANCE)

Summary

The elastoplastic behaviour of plates and shells is usually obtained in two different ways. First, the so-called global method assumes that the yield function can be expressed in terms of resultant in-plane stresses and bending moments. The success of such a formulation is due principally to its simplicity as a natural generalization of the elastic theory of plates and shells. Besides the well-known difficulty on an adequate expression of the adopted criterion, such a model leads to a linear variation of in-plane stresses through the thickness of the structure. If this description is widely used and seems to give entire satisfaction in limit analysis, on the contrary, in the context of elastoplastic analysis, a more refined model is sometimes necessary. The second possibility to derive the inelastic behaviour is to take account of the distribution of plastic strains along the thickness. This more complicated formulation is often preferred to the first one in computation by finite-element programs. The goal of our communication is to give a mathematical justification of the second formulation by asymptotic analysis. It consists in considering the thickness of the shell as a small parameter governing equations for plates and shells are then derived exactly from the three-dimensional description by asymptotic development with respect to the thickness. In the case of elastoplasticity, this leads to a two dimensional model with respect to the displacements of the middle surface. The plastic behaviour is characterized by multiple plastic potentials in the sense of Koiter-Mandel, with hardening parameters which are residual stresses along the thickness. The in-plane stresses are no more linear through the thickness. The derivation of the bidimensional behaviour is presented in detail in the case of standard materials for both deformation theory and incremental theory of plasticity. A comparison-between global models and the ones obtained in this paper is given for particular choice of yield criteria. A section is devoted to composite materials. It is assumed that a plasticity phenomenon can appear in the epoxy layers sticking together two layers of fibers. Then using asymptotic methods we deduce a plate model which takes into account the effect of a gliding between two layers. Finally a buckling model is suggested.

I. Three dimensional formulation for plates

Let us consider an open set Ω^{ε} such that :

$$\Omega^{\varepsilon} = \omega \times]-\varepsilon,\varepsilon[$$

The plate we are dealing with, in this paper, occupies in space the set Ω^{ε},

Inelastic Behaviour of Plates and Shells
IUTAM Symposium Rio de Janeiro 1985
Editors: L. Bevilacqua, R. Feijóo and R. Valid

where ω is the medium surface and $2\ \varepsilon$ the thickness. The lateral boundary is denoted by $\Gamma_o^\varepsilon = \gamma_o \times]-\varepsilon,\varepsilon[$ where γ is the boundary of ω. The upper and lower boundaries are $\Gamma_\pm^\varepsilon = \omega \times \{\pm\varepsilon\}$ (see figure 1).

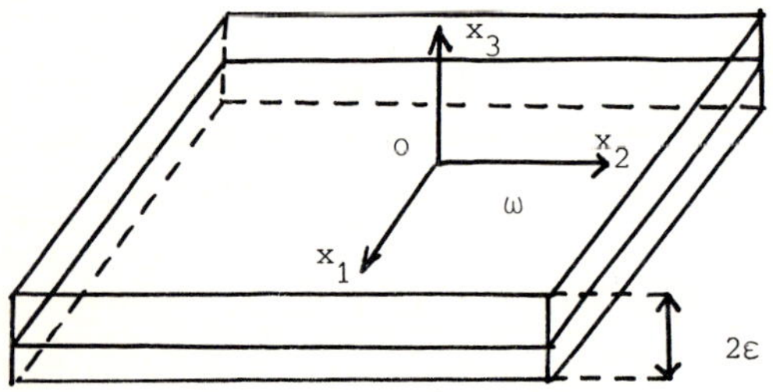

$\Gamma_o^\varepsilon = \gamma_o \times]-\varepsilon,\varepsilon[$

$\Omega^\varepsilon = \omega \times]-\varepsilon,\varepsilon[$

Figure 1

The plate is submitted to a system of loading. For sake of brevity in the notations, we only consider body forces the density components of which are denoted by f_i. The displacement field is $u = (u_i)$ and the stress field $\sigma = (\sigma_{ij})$. The constitutive relationship is

$$\sigma_{ij} = R_{ijk\ell}\ (\gamma_{k\ell} - \gamma^p_{k\ell}),\quad i,j,k,\ell \in \{1,2,3\}, \tag{1}$$

where

$$\gamma_{ij} = \frac{1}{2}\ (\partial_i u_j + \partial_j u_i)$$

is the total strain and γ^p_{ij} the plastic strain. The stiffness tensor denoted $R_{ijk\ell}$ would be such that :

$$R_{\alpha 333} = R_{\alpha\beta\mu 3} = 0\ ;\ \alpha,\beta,\mu \in \{1,2\}$$

The inverse of $R_{ijk\ell}$ is the compliance and is denoted $S_{ijk\ell}$. The yield criterion is written :

$$f(\sigma) \leqslant k$$

$f(.)$ being a convex function and k a given positive constant. The evolution of the plastic strain γ^p_{ij} is governed by Hill's Principle :

$$\forall\ \tau_{ij} \in K^\varepsilon\ ;\ \dot{\gamma}^p_{ij}\ (\tau_{ij} - \sigma_{ij}) \leqslant 0 \tag{2}$$

with

$$K^\varepsilon = \{\tau = (\tau_{ij})\ ;\ \tau_{ij} = \tau_{ji}\ ;\ f(\tau) \leqslant k \text{ on } \Omega^\varepsilon\} \tag{3}$$

Finally the principle of virtual work can be written :

$$\forall\, v \in V^{\varepsilon} \quad \int_{\Omega^{\varepsilon}} \sigma_{ij}\, \partial_i v_j \;=\; \int_{\Omega^{\varepsilon}} f_i\, v_i \tag{4}$$

where V^{ε} denotes the space of admissible displacement fields.

As a matter of fact relations (1), (2), (4) are the Prandtl-Reuss model. Our goal in this paper is to construct an approximation of the three dimensional solution based on asymptotic methods. The mathematical proof will be omitted and only the method and the results obtained will be pointed out. There are five parts. The first one presents the so-called asymptotic method. It is applied to Prandtl-Reuss model in the second part. A discussion about limit load analysis is carried out in section three and extension to composite materials is given in the fourth one.

II. The asymptotic method [1], [7]

The thickness of the plate, ie 2ε, is supposed to be small in comparison with the other dimensions. This geometrical property allows simplifications in the three dimensional model. Basically this leads to the Kirchhoff-Love assumptions stating that the normal to the medium surface is un-strained and remains - during deformation - normal to the deformed medium surface. But some complementary assumptions concerning the constitutive relationship are also needed. As a matter of fact one assumes that the normal stress can be neglected when it is compared to the in-plane stresses. Asymptotic methods provide a general framework for obtaining such simplifications from the three dimensional. It is worth noticing that the models obtained through asymptotic procedure are justified only from the energy point of view. It means that no confidence should be conceded to local interpretation of the solution of the plate model. An interesting example which emphasizes this remark is the case of a transverse crack in a plate. One can prove that even if the local behaviour is completely different from the one of the plate model solution, this last one enables to compute a correct value of the energy release rate [5]. Let us introduce now a new open set Ω by :

$$\Omega = \omega \times \,]-1,1[$$

and let us consider the mapping from Ω into Ω^{ε} by :

$$X = (x_1,x_2,x_3) \in \Omega \longrightarrow X^{\varepsilon} = F^{\varepsilon}(X) = (x_1,x_2,\varepsilon x_3) \in \Omega^{\varepsilon} \tag{5}$$

To this change of coordinates one associates a change of unknowns which is very convenient in the sequel :

$$\begin{cases} \sigma^{\varepsilon}_{\alpha\beta}(X) = \sigma_{\alpha\beta} \circ F^{\varepsilon}(X), & u^{\varepsilon}_{\alpha}(X) = u_{\alpha} \circ F^{\varepsilon}(X), \\ \sigma^{\varepsilon}_{\alpha 3}(X) = \varepsilon^{-1}\sigma_{\alpha 3} \circ F^{\varepsilon}(X), & u^{\varepsilon}_{3}(X) = \varepsilon u_{3} \circ F^{\varepsilon}(X), \\ \sigma^{\varepsilon}_{33}(X) = \varepsilon^{-2}\sigma_{33} \circ F^{\varepsilon}(X). & \end{cases} \quad (6)$$

Furthermore it is necessary to assume that the applied forces are such that (there are only body forces)

$$f_3 = \varepsilon f^{0}_{3}, \quad f_{\alpha} = f^{0}_{\alpha}$$

in order to have finite displacements. Then a simple computation enables to transform problem (1),(2),(4) into an equivalent one, but now set over the open set Ω. First of all, let us introduce the sets :

$$\begin{cases} K(\varepsilon) = \{\tau = (\tau_{ij}) \; ; \; \tau_{ij} = \tau_{ij} \; ; \\ \qquad \tau_{\alpha\beta}\tau_{\alpha\beta} - \frac{1}{3}(\tau_{\mu\mu})^2 + 2\varepsilon^2(2\tau_{\alpha 3}\tau_{\alpha 3} - \frac{1}{3}\tau_{\mu\mu}\tau_{33}) + \frac{2}{3}\varepsilon^4\tau^2_{33} \leqslant k\} \\ V = V^{\varepsilon} \text{ for } \varepsilon = 1. \end{cases} \quad (7)$$

Then the element $(\sigma^{\varepsilon},u^{\varepsilon})$ constructed from (σ,u) solution of (1),(2),(4) with formulae (6) satisfies :

$$\sigma^{\varepsilon} \in K(\varepsilon), \; u^{\varepsilon} \in V, \text{ and :}$$

$$\boxed{\begin{aligned} &\forall \tau \in K(\varepsilon) \quad S_{\alpha\beta\mu\nu}\,\dot{\sigma}^{\varepsilon}_{\mu\nu}(\tau_{\alpha\beta} - \sigma^{\varepsilon}_{\alpha\beta}) + 4\,\varepsilon^2 S_{\alpha 3\beta 3}\,\dot{\sigma}^{\varepsilon}_{\alpha 3}(\tau_{\beta 3} - \sigma^{\varepsilon}_{\beta 3}) \\ &+ \varepsilon^2 S_{33\alpha\beta}(\dot{\sigma}^{\varepsilon}_{\alpha\beta}(\tau_{33}-\sigma^{\varepsilon}_{33}) + \dot{\sigma}^{\varepsilon}_{33}(\tau_{\alpha\beta}-\sigma^{\varepsilon}_{\alpha\beta})) + \varepsilon^4 S_{3333}\dot{\sigma}^{\varepsilon}_{33}(\tau_{\alpha\beta}-\sigma^{\varepsilon}_{\alpha\beta}) \\ &\qquad\qquad \geq (\tau_{ij}-\sigma^{\varepsilon}_{ij})\,\gamma_{ij}(\dot{u}) \\ &\forall v \in V, \quad \int_{\Omega}\sigma^{\varepsilon}_{ij}\partial_i v_j = \int_{\Omega} f^{0}_{i} v_i \end{aligned}} \quad (8)$$

and for $\varepsilon = 0$ it is clear that the limit model consists in finding (σ^{0},u^{0}), element of the set $K(0) \times V$ such that :

$$\boxed{\begin{aligned} &\forall \tau \in K(0), \quad S_{\alpha\beta\mu\nu}\,\dot{\sigma}^{0}_{\mu\nu}(\tau_{\alpha\beta} - \sigma^{0}_{\alpha\beta}) \geqslant (\tau_{ij} - \sigma^{0}_{ij})\,\gamma_{ij}(\dot{u}^{0}), \\ &\forall v \in V, \quad \int_{\Omega}\sigma^{0}_{ij}\,\partial_i v_j = \int_{\Omega} f^{0}_{i} v_i . \end{aligned}} \quad (9)$$

The goal of the next chapters will be to discuss the various simplified formulations of this model. An extension to geometrical non linearities is presented in the last one.

III. Formulation of the Prandtl-Reuss model for plates

There are five steps in the characterization of (σ^o, u^o) solution of (9). Finally the model obtained will be the same as the one introduced in [9].

Step 1. In the first relation (9), it is possible to choose arbitrarily the components τ_{i3} because K(0) involves only the components $\tau_{\alpha\beta}$. Hence :

$$\gamma_{i3}(\dot{u}^o) = 0 \qquad i = 1,2,3,$$

or else :

$$\begin{cases} \partial_3 \dot{u}_3^o = 0, \\ \partial_\alpha \dot{u}_3^o + \partial_3 \dot{u}_\alpha^o = 0 \quad \alpha = 1,2. \end{cases}$$

This leads to a Kirchhoff-Love velocity field :

$$\begin{cases} \dot{u}_3^o \quad (x_1, x_3), \\ \dot{u}_\alpha^o = \dot{\underline{u}}_\alpha^o \quad (x_1, x_2) - x_3 \, \partial_\alpha \, \dot{u}_3^o(x_1, x_2), \qquad \alpha = 1,2. \end{cases} \tag{10}$$

Step 2. The first relation (9) is now restricted to in-plane stresses, we obtain :

$$\forall \, \tau_{\alpha\beta} \in K(0) \; ; \; S_{\alpha\beta\mu\nu} \, \dot{\sigma}_{\mu\nu}^o \, (\tau_{\alpha\beta} - \sigma_{\alpha\beta}) \geqslant (\tau_{\alpha\beta} - \sigma_{\alpha\beta}^o) \, \gamma_{\alpha\beta}(\dot{\underline{u}}_\alpha^o) - x_3 \, (\tau_{\alpha\beta} - \sigma_{\alpha\beta}^o) \, \partial_{\alpha\beta} \, \dot{u}_3^o \tag{11}$$

and if we restrict the virtual displacement v, in the second relation (9), to Kirchhoff-Love fields, we obtain :

$$\boxed{\forall \, v_\alpha \, (u_1, u_2), \text{ S.B.C.}^* \int_\omega n_{\alpha\beta}^o \, \partial_\alpha \, v_\beta = \int_\omega F_\alpha \, v_\alpha} \tag{12}$$

where

$$n_{\alpha\beta}^o = \int_{-1}^{1} \sigma_{\alpha\beta}^o \, dx_3 \qquad \text{and} \qquad F_\alpha = \int_{-1}^{1} f_\alpha^o dx_3,$$

and

$$\boxed{\forall \, v_3 \, (u_1, u_2), \text{ S.B.C.} \; -\int_\omega m_{\alpha\beta}^o \, \partial_{\alpha\beta} \, v_3 = \int_\omega F_3 \, v_3 - \int_\omega M_\alpha \, \partial_\alpha \, v_3} \tag{13}$$

where

$$m_{\alpha\beta}^o = \int_{-1}^{1} x_3 \, \sigma_{\alpha\beta}^o \quad \text{and} \quad F_3 = \int_{-1}^{1} f_3 dx_3, \; M_\alpha = \int_{-1}^{1} x_3 f_\alpha dx_3$$

There is a simple way to solve the obtained model. It consists in an incremental formulation, substituting the time derivative by differential ratio. Hence let us denote by $\sigma_{\alpha\beta}^j$, $n_{\alpha\beta}^j$, $m_{\alpha\beta}^j$ and u_i^j the various unknowns at time

* S.B.C. = Satisfying Boundary Conditions.

tj = j Δt. Then the relation (11) is replaced by : [8]

$$\forall \tau \in K(0),\ S_{\alpha\beta\mu\nu}\, \sigma^{j+1}_{\mu\nu}\, (\tau_{\alpha\beta} - \sigma^{j+1}_{\alpha\beta}) \geqslant (\tau_{\alpha\beta} - \sigma^{j+1}_{\alpha\beta})\, \gamma_{\alpha\beta}(u^{j+1}) + g_{\alpha\beta}\, (\tau_{\alpha\beta} - \sigma^{j+1}_{\alpha\beta}) \tag{14}$$

where

$$g_{\alpha\beta} = S_{\alpha\beta\mu\nu}\, \sigma^{j}_{\mu\nu} - \gamma_{\alpha\beta}(u^j)\ .$$

If $P_{K(0)}$ denotes the orthogonal projection on the set K(0) with respect to the scalar product induced by the compliance tensor $S_{\alpha\beta\mu\nu}$, one has from (14):

$$\boxed{\sigma^{j+1}_{\alpha\beta} = P_{K(0)}\, [\bar{R}_{\alpha\beta\mu\nu}\, \gamma_{\mu\nu}(u^{j+1}) - g_{\alpha\beta}]} \tag{15}$$

In the sequel we use the notation :

$$\bar{R}_{\alpha\beta\mu\nu} = [S_{\alpha\beta\mu\nu}]^{-1} \quad (\neq R_{\alpha\beta\mu\nu})\ .$$

<u>Step 3</u>. Substituting (15) into (12) and (13), one obtains :

$$\forall v_\alpha(x_1,x_2),\ \int_\omega \left\{ \int_{-1}^{1} P_{K(0)}\, [\bar{R}_{\alpha\beta\mu\nu}\gamma_{\mu\nu}(\underline{u}^{j+1}) - x_3\bar{R}_{\alpha\beta\mu\nu}\partial_{\mu\nu}u_3^{j+1} - g_{\alpha\beta}] \right\} \cdot \gamma_{\alpha\beta}(v) = \int_\omega F_\alpha v_\alpha \tag{16}$$

$$\forall v_3(x_1,x_2),\ -\int_\omega \left\{ \int_{-1}^{1} x_3 P_{K(0)} \left[\bar{R}_{\alpha\beta\mu\nu}\gamma_{\mu\nu}(\underline{u}^{j+1}) - x_3\bar{R}_{\alpha\beta\mu\nu}\partial_{\mu\nu}u_3^{j+1} - g_{\alpha\beta}\right] \right\} \partial_{\alpha\beta}\, v_3 = \int_\omega F_3 v_3 - \int_\omega M_\alpha \partial_\alpha v_3\ . \tag{17}$$

The set of equations (16),(17) is obviously non linear because of the projection $P_{K(0)}$. It is worth noticing that the model is two-dimensional in the sense that the unknowns u^{j+1}_α, u^{j+1}_3 are only dependent on the coordinates x_α. Hence it is not necessary to introduce multilayered finite element schemes to solve the Prandtl-Reuss model. A simple way for solving numerically equations (16),(17) consists in constructing a sequence u^P, which converges to u^{j+1} when p tends to the infinity, and which is defined by :

$$\left\{ \begin{aligned} &\forall v_\alpha(x_1,x_2),\ \text{S.B.C.},\ \int_\omega R^H_{\alpha\beta\mu\nu}\, \gamma_{\mu\nu}(\underline{u}^{p+1})\, \gamma_{\alpha\beta}(v) \\ &\quad = \int_\omega R^H_{\alpha\beta\mu\nu}\, [\gamma_{\mu\nu}(\underline{u}^p) - X^p_{\mu\nu}]\, \gamma_{\alpha\beta}(v) + \int_\omega F_\alpha v_\alpha, \\ &\forall v_3\ (x_1,x_2),\ \text{S.B.C.},\ \int_\omega I_{\alpha\beta\mu\nu}\, \partial_{\mu\nu}\, u_3^{p+1}\, \partial_{\alpha\beta}\, v_3 \\ &\quad = \int_\omega I_{\alpha\beta\mu\nu}\, [\partial_{\mu\nu}\, u_3^p - Q^P_{\mu\nu}]\, \partial_{\alpha\beta}\, v_3 + \int_\omega F_3 v_3 - \int_\omega M_\alpha\, \partial_\alpha\, v_3 \end{aligned} \right. \tag{18}$$

where

$$\begin{cases} X^P_{\alpha\beta} = \int_{-1}^{1} P_{K(0)} \, [\bar{R}_{\alpha\beta\mu\nu} \, \gamma_{\mu\nu}(\underline{u}^P) - x_3 \, \bar{R}_{\alpha\beta\mu\nu} \, \partial_{\mu\nu} \, u^P_3 - g_{\alpha\beta}] \\ Q^P_{\alpha\beta} = - \int_{-1}^{1} x_3 \, P_{K(0)} \, [\bar{R}_{\alpha\beta\mu\nu} \, \gamma_{\mu\nu}(\underline{u}^P) - x_3 \, \bar{R}_{\alpha\beta\mu\nu} \, \partial_{\mu\nu} \, u^P_3 - g_{\alpha\beta}], \end{cases}$$

$P_{K(0)}$ being the orthogonal projection on the convex K(0) with help of the scalar product induced by the compliance tensor $S_{\alpha\beta\mu\nu}$, and

$$R^H_{\alpha\beta\mu\nu} = \int_{-1}^{+1} \bar{R}_{\alpha\beta\mu\nu} \, dx_3, \quad I_{\alpha\beta\mu\nu} = \int_{-1}^{+1} x_3^2 \, \bar{R}_{\alpha\beta\mu\nu} \, dx_3$$

Remark 1. The projection $P_{K(0)}$ on K(0) is a non linear problem which should be solved a large number of times through the thickness of the plate in order to have an accurate expression of $X^P_{\alpha\beta}$ and $Q^P_{\alpha\beta}$. A nice algorithm can be formulated as follows. Let x be a symmetric matrix of $\mathbb{R}^2$. Then there exists a positive constant, say μ, such that :

$$x - P_{K(0)} \, x = \mu \, n(P_{K(0)} x)$$

where n is the outward normal to K(0) at the point $P_{K(0)} x$. If we assume that the strength yield function f is such that :

$$f(x) = \frac{1}{2} \, (Bx, x),$$

then

$$x - P_{K(0)} x = \mu \, B$$

or else

$$P_{K(0)} x = (I + \mu \, B)^{-1} \, x$$

But $P_{K(0)} x$ is on the boundary of K(0), hence :

$$\left(B \, (I + \mu \, B)^{-1} \, x, \; (I + \mu \, B)^{-1} x \right) = k$$

The value of μ can then be computed by Newton's algorithm. Practically it appears to be very fast. ⊡

Step 4. Computation of $\overset{o}{\sigma}_{\alpha 3}$

The second equations (9) can be explicited as follows :

$$\begin{cases} \partial_\beta \, \overset{o}{\sigma}_{\alpha\beta} + \partial_3 \, \overset{o}{\sigma}_{\alpha 3} + f_\alpha = 0, & \alpha = 1,2, \\ \sigma_{\alpha 3} = 0 \quad \text{for } x_3 = \pm 1 & \alpha = 1,2. \end{cases}$$

The solution of this system is :

$$\boxed{\overset{o}{\sigma}_{\alpha 3} = - \int_{-1}^{x_3} \partial_\beta \, \overset{o}{\sigma}_{\alpha\beta} - \int_{-1}^{x_3} f_\alpha} \tag{19}$$

Step 5. Computation of $\overset{o}{\sigma}_{33}$

The last equation (9) (not yet used) is :

$$\begin{cases} \partial_3 \, \sigma^o_{33} + \partial_\alpha \, \sigma^o_{\alpha 3} + f_3 = 0, \\ \sigma^o_{33} = 0 \text{ for } x_3 = \pm 1, \end{cases}$$

the solution of which is :

$$\boxed{\sigma^{o}_{33} = - \int_{-1}^{x_3} \partial_\alpha \sigma^{o}_{\alpha 3} - \int_{-1}^{x_3} f_3} \tag{20}$$

Remark 2. The elastoplastic model for plate, that we have obtained, is two-dimensional. It means that it can be solved with the help of unknown functions, only defined on the medium surface of the plate. As a matter of fact, they are the components of the displacement field $(\underline{u}_\alpha, u_3)$. But the stress field is generally three dimensional. ⊡

Remark 3. One can obtain a simple characterization of the stress field $\sigma^{j+1}_{\alpha\beta}$ solution of (14),(12) and (13). Integrating (14) over Ω, leads to :

$$\forall \tau \in K(0), \int_\Omega S_{\alpha\beta\mu\nu} \sigma^{j+1}_{\mu\nu} (\tau_{\alpha\beta} - \sigma^{j+1}_{\alpha\beta})$$

$$\geqslant \int_\Omega (\tau_{\alpha\beta} - \sigma^{j+1}_{\alpha\beta}) \gamma_{\alpha\beta}(u^{j+1}) + \int_\Omega g_{\alpha\beta}(\tau_{\alpha\beta} - \sigma^{j+1}_{\alpha\beta})$$

Then, if we define the space :

$$H_F = \Big\{ \tau = (\tau_{\alpha\beta}) \ ; \ \tau_{12} = \tau_{21} \ ; \text{ such that :}$$

$$\forall \ v_\alpha(x_1,x_2) \text{ S.B.C.}, \int_\omega n_{\alpha\beta}\, \partial_\alpha v_\beta = \int_\omega F_\alpha v_\alpha$$

and

$$\forall \ v_3(x_1,x_2) \text{ S.B.C.}, -\int_\omega m_{\alpha\beta}\, \partial_{\alpha\beta} v_3 = \int_\omega F_3 v_3 - \int_\omega M_\alpha \partial_\alpha v_3 \tag{14}$$

with

$$n_{\alpha\beta} = \int_{-1}^{+1} \tau_{\alpha\beta} \quad \text{and} \quad m_{\alpha\beta} = \int_{-1}^{1} x_3 \tau_{\alpha\beta} \Big\},$$

it appears that $\sigma^{j+1}_{\alpha\beta}$ is solution of :

$$\begin{cases} \sigma^{j+1} \in K(0) \cap H_F \\ \forall \tau \in K(0) \cap H_F, a(\sigma^{j+1}, \tau - \sigma^{j+1}) \geqslant g^j(\tau - \sigma^{j+1}) \end{cases}$$

where

$$a(\sigma,\tau) = \int_\Omega S_{\alpha\beta\mu\nu}\, \sigma_{\alpha\beta}\, \tau_{\mu\nu}, \quad g^j(\tau) = a(\sigma^j - \gamma(u^j), \tau)$$

Hence, σ^{j+1} is also solution of the following problem :

$$\boxed{\begin{array}{l} \text{minimize } \frac{1}{2} a(\tau,\tau) - g^j(\tau) \\ \tau \in K(0) \cap H_F. \end{array}} \tag{15}$$

The solution exists and is unique as soon as $K(0) \cap H_F$ is not empty. An important consequence - but quite obvious from the physical point of view - is that when only membrane loadings are involved, then the in-plane stress is reduced to $\frac{n}{2}\alpha\beta$. This result is obtained with the help of (15). Indeed let σ^{j+1}, element of $K(0) \cap H_F$, be solution of (15). We associate the stress resultant by :

$$n_{\alpha\beta}^{j+1} = \int_{-1}^{+1} \sigma_{\alpha\beta}^{j+1} ,$$

and we consider the in-plane stress field $\frac{n}{2}{}_{\alpha\beta}^{j+1}$. Because there are only in-plane loading, it is clear that $\frac{n^{j+1}}{2}{}_{\alpha\beta} \in H_F$. In other respects, from :

$$\sigma^{j+1} \in K(0) \Rightarrow f(\sigma^{j+1}) \leqslant k \Rightarrow \int_{-1}^{1} f(\sigma^{j+1}) \leqslant 2\ k$$

one deduces that :

$$f\left(\frac{n}{2}^{j+1}\right) \leqslant k \Rightarrow \frac{n}{2}^{j+1} \in K(0)$$

Then using the expression :

$$\sigma_{\alpha\beta}^{j+1} = \frac{n}{2}{}_{\alpha\beta}^{j+1} + \frac{3x_3}{2}\ m_{\alpha\beta}^{j+1} + T_{\alpha\beta}^{j+1}$$

where :

$$\int_{-1}^{1} T_{\alpha\beta}^{j+1} = 0 \qquad \text{and} \qquad \int_{-1}^{1} x_3\ T_{\alpha\beta}^{j+1} = 0,$$

one obtains :

$$a(\sigma^{j+1},\sigma^{j+1}) = a\left(\frac{n}{2}^{j+1}, \frac{n}{2}^{j+1}\right) + a\left(\frac{3x_3}{2}m^{j+1}, \frac{3x_3}{2}\ m^{j+1}\right) + a\left(T^{j+1},T^{j+1}\right)$$

which implies that :

$$a(\sigma^{j+1},\sigma^{j+1}) \geqslant a\left(\frac{n}{2}^{j+1}, \frac{n}{2}^{j+1}\right)$$

If we assume, by induction, that σ^j is x_3 independent and that $u_3^j = 0$, we have :

$$g(\sigma^{j+1}) = (S\sigma^j - \gamma(u^j),\ \sigma^{j+1}) = \left(S\sigma^j - \gamma(u^j),\ \frac{n}{2}^{j+1}\right) = g\left(\frac{n}{2}^{j+1}\right)$$

Finally we deduce that :

$$\frac{1}{2}\ a(\sigma^{j+1},\sigma^{j+1}) - g(\sigma^{j+1}) \geqslant \frac{1}{2}\ a\left(\frac{n}{2}^{j+1}, \frac{n}{2}^{j+1}\right) - g\left(\frac{n}{2}^{j+1}\right)$$

which proves that :

$$\frac{n}{2}^{j+1} = \sigma^{j+1} \tag{16}$$

⊡

Remark 4. For more general loading cases it is not possible to simplify the elastoplastic model for plates and a coupling appears between membrane and bending effects. The in-plane stress remains three dimensional. It means that plastic zones can have quite arbitrary shapes. In particular the plasticity does not cross necessarily the thickness of the plate.

IV. Global models in elastoplasticity

An important problem arising in plasticity is the determination of limit load (limit analysis). Because of the three dimensional behaviour of the in-plane stress, it is very convenient to replace the convex set K(0) by a new one, which just involves the stress resultant $n_{\alpha\beta}$ and the bending moments $m_{\alpha\beta}$. As a matter of fact we define two new convex sets, one being including in K(0) the other one being larger This leads to a lower and an upper bound of the limit load. Let us consider a splitting of the stress field into three components as follows :

$$\boxed{\tau_{\alpha\beta} = \frac{n}{2}\alpha\beta + \frac{3x_3}{2} m_{\alpha\beta} + T_{\alpha\beta}}$$

with

$$\int_{-1}^{1} T_{\alpha\beta} = \int_{-1}^{1} x_3 \, T_{\alpha\beta} = 0$$

Hence, we deduce that (f being a quadratic function) :

$$\int_{-1}^{1} f(\tau) = \frac{1}{2} f(n) + \frac{3}{2} f(m) + \int_{-1}^{1} f(T)$$

and the inequality :

$$f(\tau) \leqslant k$$

implies :

$$f\left(\frac{n}{2}\right) + \frac{1}{3} f\left(\frac{3m}{2}\right) \leqslant k$$

If we set :

$$\boxed{K_{GM} = \left\{ \tau = (\tau_{\alpha\beta}),\ \tau_{12} = \tau_{21},\ \text{such that on } \omega \quad f\left(\frac{n}{2}\right) + \frac{1}{3} f\left(\frac{3m}{2}\right) \leqslant k \right\}} \tag{17}$$

(where :

$$n_{\alpha\beta} = \int_{-1}^{1} \tau_{\alpha\beta} \quad \text{and} \quad m_{\alpha\beta} = \int_{-1}^{1} x_3 \, \tau_{\alpha\beta}),$$

it appears that :

$$K(0) \subset K_{GM}.$$

Let us now introduce the convex set :

$$\boxed{K_{Gm} = \left\{ \tau = (\tau_{\alpha\beta}),\ \tau_{12} = \tau_{21},\ \text{and on } \Omega : \quad f\left(\frac{n}{2} + \frac{3x_3}{2} m\right) \leqslant k \right\}}$$

Hence if the stress field :

$$\tau_{\alpha\beta} = \frac{n}{2}\alpha\beta + \frac{3x_3}{2} m_{\alpha\beta}$$

belongs to the convex set K_{Gm}, it is also in K(0). It means that :

$$K_{Gm} \subset K(0) \subset K_{GM}$$

These inclusions enable to compare solutions obtained with help of the various convex sets defined above. Let us define, for instance, σ^m, σ^o and σ^M the three minimae of the functional (15) over the convex sets, K_{Gm}, K(0) and K_{GM}. For sake of brevity we assume that $g^j = 0$ in (15). Then we have :

$$\|\sigma - \sigma^M\|^2 \leqslant a(\sigma - \sigma^M, \sigma - \sigma^M) \leqslant a(\sigma, \sigma - \sigma^M)$$
$$\leqslant a(\sigma, \sigma - \tau) + a(\sigma, \tau - \sigma^M)$$

and for any τ in $K(0)$; $a(\sigma, \sigma - \tau) \leqslant 0$, hence :

$$\|\sigma - \sigma^M\|^2 \leqslant c \, [\inf_{\tau \in K(0)} \|\tau - \sigma^M\|] \, \|\sigma\|$$

In a similar way one can prove that :

$$\|\sigma - \sigma^m\|^2 \leqslant c \, \|\sigma^m\| \inf_{\tau \in K_{Gm}} \|\tau - \sigma\|$$

Finally the error between solutions obtained for different convex sets is bounded by the distance between K_{Gm} and K_{GM}. The error is null when there is no bending effect and is maximum for a pure bending state of stresses.

V. Plastic behaviour of the junctions between two layers in composite laminates

Let us consider a thin plate mode of several layers sticked together by a thin layer of epoxy. We assume that this layer can have a plastic behaviour. Because the thickness is very small compared to the other dimensions of the plate (even the thickness) it is possible to approximate the three dimensional elastoplastic model by a simplified one where the non linearity (behaviour) is limited to a surface (or several surfaces in case of multilayered plates).

V.1. The plastic surface model

Let us briefly recall the formulation of the plastic surface model introduced in [3]. For sake of brevity we just consider one plastic surface. The junction between the two layers is at the height εz from the medium surface. The thickness of it is 2η and it occupies in space the open set :

$$B^\eta = \omega \times \,]\varepsilon z - \eta, \varepsilon z + \eta[\, .$$

The strength yield convex is defined by :

$$K^\eta = \{\tau = (\tau_{ij}) \; ; \; \tau_{ij} = \tau_{ji} \text{ and } f(\tau) \leqslant k \text{ on } B^\eta\}$$

where $f(.)$ is the yield function. It is clear that for $\eta_1 < \eta_2$, we have : $K^{\eta_1} \supset K^{\eta_2}$. Hence the sequence K^η is increasing. Let us denote by H_F the stress fields which equilibrate the applied forces (see chapter III). Then the Prandtl Reuss' model consists in finding an element σ^η in the space $K^\eta \cap H_F$ such that :

$$\forall \tau \in K^\eta \cap H_F, \; a(\dot\sigma^\eta, \tau - \sigma^\eta) \geqslant 0$$

A basic point in our analysis is to characterize the limit set $K^\eta \cap H_F$ when

η tends to zero. This result can be obtained with the help of Stokes formula. We proved in [3] that this limit set, denoted $K^o \cap H_F$ is deduced from the strength yield function f(.). As a matter of fact K^o depends on the normal components of the stress to the surface ω. If we denote by $P\tau$ the stress deduced from τ by :

$$(P\tau)_{ij} = (\tau_{i3}\,\delta_{j3} + \tau_{j3}\,\delta_{i3})/2$$

then K^o is defined by :

$$K^o = \left\{\tau - (\tau_{ij})\ ;\ \tau_{ij} = \tau_{ji}\ ;\ f(P\tau) \leq k \text{ on } \omega \times \{\varepsilon z\}\right\}$$

From the mathematical point of view, it should be noticed that $P\tau$ can be defined on the surface $\omega \times \{\varepsilon z\}$ as soon as $\tau \in H_F$. In the sequel it is very convenient to use the convex set $\underset{\sim}{K}^o$ which just involves vector fields defined on $\omega \times \{\varepsilon z\}$ and which can be interpreted as the normal components of stress field of K^o. Finally the limit model when η tends to zero can be formulated as follows :

$$\text{find } \sigma(t) \in K^o \cap H_F \text{ such that :}$$

$$\forall\, \tau \in K^o \cap H_F,\ a(\dot{\sigma}, \tau - \sigma) \geq 0$$

One can prove that there exists a unique displacement field u in the space:

$$V_z^\varepsilon = \{v = (v_i)\ ;\ v_i \in H^1\ (\Omega^\varepsilon/\omega \times \{\varepsilon z\})\ ;\ v_i = 0 \text{ on } \Gamma_0^\varepsilon\},$$

(ie v can be discontinuous across $\omega \times \{\varepsilon z\}$). Furthermore, there is a unique vector field $\lambda = (\lambda_i)$ in the convex set $\underset{\sim}{K}^o$ such that :

$$\left\{\begin{array}{ll} \text{i)} & \sigma_{ij} = R_{ijk\ell}\,\gamma_{k\ell}(u) \text{ on } \Omega^\varepsilon/\omega \times \{\varepsilon z\}, \\ \text{ii)} & \forall\ v \in V_z^\varepsilon, \displaystyle\int_{\Omega^\varepsilon/\omega\times\{\varepsilon z\}} \sigma_{ij}\,\partial_i\, v_j - \int_{\omega\times\{\varepsilon z\}} \lambda_i\,[v_i] = \int_{\Omega^\varepsilon} f_i\, v_i \\ \text{iii)} & \forall\, \mu \in \underset{\sim}{K}^o,\ (\mu_i - \lambda_i)\ [\dot{u}_i] \leq 0 \text{ on } \omega \times \{\varepsilon z\} \end{array}\right.$$

were $[\phi]$ denotes the jump of the function ϕ across the surface $\omega \times \{\varepsilon z\}$. This model is simple to be solved because it is basically linear, the non linearity being restricted to the surface $\omega \times \{\varepsilon z\}$. It can be solved numerically with a duality algorithm (for instance Uzawa's method). Our purpose now is to deduce from (20) a plate model.

V.2. The elastoplastic plate model

Using the same method as the one described in chapter II, we can transform problem (20) into an equivalent one but set over Ω instead of Ω^ε. Let us set :

$$V_z = V_z^\varepsilon \qquad \text{for } \varepsilon = 1$$

Then the limit model (for $\varepsilon = 0$) can be formulated as follows :

find $(\lambda^o, u^o) \in \underset{\sim}{K}^{oo} \times V_z$ such that :

$$\begin{cases} S_{\alpha\beta\mu\nu}\, \sigma^o_{\mu\nu} = \gamma_{\alpha\beta}(u^o), \\ \gamma_{i3}(u^o) = 0, \\ \forall\, \mu \in \underset{\sim}{K}^{oo} \quad (\mu_\alpha - \lambda^o_\alpha)\, [u^{-o}_\alpha] \leqslant 0 \text{ on } \omega \times \{z\}, \\ \forall\, v \in V_z, \displaystyle\int_{\Omega/\omega\times\{z\}} \sigma_{ij}\, \partial_i v_j - \int_{\omega\times\{z\}} \lambda^o_\alpha\, [v_\alpha] = \int_\Omega f^o_i\, v_i, \end{cases}$$

where $\underset{\sim}{K}^{oo}$ is the limit convex for $\varepsilon = 0$ (see section II). As a matter of fact, let us assume that :

$$\underset{\sim}{K}^o = \left\{\lambda = (\lambda_i)\ ;\ \sum_{i,j=1,E} \alpha_{ij}\, \lambda_i\, \lambda_j \leqslant k \text{ on } \omega \times \{z\}\right\},$$

then

$$\boxed{\underset{\sim}{K}^{oo} = \left\{\lambda = (\lambda_\alpha) \quad \sum_{\alpha,\beta=1,2} a_{\alpha\beta}\, \lambda_\alpha\, \lambda_\beta \leqslant k \text{ on } \omega \times \{z\}\right\}} \,. \tag{22}$$

This notation implies in particular that $\lambda_3 = 0$. Indeed the solution of (21) can be explicited with the help of two models, the non linearity being limited to one of these two. The displacement u^o_3 is independent on x_3. The components u^o_α are piecewise linear with respect to x_3 such that :

$$\boxed{\begin{cases} u^o_\alpha = \underline{u}^+_\alpha (x_1, x_2) - (x_3 - \xi^+)\, \partial_\alpha u_3 & x_3 > z, \\ u^o_\alpha = u^-_\alpha (x_1, x_2) - (x_3 - \xi^-)\, \partial_\alpha u_3 \,. & x_3 < z, \end{cases}} \tag{23}$$

($\xi^\pm$ being the medium height of the two portions of the plate). The in-plane stress $\sigma^o_{\alpha\beta}$ is also piecewise linear in x_3. We set :

$$\boxed{\sigma^o_{\alpha\beta} = \frac{n^\pm}{h^\pm}\alpha\beta + \frac{12}{(h^\pm)^3}(x_3 - \xi^\pm)\, m_{\alpha\beta}} \tag{24}$$

where

$$h^+ = (1 - z), \quad h^- = (1 + z).$$

Hence the behaviour relationships are :

$$\boxed{\begin{aligned} n^+_{\alpha\beta} &= \left[\int_z^1 \overline{R}_{\alpha\beta\mu\nu}\right] \gamma_{\mu\nu}(\underline{u}^+) - \left[\int_z^1 (x_3 - \xi^+)\, \overline{R}_{\alpha\beta\mu\nu}\right] \partial_{\mu\nu}\, u^o_3\,, \\ n^-_{\alpha\beta} &= \left[\int_{-1}^z \overline{R}_{\alpha\beta\mu\nu}\right] \gamma_{\mu\nu}(\underline{u}^-) - \left[\int_{-1}^z (x_3 - \xi^-)\, \overline{R}_{\alpha\beta\mu\nu}\right] \partial_{\mu\nu}\, u^o_3\,, \\ m_{\alpha\beta} &= -\left[\int_{-1}^z (x_3 - \xi^-)^2\, \overline{R}_{\alpha\beta\mu\nu} + \int_z^1 (x_3 - \xi^+)^2\, \overline{R}_{\alpha\beta\mu\nu}\right] \partial_{\mu\nu}\, u^o_3\,, \\ &\quad + \left[\int_{-1}^z (x_3 - \xi^-)\, \overline{R}_{\alpha\beta\mu\nu}\right] \gamma_{\mu\nu}(\underline{u}^-) + \left[\int_z^1 (x_3 - \xi^+)\, \overline{R}_{\alpha\beta\mu\nu}\right] \gamma_{\mu\nu}(\underline{u}^+) \,. \end{aligned}} \tag{25}$$

And from the equilibrium equations we obtain (with $F^+_\alpha = \int_z^1 f^0_\alpha$ and $F^-_\alpha \int_{-1}^z f^0_\alpha$):

$$\boxed{\begin{aligned}
&\forall\, v^+_\alpha,\ \int_\omega n^+_{\alpha\beta}\,\partial_\alpha\, v^+_\beta - \int_\omega \lambda_\alpha\, v^+_\alpha = \int_\omega F^+_\alpha\, v^+_\alpha \\
&\forall\, v^-_\alpha,\ \int_\omega n^-_{\alpha\beta}\,\partial_\alpha\, v^-_\beta + \int_\omega \lambda_\alpha\, v^-_\alpha = \int_\omega F^-_\alpha\, v^-_\alpha \\
&\forall\, v_3,\ -\int_\omega m_{\alpha\beta}\,\partial_{\alpha\beta}\, v_3 = \int_\omega F_3\, v_3 - \int_\omega M_\alpha\,\partial_\alpha\, v_3
\end{aligned}} \qquad (26)$$

Finally the discontinuity of $\dot{u}^0_\alpha$ across the surface $\omega \times \{z\}$ is governed by (on ω) :

$$\boxed{\forall\, \mu \in \underset{\sim}{K}^{00},\ (\mu_\alpha - \lambda^0_\alpha)\,(\dot{u}^+_\alpha - \dot{u}^-_\alpha) \leqslant 0.} \qquad (27)$$

The transverse shear stress is then deduced from $\sigma^0_{\alpha\beta}$ and λ^0_α by integration of the three dimensional equilibrium relations. The same is true for σ^0_{33}.

Remark 5. The model obtained is obviously non linear. Indeed there are seven unknown fields : $u^\pm_\alpha$, λ^0_α and u^0_3. The solution can be performed as for Prandtl and Reuss' model, using an incremental procedure. It is clear from (27) that if λ^0_α belongs to the inside of $\underset{\sim}{K}^{00}$, then the jump $[\dot{u}^0_\alpha]$ across $\omega \times \{z\}$ is null. In such a case the model is linear and there is a decoupling between membrane and bending effects. When $[\dot{u}^0_\alpha] \neq 0$, a dislocation appears between two layers. It is realistic to imagine that a delamination crack can appear for $[\dot{u}^0_\alpha]$ large enough. Hence a damage model would be to state a limitation on $[u^0]_\alpha$ above which the delamination increases. Another important point is that the crack (delamination) is submitted to shear stress (no opening mode). Hence a bifurcation would certainly appear depending of course on the toughness of each layer in the direction of the crack orientation. For instance it is clear that if the fibers of a layer are parallel to the crack tip, the delamination will certainly jump across this layer (see figure 2).

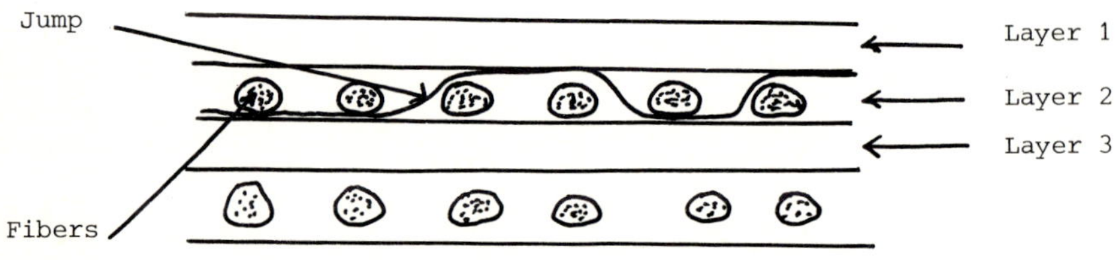

Figure 2

VI. A MODEL FOR THE ANALYSIS OF ELASTOPLASTIC BUCKLING [8],[9],[10],[11]

Let us consider in this section that the displacements of the structure are such that it is necessary to keep the complete expression of the strain tensor. Our goal is then to derive from the three dimensional a non linear plate model which permits to study both plastic and geometrical stability. The three dimensional formulation consists in finding an element (σ,u) such that on the one hand :

$$\sigma_{ij} = R_{ijk\ell}\,(\underset{\sim}{\gamma}_{k\ell} - \gamma^{p}_{k\ell}), \tag{28}$$

where

$$\underset{\sim}{\gamma}_{k\ell} = \frac{1}{2}\,(\partial_k u_\ell + \partial_\ell u_k + \partial_\ell u_j\, \partial_k u_j) \tag{29}$$

and on the other hand :

$$\forall v \in \underset{\sim}{V},\ \int_{\Omega^\varepsilon} \sigma_{ij}\, \partial_i v_j + \int_{\Omega^\varepsilon} \sigma_{ik}\, \partial_k u_\ell\, \partial_i v_\ell = \int_{\Omega^\varepsilon} f_i\, v_i\,. \tag{30}$$

Furthermore the evolution of the plastic strain γ^p is supposed to be governed by Hill's principle :

$$\forall \tau \in K^\varepsilon,\ (\tau_{ij} - \sigma_{ij}) \cdot \dot{\gamma}^{p}_{ij} \leqslant 0\,. \tag{31}$$

The question of existence of a solution is still out of reach. But using the asymptotic method, it is possible to derive from (28),(31), a limit model the solution of which approximates the assumed three dimensional one. The process is the following one :

let $(\sigma^\varepsilon,u^\varepsilon)$ be the element constructed from (σ,u) through formula (6). Then it is solution of the following set of relations :

$$\begin{aligned}\forall \tau \in K(\varepsilon),\ & S_{\alpha\beta\mu\nu}\,\dot{\sigma}^{\varepsilon}_{\mu\nu}\,(\tau_{\alpha\beta} - \sigma^{\varepsilon}_{\alpha\beta})\\ & + 4\,\varepsilon^2 S_{\alpha 3\beta 3}\,\dot{\sigma}^{\varepsilon}_{\alpha 3}\,(\tau_{\beta 3} - \sigma^{\varepsilon}_{\beta 3}) + \varepsilon^2 S_{33\mu\nu}\,\dot{\sigma}^{\varepsilon}_{33}\,(\tau_{\mu\nu} - \sigma^{\varepsilon}_{\mu\nu})\\ & + \varepsilon^2 S_{33\mu\nu}\,\dot{\sigma}^{\varepsilon}_{\mu\nu}\,(\tau_{33} - \sigma^{\varepsilon}_{33}) + \varepsilon^4 S_{3333}\,\dot{\sigma}^{\varepsilon}_{33}(\tau_{33} - \sigma^{\varepsilon}_{33})\\ & \geqslant (\tau_{ij} - \sigma^{\varepsilon}_{ij})\,(\gamma_{ij}(\dot{u}^\varepsilon) + \partial_i u^3_\alpha\, \partial_j \dot{u}^\varepsilon_3)\\ & + \varepsilon^{-2}\,(\tau_{ij} - \sigma^{\varepsilon}_{ij})\,\partial_i u^\varepsilon_3\, \partial_i \dot{u}_3,\end{aligned} \tag{32}$$

$$\begin{aligned}\forall v \in V, & \int_\Omega \sigma^\varepsilon_{ij}\, \partial_i v_j + \int_\Omega \sigma^\varepsilon_{ij}\, \partial_i u^\varepsilon_\alpha\, \partial_j v_\alpha\\ & + \varepsilon^{-2}\int_\Omega \sigma^\varepsilon_{ij}\, \partial_i u^\varepsilon_3\, \partial_j v_3 = \int_\Omega f_i\, v_i\end{aligned} \tag{33}$$

In order to construct an asymptotic expression of $(\sigma^\varepsilon,u^\varepsilon)$, with respect to ε, it is necessary to assume that the external loads are such that :

$$f_\alpha = \varepsilon^2 f^o_\alpha \quad \text{and } f_3 = \varepsilon^3 f^o_3$$

Otherwise the displacements would increase without limit when ε tends to zero. Then if we set a priori :

$$(\sigma^{\varepsilon},u^{\varepsilon}) = \varepsilon^2 (\sigma^2,u^2) + \varepsilon^4 (\sigma^4,u^4) + \ldots\ldots \text{ etc}$$

and introducing this expression in (32),(33) and by equating the predominant terms in the resulting expression, we obtain :

$$\forall \tau \in K(0),\ S_{\alpha\beta\mu\nu}\,\dot{\sigma}^2_{\mu\nu}(\tau_{\alpha\beta} - \sigma^2_{\alpha\beta}) \geqslant (\tau_{ij} - \sigma^2_{ij})\,\gamma_{ij}(\dot{u}^2) \tag{34}$$

$$+ (\tau_{ij} - \sigma^2_{ij})\,\partial_i u^2_3\,\partial_j \dot{u}^2_3,$$

(it is necessary to set $k = \varepsilon^2 k^0$ and this is an important assumption on the strength limit),

$$\forall v \in V,\ \int_{\Omega} \sigma^2_{ij}\,\partial_i v_j + \int_{\Omega} \sigma^2_{ij}\,\partial_i u^2_3\,\partial_j v_3 = \int_{\Omega} f^0_i v_i \tag{35}$$

The solution of this limit model is similar to the one described in section III. An extended justification is given in [2]. As a matter of fact, (σ^2,u^2) can be computed as follows :

The displacement u^2_3 is only dependent on the coordinates x_1 and x_2, and the other components are such that :

$$u^2_{\alpha} = \underline{u}^2_{\alpha}(x_1,x_2) - x_3\,\partial_{\alpha} u^2_3.$$

The in-plane stress field velocity $\dot{\sigma}^2_{\alpha\beta}$ is given by :

$$\forall \tau_{\alpha\beta} \in K(0),\ S_{\alpha\beta\mu\nu}\,\dot{\sigma}_{\mu\nu}(\tau_{\alpha\beta} - \sigma^2_{\alpha\beta}) \geqslant \tag{36}$$

$$(\tau_{\alpha\beta} - \sigma^2_{\alpha\beta})\,(\gamma_{\alpha\beta}(\underline{\dot{u}}^2) - x_3\,\partial_{\alpha\beta}\dot{u}_3 + \partial_{\alpha} u^2_3\,\partial_{\beta}\dot{u}^2_3)$$

and if $n^2_{\alpha\beta}$ (respectively $m^2_{\alpha\beta}$) is the stress resultant (respectively the bending moment), then (F_{α}, F_3 and M_{α} were defined in (13)) :

$$\begin{cases} \forall v_{\alpha}(x_1,x_2),\ \text{S.B.C.},\ \int_{\omega} n^2_{\alpha\beta}\,\partial_{\alpha} v_{\beta} = \int_{\omega} F_{\alpha} v_{\alpha}, \\ \forall v_3(x_1,x_2),\ \text{S.B.C.},\ -\int_{\omega} m^2_{\alpha\beta}\,\partial_{\alpha\beta} v_3 + \int_{\omega} n^2_{\alpha\beta}\,\partial_{\alpha} u^2_3\,\partial_{\beta} v_3 = \int_{\omega} F_3 v_3 - \int_{\omega} M_{\alpha}\,\partial_{\alpha} v_3\,. \end{cases} \tag{37}$$

The transverse shear stress $\sigma^2_{\alpha 3}$ and the normal stress σ^2_{33} are given by the analogous with formulae (19) and (20). The stability problem can be studied with the help of model (36),(37). As a matter of fact, it is more convenient to formulate the plate model only with respect to the displacement field. Then the stability depends on the sign of the tangent stiffness. There is an interesting case where it is possible to make more explicit the non linear model. It is obtained for an in-plane loading (ie $F_3 = M_{\alpha} = 0$). Then $u^2_3 = 0$ up to the buckling, because of the uniqueness of the solution. The displacement $\underline{u}^2_{\alpha}$ is solution of Prandtl and Reuss' model for plate ; (we have seen at remark 3 that in this case the in plane stress, $\sigma^2_{\alpha\beta}$, is reduced to the stress resultant $\frac{n^2}{2}\alpha\beta$). Hence, one can compute

$(n^2_{\alpha\beta}, \underline{u}^2_\alpha)$ up to the buckling with a simple model without geometrical non linearity. Let us assume that the loading is proportional to a real parameter, say q. The limit load is supposed to be obtained for q_ℓ and the plasticity appears in the structure for q_p. Obviously one has :

$$q_p \leqslant q_\ell$$

Let us discuss the stability model with respect to the value of q. First of all, if $q < q_p$, (q is assumed to be positive), the stress resultant $n^2_{\alpha\beta}$ is proportional to q. Let us set :

$$n^2_{\alpha\beta} = q \, \overset{o}{n}_{\alpha\beta}$$

If f (.) is the strength yielding function we have at each point of ω:

$$f(n^2_{\alpha\beta}) < k .$$

Hence the stability can be discussed in the elasticity framework. More precisely the buckling can occur if there exists an element $\dot{u}^2_3$ different from zero and satisfying the boundary conditions, such that :

$$\forall\, v_3 \text{ S.B.C.}, \int_\omega I_{\alpha\beta\mu\nu}\, \partial_{\alpha\beta}\, \dot{u}^2_3\, \partial_{\mu\nu}\, v_3 + q \int_\omega \overset{o}{n}_{\alpha\beta}\, \partial_\alpha\, \dot{u}^2_3\, \partial_\beta\, v_3 = 0 \qquad (38)$$

(indeed the non linear term $\partial_\alpha\, u^2_3\, \partial_\beta\, \dot{u}^2_3$ is null up to the buckling). The model (38) is a classical eigenvalue problem. After solving it one should check that the smallest eigenvalue is smaller than q_p. The second possibility is $q_p < q < q_\ell$. Let us denote by ω_p the plastic area defined by :

$$\omega_p = \{m \in \omega \text{ : such that } f\left(\frac{n^2}{2}\right) = k\}$$

The inequation (36) can be written on ω_p :

$$\forall\, \tau \in K(0),\ S_{\alpha\beta\mu\nu}\, \dot{\sigma}_{\mu\nu}(\tau_{\alpha\beta} - \frac{n^2}{2}\alpha\beta) \geqslant (\tau_{\alpha\beta} - \frac{n^2}{2}\alpha\beta)\ (\gamma_{\alpha\beta}(\dot{\underline{u}}^2) - x_3\, \partial_{\alpha\beta}\, \dot{u}^2_3) \qquad (39)$$

(because $\sigma^2_{\alpha\beta} = \frac{n^2}{2}\alpha\beta$ and $u^2_3 = 0$ up to the buckling). From (39) we deduce for any τ in K(0) and independent on x_3, that :

$$S_{\alpha\beta\mu\nu}\, \frac{\dot{n}}{2}\mu\nu\ (\tau_{\alpha\beta} - \frac{n^2}{2}\alpha\beta) \geqslant (\tau_{\alpha\beta} - \frac{n^2}{2}\alpha\beta)\ \gamma_{\alpha\beta}(\dot{\underline{u}}^2)$$

Hence $(\dot{n}^2, \dot{\underline{u}}^2)$ is solution of the same model as if there were no buckling and can be computed separately. In other respects, (39) implies that there exists a positive real function $\xi(x_3)$ such that :

$$\overline{R}_{\alpha\beta\mu\nu}\, [\gamma_{\mu\nu}(\dot{\underline{u}}^2) - x_3\, \partial_{\mu\nu}\, \dot{u}^2_3] = \xi\, \overline{R}_{\alpha\beta\mu\nu}\, \frac{\partial f}{\partial n_{\mu\nu}}\left(\frac{n^2}{2}\right) + \dot{\sigma}^2_{\alpha\beta},$$

and because ξ should be positive :

$$\xi = \left[\bar{R}_{\alpha\beta\mu\nu}(\gamma_{\mu\nu}(\underline{\dot{u}}^2) - x_3\partial_{\mu\nu}\dot{u}_3^2)\frac{\partial f}{\partial n_{\alpha\beta}}\right]^+ \Big/ \left(\bar{R}_{\alpha\beta\mu\nu}\frac{\partial f}{\partial n_{\alpha\beta}}\frac{\partial f}{\partial n_{\mu\nu}}\right).$$

Hence, taking the time derivative of (37) :

$$\forall\, v_3 \text{ S.B.C.}, \int_\omega I_{\alpha\beta\mu\nu}\,\partial_{\alpha\beta}\dot{u}_3^2\,\partial_{\mu\nu}v_3 + \int_\omega \xi^1 \bar{R}_{\alpha\beta\mu\nu}\frac{\partial f}{\partial n_{\mu\nu}}\partial_{\alpha\beta}v_3 \tag{40}$$

$$+ \int_\omega n^2_{\alpha\beta}\,\partial_\alpha \dot{u}^2_3\,\partial_\beta v_3 = 0$$

where ($\bar{R}_{\alpha\beta\mu\nu}$ has be assumed to be constant in x_3)

$$\xi^1 = \int_{-1}^{+1} x_3\,\xi\,.$$

Another formulation can be deduced from (40) if we set :

$$\omega_{p1} = \left\{m \in \omega_p \quad \text{where} \quad \bar{R}_{\alpha\beta\mu\nu}\,\gamma_{\mu\nu}(\underline{\dot{u}}^2)\frac{\partial f}{\partial n_{\alpha\beta}} < 0\right\},$$

$$\omega_{p2} = \left\{m \in \omega_p \quad \text{where} \quad \bar{R}_{\alpha\beta\mu\nu}\,\gamma_{\mu\nu}(\underline{\dot{u}}^2)\frac{\partial f}{\partial n_{\alpha\beta}} = 0\right\},$$

$$\omega_{p3} = \left\{m \in \omega_p \quad \text{where} \quad \bar{R}_{\alpha\beta\mu\nu}\,\gamma_{\mu\nu}(\underline{\dot{u}}^2)\frac{\partial f}{\partial n_{\alpha\beta}} > 0\right\}.$$

Then

$$\xi^1 = \begin{cases} 0 \quad \text{on } \omega_{p1}\,, \\ 0 \text{ on } \omega_{p2}\,, \\ -\left[\bar{R}_{\alpha\beta\mu\nu}\dfrac{\partial f}{\partial n_{\alpha\beta}}\partial_{\mu\nu}\dot{u}^2_3\right] \Big/ \left(\bar{R}_{\alpha\beta\mu\nu}\dfrac{\partial f}{\partial n_{\alpha\beta}}\dfrac{\partial f}{\partial n_{\mu\nu}}\right), \text{ on } \omega_{p3}\,. \end{cases}$$

Finally problem (41) is reduced to a linear eigenvalue problem at each loading step. It consists in finding an element $\dot{u}_3^2$ different from zero, satisfying the boundary conditions and such that :

$$\boxed{\begin{aligned} &\forall\, v3, \text{ S.B.C.}, \int_\omega I_{\alpha\beta\mu\nu}\,\partial_{\alpha\beta}\dot{u}^2_3\,\partial_{\mu\nu}v_3 + \int_\omega n^2_{\alpha\beta}\,\partial_\alpha\dot{u}^2_3\,\partial_\beta v_3 \\ &-\int_{\omega_{p3}} \frac{I_{\alpha\beta\mu\nu}\,\partial_{\mu\nu}\dot{u}^2_3\,\dfrac{\partial f}{\partial n_{\alpha\beta}}\,\bar{R}_{\alpha\beta\mu\nu}\dfrac{\partial f}{\partial n_{\mu\nu}}\,\partial_{\alpha\beta}v_3}{\bar{R}_{\alpha\beta\mu\nu}\dfrac{\partial f}{\partial n_{\alpha\beta}}\dfrac{\partial f}{\partial n_{\mu\nu}}} = 0. \end{aligned}} \tag{41}$$

The basic point in the obtention of (41) is that the evolution of the membrane plastic strain is independent on the buckling mode. Furthermore the plastic behaviour of the branching solution is governed by membrane stresses. This particular model seems to be a nice candidate for studying the elastoplastic buckling of plate under in-plane loading.

References

1. Ciarlet, P.G.; Destuynder, P. : A justification of the two dimensional linear plate model. J. Mécan., Vol. 18, N° 2, 1979.

2. Ciarlet, P.G.; Destuynder, P. : A justification of a non linear model in plate theory. Comp. Meth. Appl. Mechs. Eng., Vol. 17/18, p. 227-258, 1979.

3. Destuynder, P.; Neveu, D. : Sur les modèles de lignes plastiques en mécanique de la rupture. To appear in R.A.I.R.O. Analyse Numérique (Paris).

4. Destuynder, P. : Sur les modèles de plaques minces en élasto-plasticité. J. Mécan. Théo. Appl., Vol. 1, N° 1, p. 73-80, 1982.

5. Destuynder, P. : Sur la propagation des fissures dans les plaques minces en flexion. J. Mécan. Théor. et Appl., Vol. 1, N° 4, p. 579-594, 1982.

6. Duvaut, G.; Lions, J.L. : Les inéquations en mécanique et en physique. Dunod, Paris, 1972.

7. Gol'denveizer, A.L. : Derivation of an approximate theory of bending of a plate by a method of asymptotic integration of the equations of the theory of elasticity. J. Appl. Math., Vol. 19, p. 1000-1025, 1963.

8. Nguyen, Q.S.; Gary, G. : Flambage par déformations plastiques cumulées sous charges cycliques additionnelles. J. Mécan. Théor. et Appl., Vol. 2, N° 3, p. 351-373, 1983.

9. Nguyen, Q.S. : Loi de comportement élasto-plastique des plaques et des coques minces. Actes du colloque franco-polonais. Problèmes non linéaires de mécanique, p. 413-422, 1978.

10. Nguyen, Q. S. Eifurcation et stabilité des systèmes irréversibles obéissant au principe de travail maximal. J. de Mécanique Théorique et appliquée, vol 3, n°1, pp. 41-61, 1984.

11 Nguyen; Q. S. Bifurcation and stability in plasticity. Seminar of the college of engng., University of Michigan, Ann Arbor, (U.S.A.), 1984.

12 Valid R. Lectures on non linear shell theory. INRIA,EDF,CEA 1983-1984 published by INRIA,Le chesnay Rocquencourt,78150 FRANCE.

Analysis of Large Plastic Deformations in Shell Structures

C.R. CALLADINE

Department of Engineering,
University of Cambridge,
Cambridge CB2 1PZ, U.K.

Summary

Four specific problems of large plastic deformation in shell structures are analysed and discussed. They are:- inversion of a spherical shell; formation of "flaps" in long-running ductile fracture of a high-pressure pipeline; inversion of a tube; and propagating collapse of a confined tube under external pressure. All of these examples involve travelling plastic hinges; and indeed such hinges seem to be a recurrent feature of large plastic deformations of shells. Two different kinds of travelling hinge are encountered, and analysed in simple ways. The first is a sort of rolling crease, while the second is almost purely extensional in character.

I. Introduction

In this paper I shall describe some problems which involve large plastic deformation of shell structures. Each problem will illustrate different aspects of what can happen when a shell undergoes gross distortion in the plastic range, and how we may study the behaviour. I adopt this approach for several reasons. First, it seems to me that this scheme reflects the nature of the subject: engineers encounter definite problems when they use shell structures in various applications — vehicles in collision, submarine pipes, general-purpose energy absorbers, etc. — and it is the task of applied mechanicians to explain the phenomena and thus to help their colleagues to set up rational design procedures. This kind of activity is characteristic of a scientific subject at an early stage in its development. In the later stages, of course, it may prove possible to condense our knowledge and understanding into a few general statements and theorems; which eventually may even turn into axioms for future workers.

Inelastic Behaviour of Plates and Shells
IUTAM Symposium Rio de Janeiro 1985
Editors: L. Bevilacqua, R. Feijóo and R. Valid

Our subject may be taken as beginning around 1960. In the preceding decade — which began with the publication of Hill's famous book [1] on the theory of plasticity — workers in the Soviet Union, Europe and the United States had established what is now known as the "simple plastic theory" of structures; which provided the key to rational design of many structures in the field of beams and frames, plates and shells.

That theory is based on two major idealisations: the material is regarded as being non-hardening in the plastic range ("perfectly plastic"); and the kinematics of plastic distortion of the structure is regarded as involving only "small" changes from the original geometry. Under these restrictions it is possible to establish, as we all know, the central "limit theorems" of simple plastic theory, which are of proven value in many fields of engineering, as I have said.

By 1960, however, it was becoming clear that many interesting practical problems lay beyond the range of "simple plastic theory", precisely because they did not fall within the scope of the assumption about changes in geometry being "small". These problems included the plastic collapse of flat plates and arches, which can undergo major geometrical changes in the course of deformation. These problems also posed questions about the stability of equilibrium in the plastic range [2].

In 1960 there appeared two papers, by Pugsley & Macaulay [3] and Alexander [4], on the energy-absorbing characteristics of metal tubes as they crumple under axial loading. The impetus for this work came both from the collision-performance of railway vehicle bodies, and from the nuclear power industry, where it was necessary to equip heavy control rods with energy-absorbing devices in case of accidental dropping. It turned out that although the experimentally observed concertina-folding patterns of the tubes were rather complicated, it was nevertheless possible to do some relatively simple, crude and non-rigorous calculations which gave a satisfactory "first approximation" to the gross behaviour, furnished the appropriate dimensionless groups and provided useful design formulas.

I said at the beginning that we are still in the early stages of development of a plastic theory for large deflections of shell structures. Thus, for instance, the tube-crushing phenomena first studied in 1960 have still not been fully explained, in spite of some good work by Wierzbicki [5], [6] and others.

Throughout the paper I shall assume, for the sake of brevity, that elastic effects may be disregarded.

II. The ring-loaded cylindrical shell

My first example lies within the scope of "simple" plastic theory; and I give it because it establishes quickly some ideas which will be useful in discussion later on.

A long cylindrical shell of radius a and thickness t is made of material which is perfectly plastic with yield stress Y in simple tension. It is subjected to a ring load of uniform intensity F applied radially at a given cross-sectional plane, as shown in Fig. 1. We enquire at what value of F plastic collapse takes place, according to simple plastic theory.

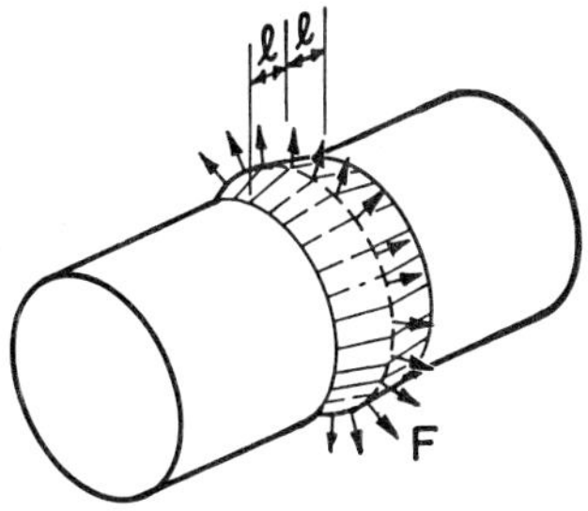

Fig. 1. Ring-loaded cylindrical shell (radius a, thickness t) in state of plastic collapse.

As in the elastic version of this problem, we are concerned primarily with a structural interaction between longitudinal bending and circumferential stretching. This kind of interaction, between bending and stretching in orthogonal directions, seems to be an ever-recurring theme in shell structures of all types [7]. A detailed analysis of the full-plastic behaviour of an element subjected to circumferential stretching stress resultant N_θ and longitudinal bending stress resultant M_ϕ gives a curved, convex interaction diagram between the two (e.g. [8], [7] chapter 18); but for most purposes of this paper it will be satisfactory to use the well-known, crude "circumscribing" yield locus for the element:

$$|N_\theta| = N_o = Yt \quad ; \quad |M_\phi| = M_o = Yt^2/4, \tag{1}$$

where t is the thickness of the shell; at yield one or both of these relations is satisfied.

We can obtain an upper bound, F^u, on the collapse value of F by doing an energy balance for the hypothetical mode of collapse shown in Fig. 1: there are three hinges, at an unknown spacing ℓ in the axial direction, and the generators remain straight between them. Let the central hinge be displaced a small distance w in the radial direction. Then the hinges rotate through angles of magnitude w/ℓ, $2w/\ell$, w/ℓ respectively, and the mean circumferential strain in the plastically deforming zone is equal to w/2a. Equating the work done by the external load to the energy dissipated in plastic bending and stretching, we have:

$$F^u.2\pi a.w = M_o.2\pi a.4w/\ell + N_o.2\pi a.2\ell.w/2a.$$

Hence

$$\frac{F^u}{Yt} = \frac{t}{\ell} + \frac{\ell}{a}\;. \tag{2}$$

We obtain the best upper-bound by minimising the RHS with respect to ℓ: thus

$$\ell = (at)^{\frac{1}{2}} \tag{3}$$

and

$$F^u = 2Yt^{\frac{3}{2}}/a^{\frac{1}{2}}. \tag{4}$$

This calculation agrees to within about 10% with the results of experiments [9], which is very satisfactory. The agreement is much better if a more exact yield locus is used, but for present purposes the crude yield locus (1) is of more interest.

Several points emerge clearly from this simple piece of analysis. (1) The optimum mode involves plastic deformation over an axial length of order $(at)^{\frac{1}{2}}$.

(2) In the optimum mode the energies dissipated in bending and stretching are equal.
(3) The mean shearing stress resultant on planes adjacent to the plane of loading is given by $F^u/2t = Y(t/a)^{\frac{1}{2}}$; and this is small in comparison with the yield stress of the material in pure shear (viz. $\frac{1}{2}Y$), provided $a/t > 10$, say. This justifies the neglect of transverse shear effects in the setting up of the yield condition for the shell element.

III. Inversion of a spherical shell

Figure 2 shows, schematically, the cross-section of a spherical shell having radius R and thickness t, which is being inverted by an inward-directed force P. (The shell is reinforced locally so that purely local deformation or "punch through" failure does not occur.)

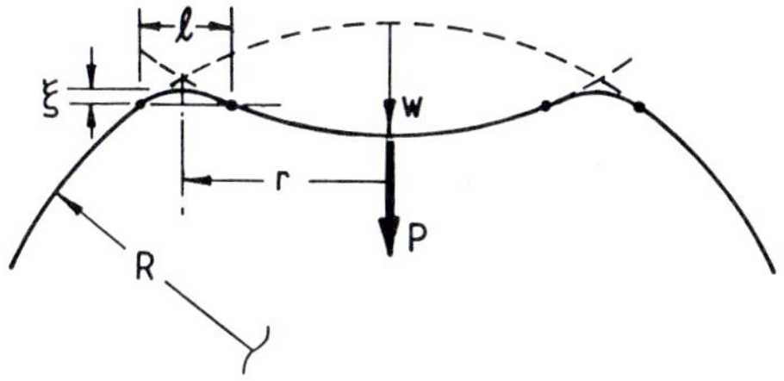

Fig. 2. Schematic cross-section of spherical shell (thickness t) being inverted by a radial force P. Hinge circles are marked by spots.

This problem was first investigated experimentally and theoretically by Wasti [10], a student of F.A. Leckie; and it has been tackled in a variety of ways subsequently [11]-[14].

At any stage in the process, the inverted region is separated from the outer portion of the sphere by a narrow toroidal "knuckle", which moves outwards as deformation proceeds. The knuckle region is connected to the two spherical portions by two travelling circumferential plastic hinge circles, of radius $r \pm \frac{1}{2}\ell$. In a frame of reference which moves with the knuckle, the material enters the knuckle at the outer hinge, which imparts the toroidal curvature to the meridian; and when it leaves by the inner hinge, curvature in the opposite sense is imposed. The meridional bending moment in the knuckle region changes between $+M_o$ at one plastic hinge and $-M_o$ at the other. Since the equations of equilibrium require M_ϕ to vary continuously along the meridian, it is reasonable to assume that during an increment of deformation the meridian of the entire toroidal part between the two hinge circles rotates as a rigid body in the plane of the meridian.

It follows immediately from kinematics (e.g. [15]) that the entry and exit hinge-circles lie in the same plane, as indicated in Fig. 2.

One of the aims of our analysis will be to determine ℓ in terms of R, t and r. Since the current configuration, Fig. 2, is a consequence of the entire preceding history of deformation, it is not immediately apparent that ℓ can be determined by a calculation related to a particular stage of the process. Here we shall treat the problem as if an "almost steady state" had been reached; this will involve, in particular, an assumption that the toroidal knuckle has uniform curvature, even though it will be found later that the mean curvature increases as the knuckle moves radially outwards.

One further point follows from the kinematical description given above. During an incremental rotation of the knuckle as a rigid body about the outer hinge, every elementary hoop of it moves inwards, towards the axis. Thus the entire knuckle region undergoes compressive hoop straining; and so we may deduce from the chosen yield locus that the entire knuckle region is at full circumferential yield stress.

Let us now turn to some conditions of equilibrium within the structure at the point of incipient plastic deformation. For the sake of simplicity we shall assume not only that the knuckle is "shallow", i.e. $r/R \ll 1$, but also that $\ell/r \ll 1$. (In fact there is no particular difficulty in relaxing either of these restrictions.)

Thus, we take the slope of the meridian at the ends of the

knuckle as $\pm r/R$; and by regarding the profile of the knuckle as parabolic we find that its "rise" ξ (Fig. 2) is given by

$$\xi = r\ell/4R. \tag{5}$$

Figure 3(a) shows the force and stress-resultants which act on the already-inverted central portion of the shell. At the outer edge we have a bending-stress resultant $M_\phi = M_o$, together with a tensile stress resultant N_ϕ. The latter is shown as being tangential to the meridian, since otherwise M_ϕ would not be a <u>maximum</u> at the hinge-circle, as it evidently must be. In Fig. 3(b) this "membrane" resultant has been replaced by its vertical and horizontal components V and H. V is determined from equilibrium in the axial direction:

$$V = P/2\pi r \tag{6}$$

and H follows from the known direction of the resultant:

$$H = VR/r = PR/2\pi r^2. \tag{7}$$

(These expressions incorporate both of the simplifications described above.) A similar analysis provides the stress-resultants acting at the outer hinge circle.

We are now in a position to to consider the equilibrium of an elementary circumferential piece of the knuckle. First we investigate the radial force-equilibrium of a piece subtending a small angle θ, as shown in Fig. 3(c). The forces transmitted across the circumferential edges are both inwards, and the circumferential forces represent full plastic compression. Radial equilibrium gives

$$2Hr\theta = Yt\ell\theta$$

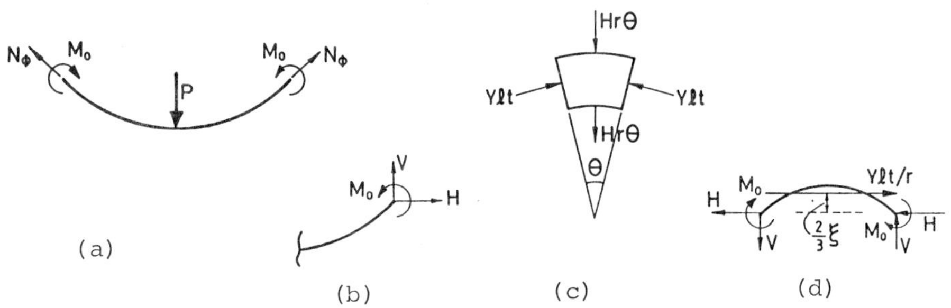

Fig. 3. Stages of the analysis of the shell shown in Fig. 2.

and hence, by (7)

$$V = Yt\ell/2R. \tag{8}$$

Next we consider moment-equilibrium about an axis tangential to the circumference. Figure 3(b) shows the stress resultants acting on the circumferential edges of a piece which has unit circumference at radius r for the sake of simplicity (i.e. $\theta = 1/r$), together with the resultant of the two forces $Yt\ell$ of Fig. 3(c). This resultant acts at the centroid of the (shallow) toroidal arc, i.e. at an elevation of $2\xi/3$ above the plane of the hinge circles. Taking moments about a point in that plane we obtain

$$V\ell = 2M_o + 2\xi Yt\ell/3r, \tag{9}$$

and hence, by (5)

$$V = 2M_o/\ell + Yt\ell/6R. \tag{10}$$

Now by (8), the last term of (10) is equal to V/3; and so (10) becomes, finally (and with (1)):

$$V = 3M_o/\ell = 3Yt^2/4\ell. \tag{11}$$

The two equilibrium equations (8), (11) may now be solved for ℓ and V:

$$\ell = (3/2)^{\frac{1}{2}}(Rt)^{\frac{1}{2}} = 1.22\ (Rt)^{\frac{1}{2}} \tag{12}$$

$$V = (3/8)^{\frac{1}{2}}Yt^{\frac{3}{2}}/R^{\frac{1}{2}} = 0.61\ Yt^{\frac{3}{2}}/R^{\frac{1}{2}}\ . \tag{13}$$

Thus we find that both ℓ and V turn out to be independent of r, which is an interesting result. Expressions (12), (13) are evidently cousins of (3), (4).

Within the assumption of shallow-shell geometry, the radial displacement w of the apex of the sphere is given by

$$w = r^2/R. \tag{14}$$

Equations (6) and (14) may thus be used to express P in terms of w:

$$P = \pi(3/2)^{\frac{1}{2}}Yt^{\frac{3}{2}}w^{\frac{1}{2}}; \tag{15}$$

or

$$P/4\pi M_o = 1.22(w/t)^{\frac{1}{2}}\ . \tag{16}$$

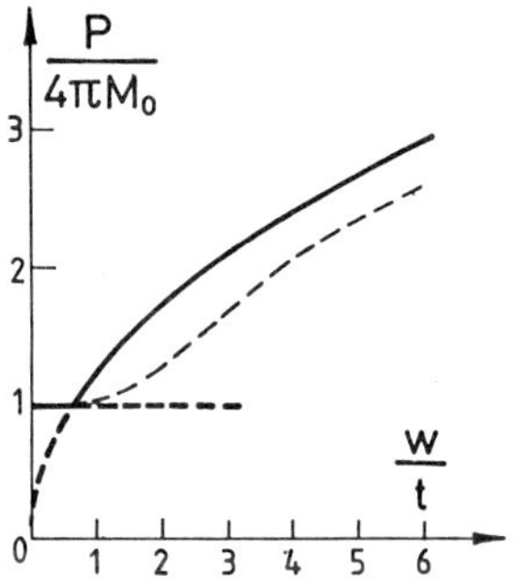

Fig. 4. Main curve, eqn (16); horizontal line, eqn (17); broken curve from ref. [12].

Formula (16) is plotted in Fig. 4. Since it is the result of a calculation in which ℓ/r is assumed to be small, it clearly cannot be considered reliable near the origin. Now when the load P is first applied, the spherical shell acts locally like a flat plate with clamped edges; for which the plastic collapse load [15] is given by

$$P = 4\pi M_o. \tag{17}$$

This line is also shown in Fig. 4, and we must regard it as cutting off the lower portion of (16).

Expression (16) agrees exactly with equation (68) of [13], when allowance is made for the fact that Updike considered the shell as being loaded through a flat surface pressed against the knuckle. In fact, Updike's analysis follows along very similar lines to mine: he uses the same yield condition , and invokes essentially the same statical condition (Fig. 3(a)) at the inner hinge. He also has an equation equivalent to (10), although he derives it in a different way by means of an energy-balance.

It is interesting to compare our results with those of Morris & Calladine [12], which are also plotted in Fig. 4. Taking a perfectly-plastic, shallow-shell approach, they followed the development of the meridional profile in detail, step by step, using an accurate "upper bound" calculation at each stage. The main overall difference between the two curves in Fig. 4 is a shift to the right equal to approximately one thickness. This is attributable to localised deformation near the apex in the early stages of deformation according to [12], which is ruled out by assumption in our analysis.

An examination of the detailed results on which [12] is based shows that when $r > 2\ell$,

$$\ell \simeq 1.3(Rt)^{\frac{1}{2}} \tag{18}$$

and

$$V \simeq 0.78Yt^{\frac{3}{2}}/R^{\frac{1}{2}}; \tag{19}$$

which are both in fairly good agreement with (12), (13). It is re-assuring to find that our value of ℓ corresponds closely to that which evolves in [12]; and indeed that both agree with photographs of deformed shells in [10]. The agreement between the present analysis and that of [12] is not complete, because we have used here a simplified yield locus (1), whereas in [12] a proper interaction between plastic bending and stretching effects was incorporated. It is comforting to know that a crude treatment of the yield locus can nevertheless give good results.

It is also interesting to note that de Oliverira & Wierzbicki [14] obtain by means of an energy-balance essentially our equation (10). But they do not use anything equivalent to (8), and instead find ℓ by minimisation of V in (10). This gives ℓ about 40% higher than our value, which is not in agreement with [12]. The lesson to be learned here is that it may be unwise to expect a reliable equilibrium condition to emerge from the minimisation of an energy expression based on an approximate yield locus.

IV. Long-running ductile fracture of high-pressure gas pipelines.

The next example concernes an aspect of the behaviour of high-pressure gas pipelines when a ductile fracture proceeds along the pipe at high speed [16], [17]. This is a very complicated problem indeed. It involves a large number of different considerations including non-steady trans-sonic gas dynamics of a non-perfect gas, both local to the burst and in the pipe as a whole; fracture mechanics; and general dynamics. I shall not describe here any of these aspects of the problem: instead, I shall consider merely the problem of determining, by means of shell theory, the shape of the "flaps" which open up behind the

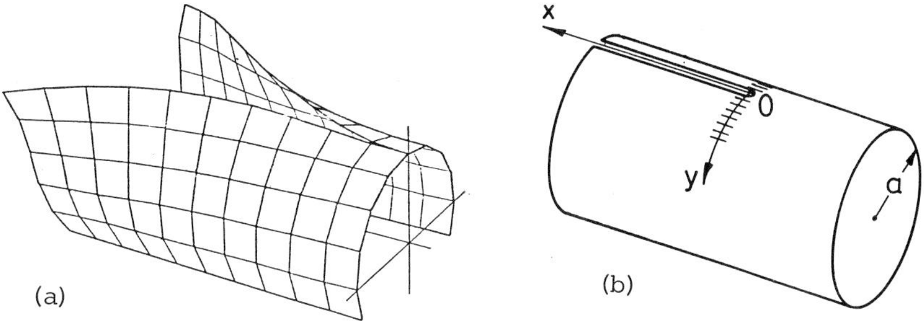

Fig. 5. (a) Computed shape of flap-opening in pipeline-fracture problem for ε_o = 0.1, b/a = 0.5 (eqn (20)). (b) Underformed pipe showing longitudinal crack; surface co-ordinate system; and the narrow zone in which plastic stretching occurs.

crack tip as it proceeds: see Fig. 5(a). These flaps form as a consequence of longitudinal plastic stretching within the centre-surface of the pipe wall. Although the flaps absorb considerable energy on account of the necessary plastic stretching (my student Abbassian concludes that this is typically of the order of 5 times the work of ductile tearing at the crack tip), they do on the other hand provide a way in which the escaping gas can do sufficient work on the walls of the pipe to maintain an "energy balance" in steady-state propagation in certain circumstances.

Consider the steady-state propagation of a crack, with its associated flaps, in a frame of reference which moves forward with the crack tip. It is reasonable to suppose that the material is given some plastic stretching strain as it passes through a narrow circumferential zone straddling the top of the pipe, as shown schematically by the shaded region of Fig. 5(b); and that thereafter, within the "flaps", it experiences no further plastic deformation apart from circumferential flexure as the flaps unfold. Consequently, along any generating line of the cylinder there is a constant pattern of strain within the region of the flaps. It is obvious that the geometrical form of the "flaps" is constrained by the manner in which the material receives its plastic strain by the passage of a "travelling stretching hinge".

In order to fix ideas, let us consider first a "small-deflection" version of a flap, in the context of an x, y cartesian coordinate system in the surface of the original pipe, as shown in Fig.5(b). Let the origin of coordinates be at the tip of the crack, with the flaps opening up on the side $x > 0$.

Again, for the sake of simplicity let us suppose that the zone of active plastic stretching lies exactly along the y-axis, and that the strain imparted to the material as it flows through this zone is

$$\varepsilon_x = \varepsilon_o \exp(-y/b), \quad x > 0. \tag{20}$$

Here, ε_o is the longitudinal strain adjacent to the torn edge, and b is a circumferential "decay length": these two parameters thus characterize the deformation.

In the small-deflection theory of cylindrical shells there is a useful geometrical expression which may be written as follows ([7], p.160):

$$-\frac{\partial^2 \varepsilon_y}{\partial x^2} + \frac{\partial^2 \gamma_{xy}}{\partial x \partial y} - \frac{\partial^2 \varepsilon_x}{\partial y^2} = -\frac{1}{a}\frac{\partial^2 w}{\partial x^2} \tag{21}$$

The LHS is an expression for the Gaussian curvature of the distorted surface in terms of the various components of surface strain, while the RHS is an expression for the Gaussian curvature in terms of principal curvatures of the surface: the negative sign on the RHS is needed when the radial component, w, of displacement is measured outwards.

In the steady state (20) will apply to all material which has passed through the plastic zone, i.e. in the region $x > 0$: in the region $x < 0$, of course, $\varepsilon_x = 0$ and the pipe is undistorted. Equation (21) gives

$$\frac{\partial^2 w}{\partial x^2} = (a/b^2)\ \varepsilon_o \exp(-y/b),$$

and integration subject to the obvious boundary conditions

$$w = \partial w/\partial x = 0 \text{ at } x = 0 \tag{22}$$

gives

$$w = (a\varepsilon_o/2b^2)\ x^2 \exp(-y/b) \tag{23}$$

as the deformed shape of the flaps for $x > 0$.

This analysis is very satisfactory in the "small deflection" range; but in practice we are concerned with larger deflections of the kind shown in Fig. 5(a). The most obvious way of proceeding would be to replace (21) by a suitable "large displacement" version, and then to integrate numerically forwards from the plane $x = 0$. This turns out to be a rather difficult task. The trouble seems to be that the solution is not necessarily unique over the entire zone $x > 0$ on account of the possibility of creases forming in the characteristic directions of the twisted surface. These difficulties may be seen as a direct consequence of our modelling the shell by a mathematical surface: such creases would not occur, of course, in a shell of finite thickness.

Abbassian has overcome this difficulty by making physical models out of thin cardboard. The material was sufficiently thin to act as a "surface", but it had sufficient flexural stiffness to avoid the ambiguities encountered in purely numerical studies of the surface. The first step was to make a model pipe of diameter 200 mm out of the cardboard. The surface of this underformed pipe was divided into 40 mm squares, and the cardboard was folded along the generators. In the "flap" region the surface was made from longitudinal strips of cardboard, composed of trapezia of height 40 mm which were related to the squares of the "underformed" part of the model by application of the straining pattern (20) at their edges. These figures were put onto the card by computer-graphics, cut out with "tabs" along the edges, scored along the edges and across the diagonals of the trapezia, and glued together. In most cases the models were made to extend 3 diameters beyond the tip of the crack. When each model was complete, the cartesian coordinates of the grid points of the flaps were recorded; then the data were read back into the computer and pictures such as Fig. 5(a) plotted out. In this way the geometry of the flaps has been determined for several values of the parameters (ε_0, b/a); and in turn these have been analysed in relation to factors which affect the distribution of pressure over the surface, and hence the input of energy to the pipe wall in a steadily-running crack.

Several points emerge from this large-deflection study of the shape of the flaps which are formed by plastic straining in the pipe wall as described above.

(1) The flaps are formed by a travelling zone of surface stretching. The bending strains — which can be determined, of course, for a given configuration when the thickness has been specified — are relatively small, and little energy of plastic deformation is absorbed in this way.

(2) Construction of a physical model turned out to be the most expeditious way of establishing the geometry of the flaps in the large-deflection range. This kind of technique, using physical models, can be useful when the geometry is awkward.

(3) The factors which determine the geometry of the flaps are not "self-contained", as in the problem of inversion of a spherical shell, for example: they depend on complicated gas-dynamical interactions which are not yet fully understood. This kind of problem, in which a shell structure is coupled with another branch of mechanics, is to be expected in future applications of shell theory to engineering problems; and it provides both a challenge and a stimulus to our thinking.

V. Axi-symmetric inversion of a tube

The large-scale inversion of a cylindrical tube of ductile metal, in the manner of Fig. 6(a), is the central feature of widely used devices for absorbing kinetic energy (e.g. [18], [19]). Figure 6 is adapted from a diagram in the first paper on the subject, by Guist & Marble, in 1966 [20]. This excellent paper raised a curious paradox which seems to have been overlooked by subsequent workers (e.g. [21], [22]) and which I shall try to resolve.

The basic analysis of [20] is extremely simple. It may be done in several different ways, and the following is perhaps the most straightforward.

As the outer part of the tube (Fig. 6(a)) moves downwards relative to the inner part, each elementary piece of the original tube in turn enters the moving toroidal region, passes through

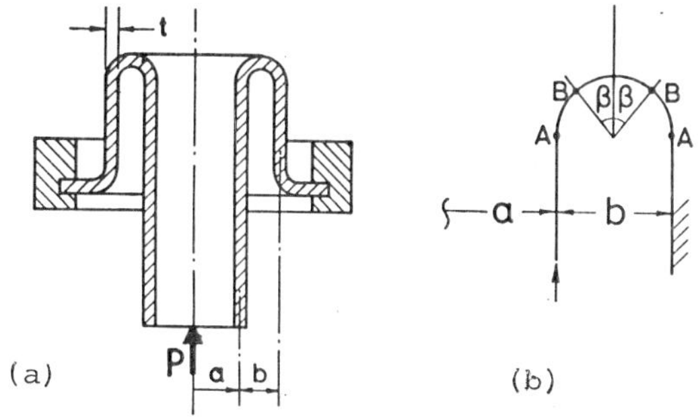

Fig. 6. (a) Cross-section of tube during inversion. As the lower end of the inner tube is pushed upwards, the tube is inverted. (b) Schematic arrangement of the knuckle region in (a).

it, and emerges at the other side. It enters as part of a tube of radius a, and it emerges as part of a tube of radius a + b. It thus experiences a circumferential strain of magnitude b/a as it passes through the toroidal region; and, if this component of strain has the largest magnitude — which is reasonable in view of the experimental observation that in this type of inversion the wall thickness changes little — then the energy absorbed per unit surface area of material is given by Ytb/a for a perfectly plastic material which obeys Tresca's yield condition ([23], §4.6).

The same piece of material also goes through a history of bending. When it passes through the "entry" travelling-hinge circle it undergoes a change of curvature in the meridional direction. The curvature imparted is equal to 2/b (i.e. the reciprocal of the radius of curvature), and this curvature remains unchanged until the material flows out through the travelling exit-hinge, where the meridional curvature is removed. The energy absorbed in bending is thus that which would be absorbed in a single change of curvature equal to 4/b. Since M_o, the full plastic moment per unit length, is equal to $Yt^2/4$, the total energy absorbed in bending is equal to Yt^2/b.

When unit area of surface is transferred in this way between the two tubes, the inner tube is shortened, relative to the toroidal region, by $1/2\pi a$, while the outer tube is lengthened by an equal amount — to a first approximation, at least, if we assume that

$b << a$. It follows that if P is the axial force required to perform the inversion, it does work equal to $P/\pi a$ during the process. Equating this to the energy absorbed internally, we have

$$P/\pi a = Ytb/a + Yt^2/b,$$

which may be re-arranged as

$$\frac{\sigma^*}{Y} = \frac{P}{2\pi atY} = \frac{1}{2}\left[\frac{b}{a} + \frac{t}{b}\right]. \qquad (24)$$

Here, on the left, the mean axial stress level σ^* has been expressed as a fraction of the uniaxial yield stress of the material.

So far, the diameter b of the toroid has been regarded as a variable. It is clear that σ^*/Y is minimum with respect to b when

$$b = (at)^{\frac{1}{2}}; \qquad (25)$$

in which case

$$\sigma^*/Y = (t/a)^{\frac{1}{2}}. \qquad (26)$$

This is a result first obtained in [20], and subsequently refined in various ways in [21], [22].

At this point it is interesting to compare the analysis with that of Alexander [4] for the collapse of a tube by the formation of axisymmetric folds in the manner of Fig. 7. Experiments show that the axial force rises to a peak as each new fold is initiated, and then falls as the lobe flattens. Alexander made an estimate of the mean axial crushing stress by equating the work done by the external load during the collapse of a single lobe to the total energy absorbed in plastic deformation within the lobe. Thinking again in terms of unit area of surface, we find that the load moves axially through a distance $1/2\pi a$; that the mean circumferential strain is equal to $c/2a$ (see Fig. 7); and that, since

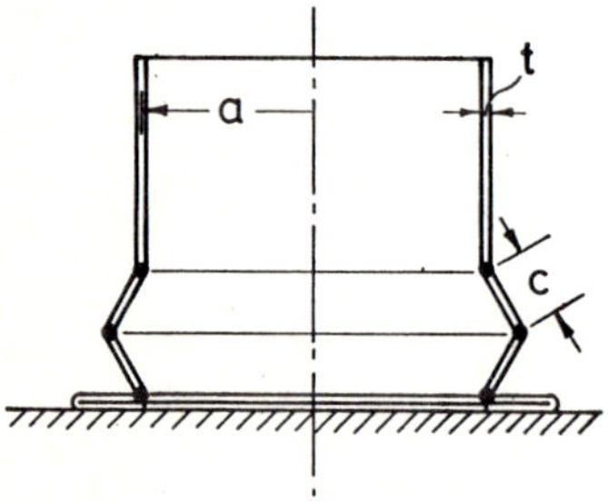

Fig. 7. Cross-section of tube under axially symmetric crumpling (after Alexander [4]).

a length c of meridian is associated with a hinge rotation π, each piece of material undergoes, on average, a change of curvature equal to π/c. In place of (24) we therefore have

$$\frac{\sigma^*}{Y} = \frac{1}{2}\left[\frac{c}{a} + \frac{\pi}{2}\frac{t}{c}\right] . \qquad (27)$$

The RHS is minimum with respect to c when

$$c = 1.25(at)^{\frac{1}{2}} , \qquad (28)$$

and with this value of c,

$$\sigma^*/Y = 1.25(t/a)^{\frac{1}{2}} . \qquad (29)$$

These two analyses of tube inversion and tube crushing are both very simple. The aim of both, of course, is to describe experimental behaviour and to provide a predictive tool for the engineer. In the absence of experimental data we might well aver that the analysis of tube-crushing is likely to be less reliable than that of tube inversion, since it involves a non-steady process, and the axial spacing c of the hinge circles is more likely to be determined at the <u>initial</u> formation of the lobe than by a minimisation of the <u>total</u> energy, as in (29). Paradoxically, the opposite is true: Alexander's formula (28) for the lobe-length agrees better with experiments that formula (25) for the toroidal diameter. And indeed, Guist and Marble found from experiments on aluminium tubes whose values of a/t ranged from 10 to 100 that the measured value of b was in the region of one-half of that given by (25); and that when the measured of b was used in (24) the calculated value of σ/Y was only about 15% lower than the experimental one.

This curious discrepancy over the value of b constitutes an interesting paradox, which warrants further study.

Suppose that we were to manufacture a partly-inverted specimen, as in Fig. 6(a), from a perfectly plastic material, and with the value of b given by (25). What will happen when we apply an upward force to the inner tube? What will be the incipient mode of plastic deformation?

One possibility, of course, is that hinge circles will form at locations shown as AA in Fig. 6(b), and that when infinitisimal plastic deformation occurs the semi-circular cross-section of the toroid will rotate as a rigid body about A on the right, thereby giving an incremental hoop strain to every hoop of material within the toroid. An energy-balance performed for this mode of incremental deformation gives precisely expression (26) again.

But it is also possible to have a different incipient mode, in which the two hinge circles BB appear in a different plane, as shown in Fig. 6(b). If incremental hinges were to occur in this way, of course, the cross-section of the toroid would not remain circular, and so we would no longer be dealing with a "self-reproducing", steady-state geometry. But there is no good reason, of course, for us to suppose that a toroidal region with an arbitrary radius would set up a self-reproducing incremental mode of deformation: this is precisely the point at which the original analysis conceals an unwarranted assumption.

Let us analyse the situation shown in Fig. 6(b), in which the hinge circles subtend angles $\pm\beta$ from the axial direction.

It is possible to do the calculation by applying the upper-bound theorem to a hypothetical incremental mode in which the two hinges rotate through equal small angles δ. It is equally simple to repeat the calculation of section III in relation to the equilibrium of the segment shown in Fig. 3(c, d), but now relaxing the "shallow shell" assumption; treating the curve BB as a circular arc; and using integration to obtain the moment-equilibrium relation. The results of these two types of calculation are identical:

$$\frac{\sigma}{Y} = \frac{1}{2}\left[\frac{t}{b}[\operatorname{cosec}\beta] + \frac{b}{a}\,[1 - \beta\cot\beta]\right]. \qquad (30)$$

When $\beta = 90^{\circ}$ we recover (24), as expected.

It is easy to show that for given values of a, t and b, the RHS of (30) is minimum when

$$\cos\beta(1 + \sin\beta)/\beta = b^2/at; \qquad (31)$$

and we must therefore expect that the value of β which actually occurs will satisfy this relation. For example, in the case

$b = (at)^{\frac{1}{2}}$ (31) gives $\beta = 57^{\circ}$, and then (30) gives

$$\sigma_m/Y = 0.77\sigma^*/Y: \tag{32}$$

here subscript m refers to the minimisation w.r.t. β.

This calculation indicates that a specimen made in this way would begin to deform plastically with hinges at $\beta = 57^{\circ}$ rather than at $\beta = 90^{\circ}$, as assumed in the previous analysis. As I have already noted, the preferred incremental mode alters the shape of the toroidal profile; and in order to follow the deformation further it would be necessary to do some detailed, step-by-step numerical calculations.

It is a straightforward matter to repeat the incremental calculation for pre-formed toroidal regions having other arbitrary values of the diameter b. Figure 8 shows the results of the calculation as a plot of σ_m/σ^* against $b/(at)^{\frac{1}{2}}$, and marks some values of β at which the minimum, σ_m, is obtained. As $b/(at)^{\frac{1}{2}}$ decreases, σ_m/σ^* and β come closer to 1 and 90°, respectively.

Now the only value of β which gives a truly steady-state mode of deformation is $\beta = 90^{\circ}$. Figure 8 is thus disappointing in relation to the observations in [20] that steady-state inversion takes place at $b/(at)^{\frac{1}{2}} \simeq 0.5$.

There is, however, another factor which we can introduce at this point. So far we have analysed an inverted tube made from a non-

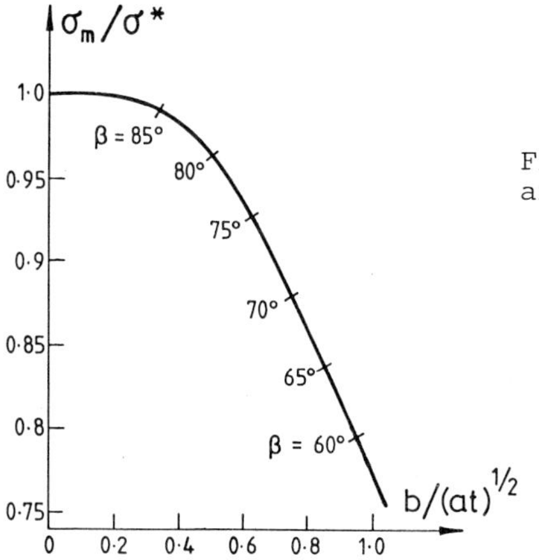

Fig. 8. Results of detailed analysis of tube-inversion.

hardening, perfectly plastic material. Real materials, such as aluminium alloys, harden when they undergo plastic deformation. It is obvious that considerable plastic strain occurs during the process of tube-inversion. For example, the measurements of Guist & Marble [20] indicate that the meridional bending strain in the knuckle region varies from about 0.2 at a/t = 100 to about 0.4 for a/t = 10. Material strained by amounts such as these will have its yield stress raised by, say, 10% above its original value. These remarks lead to the observation that although the hinge at entry to the toroidal region is formed in "fresh" material, all other plastic straining takes place in hardened material. Therefore, in terms of our previous analysis of a non-hardening material, we should consider an entry-hinge at $\beta = 90^{\circ}$ to have a specially low local value of Y. This consideration will give a slight preference to the case $\beta = 90^{\circ}$; and so we might expect the steady-state mode to occur not when $\sigma_m/\sigma^* = 1$, as previously suggested, but when $\sigma_m/\sigma^* = 0.95$, say. (This figure is a rough estimate, based on a notional 10% strain hardening, and taking account of the fact that in the region $b/(at)^{\frac{1}{2}} \simeq 0.5$ the "bending work" is dominant.)

This calculation thus provides a way of reconciling Fig. 8 with the observation in [20] that steady-state inversion takes place when $b/(at)^{\frac{1}{2}} \simeq 0.5$.

It would be possible, of course, to do a more refined calculation on the assumption of various models of hardening of the plastic material. The important point, however, is that the present analysis leading to Fig. 8, coupled with the notion of some strain hardening, successfully resolves the paradox posed by the work of Guist & Marble.

VI. Propagating buckles in externally pressurised pipes

Palmer & Martin [24] were the first to provide a satisfactory formula for the pressure required to propagate a flattened, buckled region in a submarine pipe, after a buckle has been initiated. They argued that the primary mode of deformation in the change of the cross-section of the pipe from circular to the observed "dog-bone" flattened shape was one of plastic bending of a ring; and by means of an energy-balance for an extremely simple hypothetical four-hinge mode (Fig. 9(a, b)) they obtained an estimate of the pressure required to propagate the buckle.

The calculation is very straightforward. Consider unit length of pipe. During the course of collapse the pressure does work on a volume corresponding to the shaded area in Fig. 9(a); and this work is dissipated by the rotation of four plastic hinges through 90° each. Thus

$$p.2a^2 = M_o.2\pi ;$$

hence

$$p_p/Y = (\pi/4)(t/a)^2. \qquad (33)$$

Here, subscript p stands for Palmer: p_p is "Palmer's theoretical pressure". Now that experimental data from many tests is available (e.g. [25]) over the range

$$8 < a/t < 50, \qquad (34)$$

it is clear that (33) gives good agreement with experimental observations at the upper end of the range, but that it underestimates the observed pressures for the lower values of a/t (as in [24]) by as much as about 40%: the experimental observations are fitted well by a formula [25]

$$p/Y = c\ (t/a)^{2.3}. \qquad (35)$$

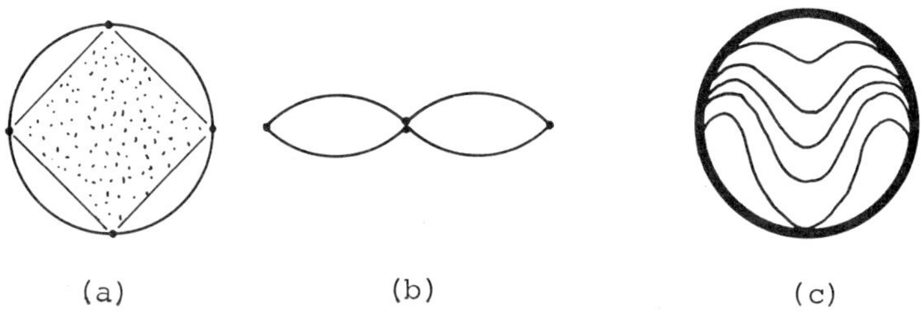

Fig. 9. Propagating buckles in externally pressurised pipelines. (a), (b) Simple hypothetical mode [24] for plastic collapse of unconfined tube. (c) Observed cross-sections of collapse of a pipe contained within a rigid cavity (c.f. Fig. 10): after [25], Fig. 5.12, as far as the point of first touch-down.

Several authors (e.g. [26], [27], [28]) have sought to explain this discrepancy by invoking strain-hardening as an explanation, along the lines suggested by Reid and Reddy [29] for the (diametral) crushing of a tube between rigid plates. It is also possible that part of the discrepancy is attributable to longitudinal stretching effects; which are unavoidable, of course, in the transition region between the circular and "dogbone" cylindrical portions of the pipe.

Kyriakides & Babcock [25] have more recently investigated another type of propagating buckle, in a pipe confined within a rigid cavity; and this will be my last example.

The mode of deformation (Figs 9(c), 10) is clearly more complicated than that of an unconfined tube: the deformation of cross-sectional "rings" evidently involves travelling hinges, and it seems obvious that the longitudinal stretching of the material will also play an important part.

The experimentally-observed propagation pressures are well fitted by the formula

$$p/Y = 1.13\pi(t/a)^2 \qquad (36)$$

in the range $8 < a/t < 50$. Thus the pressure is equal to about 4.5 times Palmer's theoretical pressure (33) for an unconfined pipe, irrespective of the value of (a/t).

I shall now attempt a preliminary analysis of the mechanics of this type of propagating buckle. The analysis will be in the form of an energy-balance, with the aim of checking that a pressure $p = 4.5p_p$ is sufficient to provide the energy absorbed in plastic deformation. In performing the calculation I shall make use of the experimentally observed cross-sectional profiles shown as Fig. 9(c).

We start with the hypothesis that energy is absorbed by two main types of structural action, namely circumferential bending and longitudinal stretching. This is reasonable in any "long-wave" deformation of a cylindrical shell: see, for example [7], chapter 9.

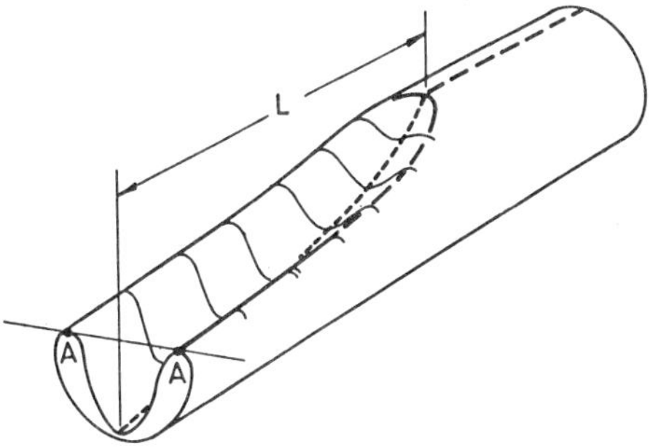

Fig. 10. Sketch of collapse of confined pipe, after photographs in [25]. The near cross-section ("exit plane") is at the point where the principal longitudinal strip (broken curve) first touches the bottom of the pipe.

Let us analyse the work done in circumferential bending. Returning first to the simple mode of Fig. 9(b), for the unconfined tube, we have a swept area of $2a^2$, and a total hinge rotation of 2π. It is useful to express the circumferential bending in terms of a mean change in curvature over the whole pipe. Since the circumference is equal to $2\pi a$, the mean curvature here is equal to $1/a$: the energy absorbed in circumferential bending is the same as if each hoop were slit once and unrolled into a flat strip. For the sequence of shapes shown in Fig. 9(c), we find that the total swept area up to the point where the pipe wall touches itself is equal to $2.4a^2$, i.e. 1.2 times the swept area for the mode of Fig. 9(a, b). The changes of curvature are, of course, more complicated to follow than those of Fig. 9(a, b), since some portions of the circumference are bent first one way and then the other as travelling hinges move through the material. In terms of energy-absorption, of course, we must add together the absolute value of all such changes; and in this way we find that the mean change of curvature over the whole circumference is equal to $3.2/a$. (These calculations have been done by S. Kamalarasa.)

It follows immediately that the pressure p_{cb} required to supply the energy dissipated in circumferential bending is given by

$$p_{cb}/p_p = 3.2/1.2 = 2.7 \quad (37)$$

In this way we thus account for about 0.6 of the experimentally observed pressure of $4.5\ p_p$.

Consider next the energy absorbed in longitudinal plastic

stretching of the material. This is a much more difficult matter. The best way of beginning is to consider the narrow "principal" strip of material (Fig. 10) which starts out as the highest generator and ends up touching the lowest generator. Let the axial length of the curved part of this strip — from the apex of the knuckle to the cross-sectional plane through the point of touch down — be L. The usual "small deflection" expressions indicate that the mean strain ε_o in the curved part is given by

$$\varepsilon_o \simeq (8/3)(a/L)^2. \qquad (38)$$

Now, since the stresses in the "banana-shaped" inverted part of the tube are unlikely to be high except at the knuckle, we must conclude that a longitudinal stretching strain of ε_o is imparted to the material when the travelling knuckle moves through it.

It is also clear that the strip must be returned again to its original length in order to lie eventually along the lower generator of the tube; and hence we must conclude that a strain of $-\varepsilon_o$ is imparted to the strip as the "exit-plane" — i.e. the cross-section which includes the point of touch-down, Fig. 10 — passes through it.

The easiest way of visualising this compressive straining is to think of the way in which the curved, banana-shaped inverted tube is turned into a cylindrical form when it passes through the plane: the action in this plane is like the straightening of a curved channel-section beam. It seems clear, indeed, that the "neutral axis" for this kind of bending must be close to the line AA in Fig. 10, which passes through the points where the inner and outer portions meet. Furthermore, the compressive straining in any longitudinal strip of the "curved channel" must be proportional to the perpendicular distance of the strip from the neutral axis.

A geometrical study of the lowest curve of Fig. 9(c) along these lines indicates that the mean longitudinal strain in the inverted region between AA is equal to about 0.45 of the peak strain; and that the inner contour-length AA is equal to about 0.45 of the complete circumference. Thus we find that the mean compressive strain imparted at the exit-plane is equal to $0.2\,\varepsilon_o$.

Now the extensional strain in any particular strip which is eliminated in passing through the exit-plane must have been provided when the material passed through the knuckle region; and so it follows that the effective mean plastic strain in the tube, taking the entry and exit processes together, is equal to $0.4\,\varepsilon_o$. (Actually, conditions at the knuckle involve bending back-and-forth as well as stretching, so the effective strain may be somewhat larger, as we shall see later.)

We cannot now proceed beyond this stage of the analysis until we can determine somehow the magnitude of ε_o or, equivalently, of L/a. We shall return to this critical point later; but for the present let us be content to work out the value of ε_o which would make the energy balance come out right.

Let p_{ls} be the pressure required to provide the energy dissipated in longitudinal stretching. Since

$$p = p_{cb} + p_{ls} , \tag{39}$$

(36) and (37) provide

$$p_{ls} = 1.8\ p_p . \tag{40}$$

Now when the buckle moves forward by unit length, the energy-balance of longitudinal stretching gives

$$p_{ls}(2.4a^2) = 2\pi a t Y(0.4\varepsilon_o) . \tag{41}$$

This yields, on substitution from (40) and (36) the simple expression

$$\varepsilon_o = 1.4\ \frac{t}{a} ; \tag{42}$$

or, equivalently, by (38)

$$Lt^{\frac{1}{2}}/a^{\frac{3}{2}} = 1.4 . \tag{43}$$

For example, a tube having (a/t) = 15 would have L/a = 5.4. This seems to be consistent with photographs of buckled tubes in [25].

There now remains only the key question of what determines L/a in a given tube. Plainly we cannot invoke the idea that the value of L is that which minimises the total energy absorbed, since the energy of circumferential bending is independent of L and the energy dissipated in stretching is inversely proportional to L^2. Here then we have another example, like that of tube inversion, in which the minimisation of energy associated with an assumed steady-state mode of deformation does not help to determine the parameters of the mode-shape. The dimensionless group $(Lt^{\frac{1}{2}}/a^{\frac{3}{2}})$ in (43) is, of course, familiar in the mechanics of elastic cylindrical shells which deform in conditions where the main structural actions are circumferential bending and longitudinal stretching ([7], chapter 9), so its appearance here also is not surprising. But this does nothing to explain why the group has a particular value in the present case.

I now suggest that what determines the length L is in fact the axial strain ε_o which is imparted to the principal strip when it flows through the curved knuckle on entry to the inverted region. I claim that ε_o is determined by <u>local</u> conditions in the knuckle; and that these local conditions determine in turn the overall length L of the transition zone.

Figure 11 shows some views of the apex of the curved knuckle shown in Fig. 10. Figure 11(a) is a perspective sketch in which the knuckle is represented by a single curved plane crease of intersection between two cylindrical surfaces. The plane of the crease makes a small angle β with the plane tangential to the tube at the upper generator; and thus the upper generator is bent through angle 2β, as shown. Figure 11(b) is an enlarged view, normal to the plane of the crease, of a portion of the apex of the knuckle. The two cylindrical surfaces of Fig. 11(a) are actually jointed smoothly by a locally toroidal region of width ℓ. The arrangement is analagous to Fig. 3(c), except that here

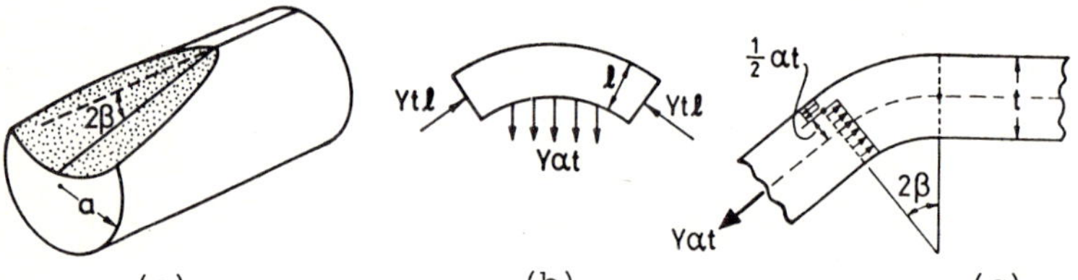

Fig. 11. Stage in the analysis of conditions at the apex of the knuckle in Fig. 10.

the knuckle is subjected to a one-sided tension per unit length equal to $\alpha N_o = \alpha Yt$: we suppose that the longitudinal tension is some unknown fraction α (< 1) of the full-plastic tension. The local radius of curvature of the apex-arc of the knuckle of Fig. 11(b) is equal to βa, from the geometry of Fig. 11(a); and this corresponds to the radius r of the knuckle shown in Figs 2 and 3(c). We assume, as before, that the arc is in full-plastic compression; and thus the equilibrium of an element at the apex requires

$$\alpha Yt.a\beta = Y\ell t ,$$

or

$$\ell = \alpha\beta a . \tag{44}$$

Figure 11(c) shows a further enlarged section (in the axial, vertical plane) of the principal strip of the tube as it flows through the knuckle region. The centre-line length of the curved portion is ℓ. When it passes through the entry hinge from right to left, the strip is given a curvature $2\beta/\ell$; and this curvature is removed again at the exit-hinge. At the entry hinge there is a pure plastic bending moment, and the "neutral surface" is at the centre of the strip: the spot marks the instantaneous centre. At the exit-hinge, however, the situation is different: the "neutral surface" is eccentric by $\frac{1}{2}\alpha t$, because the central part of the thickness of the sheet carries the longitudinal tension. Again, the spot marks the instantaneous centre, and the distribution of stress on the section is also shown. Since the neutral surface is inextensional, the centre-surface of the sheet receives an axial strain ε_o as it passes through the exit hinge, equal to the eccentricity multiplied by the curvature: thus

$$\varepsilon_o = \tfrac{1}{2}\alpha t.2\beta/\ell . \tag{45}$$

Using (44) we obtain finally the relation

$$\varepsilon_o = t/a . \tag{46}$$

This is a most unexpected and surprising result, since it indicates that ε_o has a universal value for a given tube, independent of both α and β: ε_o seems to be an invariant feature of this kind of "inverting knuckle". This result should be susceptible to simple experimental verification.

The final step in the analysis is the remark that (42) and (46), though not identical, do have the same form. We noted earlier that our estimation of the energy dissipated in material as it passed through the knuckle was rather crude, since it neglected the back-and-forth bending at the closely-spaced entry and exit hinges in Fig. 11(c). A simple calculation shows that the energy dissipated per unit area of sheet in passing through the exit-hinge always exceeds $Yt\varepsilon_o$ by an amount proportional to α^2; and we should add to this the energy dissipated in the purely flexural entry-hinge. The total energy dissipated depends on the value of α, about which we have no information. If we take, as a guess, $\alpha = 0.5$ we find that the energy dissipated per unit area in the two hinges together is equal to $2.3\ Yt\varepsilon_o$, which is $1.3\ Yt\varepsilon_o$ greater than the previous estimate. This in turn would increase the mean total extensional strain, used in expression (41), from $0.4\varepsilon_o$ to $0.4 \times (3.3/2)\varepsilon_o$, which would give

$$\varepsilon_o = 0.82(t/a) \tag{47}$$

instead of (42); which would be very satisfactory.

It is clear that more work needs to be done on the details of this calculation. But the main idea of the present analysis — that L depends on ε_o, which in turn depends on local conditions at the inverting knuckle — seems to be broadly in accord with available experimental data.

VII. Conclusions

The various examples which I have described illustrate some of the wide range of interesting problems which occur in the field of large plastic deformations of shell structures.

One of the most striking features of shell structures in general is the strong kinematic constraints which are imposed by the condition of continuity of the surface. An extreme form of kinematic constraint is that of complete inextensibility of the central surface of a shell; and indeed inextensional modes of deformation of "open" shells are well known (e.g. [7] chapter 6). Inextensional deformations are possible in some closed surfaces, also: two obvious examples are the change from a cylindrical surface into the well-known "Yoshimura" pattern of plane triangular facets, and the inversion of a spherical surface as in Fig. 2, but with a sharp circumferential crease instead of a smooth "knuckle". The sharp creases of these truly inextensional modes would act as rather complicated stress-raising features, and so we would expect to see plastic deformation there. This effect will be obvious to anyone who has applied axial load to a cylinder already folded in the "Yoshimura" pattern.

But purely inextensional deformation is not a totally satisfactory starting-point for thinking about our subject, as it immediately raises great difficulties. Thus, in particular, the flexural stiffness of sheet material inhibits that sharp, angular folding which is required by the conditions of strict inextensionality. In the problem of inversion of a spherical shell it is the knuckle region which absorbs almost all of the analytical effort; and the notion of starting with a sharp crease rather than a smooth knuckle is not helpful.

In fact the travelling plastic hinge is of the essence of each of my large-deformation examples, apart from Alexander's [4] axisymmetric crumpling of a tube. These examples, taken as a whole, illustrate two kinds of travelling plastic hinge. In the first kind there is a tightly-curved knuckle contained between two moving flexural hinges: the separation between these hinges is generally by a length of order $(Rt)^{\frac{1}{2}}$, where R is an overall radius of curvature and t is the thickness of the shell. In such an arrangement the energies dissipated in plastic deformation of bending and stretching are of the same order of magnitude; but there seems to be no over-riding condition that they should be equal, in contrast to the small-deflection, non-travelling hinge problem of section III, and in spite of the frequent occurrence of equations such as (10), (24) and (27).

The second kind of travelling plastic hinge in my examples is the "stretching" variety. This is seen in the problem of the propagating fracture of a pipeline (Fig. 5) and also, less obviously, at the exit-plane of the propagating buckle in the confined tube (Fig. 10). Hinges of this type absorb practically no energy in bending.

Another lesson to emerge from the assorted examples is that mode-forms of travelling hinge patterns cannot necessarily be determined by the minimisation of global energy functions with respect to "pattern parameters". Indeed, in the case of the propagating buckle (Fig. 10) it seems clear that the overall length of the inverting region is determined by the mechanical behaviour within a small, quite local region of the advancing knuckle. Some may argue, I suppose, that any local problem may be converted into an extremum-global problem by means of suitable mathematical manipulations. This idea seems to be not very useful in our present field of study.

All of my examples have been extremely simple geometrically, and have involved a minimum of complicating factors such as strain-hardening and dynamics (c.f. [30]-[32]). Practical engineers are interested in the deformation of structures having much more complicated geometries, such as vehicle bodies consisting of shells which are reinforced by ribs of various sorts, or which are constructed as assemblies of relatively flat plates. Moreover, even shells having a simple geometry need not necessarily deform in a simple way: for example, the process of inversion of a spherical shell (FIg. 2) can involve, in later stages, a non-circular travelling knuckle region, becoming polygonal in form [10] and involving folds somewhat reminiscent of Fig. 10.

Another important problem-area, which I have only marginally touched on, is the tearing of structures: here the strong kinematical constraints are overcome by a mixture of tearing and the passage of plastic hinges. These further problems provide us with an interesting challenge; and I venture to hope that the methods which I have expounded here will prove to be fruitful in this extended domain.

VIII. References

1. Hill, R.: The Mathematical Theory of Plasticity. Oxford: Clarendon Press 1950.

2. Onat, E.T.: The influence of geometry changes on the load-deformation behaviour of plastic solids. In Plasticity (Proceedings of Second Symposium on Naval Structural Mechanics), ed. Lee, E.H.; Symonds, P.S., pp 225-238. Oxford: Pergamon Press 1960.

3. Pugsley, Sir Alfred; Macaulay, M.: The large-scale crumpling of thin cylindrical columns. Quarterly Journal of Mechanics and Applied Mathematics 13 (1960) 1-9.

4. Alexander, J.M.: An approximate analysis of the collapse of thin cylindrical shells under axial loading. Quarterly Journal of Mechanics and Applied Mathematics 13 (1960) 10-15.

5. Wierzbicki, T.: Crushing behaviour of plate intersections. In Structural Crashworthiness, ed. Jones, N.; Wierzbicki, T., pp 66-95. London: Butterworths 1983.

6. Wierzbicki, T.; Abramowicz, W.: On the crushing mechanics of thin-walled structures. Journal of Applied Mechanics, 50 (1983) 722-734.

7. Calladine, C.R.: Theory of shell structures. Cambridge University Press 1983.

8. Onat, E.T.: The plastic collapse of cylindrical shells under axially symmetrical loading. Quarterly of Applied Mathematics 13 (1955) 63-72.

9. Demir, H.H.; Drucker, D.C.: An experimental study of cylindrical shells under ring loading. In Progress in Applied Mechanics (the Prager Anniversary Volume) pp 205-220. New York: Macmillan 1963.

10. Wasti, S.T.: Finite deformation of spherical shells. Ph.D. dissertation, University of Cambridge 1964.

11. Leckie, F.A.; Penny, R.K.: Plastic instability of a spherical shell. In Engineering Plasticity, ed. Heyman, J.; Leckie, F.A., pp 401-411. Cambridge University Press 1968.

12. Morris, A.J.; Calladine, C.R.: The local strength of a thin spherical shell loaded radially through a rigid boss. In Pressure Vessel Technology (Proceedings of the First International Conference on Pressure Vessel Technology), vol. I., ed. Berman, I., pp 35-44. New York: American Society of Mechanical Engineers 1969.

13. Updike, D.P.: On the large deformation of a rigid-plastic spherical shell compressed by a rigid plate. Journal of Engineering for Industry (Trans. A.S.M.E. (B)) 94 (1972) 949-955.

14. de Oliveira, J.G.; Wierzbicki, T.: Crushing analysis of rotationally symmetric plastic shells. Journal of Strain Analysis 17 (1982) 229-236.

15. Calladine, C.R.: Simple ideas in the large-deflection plastic theory of plates and slabs. In Engineering Plasticity, ed. Heyman, J., Leckie, F.A., pp 93-127. Cambridge University Press 1968.

16. Poynton, W.A.; Shannon, R.W.E.; Fearnehough, G.D.: The design and application of shear fracture propagation studies. Journal of Engineering Materials and Technology (Trans. A.S.M.E. (H)) 96 (1974) 323-329.

17. Popelar, C; Rosenfield, A.R.; Kanninen, M.F.: Steady-state crack propagation in pressurised pipelines. Journal of Pressure Vessel Technology (Trans. A.S.M.E. (J)) 99 (1977) 112-121.

18. Johnson, W; Reid, S.R.: Metallic energy dissipating systems. Applied Mechanics Reviews 31 (1978) 277-288.

19. Thornton, P.H.; Mahmood, H.F.; Magee, C.L.: Energy absorption by structural collapse. In Structural Crashworthiness, ed. Jones, N; Wierzbicki, T., pp 96-117. London: Butterworths 1983.

20. Guist, L.R.; Marble, D.P.: Prediction of the inversion of a circular tube. Technical Note TN-D-3622. Washington: National Aeronautics and Space Administration 1966.

21. Al-Hassani, S.T.S.; Johnson, W.; Lowe, W.T.: Characteristics of inversion tubes under axial loading. Journal of Mechanical Engineering Science 14 (1972) 370-381.

22. Kinkead, A.N.: Analysis for inversion load and energy absorption of a circular tube. Journal of Strain Analysis 18 (1983) 177-188.

23. Calladine, C.R.: Plasticity for Engineers, (2nd edition of Engineering Plasticity). Chichester: Ellis Horwood 1985.

24. Palmer, A.C.; Martin, J.H.: Buckle propagation in submarine pipelines. Nature 254 (1974) 46-48.

25. Kyriakides, S.; Babcock, C.D.: Buckle propagation phenomena in pipelines. In Collapse, ed. Thompson, J.M.T.; Hunt, G.W., pp 75-91. Cambridge University Press. 1983.

26. Steel, W.J.M.; Spence, J.: On propagating buckles and their arrest in sub-sea pipelines. Proceedings of Institution of Mechanical Engineers 194A (1983) 139-147.

27. Wierzbicki, T.; Bhat, S.U.: On the initiation and propagation of buckles in pipelines. M.I.T., Dept of Ocean Engineering, Report No. 85-2 (1985).

28. Croll, J.G.A.: Analysis of buckle propagation in marine pipelines. Journal of Constructional Steel Research (in press).

29. Reid, S.R.; Reddy, T.Y.: Effects of strain hardening on the lateral compression of tubes between rigid plates. International Journal of Solids and Structures 14 (1978) 213-225.

30. Perrone, N.; Bhadra, P.A.: A simplified method to account for plastic rate sensitivity with large deformations. Journal of Applied Mechanics, A.S.M.E., 46 (1979) 811-816.

31. Reid, S.R.: Laterally compressed metal tubes as impact energy absorbers. In Structural Crashworthiness, ed. Jones, N; Wierzbicki, T., pp 1-43. London: Butterworths 1983.

32. Calladine, C.R.; English, R.W.: Strain-rate and inertia effects in the collapse of two types of energy-absorbing structure. International Journal of Mechanical Sciences 26 (1984) 689-701.

Dynamic Behaviour of Structural Elements Liable to Buckling

M.A.SOUZA

Pontifícia Universidade Católica do Rio de Janeiro, Brasil
Departamento de Engenharia Civil

Summary

The dynamic behaviour of structural elements subjected to compressive loading is investigated by means of three simplified theoretical models. Curves of the square of the frequency versus the applied axial loading illustrate such a behaviour. The corresponding equilibrium paths are presented. The effect of the initial imperfection is taken into account. Theoretical results for a rectangular plate are also presented.

Introduction

The present work illustrates how simplified theoretical models can be used in order to get a better understanding of the complex behaviour of structural elements. Three models are presented and they reproduce the behaviour of beams, plates and cylindrical shells axially compressed. Each model has a different degree of complexity.

The adequacy of the models introduced is guaranteed by the fact that they reproduce equilibrium characteristics of beams, plates and cylindrical shells known both from experiments and theoretically. Such characteristics can be summarized by the equilibrium paths shown in Figure 1.

The great advantage of the simplified model is the capability of including different factors such as initial imperfection, for instance, in the analysis without increasing substantially the degree of complexity. And this is a very important aspect for the interpretation of experimental results since the recorded response may include the effects of different factors such as initial imperfection, residual stresses, changes in

Inelastic Behaviour of Plates and Shells
IUTAM Symposium Rio de Janeiro 1985
Editors: L. Bevilacqua, R. Feijóo and R. Valid

the boundary conditions during the loading process etc...,just to mention a few. These factors can be combined in such a way that not always the individual contributions can be quantified.

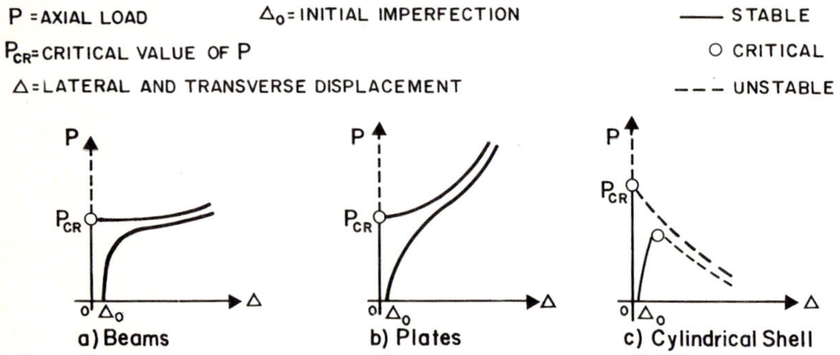

Figure 1 - Equilibrium Paths

All the difficulties mentioned up to now are related to static aspects of the problem. The degree of complexity increases substantially when we consider the dynamics of the problem: the relationship between the applied axial load and the natural frequency of vibration taking into account initial imperfection, residual stresses, etc ... .

The simplified models represent a valuable contribution for the understanding of such a complex behaviour of structural elements.

What follows is a series of theoretical results obtained for three simplified models and the rectangular plate.

The Rigid-Bars Model

The mechanical model of Figure 2 reproduces the stable symmetrical and the unstable-symmetrical bifurcation.

It consists of two rigid bars of the same length L linked by a hinge with a force and a moment spring of stiffness K and C respectively. The springs reproduce elastic characteristics of structural elements.

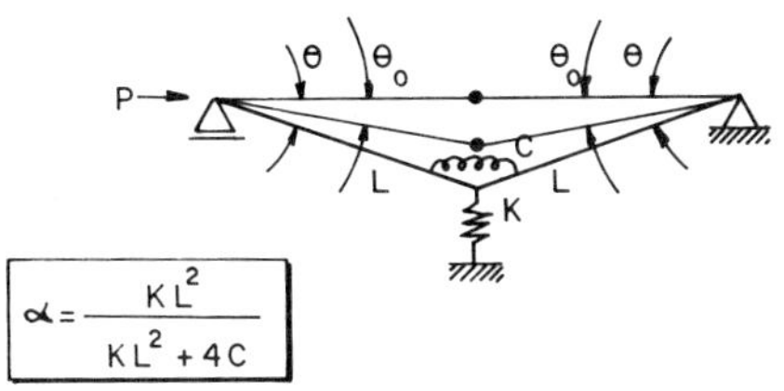

Figure 2 - The Rigid-Bars Model

Equilibrium paths for both the perfect and the imperfect model are shown in Figure 3 for different values of K and C. The equilibrium characteristics are also presented schematically .

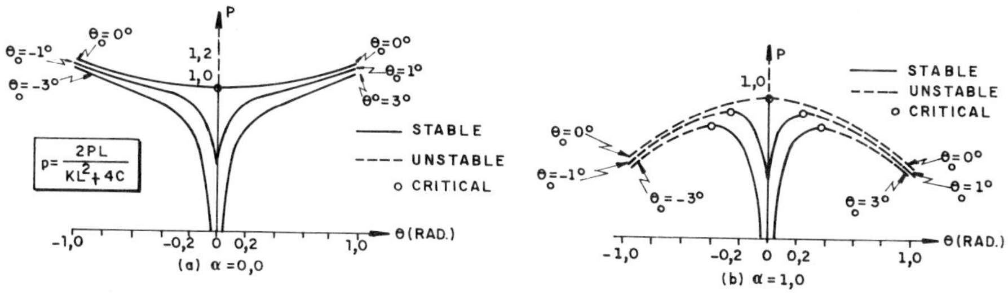

Figure 3 - Equilibrium Paths

The curves of Figure 3-a reproduce the equilibrium paths of beams and plates whereas the curves of Figure 3-b are similar to those of the cylindrical shells (the equilibrium charac - teristics).

Therefore the rigid-bars model has the capability of reproducing equilibrium characteristics which are similar to those of beams, plates and cylindrical shells (depending on the ratio between K and C).

Once the equilibrium characteristics are known the vibration characteristics can be determined and expressed in terms of the curves of axial load versus the square of the natural frequency as shown in Figure 4.

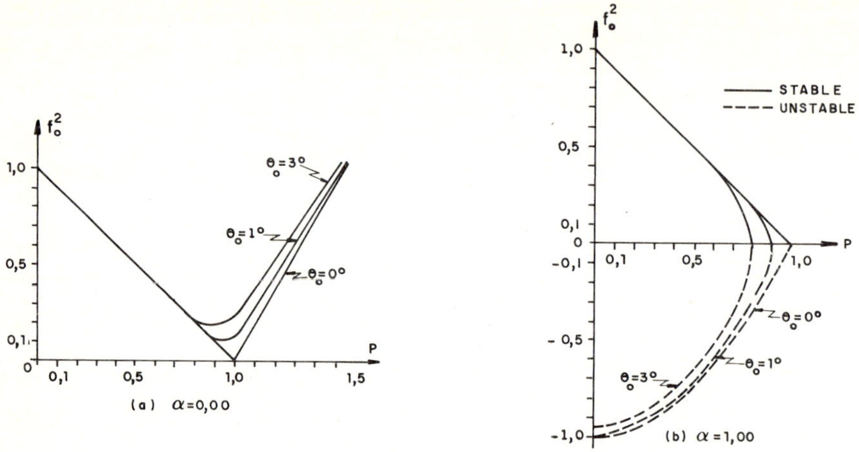

Figure 4 - $(\text{Frequency})^2$ vs. Axial Load

The curves of Figure 4-a correspond to the equilibrium paths of Figure 3-a whereas those of Figure 4-b correspond to the equilibrium paths of Figure 3-b.

Figure 4 therefore summarizes the dynamic aspect of the problem: the relationship between the axial load (and therefore the equilibrium paths) and the natural frequency of vibration.

Important conclusions are obtained from the diagrams of Figure 4 and they apply to beams, plates and cylindrical shells axially compressed :

i)Figure 4-a shows that for systems with stable equilibrium characteristics the square of the natural frequency is always positive vanishing only when the system is perfect (and in the absence of damping as it is shown by the author in the work used as reference for the present one); it also shows a decrease of the natural frequency with an increase of the axial load until a minimum is reached and then an increase of the natural frequency when the load increases (the author also shows in the above mentioned work that the rate of increase is twice that of the decrease in the vicinity of the critical load, for the perfect system). The effect of the initial imperfection in this case is to raise the natural frequency when compared to the perfect situation, for the same level of load. Figure 4-a reproduces the behaviour of beams and plates.

ii) The main characteristic of Figure 4-b is the negative value reached by the square of the natural frequency corresponding to levels of load lower than the critical. The effect of the initial imperfection in this case is to lower the natural frequency when compared to the perfect situation, for the same level of load. This is the behaviour of the cylindrical shells.

Therefore from the simple mechanical model of Figure 2 important conclusions regarding the dynamic behaviour of complex structural elements can be obtained. It will be shown next theoretical results for two specific models for rectangular plates and cylindrical shells which confirm the predictions made from the rigid-bars model.

The Rectangular Plate Model

A mechanical model suitable for the analysis of rectangular plates is shown in Figure 5. It consists of several force and moment springs which reproduce elastic characteristics of a plate.

The equilibrium and the vibration characteristics of such a model are presented in terms of the equilibrium paths and the curves of load versus the square of the frequency respectively in Figure 6. The initial imperfection is represented by a deflection of the centre section of the model and its effect is included in those diagrams.

The curves of Figure 6 are similar to those of Figures 3-a and 4-a of the rigid-bars model with $\alpha = 0.0$. Therefore the behaviour of rectangular plates previously obtained from the model of Figure 2 were confirmed.

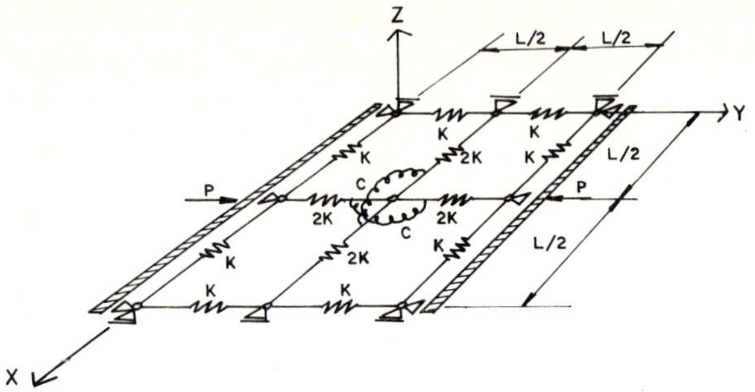

Figure 5 - The Rectangular Plate Model

Analytical results obtained by the author for a rectangular plate treated as such confirm these results and are shown by the diagrams of Figure 7. This result stresses the initial afirmative regarding the adequacy of such a simplified model.

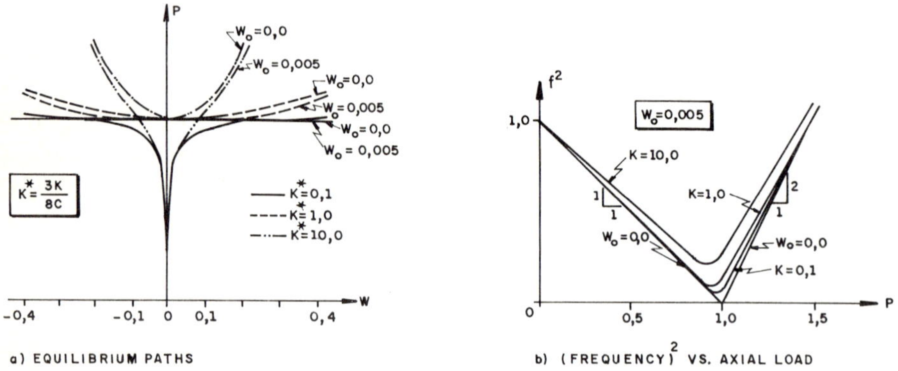

Figure 6 - Characteristic Curves - Rectangular Plate Model

<u>The Cylindrical Shell Model</u>

The cylindrical shell model consists of two elements: a strut and an arch which are associated to the main directions of the shell. This model is represented in Figure 8.

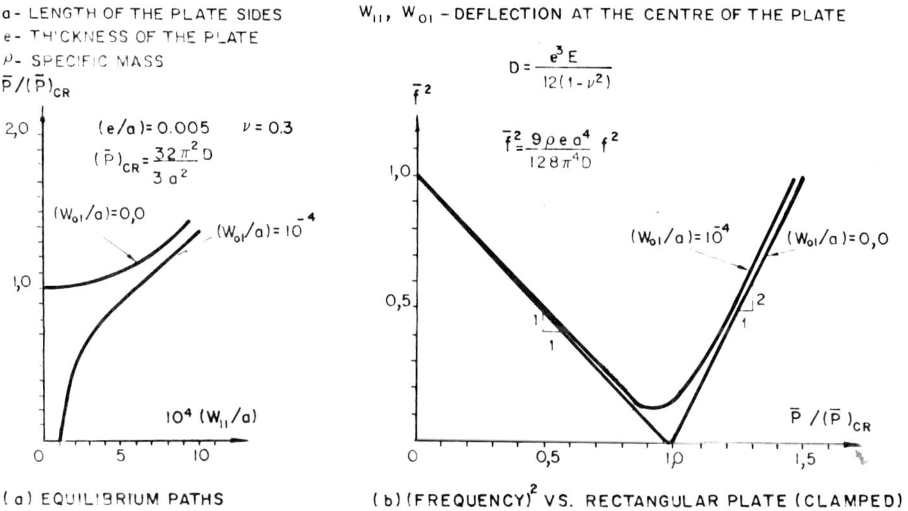

Figure 7 - Characteristic Curves - Rectangular Plate

As it can be seen from Figure 8 the cylindrical shell model has a higher degree of complexity when compared to the ones previously presented for instead of springs the elements are treated as such.

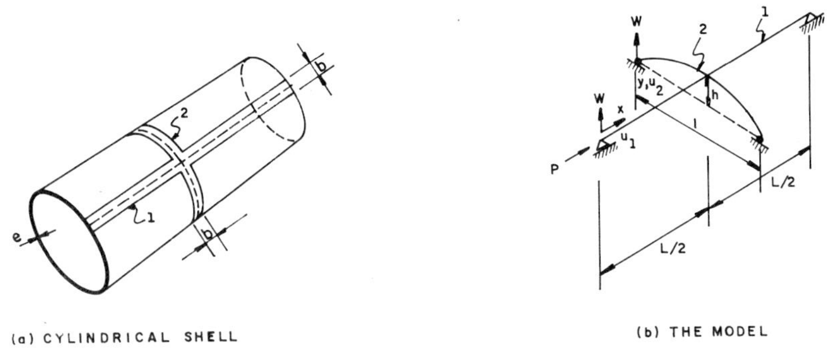

Figure 8 - The Cylindrical Shell Model

Equilibrium paths and curves of load versus the square of the frequency for the cylindrical shell model are shown in Figure 9 where different levels of initial imperfection were considered. Such an imperfection is represented by the deflection of the centre section of the model.

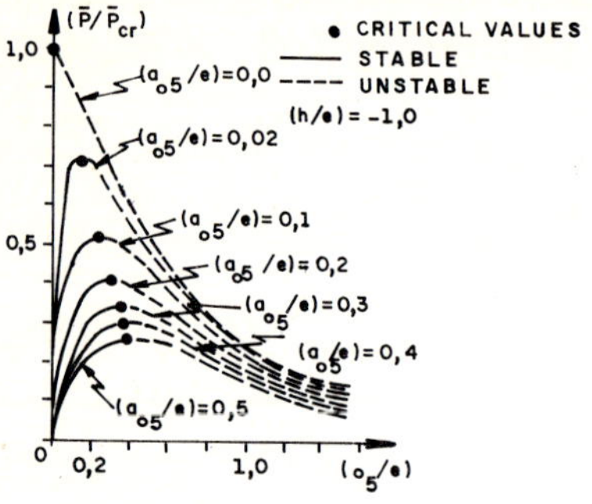

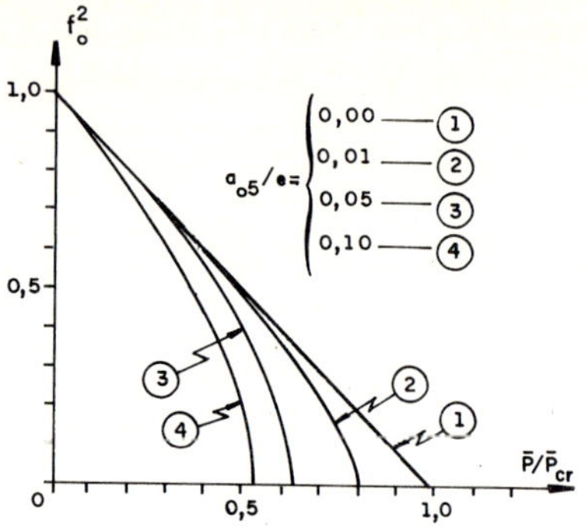

Figure 9 - Characteristic Curves of the Cylindrical Shell Model

As in the case of the rectangular plate the characteristics obtained from the rigid-bars model were reproduced, in the present case with $\alpha = 1.0$. The comparison of the diagram of Figures 3-b, 4-b with those of Figures 9-a, 9-b, respectively highlights this fact.

Therefore although the cylindrical shell model is much more complex than the rigid-bars model, in terms of the behaviour the latter one gives valuable information regarding the main characteristics of a cylindrical shell.

Conclusions

Three simplified models with different degrees of complexity were presented. The equilibrium and the vibration characteristics of each of them were summarized in terms of the equilibrium paths and the curves of load versus the square of the natural frequency.

Analytical results for the rectangular plate were also presented in terms of the same curves. They confirm the results obtained from the simplified models.

The capability of such models incorporating effects such as initial imperfection was stressed. It is hoped that the main aim of the present work was achieved: the use of simplified models for the understanding of the dynamic behaviour of structural elements liable to buckling.

Acknowledgement

This work is part of the Ph.D. thesis carried out by the author at University College, London. The author is indebted to Prof. A.C. Walker (presently at Surrey University, Guildford, England), who supervised the thesis work.

Reference

Souza, M.A., "Vibration Characteristics of Buckled Structures", Ph.D. Thesis, University College London, England, October, 1982.

Time Deflection Behaviour of Ice Plates

W. SZYSZKOWSKI, Visiting Associate Professor
P. G. GLOCKNER, Professor and Head

Department of Mechanical Engineering
The University of Calgary
Calgary, Alberta, Canada

Abstract

In analyzing ice plates it has been customary to use rather simple material models. In this paper a theory for viscoelastic plates is presented which uses a more realistic nonlinear constitutive law for ice, incorporating the most recent experimental data and including all three stages of creep and hereditary effects. In particular, the hardening phenomenon in the primary creep phase, the strain softening effect during the tertiary stage, and the inherent brittleness of the material, leading to some stress anisotropy, are taken into account. The constitutive law in this form facilitates analysis of ice structures up to failure.

An incremental formulation and numerical technique, introduced elsewhere, allows effective and efficient treatment of the governing equations, a solution technique which is demonstrated by solving the problem of a long ice plate subjected to in-plane compressive forces and undergoing cylindrical bending.

Introduction

Since even under arctic conditions ice is relatively close to its melting point, in dealing with ice structures, the viscous/time dependent properties of the material play a primary role and must be treated with extreme care in the analysis. This, in turn, suggests that the reliability of such analyses is very much a function of the accuracy of the constitutive law used. Such laws for time dependent materials are usually quite complex and general [1,2,3]* which makes their applicability to a specific material rather questionable.

A most convenient linearized form of the constitutive relation for such time dependent materials, presented in an elegant form in [1], is quite

---

* The results presented here were obtained in the course of research sponsored by the Natural Sciences and Engineering Research Council of Canada, Grant No. A-2736.

** Numbers in square brackets indicate publications listed under References.

Inelastic Behaviour of Plates and Shells
IUTAM Symposium Rio de Janeiro 1985
Editors: L. Bevilacqua, R. Feijóo and R. Valid

unacceptable for ice due to the significant nonlinearity in this material's stress/strain behaviour observed in creep tests. Despite this fact such linear approximation to the actual constitutive law was used frequently in the past [4,5,6].

A simple nonlinear constitutive law, assuming a strain rate-stress relation for the steady creep stage, was used in the analysis of ice plates in [7,8]. However, even such a constitutive relation represents a severe approximation. For example, it does not take into account the hardening effect observed in the behaviour of ice during its primary creep phase, nor does it allow for the strain-softening depicted by experiments during advanced stages of the tertiary creep phase. To remedy some of these deficiencies, a non-linear hereditary type constitutive law for non-ageing materials with fading memory was introduced in [9] for the uniaxial case and in [10] for the multiaxial case, to describe the primary (hardening) and steady creep phases of ice. An extension to this model, taking into account strain-softening effects in the tertiary phase, was presented in [11] for the uniaxial case.

Finally, to model the differences in ice response observed experimentally for samples tested in tension as opposed to compression, a damage function was introduced in [12] which was assumed to depend on the tensile stress state and which admits stress anisotropy and limits the creep process at the moment of failure.

Utilizing the constitutive law discussed in [12], this paper presents a theory of thin ice plates based on a so-called strain formulation in which the current strain depends on the current stress and stress history. In such a formulation the problem arises as to how to treat a structure in which stress varies temporally even under constant external loads. To demonstrate the relative ease with which such problems can be solved using the approach presented here, the cylindrical bending of long ice plates with initial imperfections and subjected to in-plane compressive forces applied along the longitudinal edges is treated numerically.

2. Constitutive Law for Ice

In the analysis of structures made of time-dependent materials, a proper formulation of the constitutive law is usually one of the most difficult problems. It goes without saying that such a law should be based on extensive experimental data.

Despite recent intensification in experimental studies, knowledge of the mechanical properties of ice is rather incomplete [13] in the sense that only its behaviour in uniaxial compression has been studied extensively [14,15,16]. Let us, therefore first discuss some possible forms of uniaxial constitutive laws which take into account such experimental findings. Using a so-called strain formulation, a general constitutive relation for viscoelastic materials at a constant temperature may be written in the form

$$\varepsilon(t) = \int_o^t \bar{F}[\sigma(\tau,t)]d\tau \qquad (1)$$

where ε and σ denote strain and stress, respectively, while $\bar{F}$ is a function describing the viscous properties of the material. From Equ. (1) it is clear that these properties are functions of the stress history, from the 'virgin' state (τ=o) up to the current moment (τ=t).

Some creep test data can be described reasonably well, at least within the first two stages of creep, by assuming Equ. (1) in the form

$$\varepsilon(t) = \frac{\sigma(t)}{E} + \int_o^t F[\sigma(\tau)][A_1 j(t-\tau) + A_2]d\tau \qquad (2)$$

in which the elastic response is separated, with E denoting Young's modulus, and A_1 and A_2 signifying parameters related to the recoverable and permanent creep deformations [10], respectively, $F(\sigma)$ defining the strain-stress nonlinearity of the material and $j(t)$, a monotonically decreasing function of time (with $j(t)\to 0$ as $t\to\infty$ and for convenience normalized as $j(0) = 1$), describing the hardening effects associated with the primary creep stage. A detailed discussion of the parameters and functions appearing in Equ. (2) was given in [9-12]. In particular, the nonlinearity function was assumed in the form of Norton's power law as

$$F(\sigma) = B\sigma^n \qquad (3)$$

where B and n are material constants. Using this expression for $F(\sigma)$ in

Equ. (2) and differentiating with respect to time, the strain rate, $\dot{\varepsilon}(t)$ is established as

$$\dot{\varepsilon}(t) = \frac{\dot{\sigma}}{E} + \frac{1}{\nu_1}\frac{d}{dt}\int_0^t \sigma^n(\tau)\, j(t-\tau)d\tau + \frac{1}{\nu_2}\sigma^n \tag{4}$$

where $\nu_1 = 1/(A_1B)$ and $\nu_2 = 1/(A_2B)$. The coefficients ν_1, ν_2 and n can be determined from experimental creep data by plotting the initial (maximum) and minimum creep rates against given (constant) stress, on a log-log plot. The function j(t) can also be specified from creep curves by determining two stress-independent parameters, t_o and t_1 (see Fig. 1), where t_o is proportional to the 'latent' elastic energy accumulated during the hardening process, energy which is recoverable upon unloading, and where t_1 indicates the initial rate of hardening or initial deceleration of strain rate (see [11] for details). Calculation of the derivative of the Volterra-type integral appearing in Equ. (4) will be accomplished with relative ease by means of a numerical technique introduced in [10].

For a limited range of viscous deformations restricted to the first two stages of creep and for which Equ. (4) is applicable, ice can be assumed to be an isotropic material. For this domain of behaviour, Equ. (4) can be generalized to a constitutive law for multiaxial stress states by postulating the existence of a complementary power potential, $\mathcal{P}(\sigma_{ij})$ in the form [10]

$$\mathcal{P}(\sigma_{ij}) = \int_o^{\sigma_{ij}} \dot{\varepsilon}_{ij}\, d\sigma_{ij}; \qquad \therefore\ \dot{\varepsilon}_{ij} = \frac{\partial \mathcal{P}}{\partial \sigma_{ij}} \tag{5a,b}$$

For a material obeying Norton's power law, the following specific form for $\mathcal{P}$ was assumed in [10]

$$\mathcal{P}(\sigma_{ij}) = \frac{1}{6}\frac{d}{dt}\left[\frac{\sigma_{kk}^2}{3K_b} + \frac{S^2}{G}\right] + \frac{1}{\nu_1}\frac{d}{dt}\int_o^t \frac{S^{n+1}}{(n+1)}\, j(t-\tau)d\tau + \frac{1}{\nu_2}\frac{S^{n+1}}{(n+1)} \tag{6}$$

where

$$\sigma_{kk} = \sigma_{11} + \sigma_{22} + \sigma_{33}; \quad S^2 = \frac{3}{2}s_{ij}s_{ji}; \quad s_{ij} = \sigma_{ij} - \frac{1}{3}\sigma_{kk}\delta_{ij} \tag{7a,b,c}$$

and where K_b and G denote the bulk and shear modulus, respectively. Using

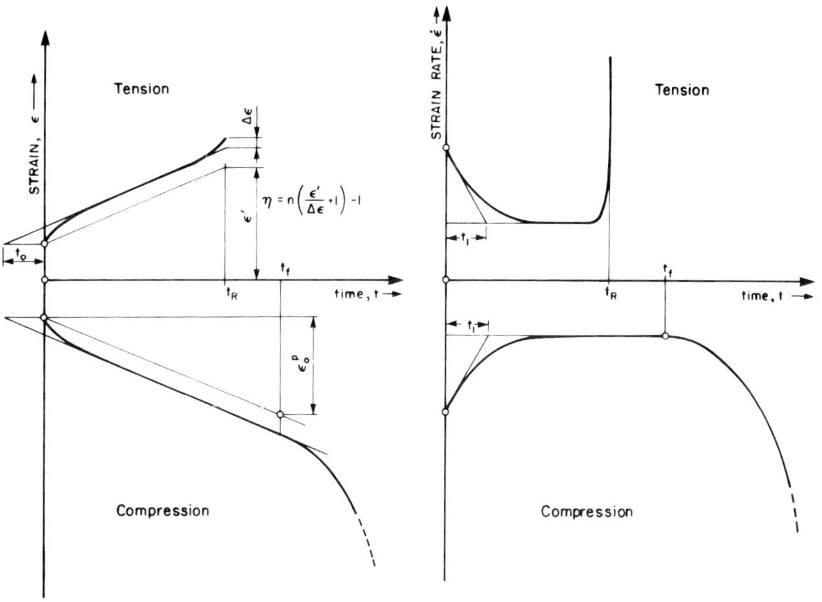

Fig. I Some Characteristic Features for Tension and Compression Creep Test Curves

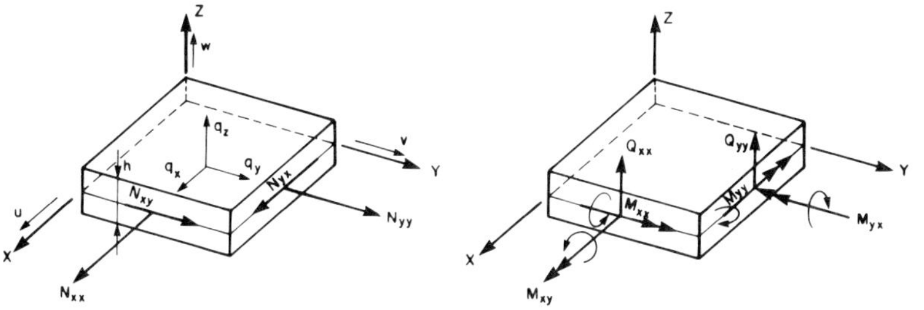

(a.) Sign Convention for Coordinates, Loads, Displacements and Membrane Stress Resultants

(b.) Sign Convention for Bending Moment and Transverse Shear Stress Resultants

Fig. 2 Plate Element Showing Positive Sense of Stress and Moment Resultants, Displacement Components and Coordinates

Equs. (5b) and (6) one obtains the strain rate as

$$\dot{\varepsilon}_{ij} = \dot{\varepsilon}^{e}_{ij} + \dot{\varepsilon}^{r}_{ij} + \dot{\varepsilon}^{p}_{ij} \tag{8a}$$

where

$$\dot{\varepsilon}^{e}_{ij} = \frac{1}{E}\,[(1+\mu)\dot{\sigma}_{ij} - \mu\,\dot{\sigma}_{kk}\,\delta_{ij}] \tag{8b}$$

$$\dot{\varepsilon}^{r}_{ij} = \frac{1}{\nu_1}\frac{d}{dt}\int_0^t \tilde{\sigma}^{n}_{ij}(\tau).j(t-\tau)\,d\tau \tag{8c}$$

$$\dot{\varepsilon}^{p}_{ij} = \frac{1}{\nu_2}\,\tilde{\sigma}^{n}_{ij} \tag{8d}$$

and where μ denotes Poisson's ratio and the 'effective viscoelastic stress', $\tilde{\sigma}_{ij}$, is defined by

$$\tilde{\sigma}_{ij} = \frac{s_{ij}}{|s_{ij}|}\left[\frac{3}{2}\,|s_{ij}|\,s^{n-1}\right]^{\frac{1}{n}} \tag{9}$$

One can show that relations (8) reduce to Equ. (4), for the uniaxial case. In addition, for a linear material for which n=1, these relations become the constitutive equations of linear viscoelasticity as given, for example, in [1]. Since $s_{kk} = 0$, the rate of dilatation, $\dot{\varepsilon}_{kk}$, is found to be

$$\dot{\varepsilon}_{kk} = \frac{1-2\mu}{E}\,\dot{\sigma}_{kk} \tag{10}$$

confirming the volume change to be creep-independent.

As indicated in the previous section, ice exhibits significant stress anisotropy at advanced stages of deformation due mainly to the development of internal microcracking resulting from inherent material brittleness. Consequently, there is substantial digression in behaviour of ice when subjected to compression as opposed to tension, in the former case the samples usually exhibiting long tertiary (accelerating) creep phases while for a tensile stress state this stage of creep is normally very short or may disappear altogether [17]. Microcracking, which disrupts the continuity of the material, is difficult to describe analytically. The effects of this deterioration can be described, qualitatively, by the introduction of a damage function, $\omega(t)$, [1], which physically represents the relative decrease in effective area available for stress transmission. Thus for a 'virgin' material, $\omega=0$, while at the instant of rupture, $\omega=1$. 'True' stress, the stress value governing the viscous process, is obtained by dividing the 'nominal' stress by $(1-\omega)$. In [12] it was assumed that microcracking affects only the permanent portion of the viscous strain rate since during advanced stages of creep, the rate of reversible viscous strain approaches zero. Thus we had [12]

$$\dot{\varepsilon}^{p}_{ij}(t) = \frac{1}{\nu_2}\left(\frac{\tilde{\sigma}_{ij}(t)}{1-\omega(t)}\right)^{n} \qquad (11)$$

To define the evolution law for the damage function, we note from [12] that tensile creep test data could reasonably be modelled, up to failure, by assuming the damage rate, $\dot{\omega}(t)$, to be a function of the 'true' stress and the current damage, in the form

$$\dot{\omega} = K\left(\frac{\sigma_{max}}{1-\omega}\right)^{\Gamma} \cdot \frac{1}{(1-\omega)^{\eta-\Gamma}} \qquad (12)$$

in which σ_{max} denotes the largest tensile stress. For a purely compressive stress state, $\sigma_{max}=0$. The material constants, K, Γ and η can be determined from tensile creep tests up to failure. In particular, η represents the ductility of the material at brittle failure (see Fig. 1). As was discussed in [12], for ice $\eta>\Gamma$ which physically means that the damage process intensifies significantly as final rupture is approached, or alternatively, that the tertiary creep phase is very short. Integrating Equ. (12) for σ_{max}=const, one obtains a relation involving the rupture time, t_R (for which $\omega=1.0$) in the form

$$(1+\eta)\; K\; t_R\; \sigma^{\Gamma}_{max} = 1 \qquad (13)$$

which allows determination of K and Γ from a log-log plot of t_R vs. σ_{max}.

In compression creep tests of ice, the tertiary creep phase is normally quite long, with strain magnitudes reaching several percent. It has been demonstrated experimentally [15,16], that the accelerating creep stage is initiated at a certain level of strain, independent of stress magnitude. The time associated with the start of the tertiary creep phase is called failure time, t_f, (see Fig. 1) which satisfies the empirical relation

$$t_f \cdot \dot{\varepsilon}_{min} = \text{constant} = C \tag{14}$$

where the constant, C, upon reflection and use of Equ. (4) for σ= const., physically clearly represents the permanent viscous strain, ε^p_o, accumulated during the primary and secondary creep stages. Analyzing test data from uniaxial compression tests, it was shown in [11] that such data for strains exceeding ε^p_o can be predicted with satisfactory accuracy by assuming the rate of permanent creep to depend on stress level as well as on the magnitude of accumulated permanent strain. In the case of a multiaxial stress state [12], the effective permanent axial strain, ε^p_{ef}, was defined as

$$\varepsilon^p_{ef}(t) = [\tfrac{2}{3}\, \varepsilon^p_{ij} \varepsilon^p_{ij}]^{\frac{1}{2}} \tag{15}$$

while the rate of permanent viscous strain, Equ. (11), is finally modified to read

$$\dot{\varepsilon}^p_{ij} = \frac{1}{\nu_2} \left(\frac{\tilde{\sigma}_{ij}(t)}{1-\omega(t)}\right)^n \cdot \left[1 + \alpha \left(\frac{\varepsilon^p_{ef}(t)}{\varepsilon^p_o} - 1\right)^a\right] \tag{16}$$

where α=0 for $\varepsilon^p_{ef}<\varepsilon^p_o$ and α>0 for $\varepsilon^p_{ef}>\varepsilon^p_o$. The material parameters α and a are determined from the accelerating stages of compression creep test curves. The strain-softening effect observed in tests on ice subjected to constant strain rate [15] is also simulated by Equ. (16) (using Equ. 8a). For the uniaxial compressive case, our constitutive model was verified against experimental data in [11].

In summary and before leaving this brief review of a multiaxial constitutive law for ice, let us note that modelling the behaviour of this material in tension and compression and up to failure requires the following constitutive constants:

2 elastic parameters, E and μ;

5 viscous parameters, ν_1, ν_2, n, t_o and t_1, determined from and defining the first two stages of creep;

3 viscous parameters, K, Γ, and η, determined from and defining the tensile tertiary creep stage;

3 viscous parameters, ε_o^p, α and a, determined from and defining the compressive tertiary creep phase.

3. Equations of a Viscoelastic Plate Theory

Consider a thin plate element of thickness h as shown in Fig. 2. Denoting the stress components at a distance z from the midsurface by $\sigma_{\alpha\beta}^z$ and $\sigma_{\alpha 3}^z$ (α=1,2), the in-plane and transverse stress resultants, $N_{\alpha\beta}$ and Q_α, and the bending moment resultants, $M_{\alpha\beta}$, are defined as

$$N_{\alpha\beta} = \int_{-h/2}^{h/2} \sigma_{\alpha\beta}^z dz; \quad Q_\alpha = \int_{-h/2}^{h/2} \sigma_{\alpha 3}^z dz; \quad M_{\alpha\beta} = \int_{-h/2}^{h/2} \sigma_{\alpha\beta}^z z dz; \qquad a=1,2 \tag{17}$$

Differentiating these expressions with respect to time, one can obtain the stress and moment resultant rates, $\dot{N}_{\alpha\beta}$, $\dot{Q}_\alpha$ and $\dot{M}_{\alpha\beta}$, as functions of $\dot{\sigma}_{\alpha\beta}^z$ and $\dot{\sigma}_{\alpha 3}^z$. The stress rates are determined using Equ. (8), with (8d) replaced by expression (16). Assuming $\sigma_{33}^z=0$, after some algebra one arrives at

$$\dot{\sigma}_{\alpha\beta}^z = (\dot{\sigma}_{\alpha\beta}^z)^e - (\dot{\sigma}_{\alpha\beta}^z)^v; \quad \dot{\sigma}_{\alpha 3}^z = (\dot{\sigma}_{\alpha 3}^z)^e - (\dot{\sigma}_{\alpha 3}^z)^v \tag{18a,b}$$

where

$$(\dot{\sigma}_{\alpha\beta}^z)^e = \frac{E}{(1-\mu^2)} [(1-\mu)\dot{\varepsilon}_{\alpha\beta}^z + \mu\, \dot{\varepsilon}_{\gamma\gamma}^z\, \delta_{\alpha\beta}]; \quad (\dot{\sigma}_{\alpha 3}^z)^e = 2G\cdot\dot{\varepsilon}_{\alpha 3}^z \tag{18c,d}$$

$$(\dot{\sigma}_{\alpha\beta}^z)^v = \frac{E}{(1-\mu^2)} \dot{\tilde{\varepsilon}}_{\alpha\beta}^z; \quad (\dot{\sigma}_{\alpha 3}^z)^v = 2G[(\dot{\varepsilon}_{\alpha 3}^z)^r + (\dot{\varepsilon}_{\alpha 3}^z)^p] \tag{18e,f}$$

and where

$$\dot{\tilde{\varepsilon}}_{\alpha\beta}^z = \frac{1}{\nu_1}\frac{d}{dt}\int_o^t (\overset{*}{\sigma}{}_{\alpha\beta}^z)^n \cdot j(t-\tau)d\tau + \frac{1}{\nu_2}\left(\frac{\overset{*}{\sigma}{}_{\alpha\beta}^z}{1-\omega}\right)^n \cdot \left[1 + \alpha\left(\frac{\varepsilon_{ef}^p}{\varepsilon_o^p} - 1\right)^a\right] \tag{18g}$$

in which

$$(\overset{*z}{\sigma}_{\alpha\beta})^n = \frac{3}{2} S^{n-1}[(1-\mu)\sigma^z_{\alpha\beta} - \frac{(1-2\mu)}{3} \sigma^z_{\gamma\gamma} \delta_{\alpha\beta}] \tag{18h}$$

In these expressions, the superscript 'e' denotes the 'elastic' portion of the stress rate while a viscous correction to this elastic rate is indicated by 'v'. The strain rate $\dot{\varepsilon}^z_{\alpha 3}$ is defined by Equ. (8) and $\dot{\varepsilon}^z_{\gamma\gamma} = \dot{\varepsilon}^z_{11} + \dot{\varepsilon}^z_{22}$. Note also that the elastic portion of the stress rate depends on the strain rate while the viscous parts are functions of the stress.

Using the Kirchhoff-Love hypothesis, the strains in a parallel plane, $\varepsilon^z_{\alpha\beta}$, are expressed in terms of midsurface tangential and bending strains, $\varepsilon_{\alpha\beta}$ and $\kappa_{\alpha\beta}$, as

$$\varepsilon^z_{\alpha\beta} = \varepsilon_{\alpha\beta} + z\,\kappa_{\alpha\beta} \tag{19}$$

which when substituted into Equs. (18) one obtains $\dot{\sigma}^z_{\alpha\beta}$ as functions of the midsurface strains. Substituting such results into Equ. (17) leads to

$$\dot{N}_{\alpha\beta} = \dot{N}^e_{\alpha\beta} - \dot{N}^v_{\alpha\beta}; \quad \dot{M}_{\alpha\beta} = \dot{M}^e_{\alpha\beta} - \dot{M}^v_{\alpha\beta} \tag{20a,b}$$

where

$$\dot{N}^e_{\alpha\beta} = \frac{Eh}{(1-\mu^2)} [(1-\mu)\dot{\varepsilon}_{\alpha\beta} + \mu\,\dot{\varepsilon}_{\gamma\gamma}\,\delta_{\alpha\beta}]; \quad \dot{N}^v_{\alpha\beta} = \frac{Eh}{(1-\mu^2)} \psi'_{\alpha\beta} \tag{20c,d}$$

$$\dot{M}^e_{\alpha\beta} = D[(1-\mu)\,\dot{\kappa}_{\alpha\beta} + \mu\,\dot{\kappa}_{\gamma\gamma}\,\delta_{\alpha\beta}]; \quad \dot{M}^v_{\alpha\beta} = D\cdot\psi''_{\alpha\beta}; \quad D = \frac{Eh^3}{12(1-\mu^2)} \tag{20e,f,g}$$

and where

$$\psi'_{\alpha\beta} = \frac{1}{h}\int_{-h/2}^{h/2} \dot{\bar{\varepsilon}}^z_{\alpha\beta}\,dz; \quad \psi''_{\alpha\beta} = \frac{12}{h^3}\int_{-h/2}^{h/2} \dot{\bar{\varepsilon}}^z_{\alpha\beta}\,z\,dz \tag{20h,j}$$

Analogous expressions can also be written for $\dot{Q}_\alpha$ in the form

$$\dot{Q}_\alpha = \dot{Q}^e_\alpha - \dot{Q}^v_\alpha \tag{21a}$$

with

$$\dot{Q}^e_\alpha = 2G \int_{-h/2}^{h/2} \dot{\varepsilon}^z_{\alpha 3} dz; \qquad \dot{Q}^v_\alpha = 2G \int_{-h/2}^{h/2} \dot{\varepsilon}^v_{\alpha 3} dz \tag{21b,c}$$

Any further details concerning the transverse shear stress resultants requires specification of the variation of $\varepsilon^z_{\alpha 3}$ across the thickness. Since in thin plate theory, Q_α is normally eliminated from the governing equations, we will not pursue this topic any further.

Equations (20) are the materially nonlinear constitutive relations for thin viscoelastic plates which when used in conjunction with standard equilibrium equations and geometrical (strain-displacement) relations lead to the governing equations for thin ice plates. Thus, for example, admitting effects of membrane stress resultants on the equilibrium of the deflected plate in the surface normal direction leads to a set of equilibrium equations in the form

$$N_{\alpha\beta,\alpha} + q_\beta = 0 \tag{22a}$$

$$Q_{\alpha,\alpha} - \kappa_{\alpha\beta} N_{\alpha\beta} + q_z = 0 \tag{22b}$$

$$M_{\alpha\beta,\alpha} - Q_\beta = 0 \tag{22c}$$

in which a comma before a subscript indicates partial differentiation with respect to that index. Using Equ. (22c) in (22b), one obtains

$$M_{\alpha\beta,\alpha\beta} - \kappa_{\alpha\beta} N_{\alpha\beta} + q_z = 0 \tag{22d}$$

thereby eliminating Q_α from the theory. Note that all inertia effects have been omitted from these equilibrium equations in view of the quasi-static nature of the slow viscous processes.

Restricting the discussion to infinitesimal displacement gradients, the strain-displacement relations are written as

$$\varepsilon_{\alpha\beta} = \frac{1}{2}(u_{\alpha,\beta} + u_{\beta,\alpha}) \tag{23a}$$

$$\kappa_{\alpha\beta} = -w_{,\alpha\beta} \tag{23b}$$

where u_α and w denote midsurface displacement components. Substituting Equ. (23) into relations (20) and the results into Equs. (22a,d) one arrives at

$$\frac{Eh}{(1-\mu^2)}\left[\left(\frac{1+\mu}{2}\right)\dot{u}_{\alpha,\alpha\beta} + \left(\frac{1-\mu}{2}\right)\dot{u}_{\beta,\alpha\alpha} - \psi'_{\alpha\beta,\alpha}\right] + \dot{q}_\beta = 0 \tag{24a}$$

$$D[\dot{w}_{,\alpha\beta\alpha\beta} + \psi''_{\alpha\beta,\alpha\beta}] - \dot{w}_{,\alpha\beta}N_{\alpha\beta} - w_{,\alpha\beta}\frac{Eh}{(1-\mu^2)}\left[\left(\frac{1-\mu}{2}\right)(\dot{u}_{\alpha,\beta} + \dot{u}_{\beta,\alpha})\right.$$

$$\left. + \mu\,\dot{u}_{\gamma,\gamma}\,\delta_{\alpha\beta} - \psi'_{\alpha\beta}\right] + \dot{q}_z = 0 \tag{24b}$$

These equations, as can be noted, are linear in the displacement rates, $\dot{u}_\alpha$, $\dot{w}$. Thus knowing the current stress and stress history at a particular instance, which allows determination of $\psi'_{\alpha\beta}$ and $\psi''_{\alpha\beta}$, together with the current displacements (which are required to determine $N_{\alpha\beta}$ and $w_{,\alpha\beta}$) the displacement rates can be determined from Equs. (24) from which, in turn, the stress, strain and displacement components for the next time interval can be obtained using some iterative scheme.

Instead of solving the above three equations for the displacement rates, one can reduce the number of unknowns to two by introducing a stress resultant function, ϕ, so as to express $N_{\alpha\beta}$ in the form

$$N_{\alpha\beta} = [\phi_{,\alpha\beta} - \phi_{,\gamma\gamma}\,\delta_{\alpha\beta}] \tag{25}$$

which for the case of $q_\beta=0$, reduces Equ. (22a) to an identity, while expression (22d) takes the form

$$D(\dot{w}_{,\alpha\beta\alpha\beta} + \psi''_{\alpha\beta,\alpha\beta}) - \dot{w}_{,\alpha\beta}(\phi_{,\alpha\beta} - \phi_{,\gamma\gamma}\,\delta_{\alpha\beta}) - w_{,\alpha\beta}(\dot{\phi}_{,\alpha\beta} - \dot{\phi}_{,\gamma\gamma}\,\delta_{\alpha\beta}) + \dot{q}_z = 0 \tag{26}$$

an equation which contains two unknowns, w and ϕ. The second equation required for the solution of this problem is obtained from compatibility, which takes the form

$$\dot{\varepsilon}_{\alpha\alpha,\beta\beta} - \dot{\varepsilon}_{\alpha\beta,\alpha\beta} = 0 \tag{27}$$

Substituting Equ. (25) into (20) and using the results to calculate $\dot{\varepsilon}_{\alpha\beta}$, substituting into Equ. (27) leads to the desired second equation in the form

$$\dot{\phi}_{,\alpha\beta\alpha\beta} - \frac{Eh}{(1-\mu^2)} [\psi'_{\alpha\alpha,\beta\beta} - (1+\mu)\, \psi'_{\alpha\beta,\alpha\beta}] = 0 \tag{28}$$

The form of the equations derived in this section suggests an incremental approach for their numerical solution, a technique which is discussed in detail in [12] and will be used in the next section to analyze the cylindrical bending of a long ice plate with initial (imperfection) displacements $w_o(x)$ and subjected to uniform in-plane compressive loads, $N_{xx}=-N_o$, applied along the longitudinal boundaries (see Fig. 3). For such a problem in which all variables are independent of the coordinate y, the governing equations reduce to

$$D\, \dot{w}_{,xxxx} + N_o\, \dot{w}_{,xx} = -D\, \psi''_{xx,xx} \tag{29}$$

while the membrane stress resultants in such a cylindrical bending problem have to satisfy the conditions

$$\dot{N}_{xx} = \dot{N}_{xy} = 0; \qquad \int_o^b \dot{N}_{yy} dx = 0 \tag{30a,b,c}$$

which when used in Equ. (20) leads to

$$\dot{\varepsilon}_{xx} + \mu\, \dot{\varepsilon}_{yy} = \psi'_{xx} \tag{31a}$$

$$\dot{\varepsilon}_{yy} + \frac{\mu}{b} \int_o^b \dot{\varepsilon}_{xx} dx = \frac{1}{b} \int_o^b \psi'_{yy} dx \tag{31b}$$

Note that this apparently uniaxial problem has to be solved in two spatial dimensions due to the coupling effects resulting from the functions ψ'_{xx}, ψ'_{yy} and ψ''_{xx}, the calculation of which requires both σ_{xx} as well as σ_{yy}.

4. Numerical Procedure and Results

The numerical solution procedure, using an incremental approach, consists of the following steps (for details see [12]):

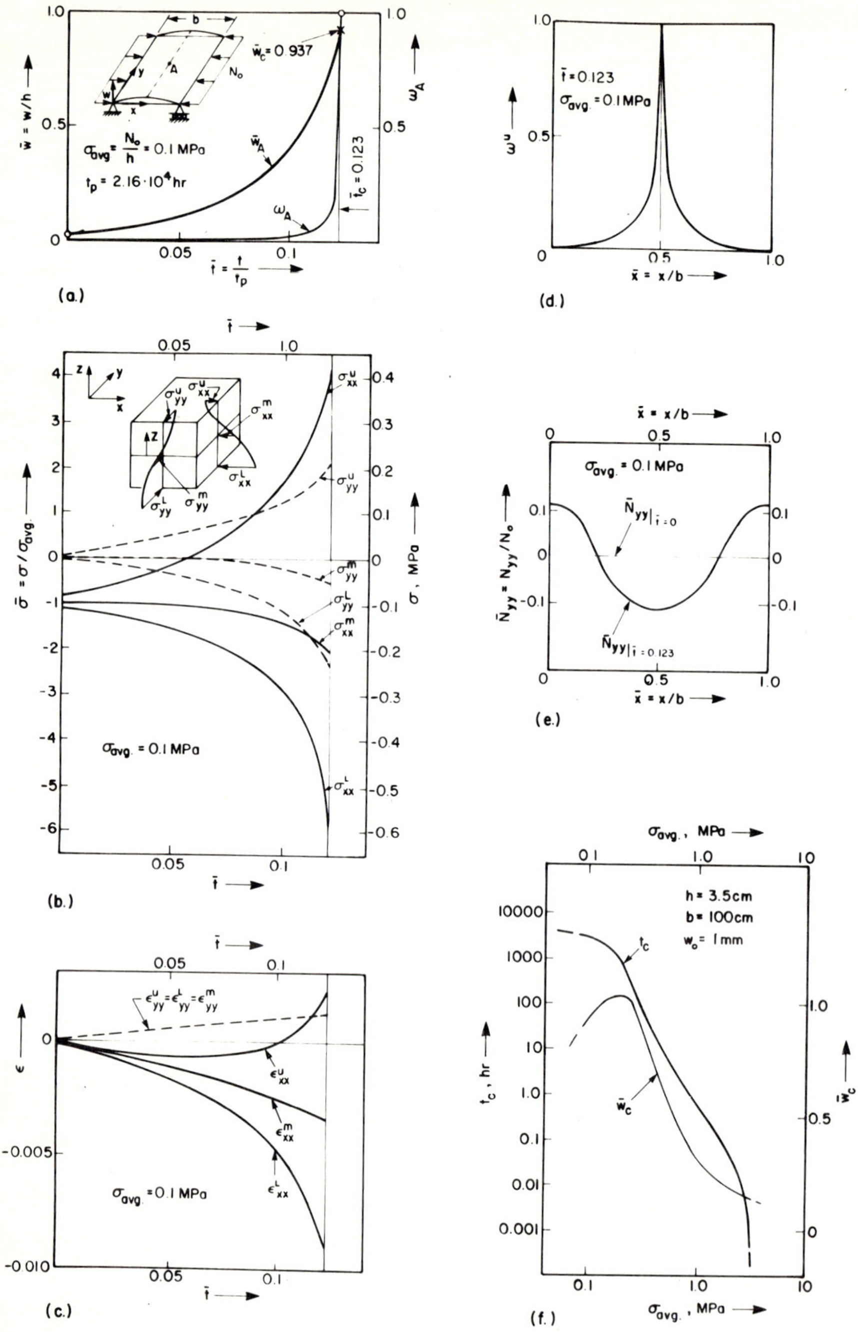

Fig. 3 Some Numerical Results

1. at a certain instant of time, the complete states of stress, deformation/deflection and damage should be known (at $t=0$ the elastic solution is taken to be the initial deformed state).
2. using the equations presented, the rates of displacements, (Equ. 24) strains (Equ. 23), stresses (Equ. 16) and damage (Equ. 12) are determined at selected points (nodes) of the plate.
3. using some time interval, δt, increments for all variables during this time interval are calculated and are added to the previous values to obtain updated values for the time $(t+\delta t)$.
4. steps (1)-(3) are repeated until the damage function reaches the rupture level $(\omega=1)$.

On the basis of experimental data presented in [14], the following visco-elastic constants are assumed for ice at a temperature of -7°C:

$$E = 4.25 \text{ GPa}; \qquad \mu = 0.33$$

$$n = 1.8; \; \nu_2 = 2.4 \times 10^4 (\text{MPa})^n \text{ hr}; \; \nu_1 = \nu_2/50; \; t_o = 100 \text{ hr}; \; t_1 = 1 \text{ hr}$$

The parameters defining the damage process are taken in accordance with [12] as

$$\Gamma = 1.8; \quad \eta = 15.0; \quad K = 6.5 \times 10^{-4} [(\text{MPa})^{\Gamma} \text{ hr}]^{-1}$$

while the material constants associated with the tertiary compressive creep stage of ice were selected as [11]

$$\alpha = 0.65; \quad \varepsilon_o^p = 0.7\%; \quad a = 1.0$$

The geometry of the plate was taken to be $h=3.5$ cm, $b=100$ cm, the initial imperfection at midspan , $w_o=1.0$ mm. The plate was divided into 10 sectors in each of the x and y directions resulting in 100 node points at each one of which the above outlined step-wise procedure was applied. Some of the results from this numerical analysis are indicated on Fig. 3.

Typical load-deflection behaviour is depicted on Fig. 3a for $\sigma_{avg}=0.1$ MPa. The time parameter, t_p, is calculated as $t_p=\nu_1/\sigma^n_{avg}$. The viscous process is terminated at the instant of collapse, designated by the collapse time,

t_c, for which the damage function at point A on the upper surface, ω_A, reaches the rupture level (ω=1.0). Variation of ω with time, for the same point, is also shown on the same figure and indicates the rapid increase in ω as $t \to t_c$.

Figs. 3b and c show stress and strain histories for selected points in the cross section containing point A, with superscripts u, m, and L denoting the upper, the middle and the lower plate surface, respectively. Fig. 3d shows the variation of the damage function across the plate width and indicates the relative concentration of damage near midspan.

As a result of the coupling effects inherent in the problem and mentioned above, the membrane stress resultant, N_{yy}, which at t=0 is equal to zero, varies across the plate width as shown in Fig. 3e. The last graph, Fig. 3f, depicts the collapse time, t_c, and the corresponding midspan deflection, $\bar{w}_c$, as a function of the average stress, σ_{avg}. Note that for relatively light loads (ie. low σ_{avg}), the collapse deflection, $\bar{w}_c$, decreases with decreasing σ_{avg}, which is a consequence of the damage process progressing in areas of tensile stress even at small stress/deflection levels, provided such stress duration is sufficiently long.

Conclusions

As opposed to previous work on ice plates, this paper presents a theory for nonlinear viscoelastic thin plates based on a constitutive model which takes into account all experimental data on ice available to the writers. The model defines the behaviour of ice in tension and compression up to failure thus including all stages of creep. To simulate the effects of micro-cracking and resulting stress anisotropy at advanced stages of deformation, a damage function is introduced which is assumed to be tensile strength dependent. The model also describes the hardening effect observed during the primary creep stage as well as the strain softening phenomenon exhibited by ice samples during their tertiary creep phase. This constitutive law is of a hereditary type. Accordingly, the current strain rate is a function of the current stress and stress history. With all these features, simulating actual ice behaviour, it is felt that this is a more realistic representation of material behaviour than previous models taken from linear viscoelasticity or the constitutive relations used in describing the time-dependent behaviour of metals at elevated temperatures.

Since the emphasis of this paper is on the treatment of material nonlinearity and the associated numerical techniques and convergence difficulties, the constitutive model was applied to a relatively simple structure, namely the cylindrical bending of a thin long plate with initial (imperfection) deflections subjected to uniform compressive loads applied along its longitudinal edges. The example was chosen because even under such constant external loading, the internal stress and strain states vary temporally. In addition, the numerical example was also selected so as to demonstrate the effectiveness of the incremental formulation introduced in [10], a formulation which facilitates the numerical solution of such time dependent problems.

The results indicate that the governing feature in the behaviour of such plates is the limited tensile strength of ice. The importance of a damage function, describing the effects of micro-cracking and stress anisotropy, was therefore brought into focus, underlining the urgency of the development of more comprehensive and precise damage theories for brittle materials as well as the need for additional experimentation designed to allow determination of parameters defining the damage process in ice. The results also confirm the fact that due to the inherent brittleness of the material, the collapse deflections of the plate are relatively small, keeping strain and displacement magnitudes well within the domain of a geometrically linear theory. From a practical viewpoint, this observation suggests reinforcing the material thereby providing tensile strength to ice, as was done for other brittle materials, such as concrete.

Finally, we note that the constitutive model presented here and in [12] should be considered a first attempt in simulating the complex behaviour of brittle materials, such as ice. Refinement of a model such as this, however, will require substantial additional experimental data.

References

1. Rabotnov, Yu N., "Creep Problems in Structural Members", North-Holland Publ. Co., Amsterdam-London, 1969.

2. Krempl, E., "On the Interaction of Rate and History Dependence in Structural Metals", Acta. Mech., 22, 1975, pp. 53-90.

3. Spring, U., Morland, L.W., "Integral Representations for the Visco-elastic Deformation of Ice", Cold Regions Sci. and Technol., 6, 1983, pp. 185-193.

4. Williams, F.M., "Time Dependent Deflections of Nonhomogeneous Ice Plates", Acta Mechanica, 25, 1976, pp. 29-44.

5. Vinogradov, A., Glockner, P.G., "Buckling of Spherical Viscoelastic Shells", Proc. ASCE, 106, J. Structural Div. No. ST1, Paper No. 15116, Jan. 1980, pp. 59-67.

6. Hutter, K., "On the Mechanics of Floating Ice Sheets", Mitteilungen der Versuchsanstalt fur Wasserbau, Hydrologie und Glaziologie, No. 28, Zurich 1978.

7. Hutter, K., Williams, F.W., "Theory of Floating Ice Sheets", Physics and Mechanics of Ice, Proc. of IUTAM Symp., Copenhagen, Aug. 1979, Springer-Verlag, Berlin-Heidelberg 1980, pp. 147-161.

8. Williams, F.W., Hutter, K., "Thermal Response of Unconfined Ice Shelves to Climatic Conditions", Acta Mechanica, 48, 1983, pp. 131-146.

9. Szyszkowski, W., Dost, S. and Glockner, P.G., "A Nonlinear Constitutive Model for Ice", Int. J. Solids Structures, 21, No. 3, 1985, pp. 307-321.

10. Szyszkwoski, W., Glockner, P.G., "On a Multi-Axial Nonlinear Hereditary Constitutive Law for Non-Ageing Materials with Fading Memory", Dept. of Mech. Engg., The University of Calgary, Report No. 321, May 1985.

11. Szyszkowski, W., Glockner, P.G., "Modelling the Time-Dependent Behaviour of Ice", Cold Regions Sci. and Technology (in press).

12. Szyszkowski, W., Glockner, P.G., "On a Multiaxial Constitutive Law for Ice", Dept. of Mech. Engg., The University of Calgary, Report No. 327, July 1985.

13. Weeks, W.F., Mellor, M., "Mechanical Properties of Ice in the Arctic Seas", Proc. on Arctic Technology and Policy, M.I.T., March 1983.

14. Voitkovski, K.F., "The Mechanical Properties of Ice", Izd. Akademii Nauk SSSR, 1960, Trans. AMS-T-R-391, Am. Met. Soc., Office of Techn. Services, U.S. Dept. of Commerce, Washington, 25.

15. Mellor, M., Cole, David M., "Stress/Strain/Time Relations for Ice Under Uniaxial Compression", Cold Regions Sci. and Technol., 6, 1983, pp. 207-230.

16. Jacka, T.H., "The Time and Strain Required for Development of Minimum Strain Rates in Ice", Cold Regions Sci. and Technol., 8, 1984, pp. 261-268.

17. Hawkes, I., Mellor, M., "Deformation and Fracture of Ice Under Uniaxial Stress", Journal of Glaciology, 11, 1972, pp. 103-131.

Fracture and Fatigue of Composite Plates and Shells

V.V. Bolotin

Academy of Sciences of USSR
Institute of Mechanical Engineering
Moscow, USSR.

Summary

A survey of the up-to-date state of the fracture and fatigue theory is presented with special emphasis on delaminations, edge effects and related phenomena typical for composite plates and shells. Equations governing the growth of delaminations are presented based on the generalized concept of energy release with an account of damage accumulated in the interlayer zone. For example, delaminations in multilayered composite plates, spherical and circular cylindrical shells are studied. The physical non-linearity due to the microdamage accumulation, as well as the geometrical non-linearity originated from the buckling of a delaminated region are taken into account. Cyclic, static and combined fatigue are considered. The effect of delaminations on the load carrying capacity of composite plates and shells is discussed.

Introduction

Properties of most of structural composite materials are unique. Firstly, they combine extremely high strength with low mass density. Secondly, manufacturing of composites requires less human labor and energy than that of most conventional structural materials. In third place, manufacturing of composites can be maintained with less environment pollution than, say, steel production. From all of these three viewpoints, composites may be considered as materials of the future.

High strength of up-to-date composites is achieved by the adequate association of components' properties, and high strength of composite structures - by coordination between the fields of mechanical properties and the stress fields under design loads and actions. But properties of composites and composite structures are ambivalent. For example, high strength of a composite in the direction of reinforcement is accompanied with a rather

Inelastic Behaviour of Plates and Shells
IUTAM Symposium Rio de Janeiro 1985
Editors: L. Bevilacqua, R. Feijóo and R. Valid

low resistance against transverse shear and transverse tension. Therefore, composite structures are more sensitive to small deviations of loads and actions from their design values and directions, compared with structures made of conventional quasi-isotropic materials. Effects accounting for the latter type of structures as second rate and even negligible, become of significance for those composites which are highly anisotropic, and/or highly non-homogeneous in the microscale.

Most composite structures are realized as plates and shells, which may be treated as laminated structures [1]. For example, each ply of a composite shell manufactured by filament winding is, in the macroscale, an anisotropic thin shell. There are also very thin matrix interlayers between neighbouring plies. These interlayers are of special significance when the transverse shear and tension are considered. Generally, a composite plate or shell has a more or less regular structure along the thickness, and "strong" layers alternate with "weak" ones [2,3]. Since the strength of matrix is relatively low, composite structures are especially sensitive with respect to interlayer flaws such as delaminations. Most of these flaws are initiated at the fabrication stage due to shrinkage of the matrix and the thermal stresses [4]. Delaminations can be created also at the transportation, storage and service stages under thermal actions, local forces, surface impacts, etc.

Two kinds of interlayer flaws are distinguished in the present study. The first ones are situated in the bulk of a structure. With application to shell structures, the position of a flaw can be characterized by the distances H_1 and H_2 from the external surfaces of a shell. If $H_1 \sim H_2 \sim H$, where H is the thickness of the shell, we say that the flaw is an internal delamination (Fig. 1a). The second kind of flaws are those which are situated near the surface. The thickness h of the delaminated part of a structure satisfies to the condition $h \ll H$ (Fig. 1b). We call these flaws the surface delaminations.

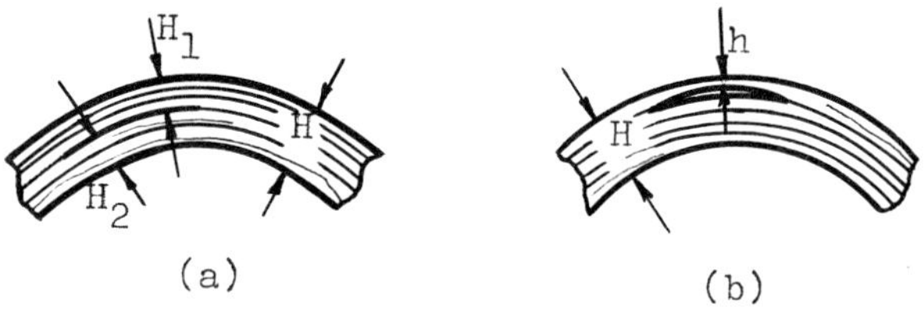

Fig. 1. Delaminations in composite shells: (a) internal; (b) surface delaminations.

Behavior of surface delaminations, generally, includes the buckling phenomena. It is typical for compressed structural components, and for non-uniform temperature fields. Due to the Poisson's effect, buckling can also be met in components in tension. It is evident that an adequate fracture mechanics of surface delaminations should include the account of buckling, and, therefore of geometrical non-linearities. Typical examples are presented in Fig. 2. Fig. 2a and b correspond to components in tension with initial surface notches. Fig. 2c corresponds to components

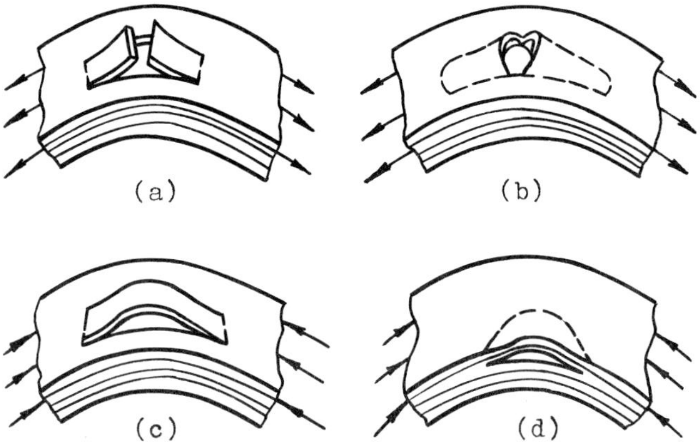

Fig. 2. Fracture of composite shells in delamination modes: (a) peeling in tension; (b) pocket-like peeling in tension; (c) buckling in compression; (d) edge buckling in compression.

in compression when the origin of a delamination is either an initial poor adhesion or a shrinkage crack. Fig. 2d shows a delamination originated from an edge effect. Only in case 2a we may not take into consideration the geometrical non-linearity.

The following groups of problems for composite plates and shells containing delaminations are of interest:

- stability of delaminations (in the sense of fracture mechanics);
- growth of delaminations under long acting and /or cyclic loading;
- ultimate loads and ultimate loads combinations for delaminated structures;
- admissible sizes of delaminations with respect to long acting and/or cyclic loading.

The behaviour of internal delamination in opening and shear fracture modes is more or less similar to that of cracks in the conventional fracture mechanics, although there are some complications due to the essential non-homogeneity and anisotropy of composites [6]. But even the basic problem of fracture of unidirectional fibrous composites in tension along fibres does not fit into the patterns of the linear fracture mechanics. To explain the "brush-like" character of fracture, additional factors should be taken into account such as the separation of fibers due to the damage of the matrix and the scale effect of strength of fibres [7]. Concerning the surface delaminations in compression, we must take into account geometrical non-linearities [8]. As to the growth of delaminations under long acting and/or cyclic loading, an essential generalization of the conventional fracture mechanics is required [9].

A United Model of Fracture and Fatigue

It is common knowledge that the fatigue cracks growth is connected closely with the microdamage accumulation in the material's structure. Initiation of fatigue cracks is the result of a local damage accumulation near the most stressed or weakest microstructural elements. Moreover, interaction between the microcracking near the crack-tip and the energy release due to the crack growth seems to be the principal mechanism governing the propagation of fatigue cracks. This conclusion is valid both for cyclic and static fatigue, for linear-elastic, elasto-plastic and visco-elastic materials.

In paper [9] a theory of fatigue crack growth was proposed. Generalizing the Griffith's energy approach, the damage accumulation near the crack-tip and on the crack's prolongation was taken into account. The essential point of the theory is the assumption that stresses near the crack-tips are bounded. Opposite to the common fracture mechanics, a crack is treated as a slit with finite curvature near the tips. This assumption introduces characteristic length parameters which are of significance for any consistent theory of crack growth. Another part of the proposed theory is the introduction of kinetic equations governing micro-damage accumulation. The proposed model can be generalized easily upon multi-axial stress-strain states, multiparameter cracks, combination of static and cyclic fatigue, corrosion cracks, etc. The theory permits to include quite naturally geometrical non-linearities, and, therefore, becomes applicable to delaminations in compressed composite plates and shells [5,9].

We treat the Griffith's concept in terms of the principle of virtual work. The central concepts of the theory are: the Griffith's variation, equilibrium and stability of a cracked body in the Griffith's sense. Following the Griffith's original idea, we consider the whole energy of the system "cracked body loading". Introduce a small isochronic increment of this functional due to virtual irreversible variations of cracks sizes and calculated under the following conditions: both the initial and disturbed states satisfy equilibrium and compatibility equations (except maybe vicinities of the crack-tips), heat flow to the body is absent, external loads and external displacements are fixed. This increment by definition is the Griffith's variation of the functional. Consider the Griffith's variation of the whole energy taken with the opposite sign:

$$\delta I = -\delta U + \delta A_p - \delta \phi - \delta A_\gamma \quad (1)$$

Here U is the potential energy of the body, δA_p is elementary work of external loads, $\delta\phi$ is elementary isochronic dissipation in the volume of the body, and δA_γ is elementary work spent on crack propagation. We say that a state of a cracked body is equilibrium in the Griffith's sense if $\delta I = 0$. We say that the body is in the sub-equilibrium state if $\delta I < 0$, and in the non-equilibrium state if $\delta I > 0$. Sub-equilibrium states are stable by definition. Stability of equilibrium states depends on the sign of the second Griffith's variation $\delta^2 I = \delta(\delta I)$. If $\delta^2 I < 0$ an equilibrium state is stable, if $\delta^2 I > 0$ it is unstable. In the case $\delta^2 I = 0$ an equilibrium is neutral, and the study of variation of higher order is required.

There is the analogy between the preceding interpretation of the Griffith's concept and general principles of analytical statics. The central notion of the Lagrange-Dirichlet principle is potential energy U of a conservative system. A system is in the equilibrium state if $\delta U = 0$. The equilibrium is stable if $\delta^2 U > 0$ and unstable if $\delta^2 U < 0$. This theorem is usually illustrated with a heavy ball placed on an ideally smooth surface. The simplest mechanical model of a cracked body is shown in Fig. 3. A heavy cylinder is placed on a geared cylindrical surface. The cylinder is equipped with a ratchet detent preventing the return motion. The notions of sub-equilibrium and non-equilibrium states, as well as of stable, neutral and unstable states are covered completely by the model presented in Fig. 3. The essential feature of fatigue crack problem is that all cracks are "incurable". Hence the condition $\delta l \geqslant 0$ holds. The cylinder with the ratchet is a mechanical system with unilateral constraints having an irreversible set of equilibrium states. In this meaning, condition $\delta I \leqslant 0$ is an analogue of the condition $\delta A \leqslant 0$ of the principle of virtual work for mechanical systems with unilateral constraints.

Consider a body containing a set of cracks given with vector $l = (l_1,\ldots,l_m)^T$ where l_j are characteristic sizes of cracks. We re-write Eq. (1) in the form

$$\delta I = \sum_{j=1}^{m} (G_j - \Gamma_j)\,\delta l_j \qquad (2)$$

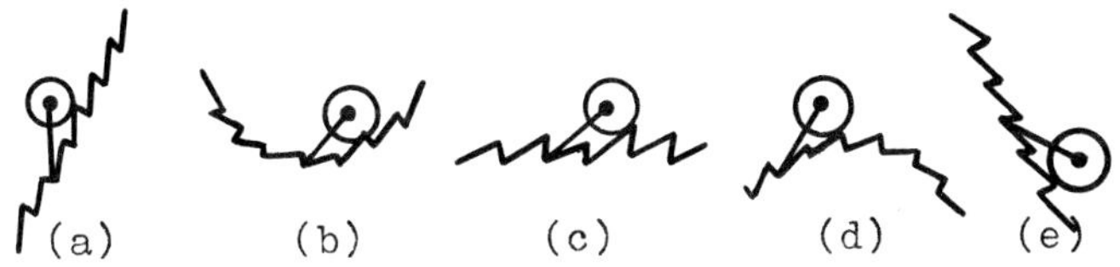

Fig. 3. The simplest mechanical model illustrating crack growth. States: (a) sub-equilibrium; (b) stable equilibrium; (c) neutral equilibrium; (d) instable equilibrium; (e) non-equilibrium.

where G_j are generalized forces moving cracks, i.e. analogues of Irwin's energy release rate G. Corresponding resistance forces Γ_j relate to elementary work δA_γ and are similar to the critical value of energy release rate G_c. The condition of equilibrium $\delta I = 0$ results in

$$G_j = \Gamma_j \quad (j = 1,\ldots,m). \tag{3}$$

Stability of the system depends on the sign of the quadratic form

$$\delta^2 I = \sum_{j=1}^{m} \sum_{k=1}^{m} \frac{\partial (G_j - \Gamma_j)}{\partial l_k} \delta l_j \delta l_k \tag{4}$$

under the condition that all $\delta l_j \geqslant 0$.

To apply the theory to growing fatigue cracks, damage accumulation in microstructure should be taken into account. Damage at the crack-tip is described with the damage vector $\psi = (\psi_1,\ldots,\psi_m)^T$. Components of this vector are associated either with crack sizes (i.e. with components of vector l) or with the type of microcracking. Damage vector ψ enters into Eqs. (3) and (4) with the resistance forces Γ_j because the latters are assumed diminishing due to microcracking. The moving forces G_j, generally, also may depend on ψ.

Consider firstly the static fatigue when loads and actions are varying very slowly. Introduce the vector s(t) characterizing loads and actions. Consider the differences of generalized forces

$$H_j(t) = G_j[l(t),s(t),\psi(t)] - \Gamma_j[l(t),s(t),\psi(t)] \qquad (j = 1,\dots,m) \tag{5}$$

If at $t=t_o$ the body is in a sub-equilibrium state, all $H_j(t_o) < 0$. One of the cracks begins to grow with respect to the generalized coordinate l_j when the equality $H_j(t) = 0$ is reached for the first time. The growth is stable if the quadratic form given with Eq. (4) is definitely negative. Then the equality $H_j(t) = 0$ will be followed at a certain time segment until the crack growth terminates because of the load drop, a non-homogeneity entered on the path of the crack, etc. The other case is the instability of the crack in the form of a jump-like growth to a new sub-equilibrium state, or to the final failure. In the absence of these phenomena, a stable fatigue crack is developing remaining approximately to be in equilibrium and stable in the Griffith's sense. This postulate was suggested firstly in [9].

To obtain a closed set of equations, additional equations should be taken governing the damage accumulation on the prolongations of cracks. We take these equations in the form

$$\frac{\partial\phi(\lambda,t)}{\partial t} = \mathop{\Phi}_{\tau=t_o}^{\tau=t}\{\lambda, l(\tau), s(\tau), \phi(\lambda,\tau)\} \tag{6}$$

where $\Phi\{.\}$ is a functional of the history of loading and damage, λ is the prolongation of vector l, and $\phi(\lambda,t)$ is the damage vector on this prolongation. Evidently, the following identity takes place

$$\psi(t) \equiv \phi[l(t),t]. \tag{7}$$

Similar equations are valid in the case of cyclic fatigue. Instead of $H_j(t)$ the functional $\sup H_j(t)$ is used. Supremum is taken at the segement (t_{n-1}, t_n) corresponding to n-th cycle of loading. Then the equation similar to Eq. (6) is to be replaced with a finite-difference analogue with respect to $\phi(\lambda,t_n)-\phi(\lambda,t_{n-1})$. A crack advances at a distance with the order of magnitude equal to the characteristic size of the material's structure, if during the n-th cycle even if once the inequality sup $H_j(t) > 0$ takes place. But, when the number of cycles up to fracture is very large, and extremum values of the components of the vector s(t) are varying slowly from one cycle to another, the cycle number n may be treated as a continuous argument. As a result, we come to Eqs. (5)-(7) replacing t by n.

The process s(t) includes extreme values of loads or nominal stresses. In the case of combined loading when the input both of quasistatic and cyclic actions is important, damage originated from both sources ought to be summarized. It should be mentioned that Eq. (1) and its analogue for the cyclic fatigue, in principle, can incorporate also temperature, concentrations of active agents, and other environmental factors which have an influence on the damage accumulation and are considered significant.

Delamination in a Notched Composite Plate in Tension.

To illustrate the application of the theory, consider an elastic laminated plate with an initial surface notch which has a depth h. Let in the vicinity of the notch, a delamination zone be originated with the length 2l and the width b (Fig. 4). Assume that the delamination is in a plane strain state, and omit all secondary details. Denote ε the nominal tensile strain in x-direction. Potential energy of the plate $U = \text{const} - \frac{1}{2} E_x bhl\varepsilon^2 (1 - \nu_{xy}\nu_{yx})^{-1}$. Here E_x is Young's modulus in x-direction and ν_{xy}, ν_{yx} are Poisson's ratios. The moving generalized force related to one of the branches of the delamination is

$$G = \frac{1}{2} \frac{E_x bh\varepsilon^2}{1-\nu_{xy}\nu_{yx}} \tag{8}$$

Virtual fracture work corresponding to the advancement of one of the branches on the distance δl is $\delta A_\gamma = \gamma_o b\delta l + \gamma_1 h\delta l$. Here

γ_o is the specific fracture work for the matrix interlayer, and γ_1 is the specific fracture work spent on splitting along the delamination. For unidirectional layers $\gamma_1 \sim \gamma_o$. Then at $h \ll b$ the generalized resistance force for the undamaged composite may be taken as

$$\Gamma_o = \gamma_o b. \tag{9}$$

The state of the delamination in undamaged composite is equilibrium at $G = \Gamma_o$. Hence the critical (in the Griffith's sense) nominal strain is

$$\varepsilon_\infty = \left[\frac{2\gamma_o(1-\nu_{xy}\nu_{yx})}{E_x h} \right]^{\frac{1}{2}} \tag{10}$$

The corresponding equilibrium state is neutral since $\delta^2 I = 0$. But this circumstance has no significance for further analysis because we are going to treat the case when $G < \Gamma_o$.

At a certain time segment $[0,t_*]$ the delamination does not grow because it is in a sub-equilibrium state. This situation corresponds to the initiation stage during which damage is accumulated at $|x| \geqslant l$. Introduce damage measure $\phi(x,t)$ for the matrix interlayer. As usual, we assume $\phi = 0$ for the undamaged interlayer, and $\phi = 1$ for the completely debonded layer. We assume that the damage is created by tangential stresses $\tau(x,t)$ in the interlayer. Let Eq. (6) take the form

$$\frac{\partial \phi}{\partial t} = \begin{cases} 0, & |\tau| \leqslant \tau_{th} \\ \dfrac{1}{t_c}\left(\dfrac{|\tau| - \tau_{th}}{\tau_\psi}\right)^m, & |\tau| > \tau_{th}. \end{cases} \tag{11}$$

Here τ_{th} is a threshold damaging stress, and τ_ψ is a material constant characterizing resistance of the interlayer against the damage accumulation. Power exponent m is analgous to exponents of static fatigue curves in standard tests, and t_c is a certain time constant. The simplest equation for the stress $\tau(x,t)$ in the membrane approximation is

$$|\tau| = \frac{G_m \lambda_o \varepsilon}{h_m} \exp\left(-\frac{x-l}{\lambda_o}\right) \tag{12}$$

with the shear modulus G_m, the matrix interlayer thickness h_m, and the characteristic length

$$\lambda_o = (hh_m)^{\frac{1}{2}} \left[\frac{E_x}{G_m(1-\nu_{xy}\nu_{yx})}\right]^{\frac{1}{2}} . \tag{13}$$

The latter can be interpreted as a characteristic length of the boundary effect [2] or an ineffective length in the vicinity of ruptured fiber or monolayer [11]. Using Eqs. (11) and (12), and the initial condition $\phi(x,0) = \phi_o(x)$, the damage measure $\psi(t) = \phi[l(t),t]$ at the tip of the delamination is

$$\psi(t) = \phi_o[l(t)] + \frac{1}{t_c} \int_0^t \left\{\frac{|\tau[l(t_1),t_1]| - \tau_{th}}{\tau_\psi}\right\}^m dt_1 . \tag{14}$$

The special notation of the integral reminds that at $\tau(t_1) < \tau_{th}$ the integral must be placed to zero.

For further discussion, the relation $\Gamma(\psi)$ ought to be specified. We assume $\Gamma = \Gamma_o(1 - \psi^\alpha)$ where Γ_o is taken from Eq. (9), and $\alpha > 0$, e.g. $\alpha = 1$. This assumption is more or less realistic. The moment of termination of the initial stage is the first positive root of the equation $G(t) = \Gamma(t)$ at $l = l_o$ = const. Using Eqs. (10) and (14) we obtain equation with respect to t_*:

$$\phi_o(l_o) + \frac{1}{t_c} \int_0^{t_*} \left[\frac{|\tau(l_o,t_1)| - \tau_{th}}{\tau_\psi}\right]^m dt_1 = \left[1 - \frac{\varepsilon^2(t_*)}{\varepsilon_\infty^2}\right]^{1/\alpha} \tag{15}$$

After the tip begins to propagate, the half-length $l(t)$ is to be found from equation

$$\phi[l(t)] + \frac{1}{t_c} \int_0^t \left\{ \frac{|\tau[l(t_1),t_1]| - \tau_{th}}{\tau_\psi} \right\}^m dt_1 =$$

$$= \left[1 - \frac{\varepsilon^2(t)}{\varepsilon_\infty^2} \right]^{1/\alpha} \qquad (16)$$

Eqs. (15) and (16) are to be solved numerically.

Consider in more details the case ε = const, $\phi_o(x) = 0$, and $\tau_{th} = 0$. An explicit formula for t_* follows from Eq. (15)

$$t_* = t_c \left(\frac{\varepsilon_\psi}{\varepsilon} \right)^m \left(1 - \frac{\varepsilon^2}{\varepsilon_\infty^2} \right)^{1/\alpha} . \qquad (17)$$

where Eq. (12) and notation $\varepsilon_\psi = (\tau_\psi h_m)/(G_m \lambda_o)$ are used.

The process of growth of the delamination is illustrated in Fig. 5a, and the process of damage accumulation at the tip in Fig. 5b. Lines 1,2,3 correspond to various nominal strain levels ε = const.

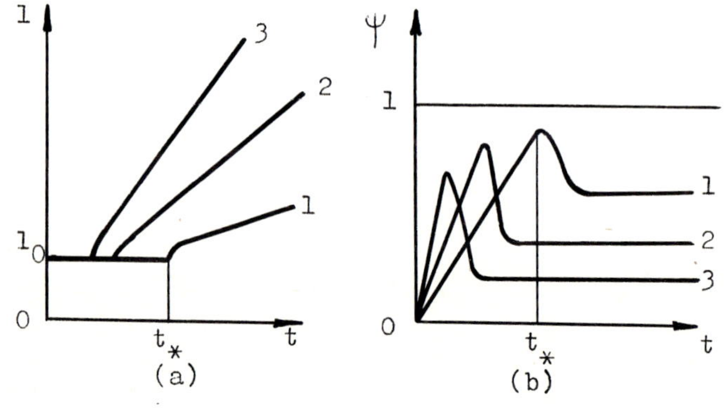

Fig. 5. Static fatigue in tension: (a) delamination growth; (b) damage accumulation in the matrix interlayer near the tip of the delamination. Lines 1,2 and 3 correspon to three ascending nominal strain levels.

At $t > t_*$ a nonstationary stage of propagation takes place. Then the growth rate dl/dt as well as the damage measure ψ approach certain asymptotic values. To estimate these values, we put in Eq. (14) $dl/dt \approx$ const, $l(t) \approx l(t_1) + (dl/dt)(t - t_1)$, and replace the lower limit of integration by $-\infty$. Then

$$\psi \approx \frac{\lambda_o}{mt_c}\left(\frac{dl}{dt}\right)^{-1}\left(\frac{\varepsilon}{\varepsilon_\psi}\right)^m , \tag{18}$$

$$\frac{dl}{dt} \approx \frac{\lambda_o}{mt_c}\left(\frac{\varepsilon}{\varepsilon_\psi}\right)^m\left(1 - \frac{\varepsilon^2}{\varepsilon_\infty^2}\right)^{-1/\alpha} . \tag{19}$$

Eq. (19) reminds the well-known semi-empirical equation of fatigue crack growth [12], and is analgous to the corresponding theoretical equations [9,10]. Eq. (19) is also valid when $\varepsilon(t)$ is varying sufficiently slowly. More precisely, the relative variation of ε should be small during a time segment of the order of magnitude $\lambda_o(dl/dt)^{-1}$. Then Eq. (19) becomes the differential equation with respect to l describing slow propagation of the delamination in static fatigue. Initial condition is $l(t_*) = l_o$ where t_* is the last moment of the initiation stage. Thus, Eqs. (15) or (17) should be applied before the use of Eq. (19).

Delamination in a Composite Plate in Compression

Let a delamination be located in the compression zone of an elastic plate or shell. The source of compression may be external loading, non-uniform temperature distribution or residual stresses created in the fabrication process. If the delamination is buckled, geometrical non-linearity has to be taken into account.

The simplest problem of this kind is shown in Fig. 6. We assume the initial state of the delamination to be plane and non-stressed. Before buckling occurs, the delamination is subjected to the same compression strain ε as the attached part of the plate. Buckling begins at $\varepsilon > \varepsilon_*$ where $\varepsilon_*(l)$ is the critical (in the Euler's sense) strain

$$\varepsilon_*(l) = \frac{\pi^2}{3}\left(\frac{h}{l}\right)^2 \tag{20}$$

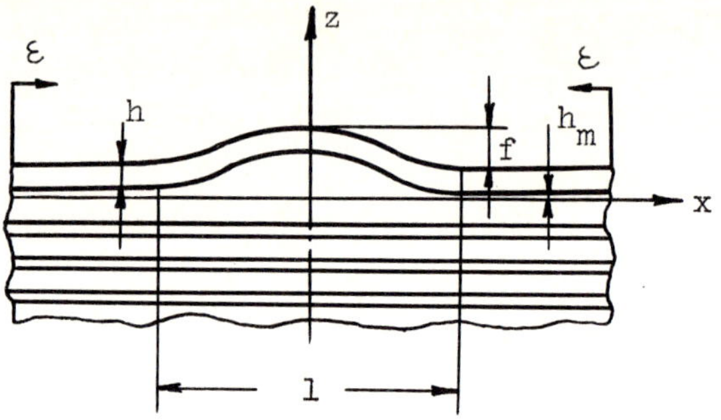

Fig. 6. Fracture of a layered plate in compression.

In this section we assume $\varepsilon > 0$ for compression. Post-buckling form of the delaminated zone is taken as

$$w(x) = f \cos^2\left(\frac{\pi x}{l}\right) \tag{21}$$

where f is the maximum deflection (Fig. 6). Relative displacement of the boundaries of the delamination is connected with the nominal strain ε. At $\varepsilon > \varepsilon_*(l)$ we obtain

$$f = \frac{2l}{\pi}\left[\varepsilon - \varepsilon_*(l)\right]^{\frac{1}{2}} \tag{22}$$

With the account of bending in the delamination zone, potential energy of the stressed plate is

$$U = \text{const} - \frac{1}{2}\frac{E_x bhl}{1-\nu_{xy}\nu_{yx}}\left[\varepsilon^2 - \varepsilon_*^2(l) - \frac{h^2}{12}\int_{-l/2}^{l/2}\left(\frac{\partial^2 w}{\partial x^2}\right)^2 dx\right]$$

Substituing Eqs. (21) and (22), we obtain $U = \text{const} - \frac{1}{2}E_x bh(1 - \nu_{xy}\nu_{yx})^{-1}[\varepsilon - \varepsilon_*(l)]^2$. If the loading is "rigid", i.e. kinematic one, virtual work of external forces is equal to zero, and

$$G = \frac{1}{2}\frac{E_x bh}{1-\nu_{xy}\nu_{yx}}\left[\varepsilon^2 + 2\varepsilon\varepsilon_*(l) - 3\varepsilon_*^2(l)\right]. \tag{23}$$

Here notation (20) for the critical strain is used.

The generalized resistance force for undamaged composite is given with Eq. (9). The condition of equilibrium $G = \Gamma_o$ with account of Eqs. (10) and (23) results in

$$\varepsilon^2 + 2\varepsilon\varepsilon_*(l) - 3\varepsilon_*^2(l) = \varepsilon_\infty^2. \tag{24}$$

Eq. (24) may be interpreted as an equation with respect to equilibrium (in the Griffith's sense) sizes of the delamination. Stability condition for an equilibrium state is

$$\frac{dl}{d\varepsilon} > 0. \tag{25}$$

Let $\varepsilon_{**}(l)$ be the value of ε satisfying Eq. (24) and the inverse relation be $l = l_{**}(\varepsilon)$. A typical stability diagram is presented in Fig. 7. According to Eq. (25), the ascending branch of

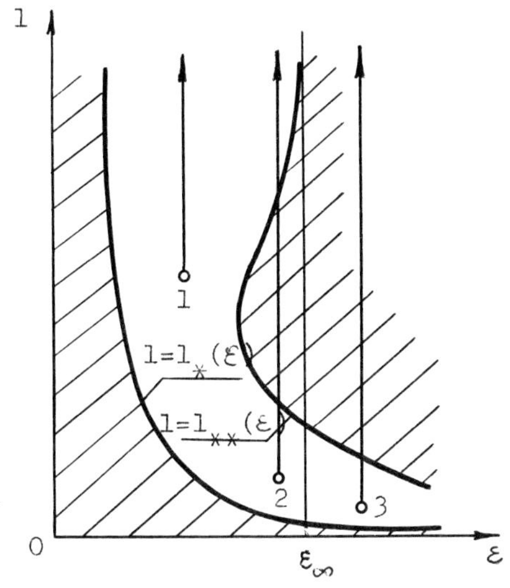

Fig. 7. The region of stability (in the Euler's sense) of the plane form of a delamination, and the region of instability of the buckled delamination (in the Griffith's sense). Stable growth of the delamination occurs in the unshaded region.

the line $l = l_{**}(\varepsilon)$ is stable, and the descending one is unstable. The relation $l = l_*(\varepsilon)$ is plotted in Fig. 7, too, corresponding to Eq. (20) for the buckling strain $\varepsilon_*(l)$. The initial plane state of the delamination remains plane and stable in the shaded area on the left hand side from the line $l - l_*(\varepsilon)$. The unshaded area corresponds to sub-equilibrium and, therefore, stable (in the Griffith's sense) buckled delaminations. The area located on the right-hand side from the line $l = l_{**}(\varepsilon)$ corresponds to non-equilibrium and, therefore, unstable delaminations.

Thin buckled delaminations have already been discussed by a number of authors [13-18] mainly from the stability viewpoint. The problem of growth of buckled delaminations due to static and/or cyclic fatigue and fracture was studied in [9]. For the sake of determination, static fatigue is considered later. Let the damage measure be the solution of Eq. (11). Condition $G(t) = \Gamma_o[1 - \psi^\alpha(t)]$ with $\psi(t)$ defined from Eq. (14) results in equations similar to Eqs. (15) and (16). In particular, duration of the initiation stage is characterized by the root t_* of equation

$$\phi_o(l_o) + \frac{1}{t_c}\int_0^{t_*}\left[\frac{|\tau(l_o,t_o)| - \tau_{th}}{\tau_\psi}\right]^m dt_1 =$$

$$\left[1 - \frac{\varepsilon^2(t_*) + 2\varepsilon(t_*)\varepsilon_*(l_o) - 3\varepsilon_*^2(l_o)}{\varepsilon_\infty^2}\right]^{1/\alpha} . \qquad (26)$$

The stable growth at $t > t_*$ is governed by equation

$$\phi_o[l(t)] + \frac{1}{t_c}\int_0^{t}\left\{\frac{|\tau[l(t_1),t_1]| - \tau_{th}}{\tau_\psi}\right\}^m dt_1 =$$

$$\left\{1 - \frac{\varepsilon^2(t) + 2\varepsilon(t)\varepsilon_*[l(t)] - 3\varepsilon_*^2[l(t)]}{\varepsilon_\infty^2}\right\}^{1/\alpha} . \qquad (27)$$

If the tip is propagating monotonically and sufficiently slowly, Eq. (27) may be reduced approximately to a differential equation similar to Eq. (19). The right-hand side becomes rather cumbersome if bending and shearing of the surface layer are incorporated. Only to keep transparency, we apply Eq. (12) which does not account for these effects. At $\tau_{th} = 0$ we obtain

$$\frac{dl}{dt} \approx \frac{\lambda_o}{mt_c} \left(\frac{\varepsilon}{\varepsilon_\psi} \right)^m \left[1 - \frac{\varepsilon^2 + 2\varepsilon\varepsilon_*(l) - 3\varepsilon_*^2(l)}{\varepsilon_\infty^2} \right]^{-1/\alpha} . \quad (28)$$

At $\varepsilon >> \varepsilon_*(l)$ all moment effects become negligible, and Eqs. (26), (27) and (28) turn into Eqs. (15), (16) and (19) correspondingly. It means that the difference between delaminations in tension and in compression becomes negligible.

Soon qualitative conclusions from Eqs. (26)-(27) are presented in Fig. 8. Lines 1, 2, and 3 correspond to initial states 1, 2

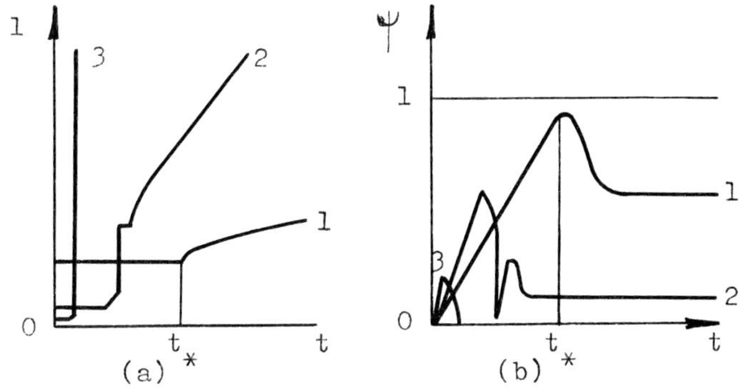

Fig. 8. Static fatigue in compression: (a) delamination growth; (b) damage accumulation in the matrix interlayer near the tip. Lines 1,2 and 3 correspond to the respective initial states from Fig. 7.

and 3 in Fig. 7, respectively. In case 1 the initial size of the delamination is comparatively large, and $\varepsilon << \varepsilon_\infty$. The initial state is sub-equilibrium. The size of the delamination begins to grow after the end of the initiation stage. Meanwhile, the size increases, the first (momentless) term in the brackets of Eq. (23) becomes dominant. The general picture in case 1 is similar to that for a notched plate in tension (Fig. 5). The initial state in case 2 is sub-equilibrium, and a comparatively short initiation stage exists too. After the growth of the delamination up to a certain unstable state, a jump to a new sub-equilibrium state occurs. The new size can be evaluated from energy balance considerations [9]. The damage measure ψ in the new state drops essentially since the tip propagates into a comparatively undamaged zone (see line 2 in Fig. 8b). Then the second initiation stage takes place resulting in the following stable growth.

The initial values l_o and ε in case 3 correspond to a point located in a narrow zone between the area where no buckling occurs, and the non-equilibrium area (Fig. 7). Very short incubation and stable growth stages terminate in total splitting of the surface layer.

Elastic limit strain for most structural composites is of the order of magnitude 10^{-3}. Hence, the typical behaviour of compressed delaminations is described by line 1 in Fig. 8. It should be mentioned, however, that delaminations in real structures are usually pre-buckled and/or pre-stressed due to the fabrication flaws and the loading prehistory. Therefore, the damage accumulation and the delamination growth begin at $\varepsilon < \varepsilon_*(l)$. The effect of initial deflections and initial stresses was studied in [5].

Elliptical delaminations

Elliptical delamination appears to be a comparatively adequate model of fabrication flaws in composite plates and shells. For example, a delamination in an orthotropic circular cylindrical shell is shown in Fig. 9. We assume that in involution or in projection on a tangential plane, the form of the delamination is close to an ellipse with semi-axes a and b. If principal directions of orthotropy and ones of the nominal strain field

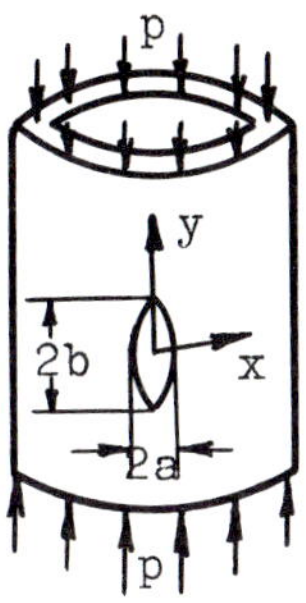

Fig. 9. Elliptical delamination in a circular cylindrical composite shell in compression.

coincide with the principal axes of the ellipse, the form of the delamination will remain close to the ellipse during the further loading history. This means that we come to a two-parametrical problem of the fracture mechanics. This problem was studied primarily in [5].

Define ε_x and ε_y, the nominal principal strains, assuming them positive in compression. Let the buckling deflection f satisfy to condition $f \ll \min\{a,b\}$. In a typical situation $f \sim h$, where h is the thickness of the delamination. Let $\max\{a,b\} \ll \min\{R_1,R_2\}$ with principal curvature radii of the shell R_1 and R_2. Then the delamination may be treated in the framework of the theory of thin shallow shells.

A variational approach was used in [5] to obtain an approximate solution. Both initial and following deflections are taken in the form

$$w(x,y) = f\left(1 - \frac{x^2}{a^2} - \frac{y^2}{b^2}\right)^2 \qquad (29)$$

The right-hand side of Eq. (29) satisfies to conditions for the delamination clamped on the boundary which is an ellipse with the semi-axes a and b. Outside of the elliptic area w=0. If $f \sim h$, membrane strains in the delamination are close to the critical buckling values. Denote the critical strains ε_x^* and

ε_y^*. In the general case, the latter depends on a and b, as well as on the nominal strains ε_x and ε_y. To express f, ε_x^* and ε_y^* through ε_x, ε_y, a and b, three equations are needed. The first equation follows from the variational principle of the theory of elastic stability under the assumption that the buckling mode is taken from Eq. (29). To obtain two remaining equations, we use relations between averaged deformation of the chords of the ellipse, the nominal and critical strains. Final equations are cumbersome, and we refer to papers [5,19,20] for more details.

Now potential deformation energy U of the delaminated shell is calculated. Treating the loading as "rigid", the generalized forces moving the delaminations are $G_a = -\partial U/\partial a$, $G_b = -\partial U/\partial b$. Virtual work of fracture for initially undamaged composite is $\delta A_\gamma = \gamma_o[\pi(a + \delta a)(b + \delta b) - \pi ab]$. Hence, the generalized resistance forces are $\Gamma_a = 2\pi\gamma_o b$, $\Gamma_b = 2\pi\gamma_o a$. The state of the delamination is a sub-equilibrium one if

$$G_a(a,b) < 2\pi\gamma_o b, \quad G_b(a,b) < 2\pi\gamma_o a. \tag{30}$$

When the equality occurs in one of Eqs. (30), it does not mean necessarily that the delamination becomes unstable. Stability of an equilibrium state depends on properties of the quadratic form (4) at m = 2, and $\delta l_1 \equiv \delta a \geqslant 0$, $\delta l_2 \equiv \delta b \geqslant 0$.

Analysis of stability of delaminations in a spherical shell in hydrostatic pressure was made in [19]. A similar study is presented in [20] for a circular cylindrical shell in longitudinal compression. A typical diagram for a spherical transversally isotropic shell is given in Fig. 10. Non-dimensional sizes a/h and b/h are plotted along horizontal axes, and nondimensional nominal deformation $\varepsilon R/h$ along the vertical axis. The line ABC is the trace of intersection between the surface corresponding to the boundary of buckling of the delamination and the boundary of a rectangular parallelepiped. The line A'B'C' is the trace of the similar surface corresponding to one of conditions $G_a(a,b) = 2\pi\gamma_o b$, $G_b(a,b) = 2\pi\gamma_o a$. Both conditions are satisfied simultaneously only at a = b. In the point B'' and in neighbouring points of the surface A'B'C' stability of an equilibrium state

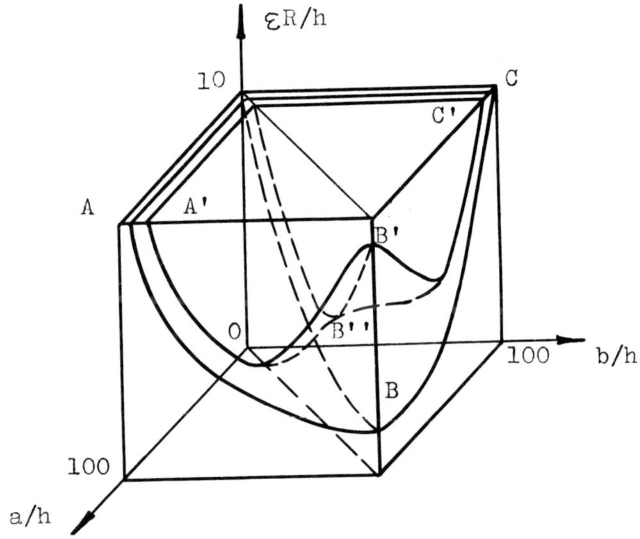

Fig. 10. States of an elliptical delamination in a spherical composite shell in external uniform pressure. The region of buckling is limited with the surface ABC, and the region of instability of the buckled delamination(in the Griffith's sense) with the surface A'B'C'.

changes into instability. From the viewpoint of fatigue, the most interesting points are in the area between the two mentioned surfaces (compare with Fig. 7).

A simpler analytical example is taken from paper [5]. Let a delamination have a circular form with the radius a. The plate is in homogeneous compression with nominal strains $\varepsilon_x = \varepsilon_y = \varepsilon$. The material of the delamination is isotropic with Young's modulus E and Poisson's ratio $\nu = 1/3$. The buckling critical strain determined with a variational method using Eq. (29) is

$$\varepsilon_*(a) = \left(\frac{h}{a}\right)^2. \tag{31}$$

The generalized moving force is

$$G = 3\pi Eha[\varepsilon^2 - \varepsilon_*^2(a)]. \tag{32}$$

To estimate damage at the tip of the delamination we use Eq.(11) at $\tau_{th} = 0$ and τ with accordance with Eq. (12). The coordinate x is measured in the radial direction. The slow growth of the delamination is governed with equation

$$\frac{da}{dt} = \frac{\lambda_o}{mt_c}\left(\frac{\varepsilon}{\varepsilon_\psi}\right)^m \left[1 - \frac{\varepsilon^2 - \varepsilon_*^2(a)}{\varepsilon_\infty^2}\right]^{-1/\alpha} \tag{33}$$

where ε_∞ is the critical (in the Griffith's sense) value of nominal strain for undamaged plate:

$$\varepsilon_\infty = \left(\frac{2\gamma_o}{3Eh}\right)^{\frac{1}{2}} . \tag{34}$$

It is evident that Eqs. (31), (32), (33) and (34) are analogous to Eqs. (20), (23), (19) and (10), respectively. The considered problem is actually, one-parametric. In more general cases, the growth rate depends on a number of factors, such as relations between magnitudes and signs of nominal strains, orthotropy and curvature parameters of a delamination, etc. Behavior of a delamination depends on its relative thickness too. For example, an initially circular isotropic delamination in a plate subjected to uni-axial tension grows in the direction of tension if the relative thickness is small. But when bending effects in the delamination become significant, the picture varies. Due to the Poisson's transverse shortening buckling of the delamination occurs, and the transverse moving force becomes dominant. The delamination begins to grow in the transverse direction.

Delaminations of arbitrary form

Following paper [21] we consider briefly a thin delamination which has an arbitrary form in the projection on a tangential x,y plane (Fig. 11). Let $h/a \ll 1$, $a/R \ll 1$, $h/H \ll 1$ where h is the thickness, a is the characteristic size of the delamination, R is the minimum curvature radius of the shell, and H is the shell thickness. Let the delaminated zone be limited with a smooth curve S. Equations of S in the polar coordinates r,θ let $r = \rho(\theta,t)$. Neglecting the potential energy of bending, we obtain

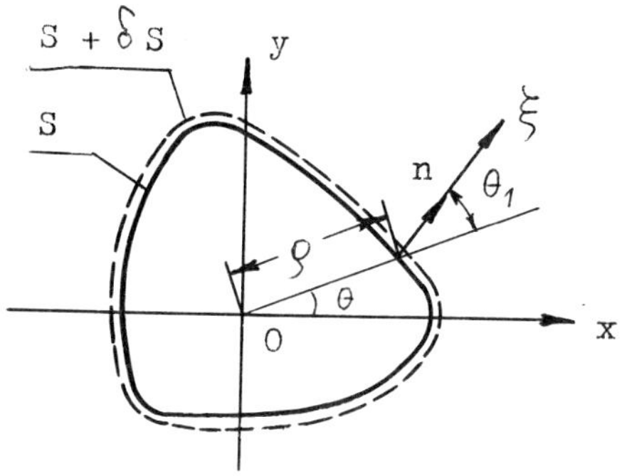

Fig. 11. Delamination of arbitrary form in the projection on a tangential plane.

$$U = \text{const} - \frac{1}{2}\iint_{\Omega} E_x h \varepsilon_r^2 d\Omega. \tag{35}$$

Here ε_r is the reduced nominal strain in the main shell. For example, in the case of orthotropy $\varepsilon_r^2 = (1 - \nu_{xy}\nu_{yx})^{-1}[\varepsilon_x^2 + 2\nu_{xy}\varepsilon_x\varepsilon_y + (E_y/E_x)\varepsilon_y^2 = (2G_{xy}/E_x)\varepsilon_{xy}^2]$ where common notations of elastic constants are used. Virtual work of fracture with account to damage is

$$\delta A_\gamma = \int_S \gamma(\psi)|\, ds \times \delta l|. \tag{36}$$

Eqs. (35) and (36) result in

$$E_x h \varepsilon_r^2 = 2\gamma(\psi). \tag{37}$$

This condition is to be satisfied in all the points of the boundary S where the delamination grows slowly. The tip of delamination remains fixed where $E_x h \varepsilon_r^2 < 2\gamma(\psi)$. Evidently Eq. (37) is a generalization of the equilibrium condition $G_j = \Gamma_j(\psi)$ upon a continuum-degree-of-freedom system. The polar angle θ with the values from the segment $[0,2\pi]$ takes the part of the index j at G_j and $\Gamma_j(\psi)$, and the polar radius $\rho(\theta)$ takes the part of the set of generalized coordinates.

Let the damage in the interlayer be produced by tangential stresses $\tau(\xi,t)$. In the momentless approximation τ is proportional to the strain ε_n in the direction of the normal vector n to the boundary S (Fig. 11). With an account of Eq. (12) we obtain the equation

$$\frac{\partial\phi}{\partial t} = \frac{1}{t_c} f\left(\frac{\varepsilon_n}{\varepsilon_\psi}\right) \exp\left[-\frac{(r-\rho)\cos\theta_1}{\lambda_o}\right] \tag{38}$$

where f(.) is a function similar to the power function (11) at $\tau_{th} = 0$; t_c, λ_o, and ε_ψ are of the same meaning as in Eqs. (12), (17) and (19). Notation θ_1 for the angle between vectors n and $r(\theta)$ is introduced in Eq. (38). To close the set of equations, we place $\psi(\theta,t) \equiv \phi[\rho(\theta,t),t]$.

Approximate equation analogous to Eq. (19) takes the form

$$\frac{\partial\rho}{\partial t} \approx \frac{\lambda_o}{mt_c} f\left(\frac{\varepsilon_n}{\varepsilon_\psi}\right)\left(1 - \frac{\varepsilon_r^2}{\varepsilon_\infty^2}\right)^{-1/\alpha}\left[1 + \left(\frac{\partial\rho}{\partial\theta}\right)^2 \frac{1}{\rho^2}\right]^{\frac{1}{2}} \tag{39}$$

The initial condition is $\rho[\theta,t_*(\theta)] = \rho_o(\theta)$ where $r = \rho_o(\theta)$ is the equation of the boundary in the initial state, and $t_*(\theta)$ is the moment of termination of the initiation stage at the point with the polar angle θ.

In principle, physical non-linearity of reinforcement layer also can be included into the theory, as was done in [9] where the effect of damage on the potential energy release, and therefore, on the generalized moving forces was taken into consideration.

Load Capacity of Delaminated Plates and Shells

Effect of delaminations on the carrying capacity of structural components in compression has been studied in [22]. Some numerical results are presented in Fig. 12 relating to a compressed layered plate with a delamination placed symmetrically with respect to the length of the plate. One-dimensional problem, i.e. cylindrical bending is considered. Fig. 12 shows the dependence of the ratio N_*/N_*^o on h/H and l/L. Here N_* and N_*^o are critical forces for the damaged and undamaged plate, h and H are the thickness of the delamination and the total thickness of the plate, l and L are the lengths of the delamination and of the span of the plate, respectively. Dotted lines are plotted using the model of local buckling. Dashed lines correspond to a monolythic plate whose equivalent stiffness varies step-wise along the x-axis. The stiffness of the delaminated part of the platc is calculated as for a packet from two unbounded plates. Such an approximate approach is used widely in engineering design. It follows from Fig. 12 that this approach fails even qualitatively.

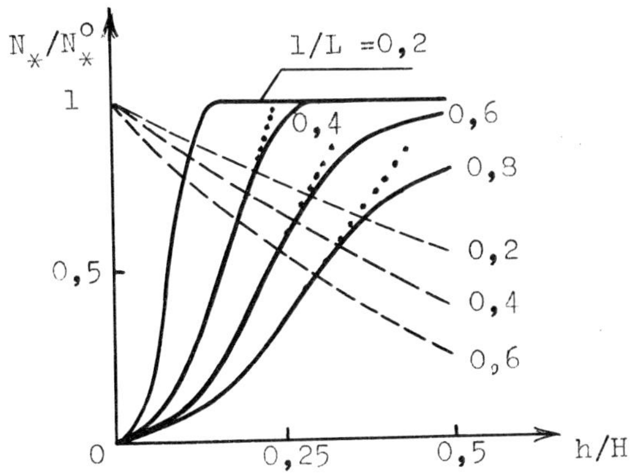

Fig. 12. Layered composite plate in compression: critical load versus the length of delamination and its position with respect to the plate thickness. Lines: solid - exact results; dotted - local buckling; dashed - step-wise varying stiffness.

A spherical transversally isotropic shell with an axisymmetrical delamination subjected to external hydrostatic pressure was considered in [23]. It is necessary to make a difference between the local buckling of surface delaminations and the global instability. Buckling modes are presented in Fig. 13. Analytical solutions of eigenvalue problems, generally, include self-intersecting, i.e. physically non-consistent modes (see the dotted line in Fig. 13b). In these cases, contact problems should be considered with an account for supports of the delaminations in the center (Fig. 13c) or along concentrical circles.

Some numerical results are presented in Fig. 14 where the ratio p_*/p_*^o is plotted against the ratio h/H and the central angle $2\theta_o$. Here p_*^o is the critical pressure for the monolithic shell, and p_* for the delaminated one. As in Fig. 12, results obtained in the buckling approximation are plotted with dotted lines.

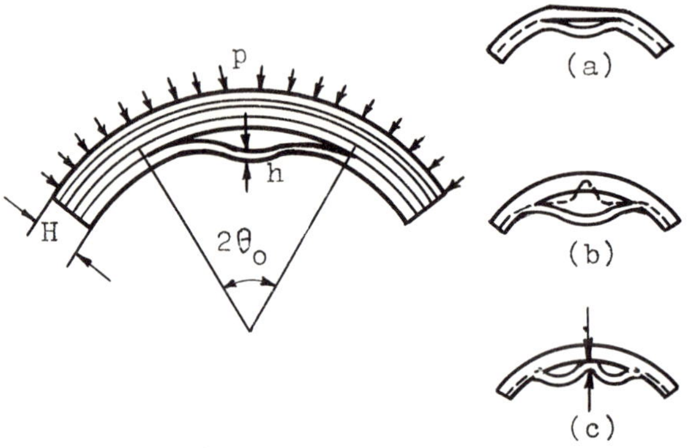

Fig. 13. Spherical layered composite shell in uniform external pressure. Modes of instability: (a) global; (b) local: (c) supported in the center of the delamination.

For one of the ratios h/H a self-intersection mode was found, and a contact problem with a support in the center was solved. Numerical results with the use of the step-wise stiffness model are presented in Fig. 15 with dashed lines. Discrepancy of results is less than for spherical shells, especially when a delam-

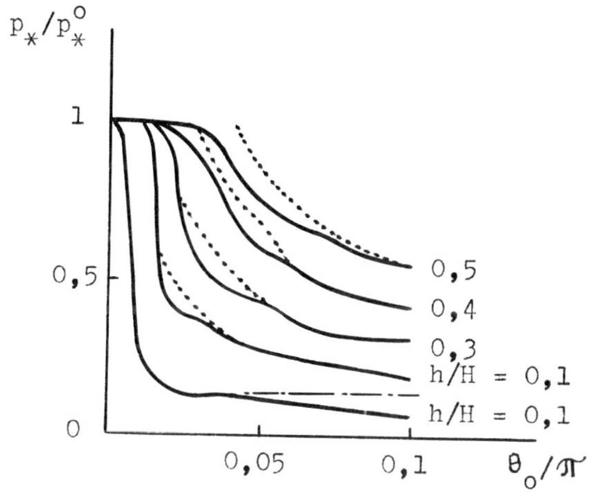

Fig. 14. Critical pressure on spherical composite shell with a circular delamination. Lines: solid - exact results; dotted - local buckling; dot-dahsed - with account of support in the center.

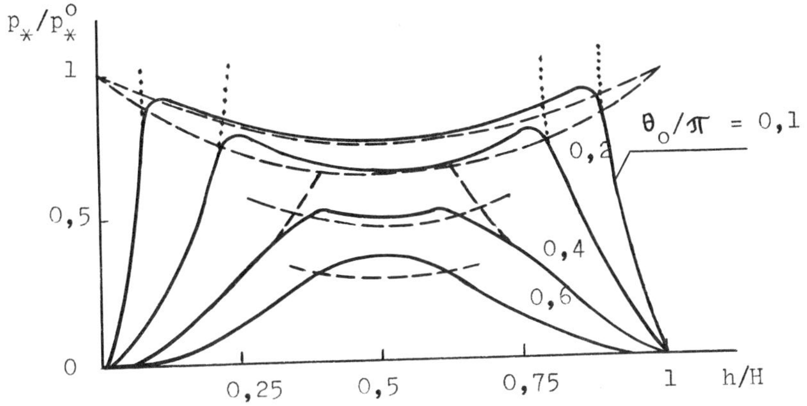

Fig. 15. The same as in Fig. 14. Dashed lines correspond to the model of step-wise varying stiffness.

ination is far from the surface, and the angle θ_o is not too large. Details are given in [23].

Instability of a structural component in the Euler's sense does not mean necessarily the total failure. It is illustrated in Fig. 16 where two plates in compression are compared with the delamination thicknesses h' and h''. Critical compressive forces are in relation N'_{*}, < N''_{*} if h' < h''. But the load carrying capacities N_{**} and N''_{**} can be in relation $N'_{**} > N''_{**}$. In Fig. 16 the carrying capacity is limited with the strength of buckled delaminations.

Structural reliability considerations should include a number of various factors: load carrying capacity of the main component, stability of delaminations in the Griffith's sense, strength of delaminations, and other requirements such as structural integrity, exterior looks, etc. The general character of restrictions is illustrated in Fig. 17, on the plane ε, a. Here ε is a characteristic nominal strain, and a is a characteristic size of the delamination. A sample ε = ε(t), a = a(t) of the loading and

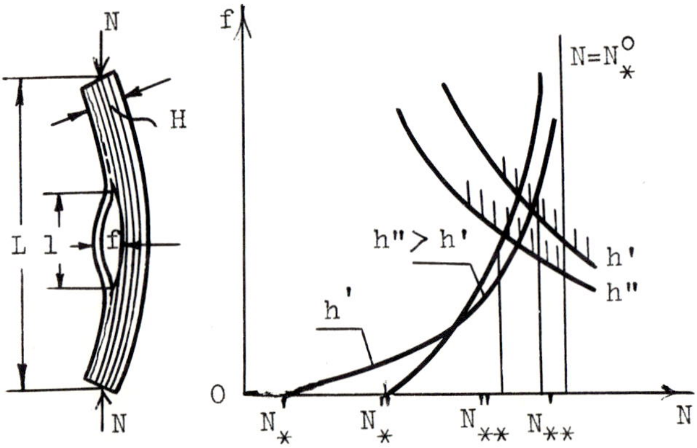

Fig. 16. Schematic relation between the instability in the Euler's sense, and the loss of load carrying capacity. Two layered plates in compression are compared with the delamination thicknesses h' and h''.

and delamination growth processes is presented in Fig. 17. The part of the plane limited with the restriction lines is an admissible region. The point corresponding to the current state of the system must not leave this region during the planned life-

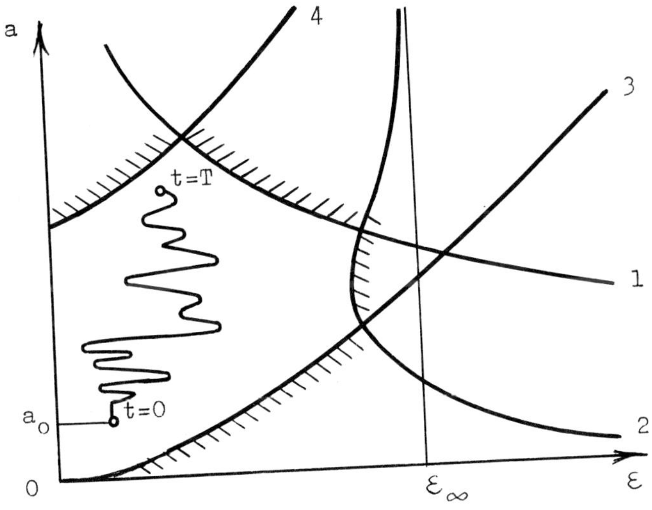

Fig. 17. Statement of the reliability problem for a structural component with a delamination. Restrictions are based on: (1) load carrying capacity of the component; (2) stability (in the Griffith's sense) of the delamination; (3) strength of the delamination; (4) other restrictions.

time T. The reliability function of a structural element is equal to the probability of staying of the process $\varepsilon = \varepsilon(t)$, $a = a(t)$ in the admissible region during the time T [24]. An inverse problem is of interest: to find such admissible initial sizes of flaws that the reliability function to the end of the planned life-time remains to be sufficiently close to unity.

References

1. Obraztsov, I.F.: Some problems of composite mechanics. In: Mechanics of Composites. Advances in Science and Technology in the USSR. Mathematics and Mechanics Series. Moscow Mir Publishers 1982, 9-19.

2. Болотин, В.В.: Основные уравнения теории армированных сред. Механика полимеров (1965) № 1 27-37.

3. Болотин, В.В.; Новичков, Ю.Н.: Механика многослойных конструкций. Москва: Машиностроение 1980.

4. Болотин, В.В.: Влияние технологических факторов на механическую надежность конструкций из композитов. Механика полимеров (1972) № 3 529-540.

5. Болотин, В.В.: Дефекты типа расслоений в конструкциях из композитных материалов. Механика композитных материалов (1984) № 2 239-255.

6. Sih, G.C.: Fracture mechanics of composite materials. In: Fracture of composite materials. Proc. of the First USA - USSR Symposium. Alphen aan den Rijn: Sijthoff and Noordhoff 1979 111-130.

7. Bolotin, V.V.: Stochastic models of fracture of unidirectional fiber components. In: Fracture of Composite Materials. Proc. of the Second USA -USSR Symposium. The Hague: Martinus Nijhoff 1982 3-16.

8. Качанов, Л.М.: Расслоение стекловолокнистых труб при внешнем давлении. Механика полимеров (1975) № 6 1106-1108.

9. Болотин, В.В.: Уравнения роста усталостных трещин. Известия АН СССР, Механика твердого тела (1983) № 4 153-160.

10. Болотин, В.В.: Объединенные модели в механике разрушения. Известия АН СССР, Механика твердого тела (1984) № 3 127-137.

11. Rosen, B.: Mechanics of reinforcement of composites. In: Fiber Composite Materials. Metals Park, Ohio: ASM 1965 37-76.

12. Панасюк, В.В.; Андрейкив, А.Е.; Ковчик, С.Е.: Методы оценки трещиностойкости конструкционных материалов. Киев: Наукова думка 1977.

13. Слепян, Л.И.: Механика трещин. Ленинград: Судостроение 1981.

14. Викторов, Е.Г.: Подрастание и излом отслоений в композитах при сжатии. В сб.: Механика материалов и конструкций. Москва: МЭИ 1982 36-40.

15. Полилов, А.Н.; Работнов, Ю.Н.: Развитие расслоений при сжатии композитов. Известия АН СССР. Механика твердого тела (1983) № 4 166-171.

16. Chai, H.; Babcock, D.; Knauss, W.: One-dimensional modelling of failure in laminated plates by delamination buckling. Internat. J. Solids and Structures 17 (1981) 1069 -1083.

17. Bottega, W.J.; Maeval, A.: Delamination buckling and growth in laminates. J. Appl. Mech. 50 (1983) 184-189.

18. Evans, A.G.; Hutchinson, J.W.: On the mechanics of delamination and spalling in compressed films. Int. J. Solids Structures 20 (1984) 455-466.

19. Мурзаханов, Г.Х.; Несин, Д.Н.: Расчет роста эллипсоидальных отслоений в сферической оболочке при циклическом нагружении. В сб.: Надежность и ресурс конструкций. Москва: МЭИ 1984 15-20.

20. Кисляков, С.А.; Нефедов, С.В.: Равновесные размеры эллиптических отслоений в ортотропной цилиндрической оболочке. В сб.: Надежность и ресурс конструкций. Москва: МЭИ 1984 29-33.

21. Болотин, В.В.: Уравнения роста отслоений в оболочках из композиционных материалов. В сб.: Надежность и ресурс конструкций. Москва: МЭИ 1984 5-10.

22. Болотин, В.В.; Зебельян, З.Х.; Курзин, А.А.: Устойчивость сжатых элементов с дефектами типа расслоений. Проблемы прочности (1980) № 7 3-8.

23. Болотин, В.В.; Зебельян, З.Х.: Устойчивость упругих сферических оболочек с расслоениями. В кн.: Расчеты на прочность. Москва: Машиностроение (1981) вып. 22 150-165.

24. Болотин, В.В.: Статистические методы в строительной механике. Москва: Стройиздат 1965; Bolotin, V.V.: Statistical methods in structural mechanics. San Francisco: Holden Day 1969.

Reference Stress Concepts for the High Temperature Deformation and Rupture of Cyclically Loaded Shell Structures

A.C.F. COCKS and F.A. LECKIE

Department of Theoretical and Applied Mechanics
University of Illinois at Urbana-Champaign

Summary

The theory of continuum damage mechanics is used to determine the properties of shells operating at temperatures sufficiently high for time-dependent deformations and material damage to be dominant factors affecting load-carrying capacity. Constitutive equations are proposed which reflect the micro-mechanical laws of damage growth, and it is demonstrated how these equations can then be used to determine results at the structural level which form the basis for consistent design procedures. The effectiveness of the method is studied by referring to the results of a number of experimental programs.

1. Introduction

When metals are subjected to stress at temperatures in excess of $T_m/3$, where T_m is the melting temperatures in ^oK, the metal suffers time-dependent creep deformations. In addition, internal damage occurs so that the metal ultimately ruptures. Consequently, when designing shell structures which operate at such elevated temperatures, consideration must be made to ensure that creep deformations do not exceed operational requirements during the life of the component. Common allowable deformations are 1% average and 5% maximum strain. In addition, the rupture conditions are that no part of the component may separate nor that local leakage can occur.

By establishing suitable constitutive equations which give the strain rates and the rate of internal damage of the material, it is possible in principle to establish by numeric means the strain, stress and damage history at all points in the shell. Such procedures tend to be very complex and it is difficult on the basis of the calculations to draw conclusions of the type which can help to reach a deeper understanding of the shell behavior. The approach to be used here is to use the results of theorems which have been established within the last ten years and which have made it possible to extend the Limit Load and Shakedown concepts to conditions when high temperature effects become important.

Inelastic Behaviour of Plates and Shells
IUTAM Symposium Rio de Janeiro 1985
Editors: L. Bevilacqua, R. Feijóo and R. Valid

For the purpose of illustration the behavior of a cylindrical shell shall be studied. The shell is loaded by cyclic internal pressure $\pm \tilde{p}$ and a constant ring load, P, and both time dependent deformation and rupture shall be taken into consideration (Fig. 1).

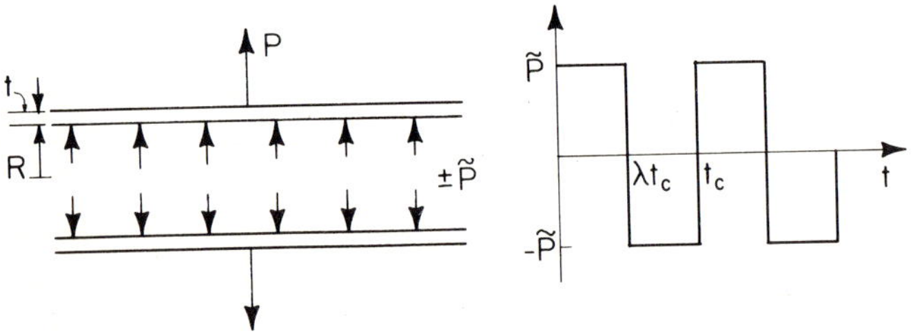

Fig. 1. Cylindrical shell of infinite length subjected to a constant ring load P and a cyclic pressure $\tilde{p}$.

2. Material Behavior

The high temperature strain/time response of a metal subjected to constant stress has the form shown in Fig. 2. After the initial time independent response, the strain rate decreases with time during the so-called transient period I when the hardening processes which occur within the material exceed the effects of thermal softening. In the region II, referred to as the steady state, the hardening and thermal rate are equal and opposite so that the strain rate is constant. In the tertiary region III, the effects of internal damage become evident so that the strain rate increases until rupture eventually occurs. In order to simplify the description of the material behavior, two sets of constitutive equations shall be used. When considering the deformations of the shell, equations shall be used which are appropriate to portions I and II of the strain/time curve when the internal hardening and softening effects are predominant. For rupture life predictions, equations are used which describe the internal damage occurring within the metal and the increasing strain rate characteristic of the tertiary behavior.

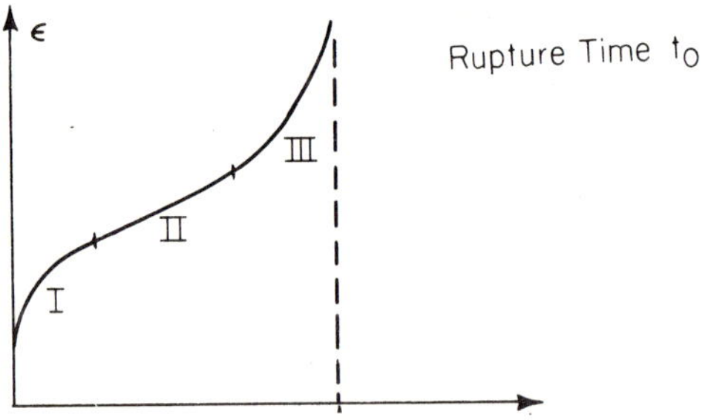

Fig. 2. Uniaxial creep curve showing the three stages of creep.

2.1 Transient Deformation Constitutive Equations

The elastic strains are defined by the equation

$$e_{ij} = C_{ijk\ell}\, \sigma_{k\ell} \tag{2.1a}$$

where $C_{ijk\ell}$ is the compliance matrix.

The time-dependent response of the material to a given multiaxial stress state can be expressed in terms of a scalar state variable s that is a measure of the present size of the yield surface in stress space. Hence

$$\dot{\varepsilon}_{ij} = f(\phi - s)\, \frac{\partial \phi}{\partial \sigma_{ij}} \tag{2.1b}$$

$$\dot{s} = h(s)\, f(\phi - s) - r(s) \tag{2.1c}$$

where $\phi(\sigma_{ij})$ is a homogeneous function of degree one in σ_{ij}. In its simplest form, f is given by the step function

$$f(\phi - s) \quad \begin{matrix} > 0 & \text{when} & \phi = s\ , \ \dot{\phi} \geqslant 0 \\ = 0 & \text{when} & \phi < s \ \text{ or } \ \phi = s \ \text{ and } \ \dot{\phi} < 0 \end{matrix} \quad . \tag{2.1d}$$

The condition $\phi > s$ cannot be achieved. The quantities $h(s)$ and $r(s)$ are the rates of strain hardening and thermal softening which depend on the value of the state variable s.

For fast loading the term $r(s)$ in Eq. 2.1c can be neglected, and $\phi = s$ so that Eqs. 2.1b and c then give

$$\dot{\varepsilon}_{ij} = \frac{\dot{s}}{h(s)}\, \frac{\partial \phi}{\partial \sigma_{ij}}$$

The function ϕ is chosen such that in a uniaxial tension test this equation reduces to

$$d\varepsilon = \frac{d\sigma}{h(\sigma)} \tag{2.2}$$

The function $h(s)$ can then be determined from the uniaxial stress-strain diagram.

Under steady state conditions uniaxial tests give the result

$$\frac{\dot{\varepsilon}}{\dot{\varepsilon}_o} = \left[\frac{\sigma}{\sigma_o}\right]^n$$

where $\dot{\varepsilon}_o$ is the strain rate corresponding to the stress σ_o and n is the so-called creep index. Also at the steady state condition $\dot{s} = 0$ in Eq. 2.1c so that

$$f(\phi - s) = \frac{r(s)}{h(s)} = \frac{r(\phi)}{h(\phi)}$$

and

$$\dot{\varepsilon}_{ij} = \frac{r(\phi)}{h(\phi)} \frac{\partial \phi}{\partial \sigma_{ij}} .$$

In uniaxial loading this gives the result,

$$\dot{\varepsilon} = \frac{r(\phi)}{h(\phi)} = \dot{\varepsilon}_o \left[\frac{\sigma}{\sigma_o}\right]^n$$

so that the recovery function $r(\phi)$ can now be deduced since the hardening function is already known.

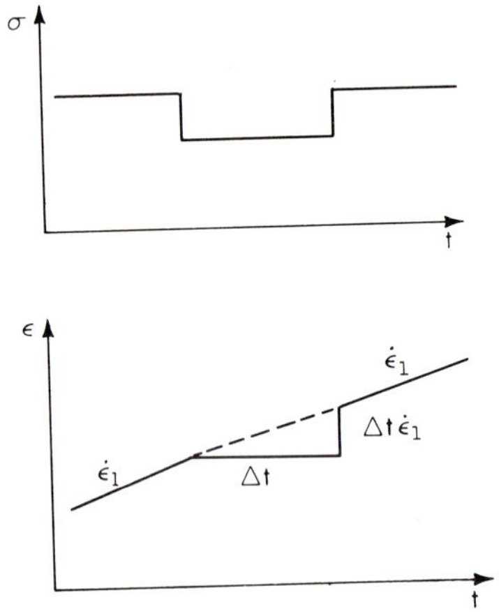

Fig. 3 Predictions of recovery model for rapid cycling

It is informative to deduce the prediction of the equations (2) to a cyclic stress loading of the form shown in Fig. 3. No additional creep strain is predicted during the period of unloading while softening continues, but on reloading the hardening is recovered with an instantaneous strain $\Delta t \, \dot{\varepsilon}_1$, so that the final strain is equal to that of a continuously loaded specimen. Such a prediction is generally close to reality, and when in error overestimates the strains so that the equations form the basis of calculations which are conservative.

2.2 Creep Rupture Constitutive Equations

The constitutive equations which have the form describing the tertiary portion of the creep curve are given by,

$$\frac{\dot{\varepsilon}_{ij}}{\dot{\varepsilon}_o} = \phi^n \left(\frac{\sigma_{ij}}{\sigma_o}\right) \frac{\partial \phi}{\partial \sigma_{ij}} g_1(\omega)$$

$$\dot{\omega} = A\, \Delta^\nu\left(\frac{\sigma_{ij}}{\sigma_o}\right) g_2(\omega) \qquad (2.4a\text{-}b)$$

where ω is a measure of the internal damage. The physical nature of the damage is not discussed here but it is sufficient to say that it is possible to identify ω rather precisely with specific forms of damage and that special forms of $g_1(\omega)$ and $g_2(\omega)$ in eqns. (2.4) can indeed be used to describe the range of mechanisms which have been identified [1]. The function of stress $\Delta(\sigma_{ij}/\sigma o)$ describes the so-called isochronous surface which is the locus of multiaxial stress states for which the rupture time is constant (Fig. 3). The constant A can be selected to give the rupture time for an applied stress σ_o. For the uniaxial loading with $\sigma = \sigma_o$, Eq. (2.4b) becomes

$$\dot{\omega} = A\, g_2(\omega)$$

which on integration gives the rupture time t_o

$$t_o = \int_0^1 \frac{1}{A} \frac{d\omega}{g_2(\omega)}$$

Two specific forms of Δ are discussed which represent the extremes of material behavior. For some materials strengthened by precipitate hardening $\Delta\left(\frac{\sigma_{ij}}{\sigma_o}\right) = \frac{\bar{\sigma}}{\sigma_o}$ where $\bar{\sigma}$ is the effective stress, and Eq. 2.4b then has the form

$$\dot{\omega} = A\left(\frac{\bar{\sigma}}{\sigma_o}\right)^\nu g_2(\omega)\,. \qquad (2.4c)$$

A material within this class shall be referred to as a $\bar{\sigma}$ material. For another class of materials the growth of damage is dictated by the maximum stress σ_I so that $\Delta\left(\frac{\sigma_{ij}}{\sigma_o}\right) = \frac{\sigma_I}{\sigma_o}$ and the damage growth equation becomes

$$\dot{\omega} = A\left(\frac{\sigma_I}{\sigma_o}\right)^\nu g_2(\omega)\,. \qquad (2.4d)$$

Materials satisfying this relationhip are referred to as σ_I materials. These extreme forms of the isochronous surface for plane stress conditions are shown in Fig. 4 and both forms shall be used in this study.

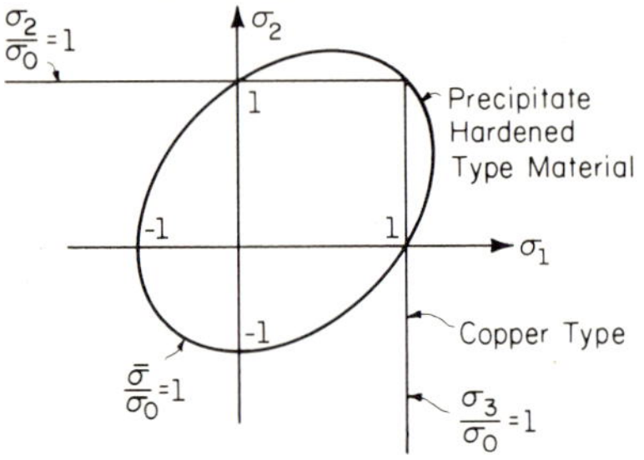

Fig. 4 Isochronous surfaces for $\bar{\sigma}$ and σ_I materials in plane stress space.

Nonproportional loading tests [2] on a precipitate hardened aluminum indicate that the damage is isotropic. Similar tests on copper indicate however that damage grows on planes independent of each other. Consequently if the stress field is rotated, the life of the material is dictated by the rupture of the plane which suffers longest exposure to maximum stress, and the failure time is independent of the damage in other directions. Consequently the concept of independent damage directions becomes useful and this property shall be used in the next section.

3. Generalized Forces and Moments

The generalized forces to be considered in this problem are the hoop stress resultant N_θ and the moment resultant M_ϕ .

From the constitutive equations (2.1) it is possible to determine the expression for the steady state creep energy dissipation rate as

$$\dot{D}(\sigma_{ij}) = \sigma_o \dot{\varepsilon}_o \phi^{n+1} \left(\frac{\sigma_{ij}}{\sigma_o}\right) . \tag{3.1}$$

To simplify the shell analysis expressions for the constant energy dissipation rate surfaces and for the isochronous surfaces should be expressed in terms of N_θ and M_ϕ . This can prove to be a lengthy exercise in itself and instead of following this route, use shall be made of the results of Hodge [3].

3.1 Constant Energy Dissipation Rate Surfaces

The constant energy dissipation rate $\sigma_o \dot{\varepsilon}_o$ per unit thickness is shown in

Fig. 4, where M_o and N_o are the normalizing factors

$$M_o = \frac{\sigma_o t^2}{4} \quad \text{and} \quad N_o = \sigma_o t \ . \tag{3.2}$$

These expressions are recognized as those used in Hodge's expressions with σ_o replacing the yield stress σ_y. In forming the surface of Fig. 5a the same through thickness stress fields as exist in plasticity are assumed to exist in creep also. This procedure gives an upper bound on energy dissipaton rates [4], which is nevertheless close to the exact value.

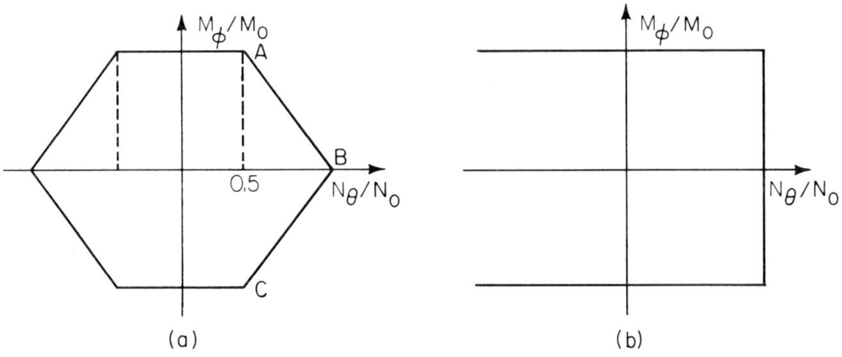

Fig. 5 (a) Surface of constant energy dissipation rate and isochronous surface for $\bar{\sigma}$ material.
(b) Isochronous surface for σ_I material.

3.2.1 Isochronous Surface for $\bar{\sigma}$ Material

For materials whose rate of creep damage is governed by the effective stress $\bar{\sigma}$, damage occurs irrespective of the sign of the stress. Experiments by Hayhurst [5] on precipitate hardened aluminum beams in bending verify the validity of the assertion. He also demonstrates that when the applied moment is

$$M_o = \frac{\sigma_o t^2}{4} \tag{3.3}$$

that the rupture time of the beam is t_o. From this behavior it can be deduced that the isochronous surface for rupture time t_o has the same form as the constant energy dissipation rate surface. The resulting isochronous surface for a $\bar{\sigma}$ material and for the given shell is then that shown in Fig. 5(a).

3.2.2 Isochronous Surface for σ_1 Material

When materials suffer the type of creep damage which is dependent on the value of the maximum stress, it is implied that the stress is tensile. Damage does not grow when the stress is compressive. This means in the case of a beam in bending that damage only occurs in those portions where stress is tensile. Hayhurst [5] has shown in this case that the moment which gives a rupture time t_o is given by

$$M_o = \frac{\sigma_o t^2}{4.2^{1/\nu}} . \qquad (3.4)$$

Since the beam is symmetric a moment of opposite sign shall give the same rupture time except that it is the opposite face that shall suffer damage.

Finally it was noted from previous experiments [2] that faces in different directions do not interact. Hence since M_ϕ and N_θ are noninteracting, and the isochronous surface shown in Fig. 5(b) is obtained. Since no damage is caused by compressive stress the surface extends infinitely far to the left.

4. Bounding Theorems for Deformation and Rupture Time

Using constitutive equations described by Eq. (2.1) Ponter [6] obtained deformation bounds for structures subjected to cyclic loading. The general deformation bound takes the form

$$\int_{S_T} T_i \dot{u}_i \, dS \leqslant \frac{1}{n} \dot{\varepsilon}_o \sigma_o \int_V \phi^{n+1} \left[\frac{n}{n+1} \frac{\sigma_{ij}}{\sigma_o} (t_m) \right] dV \qquad (4.1)$$

where

$$\sigma^*_{ij} = \sigma^{*P}_{ij} + \sigma^{*T}_{ij} + \hat{\sigma}_{ij}(t) \qquad (4.2)$$

and

σ^{*P}_{ij} is a stress field in equilibrium with the applied primary load P_i ,

σ^{*T}_{ij} is a stress distribution in equilibrium with a dummy load T_i applied in the direction of the required displacement rate $\dot{u}_i$,

$\hat{\sigma}_{ij}(t)$ is the elastic stress distribution resulting from the variation of load at time t ,

t_m is the time in the cycle when $\phi\left[\frac{\sigma^*_{ij}(t_m)}{\sigma_o}\right]$ is a maximum,

$\dot{u}_i$ is the mean displacement rate in the direction of T_i and S_T is the surface over which T_i is applied,

V is the volume of the structure.

An optimum bound can be found by making appropriate choices of the magnitude of T_i and the stress distributions σ^{*P}_{ij} and σ^{*T}_{ij}. However this shall not be attempted here and use shall be made of solutions which arise from a shakedown analysis.

Consider the situation where a structure is subjected to a cyclic history of loading $P_i(t)$. A value of the yield stress, $\sigma_y = \sigma_o$, can be chosen such that a replica structure composed of an elastic-perfectly plastic material just shakes-down. Then an equilibrium stress field exists such that

$$\phi(\sigma^{*P}_{ij} + \hat{\sigma}_{ij}(t_m)) \leqslant \sigma_o \tag{4.3}$$

When the replica structure is subjected to a dummy load plastic collapse occurs when

$$T = T_L$$

For $T = \eta T_L$ an equilibrium stress field can be found such that

$$\phi(\sigma^{*T}_{ij}) \leqslant \eta\, \sigma_o \tag{4.4}$$

Substituting eqns. (4.3) and (4.4) into eqn. (4.1) and making use of the inequality

$$\phi(\sigma^a_{ij} + \sigma^b_{ij}) \leqslant \phi(\sigma^a_{ij}) + \phi(\sigma^b_{ij})$$

gives

$$\dot{u}_T \leqslant \frac{\sigma_y}{T_L}\,\dot{\varepsilon}_o\, \frac{\{\frac{n}{n+1}(1+\eta)\}^n}{n\,\eta}\, V$$

Optimizing eqn. (3.5) w.r.t. η gives $\eta = \frac{1}{n}$ and

$$\dot{u}_T \leqslant \frac{\sigma_y}{T_L}\,\dot{\varepsilon}_o\, V \tag{4.5}$$

This bound requires a knowledge of the limit load T_L and the shakedown solution which determines σ_o and, hence, $\dot{\varepsilon}_o$.

When a bound on the mean displacement rate over an area, A, of the structure is required a dummy pressure p can be applied over this area and the bound becomes

$$\dot{u}_n \leqslant \frac{\sigma_y}{P_L} \dot{\varepsilon}_o \frac{V}{A} \tag{4.6}$$

where $\dot{u}_n$ is the mean displacement rate and P_L is the limit load for the dummy pressure. We use this result in section 6 to obtain a bound on the mean displacement rate of the section of a shell.

Care should be exercised in the use of this bound. Acceptable results can be obtained in situations where, in the shakedown and limit load solution, the entire structure deforms plastically. In the problem analyzed in this paper, however, the ratchet mechanism beyond the shakedown boundary and the collapse mechanism at the limit load are such that only a small proportion of the structure deforms plastically. We discuss the applicability of eqn. (4.6) to this type of problem in section 6.

An upper bound can be obtained on the rupture time in terms of the shakedown load associated with a yield surface of the same form as the isochronous surface for the material [7]. We follow Ponter [7] in deriving a bound on the time for initiation of rupture in a structure subjected to cyclic loading of the type shown in Fig. 1, where one extreme of loading is maintained for $0 \leqslant \tau < \lambda$ and the other extreme for $\lambda \leqslant \tau \leqslant 1$, where $0 \leqslant \lambda \leqslant 0.5$ and $\tau = t_c$, t_c being the cycle time.

From eqn. (2.4) we note that prior to failure of a material element which experiences a stress σ_{ij} that

$$\int_o^t \Delta^\nu\left(\frac{\sigma_{ij}}{\sigma_o}\right) dt = \int_o^{\bar{\omega}} \frac{1}{A} \frac{d\omega}{g_2(\omega)} \leqslant t_o \tag{4.7}$$

where $\bar{\omega} \leqslant 1$. For conditions of rapid cycling the stress at each end of the cycle remains approximately constant and the convexity condition for Δ^ν at any instant is given by

$$\Delta^\nu\left(\frac{\sigma_{ij}}{\sigma_o}\right) - \Delta^\nu\left(\frac{\sigma^s_{ij}}{\sigma_o}\right) - \frac{\partial \Delta^\nu}{\partial \sigma_{ij}{}^s}(\sigma_{ij} - \sigma^s_{ij}) \geqslant 0 \tag{4.8}$$

We identify σ_{ij} with the actual solution and σ^s_{ij} with the shakedown solution for a perfectly plastic material of yield strength σ_o, where

$$\Delta^\nu\left(\frac{\sigma_{ij}}{\sigma_o}\right) = 1 \; ; \qquad d\varepsilon^p_{ij} = \mu \frac{\partial \Delta^\nu}{\partial \sigma_{ij}} \, , \qquad \mu \geqslant 0 \tag{4.9}$$

represent the yield surface and associated flow rule, and apply eqn. (4.8) at each extreme of the cycle:

$$\left.\begin{aligned} \mu_1 \Delta^\nu\left(\frac{\sigma^1_{ij}}{\sigma_o}\right) - \mu_1 \Delta^\nu\left(\frac{\sigma^{1s}_{ij}}{\sigma_o}\right) - \mu_1 \frac{\partial \Delta^\nu}{\partial \sigma^{1s}_{ij}} \left(\sigma^1_{ij} - \sigma^{1s}_{ij}\right) \geqslant 0 \\ \mu_2 \Delta^\nu\left(\frac{\sigma^2_{ij}}{\sigma_o}\right) - \mu_2 \Delta^\nu\left(\frac{\sigma^{2s}_{ij}}{\sigma_o}\right) - \mu_2 \frac{\partial \Delta^\nu}{\partial \sigma^{2s}_{ij}} \left(\sigma^2_{ij} - \sigma^{2s}_{ij}\right) \geqslant 0 \end{aligned}\right\} \tag{4.10}$$

where the first of eqns. (4.10) applies when $0 \leqslant \tau < \lambda$ and the second when $\lambda \leqslant \tau \leqslant 1$. Combining eqns. (4.10) and noting that plastic straining can only occur when $\Delta^\nu\left(\frac{\sigma^s_{ij}}{\sigma_o}\right) = 1$ for the perfectly plastic material we obtain

$$\frac{\mu_1}{\lambda} \lambda \Delta^\nu\left(\frac{\sigma^1_{ij}}{\sigma_o}\right) + \frac{\mu_2}{(1-\lambda)} (1 - \lambda) \Delta^\nu\left(\frac{\sigma^2_{ij}}{\sigma_o}\right) - (\mu_1 + \mu_2) - d\varepsilon^p_{ij}\rho_{ij} \geqslant 0 \tag{4.11}$$

where $\rho_{ij} = \sigma^1_{ij} - \sigma^{1s}_{ij} = \sigma^2_{ij} - \sigma^{2s}_{ij}$ is a residual stress field and $d\varepsilon^p_{ij}$ is the plastic strain experienced by the element of material during the cycle at shakedown. The inequality of eqn. (4.11) is still retained if μ_1/λ and $\mu_2/(1 - \lambda)$ are replaced by $\bar{\mu}$, where $\bar{\mu}$ is the maximum of μ_1/λ and $\mu_2/(1 - \lambda)$. Integrating eqn. (4.11) over the volume then gives

$$\int_V \bar{\mu}\left\{\lambda \Delta^\nu\left(\frac{\sigma^1_{ij}}{\sigma_o}\right) + (1 - \lambda)\Delta^\nu\left(\frac{\sigma^2_{ij}}{\sigma_o}\right)\right\} dV \geqslant \int_V (\mu_1 + \mu_2) dV \tag{4.12}$$

Integrating eqn. (4.12) from $t = 0$ to $t = t_i$, the initiation time for rupture, and making use of the inequality of eqn. (4.7) yields

$$t_i \leqslant t_o \frac{\int_V \bar{\mu} \, dV}{\int_V (\mu_1 + \mu_2) dV} \leqslant \frac{t_o}{\lambda} \tag{4.13}$$

where t_o is the time to failure in a uniaxial test under a stress σ_o.

As discussed by Ponter [7] this bound can drastically overestimate the time to rupture if λ is small. In the present paper we limit our attention to situations where $\lambda = 1/2$ and eqn. (4.13) becomes

$$t_i \leqslant 2t_o \tag{4.14}$$

5. Shakedown Solutions for Shell Problem

For the present class of problems where a structure is subjected to a constant load and a cyclic load the shakedown boundary is best obtained using a method due to Gokhfeld and Cherniavsky [8]. The method involves the construction of a modified yield surface which is used in a limit load calculation for the structure subjected only to the constant load. For the shell problem considered here it proves advantageous to obtain a modified yield condition directly in terms of N_θ and M_ϕ. Fig. 6a shows the limited interaction yield surface and the range of stress experienced by an element of material during the application of the cyclic pressure loading, $\tilde{p}$. If the line representing this stress history is translated such that one end touches the yield surface and the other end remains inside or on the surface, then the locus of the centre of this line represents the modified yield surface, Fig. 6b. If a distribution of N_θ and M_ϕ can be found that is in equilibrium with the applied constant load P and nowhere violates "modified yield," then when the cyclic loading is applied the generalized force and moment always lie within the actual yield surface and the structure will shakedown [9]. We now use this method to obtain modified limit load solutions for the limited interaction and Hodge yield conditions.

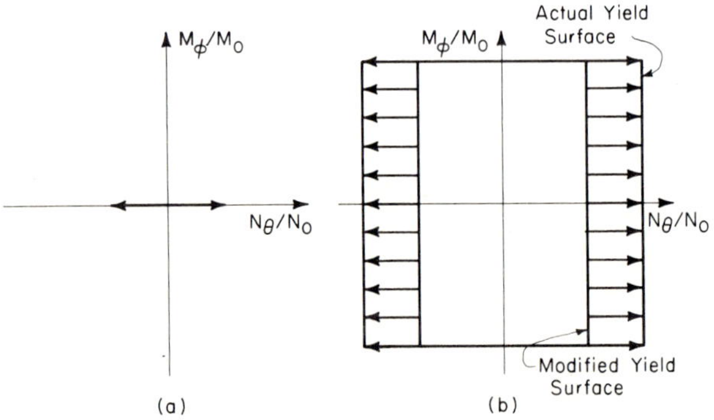

Fig. 6 Method of constructing modified yield surface

5.1 Equilibrium Equations

If the ring load is applied at $x = 0$ along the length of the tube, then for $x > 0$ the governing equilibrium equation becomes [3]

$$\frac{d^2 m_\phi}{d\bar{x}^2} + 4n_\theta = 0 \qquad (5.1)$$

where $m_\phi = \frac{M_\phi}{M_o}$, $n_\theta = \frac{N_\theta}{N_o}$ and $\bar{x} = \frac{x}{\sqrt{4M_o R/N_o}}$

The value of n_θ is limited by the yield condition. As a result both m_ϕ and $\frac{dm_\phi}{d\bar{x}}$ must be continuous along this length of tube.

5.2 Limited Interaction Surface

The modified yield surface for a given internal cyclic pressure, $p_o = \tilde{p}/\tilde{p}_L$, where

$$\tilde{p}_L = N_o/R , \qquad (5.2)$$

is shown in Fig. 6b. Before attempting to solve the equilibrium equation subject to this modified yield condition, it is instructive to inquire as to the likely mechanism of incremental collapse. In the absence of any cyclic loading the collapse mechanism is that in Fig. 7: at $x = 0$ a hinge line forms and $m_\phi = -1$; between $x = 0$ and $x = L$ the radius of the tube increases so that $n_\theta = 1$; another hinge line forms at $x = L$ and $m_\theta = 1$. We might expect a similar mechanism when the cyclic pressure is applied. Then for $|x| \leqslant L$

$$n_\theta = 1 - p_o$$

and eqn. 5.1 becomes

$$\frac{d^2 m_\phi}{d\bar{x}^2} = -4(1 - P_o)$$

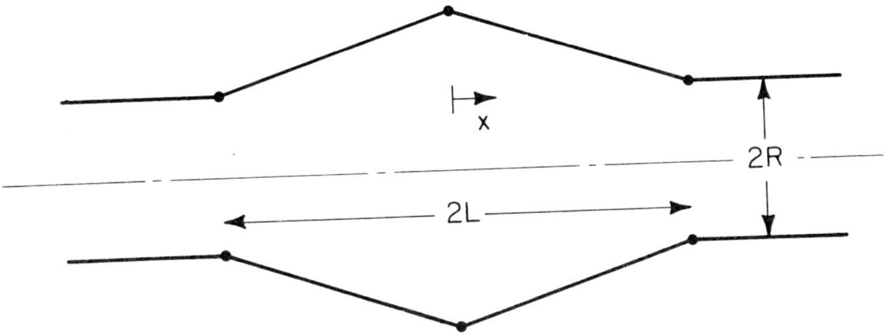

Fig. 7 Collapse mechanism for ring load alone

Solving this equation subject to the boundary condition

$$\left.\begin{aligned} m_\phi &= -1 \\ \frac{\partial m_\phi}{\partial \bar{x}} &= a\frac{P}{P_L} \end{aligned}\right\} \text{ at } \bar{x} = 0$$

$$\left.\begin{aligned} m_\phi &= 1 \\ \frac{\partial m_\phi}{\partial \bar{x}} &= 0 \end{aligned}\right\} \text{ at } \bar{x} = \bar{L} \tag{5.3}$$

where $a = 2\sqrt{2}$ and

$$P_L = \frac{8M_o}{\sqrt{4M_oR/N_o}} \tag{5.4}$$

is the limit load in the absence of cyclic loading,

gives $\bar{L} = (1 - p_o)^{-1/2}$

$$\text{and} \quad \frac{P}{P_L} = (1 - p_o)^{1/2} \tag{5.5}$$

This equation is plotted in Fig. 8.

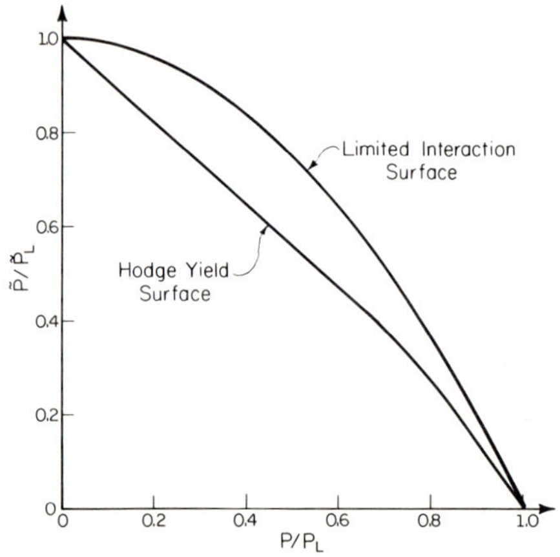

Fig. 8 Shakedown boundaries of limited interaction and Hodge yield conditions.

Before stating that this is a possible solution for the shakedown boundary we must check if a solution to eqn. (5.1) can be found beyond $\bar{x} = \bar{L}$ which does not violate the modified yield condition. A solution can be found by expressing m_ϕ in terms of a polynomial and choosing the coefficients such that m_ϕ and n_θ remain inside the yield surface. A possible solution is

$$m_\phi = 2\left(\frac{\bar{y}}{\bar{k}}\right)^3 - 3\left(\frac{\bar{y}}{\bar{k}}\right)^2 + 1 \qquad\qquad 0 \leqslant \bar{y} \leqslant \bar{k} \qquad (5.6)$$

$$n_\phi = \frac{3}{2\bar{k}^2}\left(1 - 2\left(\frac{\bar{y}}{\bar{k}}\right)\right)$$

$$n_\theta = m_\phi = 0 \qquad\qquad \bar{y} > \bar{k}$$

where $\bar{y} = \bar{x} - \bar{L}$ and $\bar{k} = \sqrt{\frac{3}{2(1 - p_o)}}$

5.3 Hodge's Yield Condition

A solution using Hodge's yield condition can be obtained in much the same way as the solution of the previous sub-section. We again assume that hinges form at $\bar{x} = 0$ and $\bar{x} = \bar{L}$ and that the values of m_ϕ and n_θ in between lie along the line ABC of the modified yield surface of Fig. A1. The analysis is more complex than the above and details are given in the Appendix. The resulting shakedown boundary is given in Fig. 8. Again the load P has been normalized using P_L, but now P_L is the limit load for the Hodge yield condition: .

$$P_L = \frac{6.92\ M_o}{\sqrt{4M_o R/N_o}} \qquad (5.7)$$

6. Deformation and Life Bounds for Shell Problem

The results of the last section can be combined with the results of section 4 to give bounds on the deformation rate and life of the structure. When applying the deformation bound the best results are obtained if in the solution of the dummy limit load problem the region of plastic deformation is similar to that in the shakedown solution. We achieve this by applying a uniform dummy pressure over the region that deforms plastically in the shakedown solution. The result then bounds the mean displacement rate over this region.

The shakedown boundary for the Hodge yield condition is given in Fig. 8. This boundary can be represented by an equation of the form

$$\sigma_y = \frac{P}{P_L}\,\sigma_y \cdot f\left(\frac{\tilde{p}/\tilde{p}_L}{P/P_L}\right) \qquad (6.1)$$

where $f\left(\frac{\tilde{P}/\tilde{P}_L}{P/P_L}\right)$ is a function of the ratio of normalized loads. Since $P_L \alpha \sigma_y$ the quantity σ_y/P_L is independent of material properties and the r.h.s. of eqn. 6.1 is only a function of the applied loading and geometry. The limit loads $\tilde{p}_L$ and P_L are given by eqns. (5.2) and (5.7), which, for the values of M_o and N_o given in section (3.1) become

$$\tilde{p}_L = \frac{\sigma_y t}{R} \; ; \quad P_L = \frac{1.73\, \sigma_y t^2}{\sqrt{Rt}}$$

For given values of P and $\tilde{p}$ a reference stress, σ_o, can be chosen such that

$$\sigma_o = \frac{P}{P_L}\, \sigma_y\, f\left(\frac{\tilde{p}/\tilde{p}_L}{P/P_L}\right) \qquad (6.2)$$

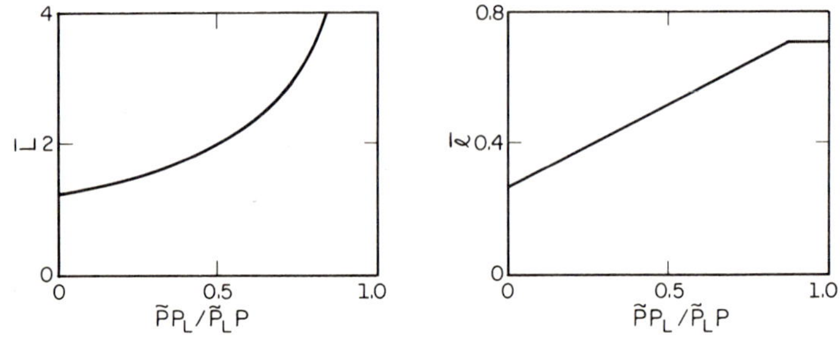

Fig. 9 Values of $\bar{L}$ and $\bar{\ell}$ obtained from the shakedown solution of Appendix A.

For a given ratio of loads the extent of the plastically deforming region is given in Fig. 9. The dimensionless length $\bar{\ell}$ is defined in Appendix A. If a uniform pressure is applied over the length 2I the resulting limit load is that shown in Fig. 10. A bound on the mean displacement rate is then given by eqn. (4.6) where $\dot{\varepsilon}_o$ is the strain rate in a uniaxial test conducted at constant stress σ_o and

$$\frac{\sigma_y}{P_L} = \frac{R}{\alpha h}$$

where α is given in Fig. 10. For the volume V we use the volume of the plastically deforming region so that

$$\frac{V}{A} = h$$

and eqn. (4.6) becomes

$$\dot{u}_m \leq \frac{R\,\dot{\varepsilon}_o}{\alpha} \tag{6.3}$$

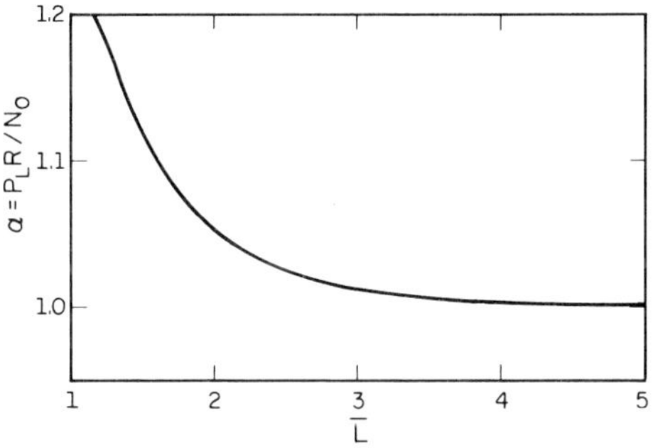

Fig. 10 Normalized limit load for pressure loading over a length 2L of the cylinder.

We would expect this bound to give a reasonable estimate of the deformation rate provided the ratchet mechanism from the shakedown solution does not extend over a significant length of tube. This limits the applicability of eqn. (6.3) to small values of $\tilde{p}\,P_L/\tilde{p}_L P$, certainly less than 0.8, Fig. 9.

The reference stress for the time to rupture is also given by the shakedown boundary solution. For an effective stress material the reference stress is the same as that used above for the deformation of the shell and the time to failure is given by eqn. (4.14). For a material that fails according to a maximum principal stress criterion the appropriate shakedown solution is that for the limited interaction surface of section 5.1 (the solution presented in section 5.1, for the complete limited interaction surface of Fig. 6, is the same as that for the surface of Fig. 5b since the normal forces always remain tensile). The shakedown boundary is then given by eqn. (5.4), or in the form of eqn. (6.1) by

$$\sigma_y = \frac{1}{2}\frac{\tilde{p}}{\tilde{p}_L}\sigma_y + \left\{\left(\frac{1}{2}\frac{\tilde{p}}{\tilde{p}_L}\sigma_y\right)^2 + \left(\frac{P}{P_L}\sigma_y\right)^2\right\}^{1/2} \tag{6.4}$$

where $\tilde{p}_L$ and P_L are obtained from eqns. (5.2) and (5.4) with N_o and M_o given by eqns. (3.2) and (3.4):

$$\tilde{p}_L = \frac{\sigma_y t}{R} \; ; \qquad P_L = \frac{2\sigma_y t^2}{2^{1/2}\nu\sqrt{Rt}}$$

The reference stress then becomes

$$\sigma_o = \frac{1}{2}\frac{\tilde{p}}{\tilde{p}_L}\sigma_y + \{(\frac{1}{2}\frac{\tilde{p}}{\tilde{p}_L}\sigma_y)^2 + (\frac{\tilde{P}}{\tilde{P}_L}\sigma_y)\}^{1/2} \tag{6.5}$$

for arbitrary values of $\tilde{p}$ and P. The time to failure is again bounded by eqn. (4.14), where t_o is the time to failure in the reference test.

Acknowledgement

Authors acknowledge support from the NSF grant MEA 82-10620.

References

1. Ashby, M. F.; Dyson, B. F.: Creep damage mechanics and micromechanisms. National Physical Laboratory, Report DMA(A) 77 1984.

2. Trampczynski, W. A.; Hayhurst, D. R.; Leckie, F. A.: Jnl. Mech. Phys. Solids. 29 (1981) 353.

3. Hodge, P. G.: Limit Analysis of Rotationally Symmetric Plates and Shells. Englewood Cliffs, N.J.: Prentice-Hall 1963.

4. Ponter, A. R. S.; Leckie, F. A.: The application of energy theorems to bodies which creep in the plastic range. J. Appl. Mech. 37 (1970) 753.

5. Hayhurst, D. R.: Estimates of the creep rupture lives of structures subjected to cyclic loading. Int. J. Mech. Sci. 18 (1976) 75.

6. Ponter, A. R. S.: Deformation bounds for the Bailey-Orowan theory of creep. J. Appl. Mech. 42 (1975) 619.

7. Ponter, A. R. S: Upper bounds on the creep rupture life of structures subjected to variable load and temp. Int. J. Mech. Sci. 19 (1977) 79.

8. Gokhfeld, D. A. and Cherniavsky, O. R.: Limit analysis of structures at thermal cycling. Sijthoff and Noordhoff 1980.

9. Koiter, W. T.: General theorems for elastic-plastic solids. Prog. Solid Mech. 1 (1960) 166.

Appendix A Shakedown Boundary for Hodge Yield Condition

First we consider the situation where $p_o < \frac{1}{2}$, and the shape of the modified yield surface is given in Fig. A1(a). We assume that the tube deforms plastically over a length $2\bar{L}$ such that the stress state moves along ABC of the yield surface as $\bar{x}$ increases.

If the stress state lies along BC of Fig. A1a

$$n_\theta = 1 - p_o + \frac{m_\phi}{2} \tag{A1}$$

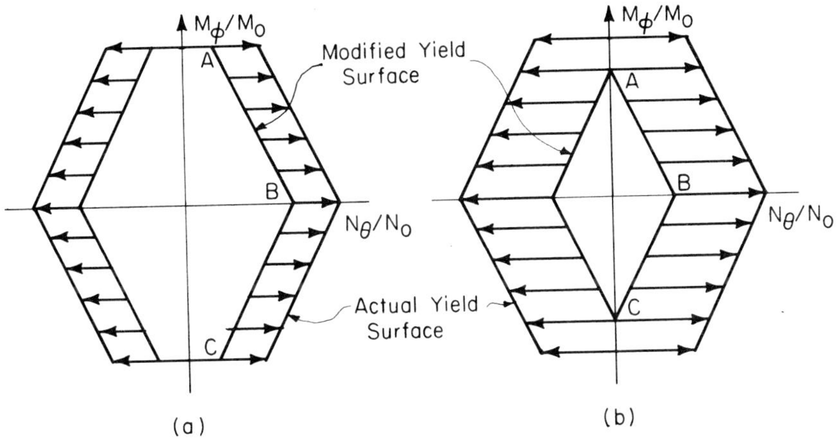

Fig. A1a. Modified yield surfaces for Hodge yield condition when (a) $p_o < \frac{1}{2}$ and (b) $p_o > \frac{1}{2}$.

and eqn. (5.1) becomes

$$\frac{\partial^2 m_\phi}{\partial \bar{x}^2} + 2m_\phi = -4(1 - p_o) \tag{A2}$$

The general solution to this equation is

$$m_\phi = A \cos \sqrt{2}\,\bar{x} + B \sin \sqrt{2}\,\bar{x} - 2(1 - p_o) \tag{A3}$$

For a stress state along AB of Fig. A1(a)

$$n_\theta = (1 - p_o) - \frac{m_\phi}{2} \tag{A4}$$

$$\frac{\partial^2 m_\phi}{\partial \bar{x}^2} - 2m_\phi = -4(1 - p_o) \tag{A5}$$

and $$m_\phi = C\exp(\sqrt{2}\,\bar{x}) + D\exp(-\sqrt{2}\,\bar{x}) + 2(1 - p_o) \tag{A6}$$

We solve these equations subject to the boundary conditions of (5.3) with $a = 2.45$ and in addition impose the conditions that m_ϕ and $\frac{\partial m_\phi}{\partial \bar{x}}$ are continuous as the solution to the governing equations changes from eqn. (A3) to eqn. (A6), i.e.,

$$\left.\begin{array}{l} m_\phi = 0 \\ \dfrac{\partial m_\phi}{\partial \bar{x}} \text{ is continuous} \end{array}\right\} \text{ at } \bar{x} = \bar{\ell} \tag{A7}$$

The result is

$$\frac{P}{P_L} = \frac{8(1 - p_o) - 4(1 - 2p_o)\cos\sqrt{2}\,\bar{\ell}}{\sin\sqrt{2}\,\bar{\ell}} \qquad \text{(A8)}$$

where

$$(1 - 2p_o) = 2(1 - p_o)\cos\sqrt{2}\,\bar{\ell} - (1 - 2p_o)\sin\sqrt{2}\,\bar{\ell}\,\sinh\sqrt{2}(\bar{L} - \bar{\ell}) \qquad \text{(A9)}$$

and $$\cosh\sqrt{2}(\bar{L} - \bar{\ell}) = \frac{2(1 - p_o)}{(1 - 2p_o)} \qquad \text{(A10)}$$

The values of $\bar{L}$ and $\bar{\ell}$ given by eqns. (A9) and (A10) are shown in Fig. 9 and the shakedown boundary of eqn. (A8) is plotted in Fig. 8.

An equilibrium stress field in the remainder of the tube which does not violate yield is given by eqn. (5.6) with $\bar{k} = \sqrt{\dfrac{3}{2(\frac{1}{2} - P_o)}}$.

When $p_o \geqslant \frac{1}{2}$ the modified yield surface is like that in Fig. A1(b). The governing equations for stress states along BC and AB of the yield surface are given by eqns. (A3) and (A6). These must now be solved subject to the boundary conditions

$$\left.\begin{aligned} m_\phi &= -\,2(1 - p_o) \\ \frac{\partial m_\phi}{\partial \bar{x}} &= 2.45\,\frac{P}{P_L} \end{aligned}\right\} \text{ at } \bar{x} = 0$$

$$\left.\begin{aligned} m_\phi &= 0 \\ \frac{\partial m_\phi}{\partial \bar{x}} &\text{ is continuous} \end{aligned}\right\} \text{ at } \bar{x} = \bar{\ell}$$

$$\left.\begin{aligned} m_\phi &= 2(1 - p_o) \\ \frac{\partial m_\phi}{\partial \bar{x}} &= 0 \end{aligned}\right\} \text{ at } \bar{x} = \bar{L}$$

We find $\bar{L} = \infty$; $\bar{\ell} = \dfrac{\pi}{2\sqrt{2}}$

and $$\frac{P}{P_L} = 1.15(1 - p_o)$$

This equation completes the shakedown boundary of Fig. 8.

Appendix B Limit Load for Shell

In this Appendix we obtain the limit load for a thin wall tube which is subjected to a constant pressure, $\tilde{p}$, over the region $-\bar{L} \leqslant \bar{x} \leqslant \bar{L}$. For $|\bar{x}| \leqslant \bar{L}$ the equilibrium equation is

$$\frac{\partial^2 m_\phi}{\partial \bar{x}^2} + 4(n_\theta - p_1) = 0 \tag{B1}$$

where $p_1 = \dfrac{\tilde{p}R}{N_o}$,

and for $|\bar{x}| > \bar{L}$ the equilibrium equation is

$$\frac{\partial^2 m_\phi}{\partial \bar{x}^2} + 4n_\theta = 0 \tag{B2}$$

If the resultant forces lie along BC of Fig. 5a

$$n_\theta = 1 + \frac{m_\phi}{2}$$

and equation (B1) becomes

$$\frac{\partial^2 m_\phi}{\partial \bar{x}^2} + 2m_\phi = 4(p_1 - 1) \tag{B3}$$

whose general solution is

$$m_\phi = A \cos \sqrt{2}\,\bar{x} + B \sin \sqrt{2}\,\bar{x} + 2(p_1 - 1) \tag{B4}$$

For a combination of forces along AB

$$n_\theta = 1 - \frac{m_\phi}{2}$$

$$\frac{\partial^2 m_\phi}{\partial \bar{x}^2} - 2m_\phi = 4(p_1 - 1) \tag{B5}$$

which has the general solution.

$$m_\phi = C\exp(\sqrt{2}\,\bar{x}) + D\exp(-\sqrt{2}\,\bar{x}) - 2(p_1 - 1) \tag{B6}$$

When equilibrium is determined by eqn. (B2) and the force resultants are along AB of the yield surface, we obtain the relationships

$$\frac{\partial^2 m_\phi}{\partial \bar{x}^2} - 2m_\phi = -4$$

and $m_\phi = E\exp(\sqrt{2}\,\bar{x}) + F\exp(-\sqrt{2}\,\bar{x}) + 2$ (B7)

We assume a solution where eqn. (B4) holds for $0 \leqslant |x| \leqslant \bar{\ell}$, eqn. (B6) for $\bar{\ell} \leqslant |x| \leqslant \bar{L}$ and eqn. (B7) for $\bar{L} < \bar{x} \leqslant \bar{k}$. These equations must be solved subject to the following boundary conditions:

$$\left.\begin{aligned} m_\phi &= -1 \\ \frac{\partial m_\phi}{\partial \bar{x}} &= 0 \end{aligned}\right\} \text{at } \bar{x} = 0$$

$$\left.\begin{aligned} m_\phi &= 0 \\ \frac{\partial m_\phi}{\partial \bar{x}} &\quad \text{continuous} \end{aligned}\right\} \text{at } \bar{x} = \bar{\ell}$$

$$\left.\begin{aligned} m_\phi &\quad \text{continuous} \\ \frac{\partial m_\phi}{\partial \bar{x}} &\quad \text{continuous} \end{aligned}\right\} \text{at } \bar{x} = \bar{L} \qquad \text{(B8)}$$

$$\left.\begin{aligned} m_\phi &= 1 \\ \frac{\partial m_\phi}{\partial \bar{x}} &= 0 \end{aligned}\right\} \text{at } \bar{x} = \bar{k}$$

The limiting pressure is then given by

$$(p_1\exp(-\sqrt{2}\,\bar{L}) - (p_1 - 1)\exp(-\sqrt{2}\,\bar{\ell})(1 + \tan(\sqrt{2}\,\bar{\ell}))$$
$$\times\ (p_1\exp(\sqrt{2}\,\bar{L}) - (p_1 - 1)\exp(\sqrt{2}\,\bar{\ell})(1 - \tan(\sqrt{2}\,\bar{\ell}))$$
$$= 0.25 \qquad \text{(B9)}$$

The value of $\bar{\ell}$ can be chosen to give the optimum value of p_1. The resulting bound is plotted in Fig. 10 as a function of $\bar{L}$.

Failure of Hyperbolicity in Soft Shells

Michail Zak

Jet Propulsion Laboratory, California Institute of Technology
Pasadena, California, USA

Introduction

The loss of hyperbolicity in elastodynamics leads to a special type of instability (wrinkling) which is accompanied by a collapse of elasticity, Refs. [1]-[3]. From the mathematical viewpoint this phenomenon is associated with occurrence of imaginary acoustic speeds and local maxima of the potential energy as a function of strains (Hadamard's instability). From the mechanical viewpoint the collapse of elasticity results from the failure of shear resistance under severe compression loading.

The criteria of the collapse of elasticity are

$$T^{ii} < \begin{cases} -\dfrac{\partial T^{ij}}{\partial \varepsilon_{ij}} & \text{if } i = j \\[2ex] -\dfrac{1}{2}\dfrac{\partial T^{ij}}{\partial \varepsilon_{ij}} & \text{if } i \neq j \end{cases} \tag{1}$$

in which the stresses T^{ij} and strains ε_{ij} are referred to a local cartesian basis.

Practically, these inequalities occur in such materials where the stiffness in one direction is significantly smaller (or larger) than the stiffness in other directions (thin films, soft shells, laminated materials, etc.). Solutions to the governing equations of elastodynamics in the region of instability have the following structure (Refs. 1, 2):

Inelastic Behaviour of Plates and Shells
IUTAM Symposium Rio de Janeiro 1985
Editors: L. Bevilacqua, R. Feijóo and R. Valid

$$\underline{u}_* = \underline{u}(r,t) + \underline{\tilde{u}}(r,t) \tag{2}$$

in which u, and $\tilde{u}$ are the basic and perturbed displacements, respectively, r is the position vector of points in space, while

$$\underline{\tilde{u}} = \underline{\tilde{u}}_o \exp\left(k\sqrt{-\lambda_o^2}\right)t \sin \underline{k}\cdot\underline{r}, \quad \underline{\tilde{u}}_o = \underline{\tilde{u}}_{t=o} \to 0, \quad k \to \infty \tag{3}$$

Here

$$\lambda_o^2 = \begin{cases} \frac{1}{\rho}\left(T^{ii} + \frac{1}{2}\frac{\partial T^{ij}}{\partial \varepsilon_{ij}}\right) & \text{if } i \neq j \\ \\ \frac{1}{\rho}\left(T^{ii} + \frac{\partial T^{ii}}{\partial \varepsilon_{ii}}\right) & \text{if } i = j \end{cases} \tag{4}$$

in which ρ is the material density.

In the case of the inequality (1) the acoustic speed λ_o given by Eq. (4) become imaginary, and the perturbed displacement (3) grows exponentially no matter how small the initial value $\underline{\tilde{u}}_o$ is chosen if the magnitude of the wave vector $\underline{k}$ is sufficiently large.

The constitutive law $T^{ii}(\varepsilon_{ij})$ fails in all the directions where $\lambda_o < 0$.

The purpose of this work is to develop a new constitutive law in the region of the collapse of elasticity describing inelastic behavior of soft shells.

2. Governing Equations for Soft Shells

A soft shell is defined as a thin shell for which the inplane strains of the middle surface are negligible in comparison to the strains in the transverse direction, i.e., the transverse rigidity is negligible in comparison to the inplane rigidity of the middle surface:

$$G_T \ll E_\ell \tag{5}$$

in which G_T is the through-the-thickness shear modulus, and E_ℓ is the longitudinal Young's modulus. Such a shell is formed by a soft material with an almost inextensible "skin." It is clearly understood that the thickness of the shell is negligible in comparison to the inplane dimensions.

In the limit $G_T \to 0$ soft shell is considered as a membrane.

The governing equation for a soft shell are given in the following form (Ref. 3):

$$\rho_o \sqrt{|g_o|} \left(\frac{\partial v^i}{\partial t} + v^i \Gamma^i_{tj} - F^i \right) = \frac{\partial}{\partial \psi^{(k)}} \left(\sqrt{|g|}\; T^{ki} \right)$$

$$+ \sqrt{|g|}\; T^{sj}\, \Gamma^i_{sj}\, T^{33} = 0, \; k = 1,2;\; i,j,s = 1,2,3 \tag{6}$$

while

$$\frac{\partial}{\partial \psi^{(3)}} \left(\sqrt{|g|}\; T^{3i} \right) = 0 \tag{7}$$

Here $\psi^{(1)}$ and $\psi^{(2)}$ are Gaussian material coordinates of the middle surface, $\psi^{(3)}$ is the out-of-plane material coordinate, g_{ij} is the metric tensor, g_{ij} is the initial state matric tensor, while $A = \det\|A_{ij}\|$, T^{sj} is the stress tensor, v^i is the velocity vector, ρ_o is the initial density, Γ^i_{sj} are the connection coefficients defined by Christoffel symbols:

$$\Gamma^n_{ij} = \frac{1}{2} g^{n\ell} \left(\frac{\partial g_{\ell i}}{\partial \psi^{(j)}} + \frac{\partial g_{\ell j}}{\partial \psi^{(i)}} - \frac{\partial g_{ij}}{\partial \psi^{(\ell)}} \right), \text{ for } i,j,n = 1,2 \tag{8}$$

$$\Gamma^3_{ij} = |g| \sqrt{g^{33}}\; b_{ij} \quad \text{for } i,j = 1,\ 2 \tag{9}$$

$$\Gamma^n_{13} = \frac{1}{2} \left(g^{n\ell} \frac{\partial g_{\ell 3}}{\partial \psi^{(i)}} - g^{nk} \frac{\partial g_{i3}}{\partial \psi^{(k)}} \right),$$

$$i,k = 1,\ 2;\ n,\ell = 1,\ 2,\ 3 \tag{10}$$

$$\Gamma_{33}{}^{n} = \frac{1}{2}\, g^{n\ell}\, \frac{\partial g_{33}}{\partial \psi^{(\ell)}} \;, \quad \ell = 1,\ 2;\ n = 1,\ 2,\ 3 \qquad (11)$$

in which b_{ij} are the coefficients of the second fundamental form of the middle surface, and Γ^{i}_{tj} are kinematical connection coefficients defined from the following compatibility equations:

$$\Gamma_{st}{}^{k} = \frac{\partial v^{k}}{\partial \psi^{(s)}} + v^{n}\, \Gamma_{sn}{}^{k} \qquad (12)$$

$$\frac{\partial \Gamma_{t\ell}{}^{n}}{\partial \psi^{(1)}} - \frac{\partial \Gamma_{1\ell}{}^{n}}{\partial t} + \Gamma_{1p}{}^{n}\Gamma_{t\ell}{}^{p} - \Gamma_{tp}{}^{n}\Gamma_{1\ell}{}^{p} \qquad (13)$$

The compatibility equations

$$\frac{\partial g_{i3}}{\partial \psi^{(i)}} = \Gamma_{ii}{}^{k}\, g_{k3} + |g|\, b_{ii},\ i,\ k=1,\ 2 \qquad (14)$$

$$b_{11}\, b_{22} - b_{12}^{2} = \Gamma_{12}{}^{\nu}\, \Gamma_{12}{}^{\delta}\, g_{\nu\delta} - \Gamma_{11}{}^{\alpha}\, \Gamma_{22}{}^{\beta}\, g_{\alpha\beta} -$$

$$- \frac{1}{2}\, \frac{\partial^{2} g_{11}}{\partial \psi^{(2)^{2}}} + \frac{\partial^{2} g_{12}}{\partial \psi^{(1)}\, \partial \psi^{(2)}} - \frac{1}{2}\, \frac{\partial^{2} g_{22}}{\partial \psi^{(1)^{2}}} \;,$$

$$(\gamma, \delta, \alpha, \beta = 1,2) \qquad (15)$$

$$\frac{\partial b_{ii}}{\partial \psi^{(2)}} - \frac{\partial b_{i2}}{\partial \psi^{(1)}} = \Gamma_{i2}{}^{1}\, B_{11} - \Gamma_{i1}{}^{2}\, b_{22}$$

$$+ \left(\Gamma_{i2}{}^{2} - \Gamma_{i1}{}^{1}\right) b_{12} \qquad (16)$$

and the constitutive equations for a Hookian material:

$$T^{ij} = C^{ijkm}\varepsilon_{km} \qquad (17)$$

where C^{ijkm} is the elastic modulus tensor, ε_{km} is the strain tensor:

$$g_{11} = 1 + \varepsilon_{11},\ g_{12} = 2\varepsilon_{12},\ g_{22} = g_{22} + 2\varepsilon_{22} \tag{18}$$

close the system of governing equations for a soft shell. It is easy to verify that all variables in these equations are expressed via geometrical invariants of the middle surface, i.e., via its shape b_{ij} and strains ε_{ij}.

One should notice that all the governing equations are two-dimensional. However, in contrast to the classical thin shell model, they do not contain bending moments. At the same time, unlike the model of a membrane, they contain transverse shear stresses. From this point of view, the soft shell model and the plane section model are at the opposite ends of the spectrum of shell models.

3. Wrinkling Criteria

Referring the criteria of the failure of hyperbolicity (1) (which are written in a local cartesian basis) to the current coordinates $\psi^{(i)}$ one arrives at the following wrinkling criteria, Ref. 3:

$$\begin{aligned} & T_{11}^{\;\;1} + G^1_{3131} + T_{22}^{\;\;1} + G^1_{3131} - \\ & \quad - \frac{1}{2}\left[\left(T^1_{11} + G^1_{3131} - T^1_{22} - G_{3232}\right)^2 \right. \\ & \quad \left. + \left(2T^1_{12} + G^1_{1213} + G^1_{3231}\right)^2\right]^{1/2} \\ & \quad = \overset{*}{T}_{ij\ \min} < 0 \end{aligned} \tag{12}$$

in which

$$T_{ij}^{\;\;1} = \alpha^s_k \alpha^\ell_q \alpha^i_{\sim s} \alpha^j_{\sim\ell}\, T^{kq},\quad i,\ j = 1,\ 2 \tag{13}$$

$$G^1_{3i3j} = \frac{1}{2}\, \alpha^s_k \alpha^\ell_q \alpha^3_s \alpha^i_\ell \alpha^s_m \alpha^\ell_n \alpha^3_s \alpha^j_\rho \frac{\partial T^{kq}}{\partial \varepsilon_{mn}},\quad i,\ j = 1,\ 2 \tag{14}$$

while

$$\frac{\partial T^{kq}}{\partial \varepsilon_{mn}} = 2G^{kqmn} = 2\sqrt{\frac{|g_o|}{g}} \; \frac{\partial^2 \Pi}{\partial \varepsilon_{kq} \; \partial \varepsilon_{mn}} \tag{15}$$

$$a^q_s = \underline{r}_s \cdot \underline{r}^q, \; \underline{r}^q = \frac{1}{2\sqrt{|g_o|}} \; e^{qks} \underline{r}_k \times \underline{r}_s, \; \tilde{a}^q_s = \underline{r}_s \cdot \underline{e}_q \tag{16}$$

where Π is the specific potential energy of the shell material, $\underline{r}_s$ are the initial state basis vectors, $\underline{r}_s$ are the current basis vectors, and $\underline{e}_k$ are the cartesian basis.

The angle α^* corresponding to the direction in which wrinkles occur is defined by:

$$\alpha^* = \text{arc tan} \frac{2[(\overset{*}{T}_{ij})_{min} - (T^1_{11} + G^1_{3131})]}{2\; T^1_{12} + G^1_{3131} + G^1_{3231}} \tag{17}$$

Clearly, the inequality (12) can lead to such a zone

$$\alpha_1 \leqslant \tilde{\alpha} \leqslant \alpha_2 \tag{18}$$

within which

$$\overset{*}{T}_{ij} < 0 \tag{19}$$

and, consequently, the instability can occur in any direction between the angles α_1 and α_2.

These angles are defined by the following expressions:

$$\alpha_1, \; \alpha_2 = \text{arctan} \left[\frac{1}{q_2} \left(-q_{12} \pm \sqrt{q^2_{12} - q_1 \; q_2} \right) \right] \tag{20}$$

where

$$q_1 = T^1_{11} + G^1_{3131},\ q_2 = T^1_{22} + G^1_{3232},$$

$$q_{12} = T^1_{12} + \frac{1}{2}(G^1_{3132} + G^1_{3231}) \ . \tag{21}$$

Four different situations can occur:

(a) $q^2_{12} > q_1 q_2$ (22)

i.e., α_1 α_2 are real and there are zones of instability given by (18)

$(b)_1$ $q_{12} = q_1 q_2,\ q_1 > 0$. (23)

In this case

$$\alpha_1 = \alpha_2 = \overset{*}{\alpha} \tag{24}$$

and the instability can occur only in the form of the cumulative effect (Ref. 3) in the direction defined by the angle $\overset{*}{\alpha}$ because

$$(\overset{*}{T}_{ij})_{min} = \overset{*}{T}_{ij}\Big|_{\alpha=\overset{*}{\alpha}} = 0 \tag{30}$$

while

$$\overset{*}{T}_{ij}\Big|_{\alpha\neq\overset{*}{\alpha}} > 0 \tag{31}$$

$(b)_2$ $q^2_{12} \leqslant q_1 q_2,\ q_1 < 0$.

Obviously that here the instability can occur in any direction.

(c) $q_{12}^2 < q_1 q_2, \quad q_1 > 0 .$ (32)

In this case the expression (30) becomes imaginary, i.e., there is no instability at all because

$$(\overset{*}{T}_{ij})_{min} > 0 . \tag{33}$$

Returning to the case (a) let the conditions defining the type of instability be recorded. As it was pointed out in Ref. 3, the wrinkling can occur only if the corresponding direction of instability coincides with an asymptotic line of the layer, i.e., if

$$b_{22}^{\beta} = 0 \tag{39}$$

where b_{22}^{β} is the coefficient of the second fundamental form referred to the axis $\underline{e}_2^{\beta}$ of the basis $\underline{e}_1^{\beta}, \underline{e}_2^{\beta}, \underline{e}_3^{\beta}, = \underline{e}_3$ turned through the angle β about the axis $\underline{e}_3$. Then:

$$b_{22}^{\beta} = b_{22}^{1} \cos^2\beta + b_{11}^{1} \sin^2\beta = b_{12}^{1} \sin 2\beta \tag{35}$$

where

$$b_{ij} = \overset{*}{\alpha}{}_k^s \overset{*}{\alpha}{}_q^{\ell} \overset{*}{\alpha}{}_{\tilde{s}}^i \overset{*}{\alpha}{}_{\tilde{\ell}}^i \, b_{ij}$$

$$\overset{*}{\alpha}{}_s^q = \dot{\underline{r}}_s \cdot \underline{r}^q , \quad \underline{r}^q = \frac{1}{2\sqrt{|g|}} e^{qks} \underline{r}_k \times \underline{r}_s \tag{36}$$

The angle $\overset{*}{\beta}$ corresponding to the equality (36) is given by the following expression:

$$\overset{*}{\beta} = \arctan\left[\frac{1}{b_{11}^{1}}\left(-b_{12}^{1} \pm \sqrt{-K}\right)\right] \tag{37}$$

where

$$K = b_{11}^{1} b_{22}^{1} - b_{12}^{1\,2} . \tag{38}$$

Thus, the existance of the asymptotic directions ($b^{\beta}_{12} = 0$) depends on the sign of the Gaussian curvature K.

There are two asymptotic directions for $K < 0$, there is only one asymptotic direction if $K=0$, and there is no such directions at all if $K > 0$.

Now the conditions of wrinkling are formulated in the following form:

$$\alpha_1 \leqslant \overset{*}{\beta} \leqslant \alpha_2 \qquad (35)$$

where α_1 and α_2 are given by equation (20).

Consequently, the wrinkling occurs only in the points with a non-positive Gauss curvature of the middle surface if the corresponding asymptotic lines are located within the zone of instability given by inequalities (39).

In all other cases the instability leads to the snap-through wrinkling, (Ref. [3] which is accompanied by strong discontinuities, i.e., by jumps of strains and stresses.

4. Constitutive Low for Inelastic Domain

Formally the solutions to the governing equations of soft shells in the domain of wrinkling are unstable and cannot be used for quantitative description of the motion. Physically in the course of the instability, the geometry of the macro-structure of the material fails and it loses all the elastic resistance in the directions of wrinkling. However, on the micro-scale the material properties remain unchanged. In other words, collapse of elasticity is caused not by changes in the rheology of the material, but rather by changes in its geometrical structure. The energy transformations in the course of the instability can be described as following: the elastic potential energy accumulated in the macrostructure is released through the kinetic energy of high frequency fluctu-ations while the latter dissipates through some friction mechanisms.

It is reasonable to assume that in solids the rate of dissipation of energy of the fluctuations is high and therefore the contribution of the fluctuations in the average motion can be neglected after some finite period of time.

This assumption allows the possibility of introducing simplified post-instability models of solids which will be discussed below. Indeed, as shown in Ref. 2, the characteristic speeds vanishes as soon as the corresponding fluctuations disappear.

Referring to the expression for the characteristic speeds in elasticity given in Ref. 2, the following constitutive law describing inelastic behavior caused by wrinkling can be introduced:

$$T_{ii} \geqslant -\frac{1}{2}\frac{\partial T_{ij}}{\partial \varepsilon_{ij}} . \tag{40}$$

Observance of this inequality guarantees that all the characteristic speeds will be positive or zero, which means that there will not be failure of hyperbolicity any more.

One should note that the inequality (40) has practical applications only if the material does not lose its elastic properties before the limit (40) occurs. Hence, the models given by Equation (40) can be applied to thin films, soft shells or laminated elastic materials, etc.

First of all let us consider the particular case of the Equation (40) when the derivative $\partial T_{ij}/\partial \varepsilon_{ij}$ does not depend on a direction selected in the body, i.e.,

$$\frac{\partial T_{ii}}{\partial \varepsilon_{ii}} = E, \quad \frac{1}{2}\frac{\partial T_{ij}}{\partial \varepsilon_{ij}} = G(i \neq j), \; E > G . \tag{41}$$

Then the Equation (40) can be written in terms of the principal stresses:

$$\overset{*}{T}_{ii} \geqslant -G . \tag{42}$$

But in general cases when the derivative $\partial T_{ij}/\partial \varepsilon_{ij}$ depends on the direction selected in the body (which can occur even in isotropic material due to pre-stresses) the Equation (40) cannot be written in an invariant form because the left-hand part of this equation belongs to a tensor of the second rank T_{ij} while the right-hand part belongs to a tensor of the fourth rank $\partial T_{ij}/\partial \varepsilon_{ks}$.

Returning to the equation (42) let us consider a thin film as a particular case of a soft shell for which:

$$G = 0 \; . \tag{43}$$

Then Equation (40) or (42) leads to a model presenting film as a no-compression material:

$$\overset{+}{T}_{11} \geqslant 0 \; , \; \overset{+}{T}_{22} \geqslant 0 \; . \tag{44}$$

5. Principal Stress Formulation of Post-Instability Models

Equation (42) formulates models in terms of principal stresses. In order to operate with principal stresses in the governing equations they must be projected on a specially selected system of coordinates, a local basis vector of which coincides with the principle directions of the corresponding local stress tensor, i.e., in the general case this system will be curvilinear, orthogonal and non-material.

Because this system is fully determined by the state of the continuum and therefore does not contain any arbitrariness, the invariant properties of the solutions are expected to be found.

Indeed, as shown in Ref. 2, where the principal stress formulation is applied to a film as a no-compression material, the coordinate lines are observed as running wrinkles.

a. Geometry of Principal Stress Coordinates

The principal stress coordinates $q_i (i = 1,2,3)$ geometrically can be introduced as orthogonal curvilinear coordinates with the metric which is supposed to be defined from dynamics:

$$\frac{\partial \underline{r}}{\partial q_i} \cdot \frac{\partial \underline{r}}{\partial q_i} = \begin{cases} g_{ij} & \text{if} \quad i=j \\ 0 & \text{if} \quad i \neq j \end{cases} \tag{45}$$

where $\underline{r}$ is the radius vector of the medium points,
g_{ij} are components of the metric tensor.

Let us introduce also a material coordinate $\psi_i (i = 1,2,3)$ by the transformation $\psi_i = \psi_i(q_1, q_2, q_3)$ and denote the coefficients:

$$a_i^j = \frac{\partial \psi_i}{\partial q_j} \tag{46}$$

while

$$\frac{\partial \underline{\underline{r}}}{\partial \psi_i} \cdot \frac{\partial \underline{\underline{r}}}{\partial \psi_i} = 1 + \varepsilon_{ii}, \quad \frac{\partial \underline{\underline{r}}}{\partial \psi_i} \cdot \frac{\partial \underline{\underline{r}}}{\partial \psi_j} = 2\varepsilon_{ij} \quad (i \neq j) \tag{47}$$

where ε_{ij} are the covariant components of the strain tensor in the material system.

It follows from (46) that

$$\frac{\partial \underline{\underline{r}}}{\partial q_i} = a_i^j \frac{\partial \underline{\underline{r}}}{\partial \psi_j}, \quad \frac{\partial \underline{\underline{r}}}{\partial \psi_i} = \tilde{a}_i^j \frac{\partial \underline{\underline{r}}}{\partial q_j}, \quad a_i^j \tilde{a}_j^k = \delta_s^k \tag{48}$$

and

$$\tilde{\varepsilon}_{ij} = a_i^\alpha a_j^\beta \varepsilon_{\alpha\beta} \tag{49}$$

where $\tilde{\varepsilon}_{ij}$ are the strain tensor components in the principal stress system.

Besides that, from (46) the geometrical equations of compatibility for the transformation coefficients a_{ij} follow:

$$\frac{\partial a_i^j}{\partial q_i} = \frac{\partial a_i^i}{\partial q_j} \; . \tag{50}$$

b) Kinematics of Principal Stress Coordinates

A velocity in the principal stress coordinates is given by:

$$\underline{v} = \frac{\partial \underline{r}}{\partial t} + \sum_{i=1}^{3} \frac{\partial \underline{r}}{\partial q_i} v_i \tag{51}$$

where $v_i = \partial q_i / \partial t$ are the relative velocities of the individual particles with respect to the system of coordinates.

Keeping in mind Equation (46) the following kinematical equations of compatibility are obtained:

$$\frac{\partial a_i^i}{\partial t} = \frac{\partial}{\partial q_i} \sum_{j=1}^{3} a_i^j v_j \; , \quad i=1,2,3 \tag{52}$$

c) Dynamics in Principal Stress Coordinates

Starting with one of the conventional forms of dynamics in orthogonal coordinates:

$$\rho \underline{a} - \underline{F} = (g_{11}g_{22}g_{33})^{-1/2} \frac{\partial}{\partial q_s} \left[(g_{11}g_{22}g_{33})^{1/2} T^{sk} \frac{\partial \underline{r}}{\partial q_k} \right] \tag{53}$$

where ρ is the density, $\underline{a}$ is the acceleration, T is the stress tensor, and $\underline{F}$ is the external force, and taking into account that

$$\underline{a} = \left(\frac{\partial}{\partial t} + \sum_{i=1}^{3} v_i \frac{\partial}{\partial q_i} \right)^2 \underline{r} \tag{55}$$

$$T^{ij} = 0 \; , \quad (i \neq j) \tag{56}$$

finally,

$$\rho \left(\frac{\partial}{\partial t} + \sum_{i=1}^{3} v_i \frac{\partial}{\partial q_i} \right)^2 \underline{r} - \underline{F} =$$

$$(g_{11}g_{22}g_{33})^{-1/2} \sum_{i=1}^{3} \frac{\partial}{\partial q_i} \left[(g_{11}g_{22}g_{33})^{1/2} \overset{+}{T}^{ii} \frac{\partial \underline{r}}{\partial q_i} \right] \tag{57}$$

where $\overset{+}{T}^{ii}$ are the principal stresses. The density ρ is expressed via the initial density ρ_0 by the formula:

$$\rho = \frac{\rho_0}{(1+2\tilde{\varepsilon}_{11})(1+2\tilde{\varepsilon}_{22})(1+2\tilde{\varepsilon}_{33})} \tag{58}$$

The Equation (57) must be completed by constitutive equations:

$$\tilde{T}^{ii} = f_i(\tilde{\varepsilon}_{11}, \tilde{\varepsilon}_{22}, \tilde{\varepsilon}_{33}, \text{etc.}) \tag{59}$$

while

$$\tilde{\varepsilon}_{ij}(i \neq j) = 0 \tag{60}$$

or because of Equation (50)

$$a_i^{\alpha} a_j^{\beta} \varepsilon_{\alpha\beta} = 0 \, , \quad (\alpha \neq \beta) \, . \tag{61}$$

It can be verified that Equations (57), (58), (59), (61) together with Equations (42), (47), (48), (49), (50), (52) form a closed system.

Indeed, the vector equation (57) with respect to unknown vector $r(q_1,q_2,q_3,t)$ contains the following additional unknowns: T^{ii}, g_{jj}, v_k. All of them by means of the Equations (48), (59), (48), (52) are expressed via $\underline{r}$, ε_{ij} and a_i^j. But strains $\tilde{\varepsilon}_{ij}$ are also expressed via $\underline{r}$ by means of Equations (49), (48), (48), while nine coefficients a_i^j are coupled by nine equations (50), (61).

Now the governing equations for the enlarged model which includes non-elastic domains are written in the following form:

$$\rho \left(\frac{\partial}{\partial t} + \sum_{i=1}^{3} v_i \frac{\partial}{\partial q_i} \right)^2 \underline{r} - \underline{F} =$$

$$\frac{1}{2} (g_{11} g_{22} g_{33})^{-1/2} \times$$

$$\left[\sum_{i=1}^{3} \frac{\partial}{\partial q_i} (g_{11} g_{22} g_{33})^{-1/2} \overset{*}{T}{}^{ii} \frac{\partial \underline{r}}{\partial q_i} \right] \tag{62}$$

where

$$\overset{*}{T}{}^{ii} = \begin{cases} \overset{+}{T}{}^{ii} & \text{if } \overset{+}{T}{}^{ii} \geqslant -G \\ -G & \text{if } \overset{+}{T}{}^{ii} < -G \end{cases} \tag{63}$$

Some applications of the theory to wrinkling films are considered in Ref. 2.

In Ref. 4 the constitutive law for inelastic domains is generalized: it includes the contributions of the kinetic energy of micro-structural fluctuations. The theory is generalized on liquid shells in Ref. 5.

Acknowledgements - The research described in this paper was carried out by the Jet Propulsion Laboratory, California Institute of Technology, under NASA Contract No. NAS7-100. The work was supported by Dr. Anthony Amos, Air Force Office of Scientific Research.

References

1. Zak, M., on the Failure of Hyperbolicity in Elasticity, Journal of Elasticity 12(2), 219-229, 1982.

2. Zak, M., Postinstability in Continuous Systems, Part I and II, Solid Mechanics Archives, 7, 467-503, 1982 and 8, 1-37, 1983.

3. Zak, M., Wrinkling Phenomenon in Structures, Parts I and II, Solid Mechanics Archives, 8, 181-216, and 279-311, 1983.

4. Zak, M., Postinstability Models in Elasticity, Acta Mechnica 52, 119-132 (1984).

5. Zak, M., Shape instability in thin viscous films and jets, Acta Mechanica, 55, 33-50, 1985.

Buckling Analysis of Reinforced Concrete Shells

E. DULÁCSKA

Chief engineer of
Institute of Architectural Development
Budapest, Hungary.

Summary

The results of the nonlinear shell buckling theory have been extended to apply to the reinforced concrete shells. A simple relation is derived for the snap through critical load of r.c. shells. The solution contains the special properties of r.c. shells, i.e. the random imperfections of the shell, the quantity of reinforcement and the quality of concrete, and the nonlinear properties of r.c. shells, i.e. cracks in the cross sections, creep of the concrete and the plasticity of concrete and reinforcement respectively. The proposed method adequately describes the results and the situation of erected domes.

1. Buckling of Shells Made of Elastic Material

The safety of a r.c. shell with respect to buckling is satisfactory if

$$p \leq p_{cr}/\gamma \tag{1}$$

Here p is the actual load acting on the shell, p_{cr} is the critical load and γ is the safety factor.

In the case of r.c. shells we have to analyse independently the critical load and the safety factor too. That is, these values differ from those of the homogeneous shells.

The behaviour of the r.c. shell is influenced by the geometric data of the shell and of the cross section, by the material properties of both the concrete and the reinforcement, by the quantity of reinforcement, and by the concrete covering.

On the whole, reinforced concrete differs from the elastic homogeneous material in the following:

Inelastic Behaviour of Plates and Shells
IUTAM Symposium Rio de Janeiro 1985
Editors: L. Bevilacqua, R. Feijóo and R. Valid

- the concrete zone under compression creeps,
- the concrete zone under tension cracks, reducing the stiffness of the cross section. At this stage the position, quantity and quality of the reinforcement play an important role,
- the materials of the shell behave elasto-plastically.

Due to the afore mentioned factors, the critical load of the r.c. shell becomes much lower than the critical load of the homogeneous shell.

The critical load of the r.c. shell can be computed from the formula:

$$p_{cr} = \zeta \beta p_{cr}(\phi). \quad (2)$$

Here, the effect of plasticity is expressed by ζ, and the effect of cracks and the reinforcement by β. The $p_{cr}(\phi)$ is the critical load of the homogeneous shell made of elastic material, including the creep effect of the concrete.

When considering a homogeneous, elastic shell with perfect geometry, the differential equations of equilibrium and compatibility restricted to small deflections lead to an eigenvalue problem. The lowest eigenvalue is the critical load of the shell (p_{cr}^{lin}), obtained with the linearized theory.

The deformation of the shell with an initial imperfection of amplitude denoted by w_o increases with the load intensity. The load intensity reaches a maximum value and, afterwards, decreases with increasing deformation ("snap-through" phenomenon). This maximum load will be called the upper critical load (p_{cr}^{u}).

The ratio of the upper critical load to the critical load corresponding to the linearized theory of the homogeneous shell is:

$$\rho = p_{cr}^{u}/p_{cr}^{lin} . \quad (3)$$

Figure 1 illustrates graphically the definitions given above. The ρ values of radially loaded spheres and axially compressed cylinders, furthermore of radially compressed short, medium-length and long cylinders are represented by the curves of Fig.2.

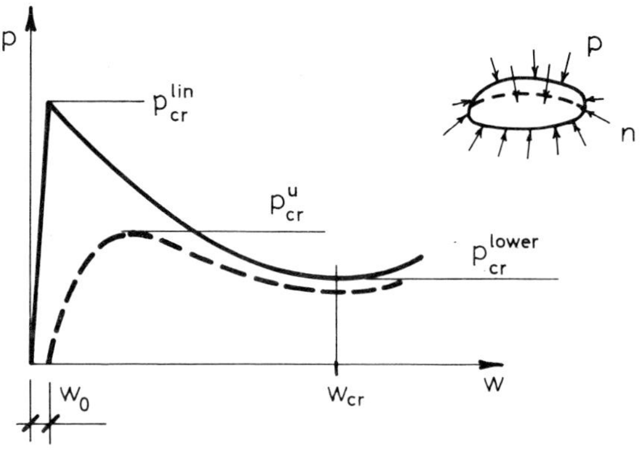

Buckling of elastic shells

Fig. 1

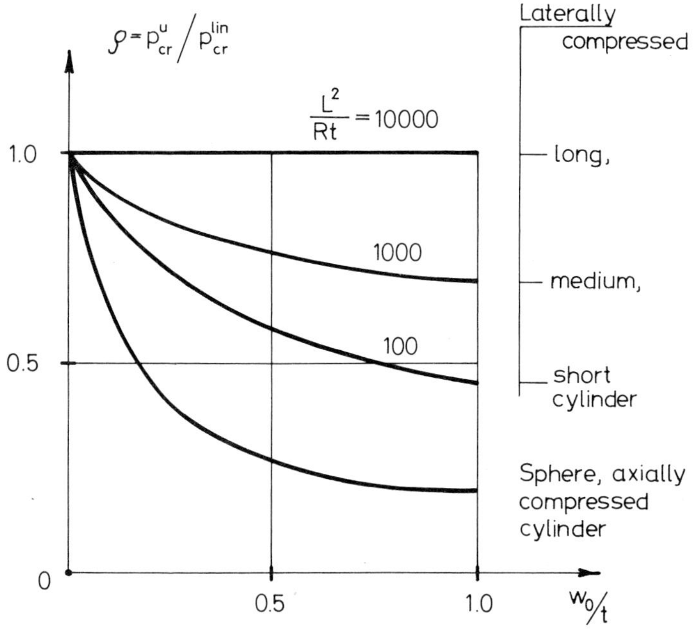

Decrease of the upper critical loads of cylindrical and spherical shells with increasing initial imperfectian w_0

Fig. 2

These curves may be closely approximated in the range $w_o/t < 1$ by the expression:

$$\rho \approx \frac{1}{1 + A(w_o/t)} \tag{4}$$

where t is the shell thickness (see in [11]). The value of A is different for various shell types. The value of ρ corresponding to $w_o/t = 0.5$, $\rho(0.5)$ (see Fig. 3) can be obtained from the literature on shell buckling. The constant A can thus be computed from the formula:

$$A = 2\left[\frac{1}{\rho(0.5)} - 1\right] . \tag{5}$$

Being aware of the value of the lower critical load (p_{cr}^{lower}) (this notation was used by Kárman), the value of $\rho(0.5)$ can be assessed from the expression:

$$\rho(0.5) \approx \frac{1}{6}\left[1 + 5\,\frac{p_{cr}^{lower}}{p_{cr}^{lin}}\right] . \tag{6}$$

Values of $\rho(0.5)$ and A are, for long or short cylinders, subjected to lateral pressure $\rho(0.5) = 0.59$ or 1.00, and $A = 1.4$ or 0.0, respectively, and for the axially compressed cylinder and radially compressed sphere $\rho(0.5) = 0.25$ and $A = 6.0$.

Finally, the critical load of the homogeneous elastic shell is determined by the formula:

$$p_{cr}^{u} = \rho\, p_{cr}^{lin}\,(E,\ t,\ R), \tag{7}$$

where E and R are the modulus of elasticity and the radius of the shell respectively.

2. The Influence of the Creep of Concrete

The effect of creep may be estimated by reducing the value of the modulus of elasticity of concrete according to the formula:

$$E_c = E_{c,o}/(1 + \phi_c) \tag{8}$$

where ρ_c is the creep factor (see [8]). The initial modulus of elasticity can be calculated from the well known expression:

$$E_{c,o} = 6750 \sqrt{f'_c} \; . \quad |N/mm^2| \tag{9}$$

Here f'_c is the cylinder strength of the concrete in N/mm^2. Ultimate compressive stress of the concrete obtained with cylindrical test samples.

The creep factor depends on the thickness of the structure, the quantity and quality of the cement used, the water cement ratio, and environmental factors. For normal circumstances ϕ_c depends only on the ultimate strength of the concrete as follows:

$$\phi_c \approx \frac{40}{10 + f'_c} \; . \tag{10}$$

If only a fraction of the load p_o is acting from the beginning, and the other fraction p_t begins to act only at a later stage t, we may reduce the creep factor ϕ_c according to Trost [13] and Zerna [14] by the formula:

$$\phi_c = \phi_c \frac{k_o + 0{,}75 \; k_t p_t / p_o}{1 + p_t / p_o} \tag{11}$$

The coefficient k takes into account the decreasing creep sensitivity of the concrete with increasing time. The value of k is 1.8 at the beginning of concrete hardening, 1.0 at the age of one month, and decreases to k = 0.5 for concretes older than one year. The value of k valid at the onset of p_o is denoted by k_o, while that valid at the onset of p_t is denoted by k_t.

The meteorological loads act only for a short term, therefore the modulus of elasticity should be only partially reduced in proportion to the ratio of short term load to all loads, but the seismic loads do not cause any creep.

3. Imperfection and Eccentricity

We have seen in Fig. 2 that the critical load of most shells sharply decrease with the initial imperfection. The imperfection consists of two parts. One part is the accidental imperfection, the amplitude of which we denote by $w_{o,accid}$. By evaluating measurements on erected cooling towers, we propose the following empirical formula for the accidental imperfection:

$$w_{o,accid} = 0.05\ t + \frac{a}{\frac{R/t}{1000} + \frac{1000}{R/t}} \left[\frac{R}{2000} \right] . \tag{12}$$

Here the factor "a" represents the influence of the accuracy of the erection method. For example, we can assume a = 1 for shells with rigid formwork, while for sliding shuttering we can take a = 6.

The other part of the imperfection can be calculated by the bending theory of shells. Its amplitude will be denoted by $w_{o,calc}$.

The superposition of the maximum values of both imperfections is rather improbable. Thus, according to the rules of the probability theory, we may take as a 95% fractile the following value for the design imperfection w_o:

$$w_o = \sqrt{w_{o,calc}^2 + 1.4\ w_{o,calc}\ w_{o,accid} + w_{o,accid}^2} \tag{13}$$

By doing so, we have taken the probability of superposition of the mean values of imperfection amplitudes into account. The standard deviation of the imperfection from their mean values will be considered in the safety factor.

We shall see later that the stiffness of the r.c. cross section depends on the eccentricity of normal forces. But only a part of the imperfection causes bending moment, i.e. eccentricity.

Therefore, the relation between imperfection and eccentricity can be analysed as follows. If we impose a small deformation w onto a shell with a given geometry and state of stress, we can determine the corresponding bending moment and the change in the membrane forces at any point and in any direction with the aid of the classical bending theory. Dividing the bending moment by the modified value of the membrane force, we arrive at the

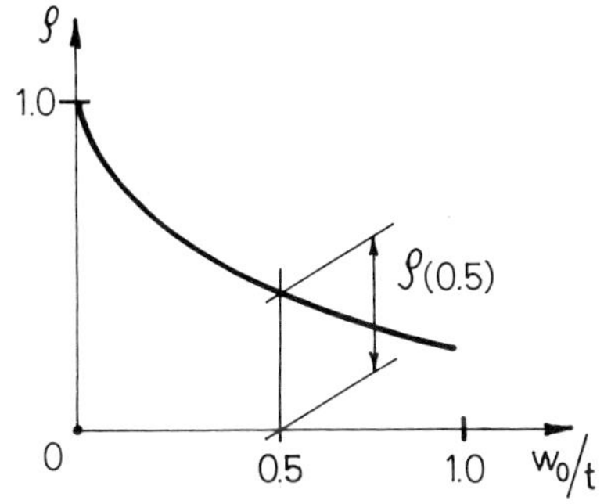

Definition of $\rho(0.5)$

Fig. 3

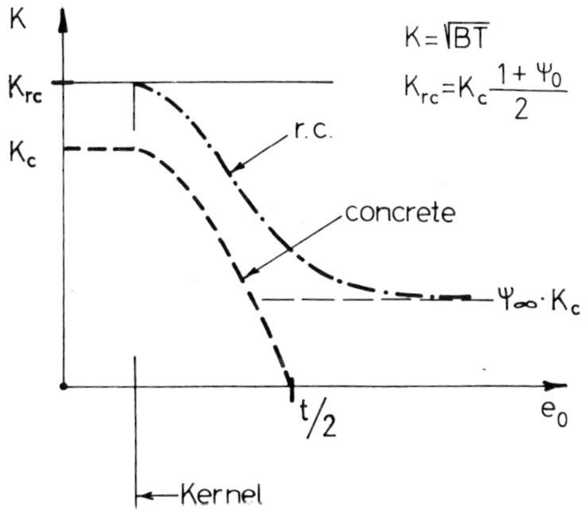

Decrease of the stiffnesses

Fig. 4

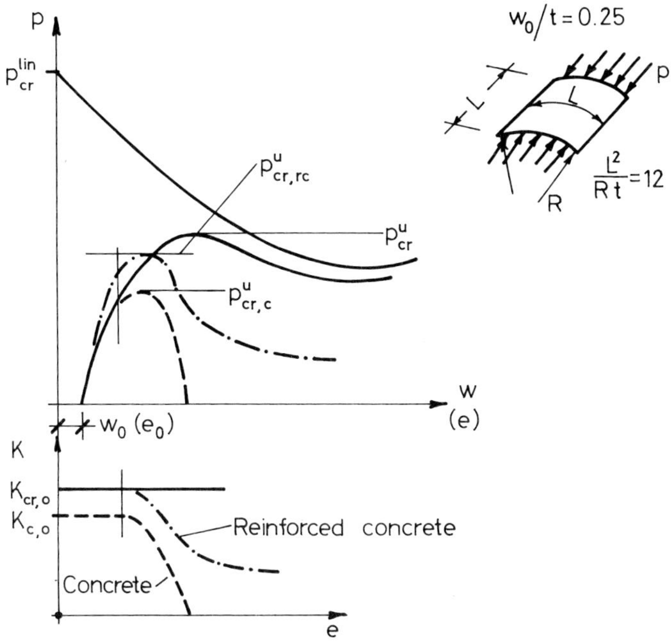

Determination of the upper critical load of the elastic reinforced concrete shell

Fig. 5

magnitude of the eccentricity. Performing this investigation for several shell surfaces and for various states of stresses the following (approximate) results for the ratio $c = e_o/w_o$ in the range $0.167 < w_{o/t} < 0.5$, was obtained:

- cylindrical shells: $c = 1.00$;
- domes: $c = 0.67$;
- hyperbolic shells: $c = 0.50$.

4. Stiffness of R.C. Cross Sections

In the following section we have to deal with the cracks of the cross section and with the reinforcement respectively.

It was shown in [5] that the critical load of shells depends on the extensional and bending rigidities of the shell, T and B respectively.

These can be taken into account as a unique, coupled "shell-buckling stiffness"

$$K = \sqrt{BT}$$

If the eccentrically compressed cross section cracks, the tensile stresses are transferred onto the reinforcement, and the rigidity of the cross section drops. The typical variation of the rigidity decreasing with increasing values of the eccentricity is to be seen in Fig. 4.

The rigidity of the uncracked cross section - which is perpendicular to the direction of the reinforcement - is $K_o = \psi_o K_c$, and in the bisector direction of the perpendicular mesh it is $K_o = \frac{1}{2}(1 + \psi_o)K_c$. After cracking, the shell buckling stiffness of the cross section perpendicular to the reinforcement direction and for very large eccentricities of the compressive force is $K_\infty = \psi_\infty K_c$, where K_c is the stiffness of the uncracked concrete cross section. In the bisector direction the stiffness is smaller, but it is increased by the effect of "tension stiffening". According to more accurate analysis we can take the K_∞ into account in all cases.

5. The Critical Load of Cracked, Elastic R.C. Shells

Fig. 5 shows the load-deformation (or - eccentricity) diagram p(e) of the homogeneous elastic cylindrical panel (see in [9]). In the lower part of Fig. 5, the shell buckling stiffnesses K of concrete and r.c. cross sections are shown for varying values of the eccentricity.

The ordinates p of the "homogeneous elastic" load-deflection curve p(w) have to be reduced proportionally to the diminishing buckling rigidity of the shell K(e). Thus, for concrete and r.c. shells respectively, the dotted and dash-dotted lines are obtained. The maximum values of these curves give the elastic critical loads belonging to the assumed imperfection amplitude for concrete and r.c. shells. It can be seen that the critical load of the r.c. shell is smaller than the critical load of the homogeneous shell, and the critical load of the concrete shell is smaller than the critical load of the r.c. shell.

Finally, we can set up the following formulae for r.c. shells:

$$\frac{e_o}{t} \leq 0.5: \; p^u_{cr,rc} = \frac{1+\psi_o}{2} \; p^u_{cr} + \psi_\infty (p^u_{cr} - p^u_{cr,c}), \qquad (14a)$$

$$\frac{e_o}{t} > 0.5: \; p^u_{cr,rc} = \psi_\infty p^u_{cr}. \qquad (14b)$$

The critical load of the concrete shell without any tensile strength can be well approximated by the formula

$$p^u_{cr,c} = p^{lin}_{cr} \left(1 - \frac{2e_o}{t}\right)^{1.5(1+w_o/e_o)} \qquad (15)$$

In r.c. shells two kinds of reinforcement are generally used: the single-layer grid (placed, as a rule, in the middle of the thickness) denoted by index 1, and the double-layer grid (placed on the two sides of the cross section), denoted by index 2.

The ψ values, calculated on the basis of the theory of r.c., are shown in Fig. 6 and in Table I.

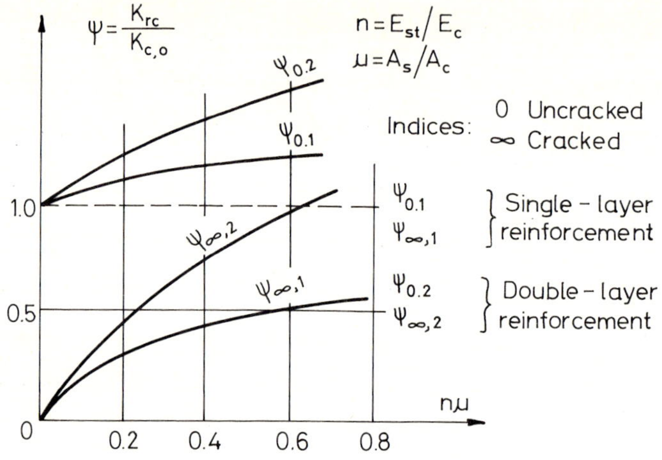

Factors ψ for calculating the shell buckling stiffness of r.c. shells plotted against the quantity of reinforcement

Fig. 6

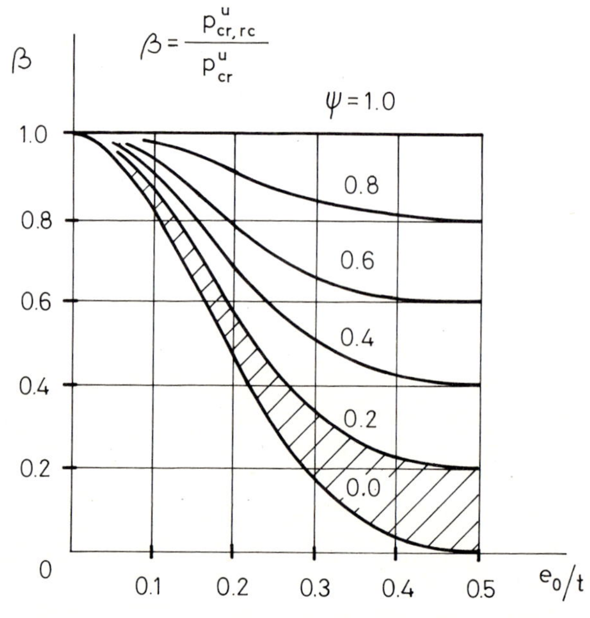

The factors β of the r.c. cylindrical shell as functions of the imperfection in the case of $\rho(0.5) = 0.25$ and $e_0 = w_0$

Fig. 7

Table I. ψ values.

$n\mu$	0	0.05	0.1	0.2	0.4	0.15	0.30	0.50
ψ_{01}	1.00	1.03	1.05	1.10	1.18	1.07	1.14	1.23
$\psi_{0,2}$	1.00	1.05	1.10	1.21	1.42	1.16	1.31	1.52
$\psi_{\infty,1}$	0	0.14	0.21	0.32	0.46	0.27	0.39	0.51
$\psi_{\infty,2}$	0	0.18	0.29	0.45	0.73	0.37	0.60	0.86

It can be seen that the rigidity of the double-layer mesh is greater than the rigidity of the single-layer mesh.

In shell structures the usual reinforcement is mostly weak, so that ψ_o only slightly exceeds unity. If we use the approximation $\psi_o \approx 1$, we can further simplify the procedure as follows (see in [10]). Introducing the ratio:

$$\beta = p^u_{cr,rc}/p^u_{cr} \tag{16}$$

we can determine the critical load of the cracked, elastic r.c. shell by the formula:

$$p^u_{cr,rc} = \beta\, p^u_{cr} = \beta\rho\, p^{lin}_{cr} \tag{17}$$

The numerical values of β for different values of e_o/t, e_o/w_o and ψ_∞ are compiled in Table II. The values corresponding to $\rho(0.5) = 0.25$ refer to shells which behave like the axially compressed cylinder. Those values have been plotted in Fig. 7. The domain between the values $\psi = 0$ and 0.2 corresponds to the cases of practice.

Looking at the horizontal lines for the values $\psi = 0.2$, it can be seen that the decrease of β is smaller at smaller values of ρ (0.5). The reason for this is that the decrease of the critical load is greater without cracks than in the other cases.

Multiplying the values of ρ and β for the axially compressed cylinder and for the value $\Psi = 0.2$, it can be seen that reinforced concrete shells are more sensitive to imperfections than homogeneous shells (Fig. 8).

Table II. Values of the factor β.

ρ(0.5)	ψ	e_o/w_o	$e_o/t=0$	0.1	0.2	0.3	0.4	≧0.5
1.00	1.0	0.5 1.0	1.0	1.0	1.0	1.0	1.0	1.0
	0.6	0.5 1.0	1.0	0.75 0.80	0.64 0.69	0.61 0.63	0.6	0.6
	0.4	0.5 1.0	1.0	0.62 0.71	0.46 0.53	0.41 0.44	0.4	0.4
	0.2	0.5 1.0	1.0	0.49 0.61	0.28 0.37	0.21 0.25	0.20 0.21	0.2
	0.0	0.5 1.0	1.0	0.37 0.51	0.10 0.22	0.02 0.06	0 0.01	0
0.75	1.0	0.5 1.0	1.0	1.0	1.0	1.0	1.0	1.0
	0.6	0.5 1.0	1.0	0.77 0.82	0.65 0.70	0.61 0.63	0.6	0.6
	0.4	0.5 1.0	1.0	0.65 0.73	0.48 0.55	0.41 0.45	0.40 0.41	0.4
	0.2	0.5 1.0	1.0	0.53 0.64	0.30 0.40	0.22 0.26	0.20 0.21	0.2
	0.0	0.5 1.0	1.0	0.42 0.55	0.13 0.24	0.02 0.08	0 0.01	0
0.50	1.0	0.5 1.0	1.0	1.0	1.0	1.0	1.0	1.0
	0.6	0.5 1.0	1.0	0.81 0.85	0.67 0.72	0.61 0.64	0.60 0.61	0.6
	0.4	0.5 1.0	1.0	0.71 0.77	0.51 0.58	0.42 0.46	0.40 0.41	0.4
	0.2	0.5 1.0	1.0	0.61 0.69	0.34 0.44	0.23 0.28	0.20 0.20	0.2
	0.0	0.5 1.0	1.0	0.51 0.61	0.18 0.30	0.04 0.10	0 0.01	0
0.25	1.0	0.5 1.0	1.0	1.0	1.0	1.0	1.0	1.0
	0.6	0.5 1.0	1.0	0.92 0.93	0.74 0.79	0.63 0.67	0.60 0.61	0.6
	0.4	0.5 1.0	1.0	0.88 0.89	0.60 0.69	0.44 0.51	0.40 0.42	0.4
	0.2	0.5 1.0	1.0	0.84 0.86	0.47 0.58	0.26 0.34	0.20 0.22	0.2
	0.0	0.5 1.0	1.0	0.81 0.82	0.34 0.48	0.07 0.18	0 0.03	0

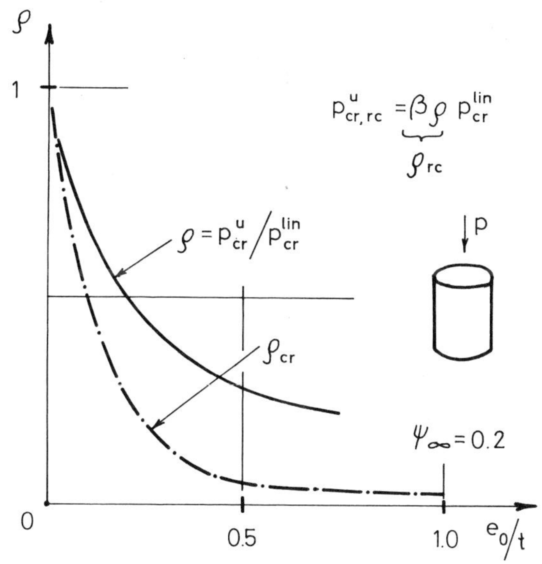

Decrease of ρ_{rc}

Fig. 8

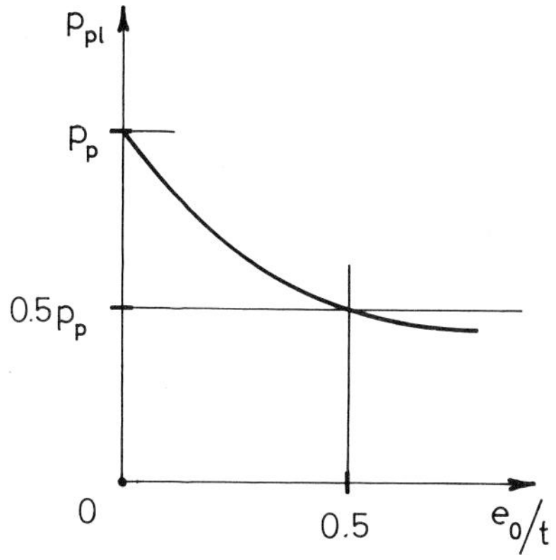

Variation of p_{pl} against e_0/t

Fig. 9

6. Effect of Plasticity

If during shell buckling, the deformations exceed the elastic limit of the r.c. shell and they become plastic, the critical load becomes less than that given by the elastic theory. Hence it is necessary to take the influence of plasticity into consideration.

More accurate investigations on r.c. shells made of approximately ideal elastic-plastic material [4], [7] showed that the critical load, which also considers plastic deformations, can be approximated within an error limit of ± 10% by the second-order Dunkerley formula:

$$\left[\frac{p_{cr}}{p_{pl}}\right]^2 + \left[\frac{p_{cr}}{p^u_{cr,rc}}\right]^2 = 1. \tag{18}$$

Here p_{cr} is the critical load which also considers plasticity, $p^u_{cr,rc}$ is the upper critical load in the case of linearly elastic material, and p_{pl} is the load which causes plastic failure assuming that the resultant force referred to in the cross-section acts with the initial eccentricity e_o, which may be calculated by the strength theory of r.c. Typical variations of p_{pl} can be seen in Fig. 9.

The factor ζ, including the effects of plastic deformation can be calculated by the formula

$$\zeta = \frac{p_{cr}}{p^u_{cr,rc}} = \sqrt{\frac{1}{1+\left[\frac{p^u_{cr,rc}}{p_{pl}}\right]^2}} \tag{19}$$

7. Critical Load for Cracked R.C. Shells of Plastic Behaviour

The plastic behaviour of cracked r.c. shells can be considered by the previous formula

$$p_{cr} = \zeta\beta p^u_{cr,rc}(\phi). \tag{2}$$

To obtain a better view, the surface of the critical load was depicted in Fig. 10.

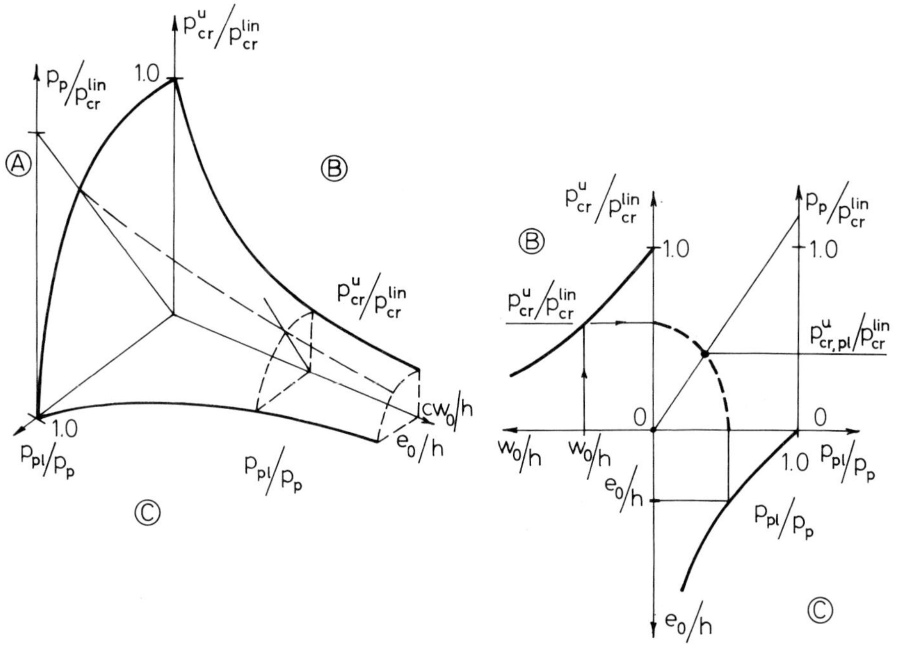

Axonometrical surface of p_{cr}^{u} of shells

Fig. 10

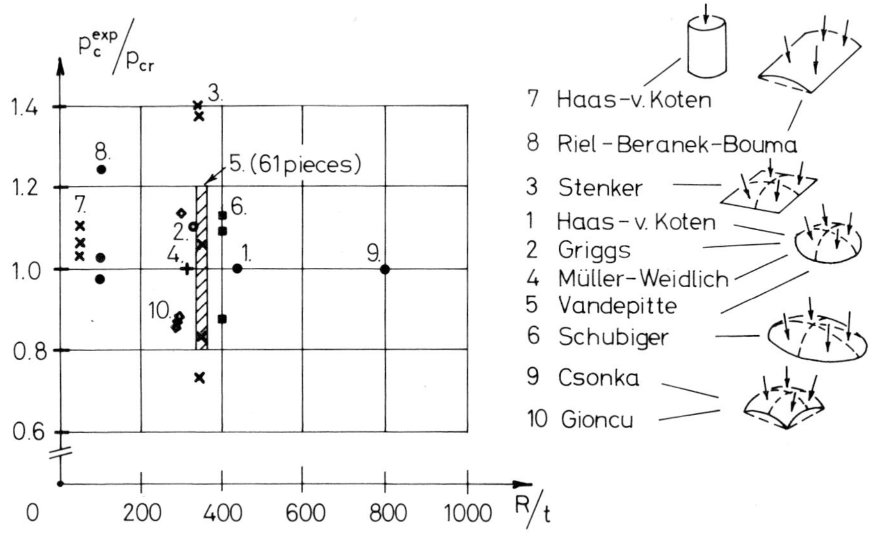

Comparison of test results with values calculated by the proposed method

Fig. 11

We computed the critical loads of all the experimental r.c. shells reported in the literature by means of Eq. (2) and compared them with the experimental values. The results are plotted against R/t in Fig. 11. Computing the probability of these results, we obtained as indicators the mean value 1.0 and the variance 11%. This shows that our procedure presented here is fairly realistic.

8. The Safety Factor [12]

According to the probability theory, the safety factor can be expressed as follows:

$$\gamma = \frac{\gamma_o}{1 - 2\,\nu_R} \; . \qquad (20)$$

Defining the safety according to the ACI Standard we obtain $\gamma_o = 1.16$. The resultant coefficient of variation can be computed from the formula:

$$\nu_R = \sqrt{\nu_{load}^2 + \nu_E^2 + \nu_\phi^2 + \nu_{cr}^2 + \nu_{pl}^2} \; . \qquad (21)$$

Here ν_{load} is the coefficient of variation of the average load; ν_E is that of the critical load of concrete or r.c. shells, taking the dispersion of the modulus of elasticity E beyond the limit of nominal strength also into account. Since the critical load is linearly proportional to E, ν_E is in fact, the coefficient of variation of E; ν_ϕ is the coefficient of variation of the critical load due to that of the creep factor ϕ_c of the concrete; ν_{cr} is the coefficient of variation of the critical load due to the change in the radius of curvature R caused by the coefficient of variation of the imperfection amplitude w_o; ν_{pl} is the coefficient of variation of the failure load due to that of the strength of the material.

We thus arrive at the values $\nu_{load} = 0.135$; $\nu_E = 0.1$; $\nu_\phi = 0.16$; $\nu_{cr} = 0.24$; $\nu_{pl} = 0.15$. We have to take $\nu_E = \nu_\phi = \nu_{cr} = 0$ for failure without buckling, and $\gamma_{el} = 3.5$ for elastic buckling.

In practice, we need a unique safety factor. This can be computed by introducing γ_{pl} and γ_{el} into formula (18), and taking into account the snap-through behavior of the shell through the factor $\rho(0.5)$. We thus obtain the unique safety factors shown in Table III.

Table III. The Values of the Unique Safety Factor.

$\rho(0.5)$	0.25	0.50	0.75	1.00	
$p_{pl}/p^{u}_{cr,rc=0}$	2.00	2.00	2.00	2.00	γ_{pl}
0,5	2.39	2.33	2.28	2.22	
1,0	2.87	2.75	2.63	2.51	
2,0	3.27	3.10	2.92	2.75	
∞	3.50	3.30	3.10	2.90	γ_{el}

We determined the critical loads of several erected large r.c. concrete domes and plotted them in Fig. 12, compared with their actual loads. On the basis of this comparison, we believe that if we compute the critical load with our method by assuming the proposed safety factors, which give an unique safety factor 2.9 at $R/_{t}$ = 500, we obtain a safety level corresponding to the practice followed until now.

9. Numerical Example

Let us perform the stability analysis of the domes of the Belgrade Exhibition Centre, consisting of a spherical surface with radius R = 56.2 m, covering a square flat area of 48 x 48 m. The shells are supported by vertical arches which rest on column rows, so that the shells cannot exert lateral thrust. The shells have a thickness of 90 mm, which increases to 200 mm in the corners. Due to lack of detailed data, we suppose that the shell wall has been thickened to 140 mm along the edges. They are made of concrete with a cube strength of 22.5 N/mm^2, reinforced by mild St37 steel (σ_{ult} = 370 N/mm^2). Ø 6/250 mm reinforcing meshes were applied, in the central zone as a single layer reinforcement, and along the edges in double-layer arrange-

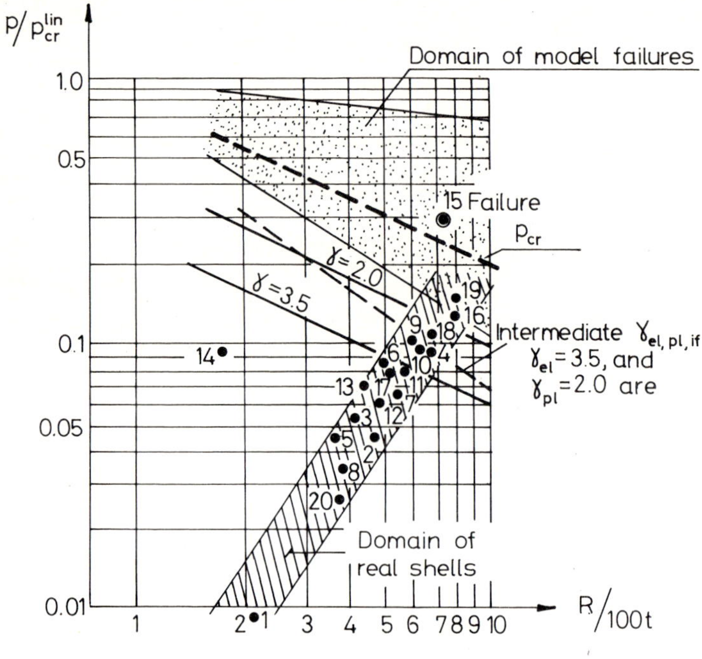

1. Jena, Germany
2. Jena, G.
3. Matsuyuma, Japan
4. Ingoviskosa Werke, G.
5. Hilling, USA
6. Hamburg, Germany
7. Windward, USA
8. Wales, GB
9. Albuquerque, USA
10. Belgrade, Yugoslavia
11. Belgrade, Yugoslavia
12. Algeciras, Spain
13. Novosibirsk, SU
14. Rome, Italy
15. Gödöllő, Hungary
16. Thessaloniki, Greece
17. Puerto Rico, USA
18. Cleadon, GB
19. Lyon, France
20. Massachusetts, USA

Safety factors of some erected r.c. domes

Fig. 12

ment near the surfaces of the shell wall. The area of reinforcement in one direction is, accordingly, 113 mm^2/m in the single-layer zone.

We perform the weight analyses with estimated data, assuming 90 mm thickness in the central part of the shell.

10 mm damp course	0.15 kN/m^2
90 mm cork heath isolation	0.14 kN/m^2
r.c. shell	2.16 kN/m^2
	2.46 kN/m^2

We further assume 0.8 kN/m^2 snow load. The cylinder strength of the concrete is:

$$f'_c = 0.8\ \sigma_{cube} = 0{,}8\ (22.5) = 18\ N/mm^2.$$

The modulus of elasticity is given by Eq. (9):

$$E_{c,o} = 6750\sqrt{18} = 28640\ N/mm^2.$$

The final value of the creep factor is determined from Eq. (10):

$$\phi_c = \frac{40}{10 + 18} = 1.43.$$

Three quarters of the load consist of dead load while the remaining 25% is the snow load occurring only at a later date for a not very long duration. Hence we set $k_t = 0{,}5$ for the snow load and, for the benefit of safety, we assume $k_o = 1$. Accordingly, we can compute the modulus of deformation from Eq. (8) with the value of the creep factor from Eq. (11):

$$E_c = \frac{28640}{1+ \frac{1 + 0.75(0.5)0.33}{1 + 0.33}\ 1.43} = 12970\ N/mm^2$$

The ratio of the moduli of deformation of steel and concrete is:

$$n = \frac{E_{st}}{E_c} = \frac{206000}{12970} = 15.9.$$

Due to the different stress states, the two kinds of reinforcement and the variable thickness, we have to investigate the stability of the shell at three places: in the middle of the shell, near the edges, and in the corners. We present here the stability analysis at the centre only.

Stability analyses at the centre of the shell

$$n\mu = 15.9 \frac{113}{9000} = 0.0199.$$

Interpolating from Fig. 6 we obtain:

$$\Psi_{0,1} = 1.000 + 0,030 \frac{0.0199}{0.050} = 1.01,$$

$$\Psi_{\infty,1} = 0.140 \frac{0.0199}{0.050} = 0.055.$$

In the middle part of the shell the bending moment is zero, hence the imperfection amplitude consists only of the accidental part: $w_{o,design} = w_{o,accid.}$. Assuming an average value, we have from Eq. (12):

$$w_{o,design} = 0.05(90) + \frac{56200}{2000}\left[\frac{1}{\frac{5620/9}{1000} + \frac{1000}{5620/9}}\right] = 17.1 \text{ mm}.$$

$$e_{o,design} = 0.67(17.1) = 11.46 \text{ mm}.$$

From Eq. (4) and (5) we obtain:

$$\rho = \frac{1}{1 + 6\frac{17.1}{90}} = 0.467.$$

The linear critical load of the sphere is:

$$p_{cr}^{lin} = 1.2\frac{Et^2}{R^2} = 1.2\frac{12970(90)^2}{(56200)^2} = 0.0395 \text{ N/mm}^2$$

$$= 39.5 \text{ kN/mm}^2,$$

Equation (7) yields:

$$p_{cr}^{u} = 0.467\ (39.5) = 18.5 \text{ kN/m}^2,$$

and Eq. (15) gives:

$$p_{cr,c}^{u} = 39.5\left[1\ \frac{2(11.46)}{90}\right]^{1.5(1+17,1/11,46)}$$

$$= 13.2 \text{ kN/m}^2.$$

Equation (14/a) yields:

$$p^{u}_{cr,rc} = \frac{1 + 1.01}{2} 13.2 + 0.055(18.5-13.2) = 13.6 \text{ kN/m}^2.$$

We compute the plastic failure load neglecting the reinforcement, assuming a 10 mm tolerance in the thickness and considering only that part of the concrete cross section on which the load is acting centrally

$$p_{pl} = \frac{2tf'_c}{R}\left[1 - \frac{2e_o}{t}\right] = \frac{2(80)18}{56200}\left[\frac{1-2(11.46)}{80}\right]$$

$$= 0.0366 \text{ N/mm}^2 = 36.6 \text{ kN/m}^2.$$

Equation (19) gives:

$$\zeta = \sqrt{\frac{1}{1+\left(\frac{13.6}{36.6}\right)^2}} = 0.94.$$

Hence, the upper critical load of the reinforced concrete shell becomes:

$$p_{cr} = 0.94\ (13.2) = 12.4 \text{ kN/m}^2.$$

The load perpendicular to the shell surface consists of the shell's own weight and of the snow load:

$$p_{actual} = 2.45 + 0.8 = 3.25 \text{ kN/m}^2.$$

The safety factor thus becomes:

$$\gamma = \frac{p_{cr}}{p_{actual}} = \frac{12.4}{3.25} = 3.8 > 3.4$$

which can be regarded as sufficient.

10. Summary.

In this paper the results of the nonlinear shell buckling theory have been generalized and applied to r.c. shells. The solution considers the special properties of r.c., i.e. cracks, plasticity, creep, and quantity of reinforcement. The proposed method adequately describes the results and the situation of erected domes.

The main results of our research are the following:
- the critical load of the r.c. shell is reduced by the plasticity, the creep, the imperfections and the cracks of the concrete;
- the critical load will be decreased by the permanent character of the load and the imperfection of the shell;
- the critical load of the r.c. shell will be increased by increasing the quantity of reinforcement and by arranging it in two layers.

References

1. Dulácska, E.: Stabilitätsuntersuchungen der Schalenkonstruktionen, Proc. Large-span Shells IASS Congr. Leningrad (1966). Tsinis, Moscow (1968).
2. Dulácska, E.: Vibration and Stability of Anisotropic Shallow Shells. Acta Techn. Acad. Sci. Hung. 65 (1969) 225-260.
3. Dulácska, E.: On the Critical Load of Shells. IASS Bulletin No. 39 (Sept. 1969) 31-36.
4. Dulácska, E.: Praktische Stabilitätsuntersuchung von zentrisch gedrückten Tragwerken aus Material mit veränderlichem Elastizitätsmodul., Bautechnik 49 (1972) 340-345.
5. Dulácska, E.: Der Steifigkeitskennwert der Schalenbeulung, Acta Techn. Acad. Sci. Hung. 87 (1978) 457-467.
6. Dulácska, E.: Die Beulung von Stahlbetonschalen, Acta Techn. Acad. Sci. Hung. 86 (1978) 93-115.
7. Dulácska, E.: Buckling of Elastic-Plastic Shells, IASS-Bulletin No. 68 (Dec. 1978) 15-20.
8. Dulácska, E.: The Influence of Creep on the Stability of Shells, Wood, Ferrocement and Plastics in Shell and Spatial Structures. Proc. IASS Symposium. University of Oulu, Finland (1980) June 87-89.
9. Dulácska, E.; Buckling of Reinforced Concrete Shells. Journ. Struct. Divis. (Proc. ASCE(107 ST. 12 (1981) 2381-2401.
10. Dulácska, E.: Explanation of the Chapter on Stability of the "Recommendations for Reinforced Concrete Shells and Folded Plates", and a Proposal to its Improvement, IASS

Bulletin (1982) 77.

11. Kollár, L.,Dulácska, E.: Buckling of Shells for Engineers. John Wiley and Sons, Chichester, New York, Brisbane, Toronto Singapore (1984).

12. Dulácska, E.: The Safety Factor to be Applied in Shell Buckling Analysis (in Hungarian), Épités - Épitészettudomány 3-4 (1984) 351-366.

13. Trost, H.: Auswirkungen des Superpositionsprinzips auf Kriech - und Relaxationsprobleme bei Beton und Spannbeton, Beton u. Stahlbetonbau 62 (1967) 230-238, 261-269.

14. Zerna, W.: Spannungs - Dehnungs - Beziehung für Beton bei einachsiger Beanspruchung.In: Aus. Theorie und Praxis des Stahlbetonbaues. W. Ernst u. Sohn. Berlin (1969).

Dynamic Buckling of a Rigid-Plastic Cylindrical Shell: A Second Order Differential Equation Subject to Four Boundary Conditions

KAREN ZAK BENBURY
Dept Math, U.S. Naval Academy, Annapolis, MD 21402

M.A. VELUSWAMI
Dept ME, Indian Inst. of Tech., Madras, INDIA, 600036

G. HORVAY
Dept MAE, North Carolina State University, Raleigh, NC 27695-7910

Summary

In the case of a Loss-Of-Coolant Accident an impulsive pressure acts on the core-support-barrel within the nuclear reactor, and the danger arises that the out-of-round shell will buckle. The dimensionless equation governing the growth $g(t)$ of initial imperfection $g(\Omega\tau_m;n) = 1$ is

$$E\{g(t;n)\} \equiv g'' - pg'\operatorname{cosec} t - qg - r = 0$$

as derived in Horvay-Veluswami [7]. Here g is amplification, τ dimensional time, τ_m time to failure, Ω natural frequency of membrane oscillation. $t = \Omega(\tau_m - \tau)$ is reversed time, i.e., $\Omega\tau_m$ represents the instant of initiating the impulse, $t = 0$ the time of collapse. $p(n)$, $q(n)$, $r(n)$ are functions of material properties, geometry (i.e., dimensions of the shell), and the buckling mode n; r in addition depends also on the loading condition. The mathematical problem: to so solve $E = 0$ that the boundary conditions be satisfied: (a) $g(\Omega\tau_m) = 1$ (normalization condition), (b) $g'(\Omega\tau_m) = 0$ (zero initial velocity), (c) $g'(0) = 0$ (zero final velocity), (d) $g(0)$ = peak value (solve the $E = 0$ problem vs n and then select the value n_{crit} for which g reaches peak value). Note that conditions (c) and (d) are redundant, because when peak value is reached, the velocity is zero. Three methods of solution are considered: I. Power series expansion in t (in conjunction with Padé approximants), exceedingly rapid convergence is observed; II. Cosine series solution, and its pair obtained by the variation of parameters method; III. Pseudo-Hell analysis . But only I is carried to completion, including a numerical confirmation that for the aluminum shell of Vaughan-Florence [4], the results coincide with those of Horvay-Veluswami-Stockton [6], based on time-consuming (and eo ipso, expensive) numerical forward integration, which also requires a passage to ∞ in the number N of integration steps. Because of the excellence of I, the traditional method II is stopped short of numerical calculations. While III constitutes a fascinating counterpart of the Hill-Mathieu theory, it would primarily reveal behavior of the solution of $E = 0$ at large negative values of t, whereas the physical problem terminates at $t = 0$; for this reason no effort was made to carry out the analysis III, only its principal features are outlined.

---

* This paper represents joint work of the authors while they were associated with the University of Massachusetts.

Inelastic Behaviour of Plates and Shells
IUTAM Symposium Rio de Janeiro 1985
Editors: L. Bevilacqua, R. Feijóo and R. Valid

Introduction

Dynamic buckling of rigid-strainhardening plastic cylindrical shells, as initiated by an inward impulse, was first studied by Abrahamson-Goodier [1]. Their work was continued by Anderson--Lindberg [2], Florence-Vaughan [3], Vaughan-Florence [4] (to be referred to hereinafter as VF), and others; for literature survey see Jones [5I, 5II]. Using VF as point of departure, Horvay-Veluswami-Stockton [6, call it HVS], and Horvay-Veluswami [7, call it HV] reconsidered the VF analysis. An inward membrane motion of velocity $\dot{u}_o$ is governed by the equation ($\dot{u} = du/d\tau$)

$$M \equiv d^2u_o/d\tau^2 + \Omega^2 u_o + 1/\tau_f = 0 \tag{1}$$

where the dimensionless quantities

$$u_o = w_o/a, \ \tau = V_o \underset{\sim}{t}/a, \ \Omega = [2(2-k)E_t/3\rho V_o^2]^{1/2} \tag{2}$$

are inward displacement (w_o = dimensional displacement, a = = shell radius), V_o = initial velocity due to impulse, τ = time ($\underset{\sim}{t}$ = dimensional time), Ω = natural frequency of oscillating membrane motion, ρ = mass density, E_t = strain-hardening coefficient, while

$$k = - \dot{\varepsilon}_x/\dot{\varepsilon}_\theta \tag{3a}$$

is the ratio of axial strain rate to hoop strain rate, having, seen VF, an experimental value (L = length of shell)

$$k \cong \frac{1}{2} e^{-L/4a} \tag{3b}$$

Finally,

$$\tau_f = \rho V_o^2 K_2/(2-k)\sigma \tag{3c}$$

is time to maximum amplitude of u_o (i.e., failure) when the Ω^2 term in (1) is neglected. The notation

$$K_1 = 2(1-k+k^2), \ K_2 = (3K_1/2)^{1/2}, \ K_3 = (2-k)/K_1 \tag{4a}$$

$$\sigma = \text{initial yield stress} \tag{4b}$$

is used. One finds that

$$u_o = \frac{\sin \Omega\tau}{\Omega} - \frac{1 - \cos\Omega\tau}{\Omega^2 \tau_f} \tag{5}$$

is the ensuing membrane displacement. (All this may be found in VF). When the Ω^2 term of (1) is not neglected, then

$$\tau_m = \Omega^{-1} \arctan \Omega\tau_f \tag{6}$$

is the time to maximum deflection; this reduces to τ_f when the series expansion of the arctan is terminated after the first term, on assumption of small $\Omega\tau_f$.

The shell is assumed to have initial imperfections (u is dimensionless bending deflection, bar denotes initial value)

$$u(0) = \sum_{o}^{\infty} \underset{\sim}{A}(0,n)\cos n\theta = \bar{u} = \bar{w}/a = \sum_{o}^{\infty} a_n \cos n\theta \tag{7}$$

(We assume no axial dependence of initial velocity, or of imperfections, or of loading; for consideration of the latter case see Veluswami-Horvay [8].) It follows that the initial out-of-roundness a_n induces a "bending response"

$$\underset{\sim}{A}_n(\tau) = a_n A(\tau;n) \tag{8}$$

which is governed by the equation HV(11);

$$E\{(\underset{\sim}{A}_n(\tau)\} \equiv \frac{d^2\underset{\sim}{A}_n}{d\tau^2} + n^4 e_5 \frac{d\underset{\sim}{A}_n/d\tau}{du_o/d\tau} - (\frac{n^2}{\tau_f} - n^4 e_6)\underset{\sim}{A}_n - n^4 e_6 a_n = 0 \tag{9}$$

where h is wall thickness, and

$$e_5 = \left\{\frac{k}{2(2-k)}\right\}^2 \frac{K_3 h^2}{a^2 \tau_f}, \quad e_6 = \frac{E_t}{\sigma} \frac{K_2 K_3}{18} \frac{h^2}{a^2 \tau_f} \tag{10}$$

It was shown in HV that equation (9) for the "amplification" A(n) may be rewritten in the form

$$E\{A(T;n)\} \equiv d^2A/dT^2 - P(dA/dT)\operatorname{cosec}\Omega\tau_m T - QA - R = 0 \tag{11}$$

where

$$T = 1 - \tau/\tau_m \tag{12}$$

and

$$t = \Omega\tau_m T \tag{13}$$

are reversed time referred to the maximum membrane deflection instant as reference zero. The coefficients in (11) have the expressions

$$P = \Omega\tau_f\tau_m e_5 n^4 (1 + \Omega^2\tau_f^2)^{-1/2}, \; Q = \tau_m^2 n^4 (n^{-2}\tau_f^{-1} - e_6), \; R = \tau_m^2 e_6 n^4 \tag{14}$$

Correspondingly,

$$T = 1 \tag{15a}$$

represents the initiation time for inward motion u_o,

$$T = 0 \tag{15b}$$

represents the instant of maximum membrane deflection (where $du_o/dT) = 0$). We cannot progress to

$$T = -1 \tag{15c}$$

which would represent the instant of resumption of zero membrane deflection if the original plastic loading line (and not the elastic line which governs) were followed also in unloading.

The equation $E = 0$ is subject to the boundary conditions

$$A(1) = 1, \quad \frac{d}{dT} A(1) = 0, \quad \frac{d}{dT} A(0) = 0 \tag{16a,b,c}$$

The critical n is to be so determined that all three boundary conditions be satisfied, and that at the same time

$$A(0) = \text{peak value} \tag{16d}$$

However one notes from (11) that all solutions bounded at $T = 0$, i.e., at $\tau = \tau_m$, obey (16c). For, if $dA(0)/dT = 0$ were violated, the second term of E would become unbounded, causing also the unboundedness of $A(0)$ (or of $d^2A(0)/dT^2$). So condition (16c) is omitted, and only (16a,b) need to be specified. It will be seen furthermore in Section 3 that (16d) is the boundary condition which determines the value of n_{crit}.

Remember that the equation (1) came about, see VF, by equating the resultant membrane hoop force to the inertia force

$$N_\theta = h\sigma_{\theta,av} = aqhd^2w_o/d\underset{\sim}{t}^2 \tag{17}$$

and noting that, in accordance with VF (17), (16)

$$\sigma_{\theta,av} = (k-2)\,\sigma\left[1 + \frac{2}{3}K_3\,u_o\,e_t/\sigma\right] \tag{18}$$

(The average is taken over the shell thickness.) Note that σ is the initial yield stress. However, in the light of VF's decision to neglect the Ω^2 term in (1), their recommendation was that mean yield stress value over the loading range be used for σ. (In HVS and HV, however, the mean yield stress value was assigned, in error, to σ in spite of the fact that the Ω^2 term was retained. In this sense the σ values used for the aluminum shell and the steel shell in HVS, HV and [8] are excessive. But the present paper will retain, for convenience of comparison, the previously used σ values.)

In the present paper our interest concentrates on establishing more compact methods for solving for $A(n)$ from (11), (16) than was developed in HVS and HV for solution from (9), and concurrently, determining the critical mode number n_{crit} at which the amplification $A(n)$ becomes most severe.

Method I. Power Series Solution of the Governing Equation; Padé

For convenience we rewrite (11), (16) in terms of the backward variable

$$t = \Omega\tau_m T = \Omega(\tau_m - \tau) \tag{19}$$

Replacing A by g, and using prime to denote t differentiation, we obtain

$$E\{g(t;n)\} \equiv g''\sin t - pg' - qg\sin t - r\sin t = 0 \qquad (20a)$$

$$p = P/\Omega\tau_m, \quad q = Q/\Omega^2\tau_m^2, \quad r = R/\Omega^2\tau_m^2 \qquad (20b)$$

(Note that the symbols p, q, r, which are used in HVS, serve a different purpose: they denote $n^{-4}P$, $n^{-4}Q$, $n^{-4}R$.)

Inasmuch as for the aluminum shell of HVS for which we shall illustrate our results (see (33))

$$\Omega\tau_m = 0.86859, \quad p = 0.88355 \qquad (21)$$

while for the steel shell of HV it is

$$\Omega\tau_m = 0.47969, \quad p = 2.2352 \qquad (22)$$

we are prompted to formulate the problem in terms of the backward variable t, so that $(\text{-forward time})^{1+p}$ in (24) below make sense. (One may safely say that p is never an integer. If it were, then the t^{1+p} factor of (24) would be replaced by a log t factor, as in the Bessel equation.) Accordingly, we specify, with respect to t, the boundary conditions

$$g(\Omega\tau_m) = 1, \quad g'(\Omega\tau_m) = 0. \quad g'(0) = 0 \qquad (23a,b,c)$$

$$g(0,n_{crit}) = \text{peak value} \qquad (23d)$$

The solution to (21) may be written in the form

$$g(t) = -r/q + AG_1(t) + Bt^{1+p}\,G_2(t) \qquad (24)$$

In the above $-r/q$ is the particular integral, and $G_1, t^{1+p}\,G_2$ represent the complementary function; A and B are coefficients to be determined, for assumed n, from the boundary conditions (23a,b).

We write the complementary function as

$$G = t^{\alpha} \sum_{o}^{\infty} c_k t^{2k} \tag{25}$$

and find, as stated in (24), that

$$\alpha = 0,\ 1 + p \tag{26}$$

are the roots of the indicial equation. The first solution we label G_1, the second $t^{1+p}\ G_2$. For the coefficients in G_1 we retain the notation c_k, while the coefficients in G_2 we label d_k.

In this fashion one obtains, on choosing

$$c_o = 1, \quad d_o = 1 \qquad \text{(27a)(26a)}$$

the expressions

$$k \geqq 1:$$

$$c_k = \frac{1}{2k(2k-1)} \sum_{j=0}^{k-1} (-1)^j c_{k-j-1} \left[\frac{q}{(2j+1)!} - \frac{(2k-2-2j)(2k-3-2j)}{(2j+3)!}\right] \tag{27b}$$

$$d_k = \frac{1}{2k(2k+1+p)} \sum_{j=0}^{k=1} (-1)^j d_{k-j-1} \left[\frac{q}{(2j+1)!} - \frac{(2k+p-1-2j)(2k+p-2-2j)}{(2j+3)!}\right] \tag{28b}$$

The Boundary Value Problem

The expressions (24), (25), (26) bear out our earlier statement that (23c) is automatically satisfied by all solutions of (20).

We satisfy (23a,b) by solving simultaneously for A(n) and B(n). Denoting

$$g'(t) = 2c_1 t G_3(t) A + (1+p) t^p\ G_4(t) B \tag{29}$$

where

$$G_3 = 1 + 2\frac{c_2}{c_1}t^2 + 3\frac{c_3}{c_1}t^4 + 4\frac{c_4}{c_1}t^6 + \dots ,$$

$$G_4 = 1 + \frac{p+3}{p+1}d_1t^2 + \frac{p+5}{p+1}d_2t^4 + \frac{p+7}{p+1}d_3t^6 + \dots \qquad (30)$$

we determine A(n), B(n) for a range of n values. Finally we seek out the n_{crit} value for which the curve $g(\Omega\tau_m;n)$ reaches a maximum.

Recognizing that $\Omega\tau_m$, in practice, is a relatively small quantity ($\lesssim 1$, see (21)), the aim is to replace the numerical forward integration of (9), as carried out in HVS and HV for a variety of n followed by drawing the A_{max} vs n curve and thereby finding the pair n_{crit}, A_{peak}, by use of the explicit solution (24). This is carried out most conveniently by the Padé approximant scheme. We shall determine A(n), B(n) and $g(\Omega\tau_m;n)$, in accordance with the Appendix, by

$$\text{Padé } [N/N] = [1/1], [2/2], [3/3] \qquad (31)$$

The Aluminum Test Shell of VF

The geometrical parameters and the material properties, see HVS (21a), are as follows:

$$a = 0.037338\text{m} = 1.47 \text{ in}, \; h = 0.001651\text{m} = 0.065 \text{ in}, \; L = = 0.076293\text{m} = 3.0037 \text{ in}$$

$$\rho = 2706\text{kg/m}^3 = 0.0^3253 \text{ lbsec}^2/\text{in}^4, \; \bar{\sigma} = 289.58\text{MPa} = = 42000 \text{ psi}, \; E_t = 2275\text{MPa} = 330000 \text{ psi} \qquad (32)$$

One obtains the derived parameters, see HVS (23), HV (20a);

$$k = 0.3, \; K_1 = 1.58, \; K_2 = 1.5395, \; K_3 = 1.07595$$

$$e_5 = 0.0^494541, \; e_6 = 0.0^281602, \; \Omega = 6.8226,$$

$$\tau_f = 0.17324, \; \tau_m = 0.12731, \; \Omega\tau_m = 0.86859, \; \underset{\sim}{t}(\text{sec}) = 0.0^32609\tau \qquad (33)$$

while for

$$n = 17 \tag{34a}$$

the coefficients of the governing equation acquire the values

$$P = 0.76744, \quad Q = 15.9919, \quad R = 11.047$$
$$p = 0.88355, \quad q = 21.197, \quad r = 14.642 \tag{34b}$$

In calculating the function values, as suggested by the results of HVS, e.g., for n = 17:

$$\begin{aligned}
G_1 = 1 &+ 91.010623t^2 + 231.037798t^4 + 203.914902t^6 \\
&+ 92.346862t^8 + 25.737342t^{10} + 4.930321t^{12} \\
&+ 0.702892t^{14} + 0.079656t^{16} + 0.0^2 7641t^{18} \\
&+ 0.03661t^{20} + \ldots
\end{aligned}$$

$$\begin{aligned}
G_2 = 1 &+ 2.764737t^2 + 2.558682t^4 + 1.197433t^6 \\
&+ 0.342110t^8 + 0.066883t^{10} + 0.0^2 9673t^{12} \\
&+ 0.0^2 1109t^{14} + 0.0^3 107t^{16} + 0.0^5 931t^{18} \\
&+ 0.0^6 772t^{20} + \ldots
\end{aligned}$$

$$\begin{aligned}
G_3 = 1 &+ 5.077161t^2 + 6.721685t^4 + 4.058729t^6 \\
&+ 1.413975t^8 + 0.325038t^{10} + 0.054058t^{12} \\
&+ 0.0^2 7002t^{14} + 0.0^3 755t^{16} + 0.0^4 726t^{18} \\
&+ 0.0^5 662t^{20} + \ldots
\end{aligned}$$

$$\begin{aligned}
G_4 = 1 &+ 5.700405t^2 + 7.992430t^4 + 5.011826t^6 \\
&+ 1.795153t^8 + 0.412596t^{10} + 0.071299t^{12} \\
&+ 0.0^2 9353t^{14} + 0.0^2 1018t^{16} + 0.0^4 983t^{18} \\
&+ 0.0^5 897t^{20} + \ldots
\end{aligned} \tag{35}$$

we Padé G_k individually, e.g.,

$$\begin{aligned}
G_1 = 1 + a_1 t^2 + \ldots + a_{K-1} t^{2(K-1)} + a_K t^{2K} [(1 + N_1 t^2 + \ldots \\
+ N_N t^{2N}) / (1 + D_1 t^2 + \ldots + D_N t^{2N})]
\end{aligned} \tag{36}$$

On using K = 0,1 we obtained unreasonable results; this must probably be attributed to the fact that the G_k commence as series with ascending (positive) coefficients, and to effectively use Padé it is desirable to have a descending series on hand. [N/N] = [1/1] gave unsatisfactory results; evidently for this problem Padé must be carried at least to [2/2], to give satisfactory results. [4/4] was also tried, using explicit formulas like (A4). However the results were unsatisfactory. This must be ascribed to round-off errors. An exploration of double precision calculations for other cases suggests that this may perhaps eliminate the difficulty. But memory storage requirements were found to be excessive, and so the effort to carry out a double precision [4/4] calculation was abandoned. It may very well be, as suggested by [13], [14] that using the Baker recurrence formulas [15, p. 77] the difficulty with round-off errors may be eliminated or at least mitigated. Because of time requirements this avenue was not pursued.

Table 1 below shows some numerical results for n = 17, and K.N = 2,2;...;5,3. (If need be one could go to higher N and then extrapolate to N = ∞ in the fashion of HV Fig. 8b. But it is found that there is no need for this.)

TABLE 1. Parameter Values for n = 17

K,N	2,2	2,3	3,2	3,3	5,2	5,3
A	103.445	20.217	20.975	20.261	20.263	20.262
B	-8476.443	-1656.565	-1718.683	-1660.233	-1660.400	-1660.327
g(0)	102.754	19.526	20.284	19.571	19.573	19.572

Making similar calculations for n = 10 through 24 one obtains the curves of Figs. 1,2. The infinities in the B(n) curve indicate that the determinant of the system for A and B vanishes at n = 17.5 and 23.1.

One thus finds for the aluminum test shell of VF at τ_m

$$\text{ZVH: } n_{crit} = 17.0: \quad A = 20.262, \quad B = -1660.327, \quad g_{peak} = 19.572 \tag{37}$$

This compares with the more time consuming and (because of truncation and round-off errors in the 4th order Runge-Kutta scheme) probably less accurate results of HVS:

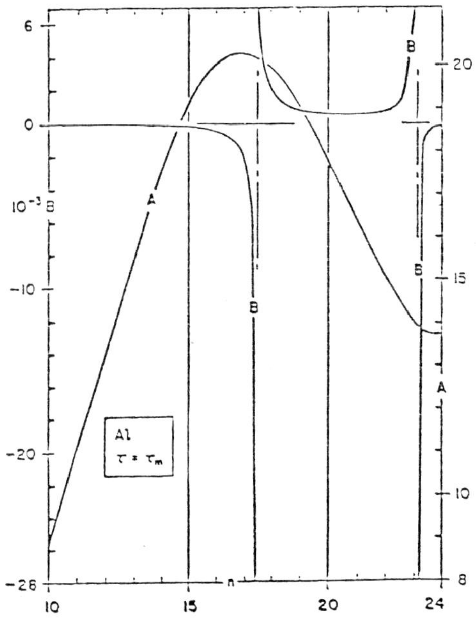

Figure 1

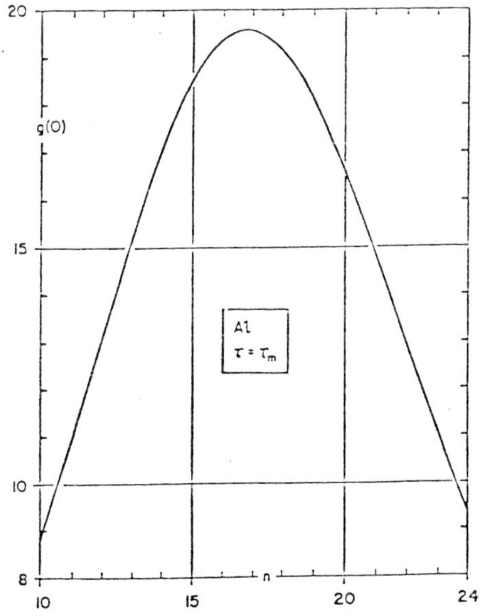

Figure 2

$$\text{HVS}: n_{crit} = 17.0: \quad g_{peak} = 19.589 \tag{38}$$

Method II. The Cosser Solution

Substitute

$$C = \sum_{o}^{\infty} A_m \cos mt \tag{39}$$

into the homogenized (20a),

$$E_h = g''\sin t - pg' - qg\sin t = 0 \tag{40}$$

One finds

$$A_o = 0, \quad A_1 = 1 \text{ (normalization)} \tag{41a,b}$$

The other coefficients of the cosine series ("cosser") solution are obtained from the recurrence formula

$$\{(M+1)^2 + q\}A_{M+1} + 2pMA_M - \{(M-1)^2 + q\}A_{M-1} = 0 \tag{42}$$

A series

$$S = \sum_{1}^{\infty} B_m \sin mt \tag{43}$$

cannot be a Second solution, for one finds from the recurrence formula that

$$B_1 = B_2 = \ldots = 0 \tag{44}$$

In the usual way S is determined by substituting

$$S = k(t)\, C(t) \tag{45}$$

into (40), which then reduces to

$$k'(2C'\sin t + C) + k''\, C = 0 \tag{46}$$

Solving for k one arrives at

$$g = c_o C + c_1 S,\ S(t) = C(t) \int_o^t dt_1 \exp\left\{-\int_o^{t_1} \frac{dt_2}{1+2\sin t_2 C'(t_2)/C(t_2)}\right\} \tag{47}$$

containing two arbitrary constants, c_o, c_1. For small t the power series solution of Section 2 appears most useful.

Method III. Pseudo-Hill Solution

We write in accordance with the familiar theory of differential equations with periodic coefficients, see Whittaker and Watson [9], Stoker [10], Strutt [11] (the yield stress symbol σ won't be used hereafter; henceforth iσ will denote the characteristic exponent),

$$g = e^{i\sigma t} \sum_{-\infty}^{\infty} A_m e^{imt} \tag{48}$$

and find, relabeling M = m-1, m, and m+1, respectively, that

$$2iE_h e^{-i\sigma t} \equiv \sum_{-\infty}^{\infty} e^{iMt} \ [\{\sigma+M+1)^2 + q\}A_{M+1} + 2p(\sigma+M)A_M$$

$$-\{(\sigma+M+1)^2 + q\}A_{M-1}] = 0 \tag{49a}$$

or written out in detail for the 5 central equations

M =

$$\begin{array}{rl}
-2: & -\{(\sigma-3)^2+q\}A_{-3}+2p(\sigma-2)A_{-2}+\{(\sigma-1)^2+q\}A_{-1} = 0 \\
-1: & -\{(\sigma-2)^2+q\}A_{-2}+2p(\sigma-1)A_{-1}+(\sigma^2+q)A_o = 0 \\
0: & -\{(\sigma-1)^2+q\}A_{-1}+2p\sigma A_o+\{(\sigma+1)^2+q\}A_1 = 0 \\
1: & -\{\sigma^2+q\}A_o+2p(\sigma+1)A_1+\{(\sigma+2)^2+q\}A_2 = 0 \\
2: & -\{(\sigma+1)^2+q\}A_1+p(\sigma+2)A_2+\{(\sigma+3)^2+q)\}A_3 = 0
\end{array} \tag{49a}$$

The value σ is determined from the requirement that the infinite system be consistent:

$$\begin{array}{cc|ccccccc|c}
M & & -3 & -2 & -1 & 0 & 1 & 2 & 3 & \\
-2 & & -\{(\sigma-3)^2+q\} & 2p(\sigma-2) & \{(\sigma-1)^2+q\} & 0 & 0 & 0 & 0 & \\
-1 & & 0 & -\{(\sigma-2)^2+q\} & 2p(\sigma-1) & \{\sigma^2+q\} & 0 & 0 & 0 & \\
0 & \Delta \equiv & 0 & 0 & -\{(\sigma-1)^2+q\} & 2p\sigma & \{(\sigma+1)^2+q & 0 & 0 & = 0 \\
1 & & 0 & 0 & 0 & -\{\sigma^2+q\} & 2p(\sigma+1) & \{(\sigma+2)^2+q\} & 0 & \\
2 & & 0 & 0 & 0 & 0 & -\{(\sigma+1)^2+q\} & 2p(\sigma+2) & \{(\sigma+3)^2+q\} & (50)
\end{array}$$

Several observations about the roots σ are immediate. (a) If $\sigma = s$ is a root, then so is $s_{\text{conjugate}}$. This, in the familiar manner, is an immediate consequence of the fact that the coefficients in (50) are real. (b) If $\sigma = s$ is a root, then so is $\sigma = s \pm k$, k = integer. This follows from the fact that by relabeling σ in (51) as $s + k$, the infinite determinant does not change; in particular, since $s = 0$ is a solution, hence so is k = any integer. (c) If $\sigma = s$ is a root, so is $-s$. Proof: relabel.

On expanding by the zeroth column we convert (50) into the more convenient from

$$
\begin{aligned}
\Delta^* \equiv 2p\sigma
&\begin{vmatrix}
2p(\sigma-1) & -\{(\sigma-2)^2+q\} & 0 & 0 & \cdots \\
\{(\sigma-1)^2+q\} & 2p(\sigma-2) & -\{(\sigma-3)^2+q\} & 0 & \cdots \\
0 & \{(\sigma-2)^2+q\} & 2p(\sigma-3) & -\{(\sigma-4)^2+q\} & \cdots \\
\vdots & \vdots & \vdots & \vdots & \vdots
\end{vmatrix} \times \\
\times &\begin{vmatrix}
2p(\sigma+1) & \{(\sigma+2)^2+q\} & 0 & 0 & \cdots \\
-\{(\sigma+1)^2+q\} & 2p(\sigma+2) & \{(\sigma+3)^2+q\} & 0 & \cdots \\
0 & -\{(\sigma+2)^2+q\} & 2p(\sigma+3) & \{(\sigma+4)^2+q\} & \cdots \\
\vdots & \vdots & \vdots & \vdots & \vdots
\end{vmatrix} + \\
+ (\sigma^2+q) \Bigg\{ &\begin{vmatrix}
2p(\sigma-1) & -\{(\sigma+2)^2+q\} & 0 & 0 & \cdots \\
\{(\sigma+1)^2+q\} & 2p(\sigma+2) & -\{(\sigma-3)^2+q\} & 0 & \cdots \\
0 & \{(\sigma-2)^2+q\} & 2p(\sigma-3) & -\{(\sigma-4)^2+q\} & \cdots \\
\vdots & \vdots & \vdots & \vdots & \vdots
\end{vmatrix} \times \\
\times &\begin{vmatrix}
\{(\sigma+1)^2+q\} & 0 & 0 & 0 & \cdots \\
-\{(\sigma+1)^2+q\} & 2p(\sigma+2) & \{(\sigma+3)^2+q\} & 0 & \cdots \\
0 & -\{(\sigma+2)^2+q\} & 2p(\sigma+3) & \{(\sigma+4)^2+q\} & \cdots \\
\vdots & \vdots & \vdots & \vdots & \vdots
\end{vmatrix} - \\
- &\begin{vmatrix}
2p(\sigma+1) & \{(\sigma+2)^2+q\} & 0 & 0 & \cdots \\
-\{(\sigma+1)^2+q\} & 2p(\sigma+2) & \{(\sigma+3)^2+q\} & 0 & \cdots \\
0 & -\{(\sigma+2)^2+q\} & 2p(\sigma+3) & \{(\sigma+4)^2+q\} & \cdots \\
\vdots & \vdots & \vdots & \vdots & \vdots
\end{vmatrix} \times \\
\times &\begin{vmatrix}
-\{(\sigma-1)^2+q\} & 0 & 0 & 0 \\
\{(\sigma-1)^2+q\} & 2p(\sigma-2) & -\{(\sigma-3)^2+q\} & 0 \\
0 & \{(\sigma-2)^2+q\} & 2p(\sigma-3) & -\{(\sigma-4)^2+q\} \\
\vdots & \vdots & \vdots & \vdots
\end{vmatrix} \Bigg\} = 0
\end{aligned}
\tag{51}
$$

The mathematical interest in equation (40) stems principally from the fact that it appears as a counterpart to the Mathieu differential equation. The Floquet theory, e.g., [9, p.412], applies unchanged. The unnatural feature is that the solution pair

$$G_I(0) = 1, \quad G_I'(0) = 0 \tag{52a}$$

$$G_{II}(0) = 0, \quad G_{II}'(0) = 1 \tag{52b}$$

constituting the fundamental pair in the Mathieu theory, e.g., [12, p.283], must be modified, inasmuch as $G'(0) \equiv 0$; hence condition (52b) cannot be prescribed.

In the Hill equation there is no first derivative term and the periodic terms all multiply only $g(t)$ in (42). Most importantly, the structure of the determinant is such that none of the diagonal terms vanish, so that all equations may be divided by the coefficient of the diagonal term, thus creating in the determinant all 1 in the main diagonal, while the off--diagonal terms become, upon division by the main diagonal term suitably small so that convergence of the infinite determinant is assured, and an explicit formula may be obtained for the roots σ. However, in our (50) we cannot reduce the diagonal terms to 1, because the divisors may take on the value 0.

The key problem becomes to establish the counterpart of the familiar explicit solution of the Hill determinant $\Delta = 0$. A further problem is to expand the solutions of the boundary value problem in terms of the pseudo-Hill functions.

While some progress has been made in exploring the solutions of $\Delta = 0$ by successive approximations, presentation of these partial results will be postponed to a later occasion, when a complete theory of $\Delta = 0$ will be available.

References

1. Abrahamson, G.R.; Goodier, J.N. "Dynamic plastic flow buckling of a cylindrical shell from uniform radial impulse," Proc. 4th U.S. Nat'l. Cong. Applied Mech. 2 (1962) 939.

2. Anderson, D.L.; Lindberg, H.E. "Dynamic pulse buckling of cylindrical shells under transient lateral pressures," AIAA J. 6 (1968) 569.

3. Florence, A.L.; Vaughan, H. "Dynamic plastic buckling of short cylindrical shells due to impulsive loading," Internat'l J. Solids Structures 4 (1968) 741.

4. Vaughan, H.; Florence, A.L. "Plastic flow buckling of cylindrical shells due to impulsive loading," J. Appl. Mech. 92 (1970) 171.

5. Jones, N. "A literature review of the dynamic plastic response of structures," Parts I and II, Shock & Vibration Digest, 1978.

6. Horvay, G.; Veluswami, M.A.; Stockton, F.D. "Dynamic plastic buckling of shells: a reconsideration of the Vaughan--Florence analysis," Proc. 5th SMIRT Conf., Berlin (1979) paper L4/8.

7. Horvay, G.; Veluswami, M.A. "Extension of the Vaughan--Florence analysis of dynamic buckling of a rigid-plastic cylindrical shell," Proc. SECTAM Conference, Knoxville (1980).

8. Veluswami, M.A.; Horvay, G. "Plastic buckling of cylindrical shells under axially varying dynamic pressure loading," Proc. 6th SMIRT Conference, Paris (1981) paper L14/2.

9. Whittaker, E.T.; Watson, G.N. A Course in Modern Analysis, Cambridge, 1946.

10. Stoker, J.J. Nonlinear Vibrations in Mechanical and Electrical Systems, Wiley-Interscience, 1950.

11. Strutt, M.J.O. Lamesche, Mathieusche u. Verwandte Funktionen in Physik u. Technik, Springer, 1932.

12. Hochstadt, H. The Functions of Mathematical Physica, Wiley--Interscience, 1971.

13. Horvay, G.; Gold, B.; Kaczenski, E.S. "Longitudinal heat propagation in three-phase laminated composites at high exciting frequencies," J. Heat Transfer 100 (1978) 281.

14. Gold, B.; Horvay, G. "Longitudinal heat propagation in three--phase laminated composites at low exciting frequencies," J. Appl. Mech. 101 (1979).

15. Baker, G.A. Essentials of Padé Approximants, Academic Press, 1975.

Appendix. Padé Approximant Scheme

We denote all series G_1, G_2, G_3, G_4 by

$$Z = 1 + a z + a z + \dots + a_n z^n \tag{A1}$$

(e.g., for $Z = G_1$: $Z = t^2$, $a_k = c_k$), and by Padé [13,14] we represent Z_n, the partial series for Z up to power z^n, by ratios of polynomials

$$\mathcal{N}_r/\mathcal{D}_c \equiv [r/c],\ r + c = n,\ \mathcal{N}_r = 1 + N_1 z + \dots + N_r z^r,$$

$$\mathcal{D}_c = 1 + D z + \dots + D_c z^c \tag{A2}$$

Using the abbreviations

$$k_4 = a_2^2 - a_1 a_3, \qquad k_6 = a_3^2 - a_2 a_4 \tag{A3}$$

and restricting ourselves to the diagonal sequence, where $r = c$, we find the approximants $N = 1,2,3$:

$N = 1$: $Z_2 \sim [1/1]$:

$$D_1 = -a_2/a_1$$
$$N_1 = a_1 + D_1$$

$N = 2$: $Z_4 \sim [2/2]$:

$$D_1 = -(a_2 a_3 - a_1 a_4)/k_4$$
$$D_2 = -(a_3 - D_1 a_2)/a_1$$
$$N_1 = a_1 + D_1$$
$$N_2 = a_2 + a_1 D_1 + D_2$$

$N = 3$: $Z_6 \sim [3/3]$:

$$D_1 = -\{(a_3 a_5 - a_2 a_6) k_4 - (a_2 a_4 - a_1 a_5) k_6\} / \{(a_3 a_4 - a_2 a_5) k_4 - (a_2 a_3 - a_1 a_4) k_6\}$$
$$D_2 = -\{a_3 a_5 - a_2 a_6 + (a_3 a_4 - a_2 a_5) D_1\}/k_6$$
$$D_3 = -\{a_5 + a_4 D_1 + a_3 D_2\}/a_2$$
$$N_1 = a_1 + D_1$$
$$N_2 = a_2 + a_1 D_1 + D_2$$
$$N_3 = a_3 + a_2 D_1 + a_1 D_2 + D_3$$

On Buckling and Post-buckling of Elastic and Inelastic Thin Shells in Primal and Mixed Formulations

R. VALID

Office National d'Etudes et de Recherches Aérospatiales
BP 72 - 92322 Châtillon Cedex, France

Summary

The great interest of mixed, hybrid, or complementary energy principles lies in the search of more refined stress-fields than by use of the primal method, in particular in zones of high stress-gradients. Moreover the accuracy of the stresses is of primary importance in buckling and post-buckling problems.
After having recalled the author's extension of the buckling criterion to various mixed principles by means of a very general method, one applies these criteria to the case of thin elastic or elastoplastic shells without shear strain, and in particular in the case of a surface principle of complementary energy with finite rotation.
It is also provided post-buckling formulae for mixed principles in the classical cases of limit points or symmetric and asymmetric bifurcation point with or without imperfections.
The criteria can be applied through a partitioning primal-mixed strategy.

I. Introduction

Since the pioneered works which marked the shell theory at its origins [1,2] and the numerous researches, discussions and controversies that they prompted [3,4], the researches have been extended in the non linear domain where problems were sharply posed as more from a fundamental than from a practical point of view [5-9]. We will then just evoke a weak part of them in relation with our goal.
Among these researches, because of the great importance of large rotation effects in shell problems with large displacements, some authors have introduced the finite rotation in the theory as independent variable [10-13]. More generally, the associated problems of structural analysis, the prodigious development of powerful computers and the modern methods of discretization [14] have thrown a strong emphasis on the variational methods [15-22] in particular in the shell domain [19,23-25]. In fact, the search of accurate stresses in the linear and non linear problems of damage, fatigue, cracks, stability, limit loads, or more simply of

Inelastic Behaviour of Plates and Shells
IUTAM Symposium Rio de Janeiro 1985
Editors: L. Bevilacqua, R. Feijóo and R. Valid

plastification, increased the interest on the utilization of mixed or hybrid variational principles, and in particular on the search of non linear complementary principles [26-28] , of which some of them could be extended to shell structures [29,13,30] .
Such investigations in the computation of as accurate as possible surface stress fields do not prevent theoretical and applied researches on three-dimensional boundary-layer phenomena like edge or interface effects, which need the introduction of additional solutions, but it goes without saying that such solutions could be justified only if they come to correct relatively good shell solutions by themselves.
One of the most important problems associated with the theory and analysis of non linear shells is the stability problem. Since the fundamental and remarkable contributions of Koiter in that field [31,32], the theory has been enriched and improved in numerous publications [33-38], and though it has been illustrated by many numerical applications [39], important difficulties remain due in particular to the great number of degrees of freedom and the very rapid increase of computational costs as soon as the structural complexity increases, but due also to numerical instabilities in the vicinity of critical points, or to unavoidable couplings between local and overall buckling [40,41], and finally to the lack of knowledges on the actual imperfections which have to be taken into account in the analysis [42-44].
These difficulties are naturally increased when some parts of the structure enter the plastic domain, and most interesting theoretical or applied works on the structural analysis and buckling load prediction of elasto-plastic shells have been published [45-56]. Likewise many fundamental works on the plastic buckling and post-buckling of general structures, and of shell structures in particular, have been devoted to this research since those of Hill [57-63].
The choice of non-linear mixed functionals and the need of refined computed buckling loads in the elastic or plastic domain led to an extension of the classical stability criterion of minimum of the total potential energy, solutions of such principles being in general saddle points of those functionals. Moreover the extension of the mixed method to post-buckling problems was necessary to make the whole method completely consistent.
In the following, one proposes to recall the mixed principles in the non-linear shell theory with an emphasis on the principle of

complementary energy with finite rotation and some variants. Then, after having recalled the general stability criterion in constrained systems [64], we will indicate its application to the preceding principles in the elastic case. The method will then be extended to the plastic case.

In a further section the results of the classical post-buckling theory will be extended to the mixed principles and the case of elasto-plastic unloading will be evoked.

One will end by an extension of the results relative to the stability of the post-buckling paths.

In all the sequel the case of Kirchhoff-Love shells will be considered alone in order to simplify the presentation, which will be made in general in an intrinsic and compact formulation.

II. Notations and primal formulation [18,9]

One considers a shell whose middle surface Σ_m is a differentiable manifold of dimension 2, embedded in the three-dimensional Euclidian space E_3. Σ_m, of generic point m, is supposed to be compact, oriented and Riemannian, namely there exists on the tangent plane $\vec{E}_2$ at m an induced scalar product, an associated topology and an orthogonal projector field π which project $\vec{E}_3$ onto $\vec{E}_2$,

$$\pi = \bar{\pi} = \pi^2 \quad \in \mathcal{L}(\vec{E}_3,\vec{E}_3),$$

where, as in the sequel, the bar designates transposition. Let O be an open set of $\mathbb{R}^2$ and a map(*)

$$\underset{\sim}{m} : m = \underset{\sim}{m}(X),\ X \in O \subset \mathbb{R}^2,$$

the natural basis at m is a line of vectors such that :

$$S = \frac{\partial m}{\partial X} = [\partial_1 m\ \partial_2 m] = [S_1\ S_2], X = \begin{bmatrix} X^1 \\ X^2 \end{bmatrix},\ \partial_\alpha m = \frac{\partial m}{\partial X^\alpha},\ \alpha = 1,2.$$

A point M of the shell on the unit normal N at m, will be given by

$$M = m + Nz$$

where z is the normal coordinate. A virtual displacement of M will be

$$\delta M = \delta m + \delta N z + N\delta z \qquad (1)$$

The Kirchhoff-Love hypotheses are summarized by

(i) the geometrical normal $\equiv$ the material normal

(ii) the normal strain energy is negligible compared with the others.

This gives

$$\left[\begin{array}{l} \bar{N}\delta N = 0 \Rightarrow \delta N \in \vec{E}_2 \ (\bar{N}\delta N = N_i \delta N^i,\ i = 1,2,3) \\ \delta[\bar{N}dm] = 0\ \forall\ dm \in \vec{E}_2, \delta m \in \vec{E}_3 \Rightarrow \delta N = -\overline{\frac{\partial \delta m}{\partial m}} N. \end{array}\right. \qquad (2)$$

(*) The underlined tilde will mean "function of".

The three-dimensional strain energy gives with (1) and (2) in the deformed state :

$$\left[\begin{array}{l} \delta W = \int_{\Sigma_m} T_r(\mathbf{n}'\ \delta\gamma + \mathbf{m}'\ \delta K),\ (=\int_{\Sigma_m}\left[\mathbf{n}'^{\alpha\beta}\ \delta\gamma_{\alpha\beta} + m'^{\alpha\beta}\ \delta K_{\alpha\beta}\right], \alpha,\beta = 1,2) \\ \mathbf{n}' = \bar{\mathbf{n}}',\ \mathbf{m}' = \bar{\mathbf{m}}',\ \delta\gamma = \delta\bar{\gamma},\ \delta K = \delta\bar{K} \in L(\vec{E}_2,\vec{E}_2), \end{array}\right. \tag{3}$$

with
$$\left[\begin{array}{l} \delta\gamma = \frac{1}{2}\left[\pi\frac{\partial\delta m}{\partial m} + \pi\overline{\frac{\partial\delta m}{\partial m}}\right] = \overline{\delta\gamma} \\ \delta K = \pi\frac{\partial\delta N}{\partial m} + \pi\overline{\frac{\partial\delta m}{\partial m}}\frac{\partial N}{\partial m} = \overline{\delta K},\ \text{with } \delta N = -\overline{\frac{\partial\delta m}{\partial m}}N. \end{array}\right. \tag{4}$$

If a former state $\mathcal{C}_o$ is taken as a reference one, (3) becomes

$$\delta W = \int_{\Sigma_{m_o}} T_r(n\ \delta\Gamma + m\ \delta K),\ (=\int_{\Sigma_{m_o}}\left[N^{\alpha\beta}\ \delta\Gamma_{\alpha\beta} + M^{\alpha\beta}\ \delta K_{\alpha\beta}\right], \alpha,\beta = 1,2) \tag{5}$$

with
$$\left[\begin{array}{l} \delta\Gamma = \frac{1}{2}\left[\overline{\frac{\partial m}{\partial m_o}}\frac{\partial\delta m}{\partial m_o} + \pi\overline{\frac{\partial\delta m}{\partial m_o}}\frac{\partial m}{\partial m_o}\right] = \overline{\delta\Gamma} \\ \delta K = \overline{\frac{\partial m}{\partial m_o}}\pi\frac{\partial\delta N}{\partial m_o} + \pi\overline{\frac{\partial\delta m}{\partial m_o}}\frac{\partial N}{\partial m_o} = \overline{\delta K} \end{array}\right\} \in L(\vec{E}_{20},\vec{E}_{20})$$

One recalls the definitions of the two fundamental forms variations of Σ_{m_o} :

$$\left[\begin{array}{l} \overline{dm_o}\ \Gamma\ dm_o \overset{\text{def.}}{=} \frac{1}{2}\left[\overline{dm}\ dm - \overline{dm_o}\ dm_o\right] \\ \overline{dm_o}\ K\ dm_o \overset{\text{def.}}{=} \overline{dm}\ \frac{\partial N}{\partial m}\ dm - \overline{dm_o}\ \frac{\partial N_o}{\partial m_o}\ dm_o \end{array}\right\} \quad m = m_o + V, \tag{6}$$

where V is the displacement of the point m_o of the middle surface, and whose virtual variations give $(5)_2$ and $(5)_3$. The principle of virtual work gives then the equilibrium equation

$$\delta W - \delta T = 0\ \forall\ \delta m\ \text{K.A.}, \tag{7}$$

where δT is the virtual work of prescribed external loads for any arbitrary kinematic admissible (K.A.) displacement δm.

Remarks

1) **n'**, **m'** are the Cauchy surface stresses, whilst **n**, **m** are the Piola-Kirchhoff surface stresses.

2) Eq. (5) and (7) give the local equations of equilibrium [9] .

III. Classical mixed formulations of Kirchhoff-Love shells

Let us pose

$$C = \begin{bmatrix}\mathbf{n}\\ \mathbf{m}\end{bmatrix},\quad D = \begin{bmatrix}\Gamma\\ K\end{bmatrix} \in \left[L(\vec{E}_{20},\vec{E}_{20})\right]^2$$

and the scalar product in $[\mathbb{L}(\vec{E}_{20},\vec{E}_{20})]^2$, where the overlined tilde means transposition,

$$\tilde{C}\,\delta D = \tilde{\mathbf{n}}\,\delta\Gamma + \tilde{\mathbf{m}}\,\delta K = T_r\,(\mathbf{n}\,\delta\Gamma + \mathbf{m}\,\delta K).$$

Let us write also (6) in the form

$$\Gamma = \underset{\sim}{\Gamma}(V), \quad K = \underset{\sim}{K}(V), \quad \text{or } D = \underset{\sim}{D}(V). \tag{8}$$

The primal principle can be written, given the constitutive equation :

$$\int_{\Sigma_{m_o}} \tilde{C}\,\delta D - \delta T = 0 \text{ with } D = \underset{\sim}{D}(V)\ \forall\ V \text{ K.A.} \tag{9}$$

We may take then Γ and K, i.e. D, as independent variables under the condition of introducing the constraint (8) in (9) by means of a Lagrange-multiplier

$$\Lambda = \begin{bmatrix} t \\ s \end{bmatrix} \in \left[\mathbb{L}(\vec{E}_{20},\vec{E}_{20})\right]^2 .$$

The result is the **five-fields principle of Hu-Washizu**

$$\left[\begin{array}{l} \int_{\Sigma_{m_o}} \left[\tilde{C}\,\delta D + \delta\left[\tilde{\Lambda}\underset{\sim}{D}(V)\right]\right] - \delta T = 0 \\ \forall\ D,\ C \in L^2(\Sigma_{m_o}),\ V \text{ K.A. s.t. } \underset{\sim}{D}(V) \in L^2(\Sigma_{m_o}). \end{array}\right. \tag{10}$$

In the case of an hyperelastic shell, $\exists$ a surface strain density α :

$$\alpha = \underset{\sim}{\alpha}(D), \quad C = \frac{\partial\underset{\sim}{\alpha}}{\partial D}, \quad \text{or } \begin{bmatrix}\tilde{\mathbf{n}} & \tilde{\mathbf{m}}\end{bmatrix} = \begin{bmatrix}\frac{\partial\alpha}{\partial\Gamma} & \frac{\partial\alpha}{\partial K}\end{bmatrix}.$$

Then (10) can be written

$$\left[\begin{array}{l} \delta\,\underset{\sim}{W}(D,\Lambda,V) = \delta\left[\int_{\Sigma_{m_o}}\left[\underset{\sim}{\alpha}(D) - \Lambda\left[\underset{\sim}{D}(V) - D\right]\right] - T\right] = 0 \\ \qquad \forall\ V \text{ K.A.},\ \underset{\sim}{\alpha}(D) \in L^1(\Sigma_{m_o}) \end{array}\right. \tag{11}$$

It should be noted that principle (11) does not need any inversion of the constitutive equation. Now (10) gives also

$$\int_{\Sigma_{m_o}} \left[\left[\tilde{C} - \tilde{\Lambda}\right]\delta D\right] + \tilde{\Lambda}\,.\,\delta\underset{\sim}{D}(V) + \delta\tilde{\Lambda}\left[\underset{\sim}{D}(V) - D\right]\right] - \delta T = 0. \tag{12}$$

If we stricly take the local Euler equation to be satisfied,

$$C = \Lambda,$$

and suppose that the constitutive equation is elastic and admits an inverse

$$C = \underset{\sim}{C}(D), \quad D = \underset{\sim}{D}(C),$$

(12) gives

$$\int_{\Sigma_{m_o}} \left[\delta\left[\tilde{C}.\underset{\sim}{D}(V)\right] - \delta\,\tilde{C}\,.\,\underset{\sim}{D}(C)\right] - \delta T = 0. \tag{13}$$

In the case of an hyperelastic shell with a strain energy density α, we introduce the complementary energy density β with the Legendre transforformation

$$\beta = \underset{\sim}{\beta}(C) = \tilde{C}\,D - \alpha, \text{ with } \alpha = \underset{\sim}{\alpha}(D),\ D = \underset{\sim}{D}(C),$$

and we easily find the **three-fields principle of Hellinger-Reissner**

$$\left[\begin{array}{l} \delta \underset{\sim}{\mathcal{H}}(C,V) = \delta\left[\int_{\Sigma_{m_o}} \left[\tilde{C}\,\underset{\sim}{D}(V) - \beta\right] - T\right] = 0 \\ \forall\ C \text{ s.t. } \beta \in L^1(\Sigma_{m_o}) \text{ and } V \text{ K.A. s.t. } \underset{\sim}{D}(V) \in L^2(\Sigma_{m_o}). \end{array}\right. \tag{14}$$

Remark : The principle (14) or (13) gives again the principle of virtual work by variation of V, and the inverted constitutive equation in a weak form, by variation of C.

If $\mathcal{C}_o$ is a prestressed state, calling C_o the prestress and δT the virtual work of the prescribed external forces, we write :

$$C = C_o + C_1,\ \delta T = \delta T_o + \delta T_1.$$

In state $\mathcal{C}_o$ the equilibrium is written

$$\int_{\Sigma_{m_o}} \left[\tilde{C}_o\ \delta\ \underset{\sim}{D}(0)\right] - \delta T_o = 0, \forall\ \delta V \text{ K.A.} \tag{15}$$

(14) and (15) give the three-field principle of Hellinger-Reissner in the updated Lagrangian formulation, namely

$$\int_{\Sigma_{m_o}} \left[\delta\left[\tilde{C}_o\left[\underset{\sim}{D}(V_1) - \underset{\sim}{D}(o)\right]\right] + \delta\left[\tilde{C}_1\,\underset{\sim}{D}(V_1)\right] - \delta\,\tilde{C}_1.\underset{\sim}{D}(C_1)\right] - \delta T_1 = 0,\ \forall\ \delta\ C_1,\ \delta\ V_1 \tag{16}$$

If the body is hyperelastic from $\mathcal{C}_o$ (16) gives :

$$\left[\begin{array}{l} \delta\underset{\sim}{\mathcal{H}}(C_1,V_1) = \delta\left[\int_{\Sigma_{m_o}} \left[\tilde{C}_o\left[\underset{\sim}{D}(V_1) - \underset{\sim}{D}(o)\right] + \tilde{C}_1\,\underset{\sim}{D}(V_1) - \underset{\sim}{\beta}\,(C_1)\right] - T_1\right] = 0 \\ \forall\ C_1,\ V_1 \text{ adm.} \end{array}\right. \tag{17}$$

If the reference state is the natural state $\mathcal{C}_{oo}$ and $\mathcal{C}_o$ a prestressed intermediate state, one obtains the total Lagrangian formulation :

$$\left[\begin{array}{l} \delta\underset{\sim}{\mathcal{H}}(C_1,V_1) = \delta\left[\int_{\Sigma_{m_{oo}}} \left[\tilde{C}_o\left[\underset{\sim}{D}(V_1) - \underset{\sim}{D}(V_o)\right] + \tilde{C}_1\,\underset{\sim}{D}(V_1) - \underset{\sim}{\beta}\,(C_1)\right] - T_1\right] = 0 \\ \text{with } V = V_o + V_1. \end{array}\right. \tag{18}$$

Principle (17) can be written in a linearized form in view of utilization of Newton-Raphson numerical procedure, and assuming $\mathbf{m}_o = 0$, it comes :

$$\left[\begin{array}{l} \delta\left[\int_{\Sigma_{m_o}}\left[\frac{1}{2}\,\tilde{\mathbf{n}}_o\,\overline{\frac{\partial V_1}{\partial m_o}}\,\frac{\partial V_1}{\partial m_o} + \tilde{\mathbf{n}}_1\,\pi_o\,\frac{\partial V_1}{\partial m_o} + \tilde{\mathbf{m}}_1\left[\pi_o\,\frac{\partial V_1}{\partial m_o}\,\frac{\partial N_o}{\partial m_o} - \pi_o\,\frac{\partial}{\partial m_o}\left[\frac{\partial V_1}{\partial m_o}N_o\right]\right] - \beta\right] - T_1\right] = 0 \;\forall\, \mathbf{n}_1 = \bar{\mathbf{n}}_1,\ \mathbf{m}_1 = \bar{\mathbf{m}}_1 \in L(\vec{E}_{20},\vec{E}_{20}) \\ \text{and } V \text{ K.A.} \end{array}\right. \tag{19}$$

<u>Remark</u>. In the Donnel-Mushtari-Vlasov (D-M-V) approximation [5], the non-linear term $\frac{1}{2}\,\tilde{n}_o\,\overline{\frac{\partial V_1}{\partial m_o}}\,\frac{\partial V_1}{\partial m_o}$ is replaced by

$\frac{1}{2}\,\tilde{n}_o\,\overline{\frac{\partial w_1}{\partial m_o}}\,\frac{\partial w_1}{\partial m_o}$ where w_1 is the deflection from state ζ_o.

Principle (19) can be modified by integrating by parts the second and third term and using the following formulae [18] :

$$\begin{array}{l} \widehat{\operatorname{div}}\, V \overset{\text{def}}{=} T_r\left(\pi_o\,\frac{\partial V}{\partial m_o}\right),\ \forall\, V \in \vec{E}_{20} \\ \int_{\Sigma_{m_o}} \widehat{\operatorname{div}}\, V = \int_{\partial\Sigma_{m_o}} \overline{\nu}_o\, V\, ds \text{ (Stokes), where } s \text{ is the boundary abscissa,} \end{array} \tag{20}$$

ν_o the external unit normal to $\partial\Sigma_{m_o}$ in $\vec{E}_{20}$.

It comes the **mixed modified three-field principle of Hellinger-Reissner**

$$\delta\left[\int_{\Sigma_{m_o}}\left[\frac{1}{2}\,\tilde{n}_o\,\overline{\frac{\partial V_1}{\partial m_o}}\,\frac{\partial V_1}{\partial m_o} - \widehat{\operatorname{div}}\left[\mathbf{n}_1 + \mathbf{m}_1\,\frac{\partial N_o}{\partial m_o} + \overline{N_o\,\widehat{\operatorname{div}}\,\mathbf{m}_1}\right]V_1 - \beta\right] + \text{boundary terms} - T_1\right] = 0 \tag{21}$$

It should be noted that V_1 must be continuous at the interfaces of Σ_{mo} and that $C_1 \in H(\widehat{\operatorname{div}}, \Sigma_{m_o})$, which imposes that the fluxes be continuous as well at the interfaces.

The stress C_1 can be taken such that

$$\widehat{\operatorname{div}}\left[\mathbf{n}_1 + \mathbf{m}_1\,\frac{\partial N_o}{\partial m_o} + N_o \operatorname{div} \mathbf{m}_1\right] + f_1 = 0 \Big/ \Sigma_{m_o} \text{ piecewise,} \tag{22}$$

where f_1 is the surface density of the prescribed forces. It is then found the **hybrid dual principle of Pian** [17,23], where the fluxes may be taken discontinuous at the interfaces, but where C_1 is not strictly statically admissible because the prestress term is not taken into account in (22).

The surface stresses can be derived from a stress function V_1^* in order to satisfy the closure condition (22) when $f_1 = 0$. It is found [18]:

$$\left[\begin{aligned} \mathbf{n} &= i_2 \frac{\partial}{\partial m_o}\left[\overline{\frac{\partial V_1^*}{\partial m_o}} N_o\right] i_2 - \frac{\partial N_o}{\partial m_o} i_2 \overline{\frac{\partial V_1^*}{\partial m_o}} i_2 \\ \mathbf{m}_1 &= \tfrac{1}{2} i_2 \left[\pi_o \frac{\partial V_1^*}{\partial m_o} + \overline{\pi_o \frac{\partial V_1^*}{\partial m_o}}\right] i_2 \end{aligned}\right. \tag{23}$$

where $i_2 \in L(\vec{E}_{20},\vec{E}_{20})$ is the +90° rotation mapping in $\vec{E}_{20}$.

IV. The complementary energy principle with finite rotation.

The finite rotation can be introduced in principle (9) by use of a surface polar decomposition of the derivative $\partial m/\partial m_o$ [13], by extension to shell problems of Fraeijs de Veubeke's method [28].

Calling h the membrane extension and R the finite rotation, one finds

$$\left[\begin{aligned} &\frac{\partial m}{\partial m_o} = R\,\pi_o\left[1_{E_{20}} + h\right],\ h = \bar{h} \in L(\vec{E}_{20},\vec{E}_{20}),\ R\ |R\ \bar{R}=\bar{R}\ R=1_{E_3}, R \in L(\vec{E}_3,\vec{E}_3) \\ &\overline{\frac{\partial m}{\partial m_o}}\frac{\partial m}{\partial m_o} = \left[1_{E_{20}} + h\right]^2 = 2\,\Gamma + 1_{E_{20}}\,. \end{aligned}\right. \tag{24}$$

It is worth noting that $\overline{R\pi_o}\,R\pi_o = 1_{E_{20}} \Rightarrow \overline{R\pi_o}$ is a "left inverse" of $R\pi_o$, and that $(24)_2$ gives a constitutive law independent of R as it must be. Introducing the constraint (24) in (9) by means of a Lagrange multiplier $t \in L(\vec{E}_2,\vec{E}_{20})$ and the constraint $(6)_2$, namely

$$\overline{\frac{\partial m}{\partial m_o}}\frac{\partial N}{\partial m_o} - \frac{\partial N_o}{\partial m_o} - K = 0,$$

by means of a multiplier $s \in L(\vec{E}_{20},\vec{E}_{20})$, the variables R, h, and K can be taken as independent variables. In the case of an hyperelastic shell and after elimination of h and K, one finds the principle

$$\left[\begin{aligned} &\delta \underset{\sim}{\mathcal{F}}(t,\mathbf{m},V,R) = \delta\left[\int_{\Sigma_{m_o}}\left[T_r\left(t\frac{\partial m}{\partial m_o} + \mathbf{m}\,K(V) - t\,R\,\pi_o\right) - \beta\right] - T\right] = 0 \\ &\text{with} = \beta = T_r(rh + \mathbf{m}K) - \alpha;\ \alpha = \underset{\sim}{\alpha}^*(h,K),\ r = \bar{r} = \frac{\partial\tilde{\alpha}}{\partial h},\ R\ |R\ \bar{R} = 1_{E_3} \end{aligned}\right. \tag{25}$$

Remarks

1) t is the surface stress of Boussinesq (or Piola-Lagrange) and is non symmetric ; r is the surface stress of Jauman, and

$$r = \frac{tR\pi_o + \overline{tR\pi_o}}{2} \tag{26}$$

2) Constraint $R\bar{R} = 1_{E_3}$ can be introduced directly in (25) [11], or perhaps more simply through a Lagrange multiplier.

From principle (25), it is possible to derive various principles [63], in particular an incremental principle, then a modified incremental principle, then an hybrid dual incremental one where the stresses would be really statically admissible, or a principle where the Jaumann stress r would be directly taken as unknown instead of the Piola-Lagrange stress, and finally a principle where the term $\underset{\sim}{N}(V)$ in $\underset{\sim}{K}(V)$ would be replaced by RN_o, a way which brings a serious simplification in the stability criterion (see § V).

The introduction of statically admissible stresses can be carried out by means of vectorial stress functions V_1^* and $\Omega_1^* \in \vec{E_3}$, such that the increments t_1 and $\mathbf{m}_1$ can be written [13]:

$$\left[\begin{array}{l} t_1\pi_o = i_2 \dfrac{\partial \Omega_1^*}{\partial m_o} - \frac{1}{2}\left[\dfrac{\partial V_1^*}{\partial m_o} + \overline{\dfrac{\partial V_1^*}{\partial m_o}}\right] i_2 \dfrac{\partial N_o}{\partial m_o} \\ \mathbf{m}_1 = \frac{1}{2}\, i_2\left[\dfrac{\partial V_1^*}{\partial m_o} + \overline{\dfrac{\partial V_1^*}{\partial m_o}}\right] i_2 \\ \forall\, V_1^*,\ \Omega_1^* \in \vec{E}_3,\ \text{s.t. } \pi_o\, \overline{N_o}\, \Omega_1^* = -\frac{1}{2}\, i_2 \left[\dfrac{\partial V_1^*}{\partial m_o} - \overline{\dfrac{\partial V_1^*}{\partial m_o}}\right], \end{array}\right. \tag{27}$$

which is equivalent to 5 scalar independent functions.

V. Stability criteria in mixed formulations

V.1. Elasticity. As all the various mixed principles were written through the introduction of new variables and some constraints, the classical primal criterion

$$\delta^2 y(x) \geqslant 0 \Rightarrow \text{Stability},\ x \in X,\ \text{a normed linear space, must be}$$

extended in case of a constraint

$$z = z(x) = 0,\ z \in \vec{E},\ \text{a normed vector space, has to be taken into}$$

account.

It is easily found with a Lagrange multiplier [64] :

$$\left[\begin{array}{ll} \delta\left[y - \Lambda z\right] = 0\ \forall\ \delta x \in \vec{X},\ \delta\Lambda \in \vec{E'} & (\Rightarrow \text{equilibrium}) \\ \delta^2\left[y - \Lambda z\right] \geqslant 0\ \forall\ \delta x \in \vec{X}\ \text{s.t. } \delta z = 0 & (\Rightarrow \text{stability}). \end{array}\right. \tag{28}$$

Remark : It should be emphasized that $(28)_2$ is calculated with $\delta^2 x = \delta^2\Lambda = 0$, at the point solution given by $(28)_1$, and with all the constraints introduced from the beginning, even if some of them have been eliminated in the final principle.

Application of (28) to **Hu-Washizu principle** (11) gives [64-25] :

$$\left[\begin{array}{l} \int_{\Sigma_{m_o}} \left[\underset{\sim}{\alpha}''(D)(\delta D)(\delta D) + \tilde{C}\, \underset{\sim}{D}''(V)(\delta V)(\delta V)\right] \geqslant 0, \text{ (dead load)}, \\ \forall\ \delta D \text{ and } \delta V \text{ K.A. s.t. } \int_{\Sigma_{m_o}} \delta\tilde{C}\left[\delta\, \underset{\sim}{D}(V) - \delta D\right] = 0\ \forall\ \delta C\ ; \end{array}\right. \tag{29}$$

and to **Hellinger-Reissner principle** (14) :

$$\left[\begin{array}{l} \int_{\Sigma_{m_o}} \left[\underset{\sim}{\beta}''(C)(\delta C)(\delta C) + \tilde{C}\, \underset{\sim}{D}''(V)(\delta V)(\delta V)\right] \geqslant 0 \\ \forall\ \delta C,\ \delta V \text{ K.A. s.t. } \int_{\Sigma_{m_o}} \left[\delta\, \tilde{\underset{\sim}{D}}(V) - \underset{\sim}{\beta}''(C)(\delta C)\right]\underline{\delta C} = 0\ \forall\ \underline{\delta C}. \end{array}\right. \tag{30}$$

As for the **complementary principle with rotation** (25), one finds :

$$\left[\begin{array}{l} \int_{\Sigma_{m_o}} \left[\underset{\sim}{\beta}''(C)(\delta C)(\delta C) + T_r(-2t\ \delta R\ \pi_o\ \delta h + \mathbf{m}\left[2\ \frac{\overline{\partial \delta V}}{\partial m_o}\ \frac{\partial \delta \underset{\sim}{N}(V)}{\partial m_o} + \right.\right.\right. \\ \left.\left.\left. + \frac{\overline{\partial m}}{\partial m_o}\ \frac{\partial \delta^2 \underset{\sim}{N}(V)}{\partial m_o}\right]\left[\underline{1} + h\right]\ t\ R\ \pi_o\ \overline{\delta\ R\ \pi_o}\ \ \delta\ R\ \pi_o)\right] \geqslant 0 \\ \forall\ \delta C,\ \delta R\ |\bar{R}\ R = 1_{E_3},\ \delta\ V \text{ K.A., s.t.} \\ \int_{\Sigma_{m_o}} T_r(\underline{\delta t}\left[\frac{\partial \delta V}{\partial m_o} - \delta\left[R\ \pi_o\left[\underline{1} + h\right]\right]\right] + \underline{\delta \mathbf{m}}\ \delta\ \frac{\overline{\partial m}}{\partial m_o}\ \frac{\partial \underset{\sim}{N}(V)}{\partial m_o} - K\ ) = 0,\ \forall\ \underline{\delta t},\ \underline{\delta \mathbf{m}} \\ \text{with } \left[\delta\tilde{h}\ \delta\tilde{K}\right] = \underset{\sim}{\beta}''(C)(\delta C)(\delta C),\ C = \begin{bmatrix} r \\ \mathbf{m} \end{bmatrix}. \end{array}\right. \tag{31}$$

<u>Remarks</u>

1) The constraint appears always linearized due to the use of a local criterion near a given state. It means the local compatibility or constitutive equation in a weak form.

2) Criteria (29)-(31) are unchanged for modified mixed principles due to a preceding remark, but functional spaces are correspondingly modified.

3) In all cases the notions of elastic and geometric stiffnesses are recovered.

4) The criteria are unchanged if linear elasticity exists only from the critical state (example hypoelasticity).

5) Extension to Koiter's sufficient condition of stability can be found for neutral equilibrium in terms of mixed formulations.

6) Quantities like $\delta \underset{\sim}{N}(V), \delta^2 \underset{\sim}{N}(V), \delta^3 \underset{\sim}{N}(V)\ldots$ are easily found in terms of δV using (2).

V.2. <u>Elasto-plasticity</u>. In that case, the analysis uses the elasto-plastic three-dimensional incremental constitutive law :

$$\underset{\sim}{\alpha}(\dot{\varepsilon}) = \tfrac{1}{2}\left[\tilde{\dot{\varepsilon}}\, a_{el}\, \dot{\varepsilon} - \frac{\langle \tilde{\sigma}\, a_{el}\, \dot{\varepsilon} \rangle^2}{\underset{\sim}{h}(\sigma^*) + \tilde{\sigma}\, a_{el}\, \sigma)}\right]$$

where
- $.a_{el}$ is the elastic constitutive mapping
- $.\langle \tilde{\sigma}\, a_{el}\, \dot{\varepsilon} \rangle = 0$ if the yield condition is not reached
- .h is an experimental function of the strain-hardening parameter σ^*.

Hill's criterion of stability [57] is applied, an extension of which is given by Nguyen Q.S. et al. [61] namely in a three-dimensional formulation based on Liapunov's definition. The result is :

$$\underset{\sim}{F}(U,\lambda) = \int_{\Omega_o}\left[\alpha_o(\dot{\varepsilon}) + \tilde{\sigma}_o \frac{\partial U}{\partial M_o}\frac{\partial U}{\partial M_o}\right] \gg \Rightarrow \text{Stability} \tag{33}$$

But following Hutchinson [60] :

$[\exists$ Bifurcation$] \Rightarrow [F = 0 +$ no unloading$]$

The preceding theorem allows to apply the standard criterion with the elasto-plastic law from a prestressed state.

In the case of shells the following identification

$$\int_{\Omega_o} \tilde{\dot{\sigma}}\, \delta\, \dot{\varepsilon} = \int_{\Sigma_{m_o}} \tilde{\dot{C}}\, \delta\, \dot{D} \tag{34}$$

with the introduction of kinematic assumptions such that $\dot{\varepsilon} = \underset{\sim}{f}(\Gamma,K,z)$ [68] gives :

$$\dot{C} = A_{ep}\, \dot{D}, \text{ or } \underset{\sim}{\alpha}(D) = \tfrac{1}{2}\, \tilde{\dot{D}}\, A_{ep}\, \dot{D} \text{ and } \underset{\sim}{\beta}(C) = \tfrac{1}{2}\, \tilde{\dot{C}}\, A_{ep}^{-1}\, \dot{C} \tag{35}$$

where A_{ep} is the shell elasto-plastic mapping. Then use is made of the preceding criteria.

Remarks

1) In general for Kirchhoff-Love (K-L) shells [18]

$$\left[\begin{array}{l} 2\,\dot{\varepsilon} = \mu^{-1}\left[2\,\dot{\gamma} + 2\,z\,\dot{K} + z^2\left[\dot{K}\,\frac{\partial N_o}{\partial m_o} + \frac{\partial N_o}{\partial m_o}\,\dot{K} + 2\,\frac{\partial N_o}{\partial m_o}\,\dot{\gamma}\,\frac{\partial N_o}{\partial m_o}\right]\mu^{-1}\right] \\ \text{where } \mu = 1_{E_{20}} + z\,\frac{\partial N_o}{\partial m_o} \end{array}\right. \tag{36}$$

2) Great simplifications are then obtained for D-M-V shells (shallow shells) [60,49,55,67]. With that assumption Nguyen Q.S. showed that the evolution of the internal parameters satisfies the general law of convexity and normality in a resulting infinitely multiple plastic potential [52].

3) Other methods for elasto-plastic analyses are based on Ilyushin approach [65,45], that is to replace all quantities in (32) by surface ones (for K-L shells) : σ by C, ε by D, ..., and the yield condition $\underset{\sim}{f}(\sigma)$ by $\underset{\sim}{\mathfrak{F}}(C)$. This way has been improved to take account of strain-hardening (Criesfield [66]), but we shall prefer the integration of the three-dimensional law with respect to z, to take account precisely (at least layer by layer) of local plastification through the thickness.

4) A_{ep}^{-1} may not be drawn from a_{ep}^{-1} because of use of kinematic assumptions.

VI. Post-buckling in mixed formulation [63]

Asymptotic expansions following Koiter's method [31-38] is used for perfect shells (P.S.) or imperfect shells (I.S.), by two ways so called "direct method" (expansion from a critical point), or "indirect method" (expansion from the fundamental path).

VI.1. Elastic shells. For perfect shells, expansion is made of the weak form of the equilibrium equation from the critical state $\underset{\sim}{\zeta}_c$:

$$\int_{\Sigma_{m_o}} C\ \delta D - \delta \underset{\sim}{T}(V,\lambda) = 0 \qquad (37)$$

where λ is the loading parameter, with

$$\left[\begin{array}{l} V = V_c = \varepsilon\ V_1 + \varepsilon^2\ V_2 + .. \\ C = C_1 + \varepsilon\ C_1 + \varepsilon^2\ C_2 + .. \\ \lambda = \lambda_c + \varepsilon\ \lambda_1 + \varepsilon^2\ \lambda_2 + ... \end{array}\right. \qquad (38)$$

Account is taken for the criterion of the critical state

$$\int_{\Sigma_{m_o}} \left[\tilde{C}_o^M.D'_c(\delta V) + \tilde{C}_c.D''_c(\delta V)(V_o^M)\right] - T''_c(\delta V)(V_o^M) = 0,\ \forall\ \delta V\ \text{K.A.}, \qquad (39)$$

where (C_o^M, V_o^M) is the mixed eigenmode, supposed to be unique.

The discussion depends on Fredholm's condition :

$$\dot{T}'_c(V_o^M)\ \lambda_1 = 0,\ \text{where}\ \dot{T} = \frac{\partial T}{\partial \lambda},\ T' = \frac{\partial T}{\partial V}\ \text{(Direct method)}.$$

The following results are found with

$$\begin{bmatrix} C \\ V \end{bmatrix} = q,\ n = \int_{\Sigma_{m_o}} \left[\tfrac{1}{2}\ \tilde{C}_c.D'''_c(V_o^M)(V_o^M)(V_o^M) + \tilde{C}_o^M.D''_c(V_o^M)(V_o^M)\right] - \qquad (40)$$

$$- \tfrac{1}{2}\ T''''_c(V_o^M)(V_o^M)(V_o^M).$$

a) **Limit point.** $\dot{T}'_c(V_o^M) \neq,\ \lambda_1 = 0$

$q = q_c + q_o^M \xi + q_2^{\perp} \xi^2 + \ldots$; $q_2^{\perp} \perp q_o^M$ given by 2d order mixed equation,

$\lambda_1 = \xi^2\ n / \dot{T}'_c(V_o^M)$

b) **Asymmetric bifurcation** : $\dot{T}'_c(V_o{}^M) = 0,\ \lambda_1 \neq 0,\ n \neq 0$

$q = q_c + q_o{}^M\xi + q_1^{\perp}\ \lambda_1,\ q_1^{\perp} \perp q_o{}^M$ given by mixed equation,

λ_1 is given by a 2d order equation.

c) **Symmetric bifurcation** : $\dot{T}'_c(V_o{}^M) = 0,\ n = 0,\ \lambda_1 = 0$

$q = q_c + q_o{}^M\xi + q_2^{\perp}\xi^2,\ \lambda = \lambda_c + \lambda_2\xi^2.$

For imperfect shells, we pose in addition to (38), with η an imperfection parameter

$$\left[\begin{array}{l} \eta = 0 + \varepsilon\,\eta_1 + \varepsilon^2\,\eta_2 \\ q_o{}^M = q_o{}^M + \varepsilon^2\, q_1{}^M + \varepsilon^2 q_2{}^M + \ldots\ldots\ldots \end{array}\right. \tag{41}$$

and expand the two equations :

$$\left[\begin{array}{l} \int_{\Sigma_{m_o}} \tilde{C}.\delta\underset{\sim}{D}(V,\eta) - \delta\underset{\sim}{T}(V,\lambda,\eta) = 0 \text{ (equilibrium)} \\ \\ \int_{\Sigma_{m_o}} \left[\tilde{C}_\eta{}^M.\underset{\sim}{D}'(V,\eta)\delta V + \tilde{C}_{c\eta}.D'_c(V_\eta{}^M)\delta V\right] - T''_c(V_\eta{}^M)\delta V = 0 \text{ (criterion),} \end{array}\right. \tag{42}$$

which expressed the fact that (C,V,λ,η) is a critical point. The results are :

a) $\dot{T}_c(V_o{}^M) = 0,\ (\underset{\sim}{C},V,\lambda,0)$ is a limit point of the P.S.

$\lambda_1 = -\left[T'^*_c(V_o{}^M) + C_c.D'_c(V_o{}^M)\right]\eta/\dot{T}'_c(V_o{}^M),\ (T^* = \frac{\partial T}{\partial \eta}).$

$q = q_c = q_o{}^M\xi + q_1^{\perp}\ \lambda_1 + \cdots;\ \lambda_1 = 0$ if $n = 0$

$q_1{}^M \perp q_o{}^M$ is known by $(42)_2$ + weak constitutive equation.

b) $\dot{T}'_c(V_o{}^M) \neq 0,\ (C_c,V_c,\lambda_c,0)$ is an asymmetric bifurcation point of the P.S.

$q = q_c = q_o{}^M\xi + q_1^{\perp}\ \lambda_1,\ \eta = \xi^2\eta^2.$

c) $\dot{T}'_c(V_o{}^n) \neq 0,\ (C_c,V_c,\lambda_c,0)$ is a symmetric bifurcation point of the P.S.

$n = 0,\ \eta_1 = \eta_2 = 0,\ \lambda_1 = 0,\ \eta = 0\ (\xi^3)\ ;\ q = q_c + q_o{}^M\xi + q_2^{\perp}\,\xi^2,$

$\lambda = \lambda_c + 0\ (\xi^2).$

Remark

The same orders of magnitude are found again as with the primal method obviously, either by the direct or by the indirect method.

VI.2. Elasto-plastic shells. The results follow those given by Hutchinson in the primal case by the indirect method for the P.S. [60] where

$V = V_F + \xi\ V_o{}^M + \xi^{1+\beta}\ V_2 + \ldots$ (V_F = fundamental path).

Shanley's condition (no unloading) is used and gives that the existence of an elasto-plastic bifurcation is subjected to no unloading as a necessary condition.

In case of unloading (U.L.), two cases may be distinguished :

a) $\lambda_1/\xi > \lambda_{1\,UL}$ standard case of hypoelasticity.

b) $\lambda_1/\xi = \lambda_{1\,UL}$ the elastic zone of U.L. spreads from a priming point. This case involves a singular perturbation technique which gives $\beta = 1/3$ for D-M-V shells. Then λ_2 can be calculated.

The case of imperfect shells has been examined by Hutchinson under some assumptions, when bifurcation occurs before or after plastification. This author confirms that no general treatment is available like Koiter's theory for conservative systems.

All the study may be looked upon with minor modifications from a mixed viewpoint.

VIII. Conclusions

The problem of mixed variational formulations, buckling and post-buckling, has been treated for Kirchhoff-Love shells, but the methods can be extended to the case of shells which undergo non negligible transverse shear strains, cases of considerable importance for multilayered composite materials. Pointless to say that for finite rotation formulations unavoidable adaptations would be necessary for the latter.

The preceding intrinsic and compact presentation has the merit, to our feelings, to be relatively simple in spite of the classical difficulties which are inherent to curved space mathematics. It may nevertheless mask some complexity in passing to detailed formulae.

In fact, considering for instance the term $\tilde{\mathbf{m}}_o \overline{\dfrac{\partial V_1}{\partial m_{oo}}}\, \dfrac{\partial M_1}{\partial m_{oo}}$,

which appears in some incremental formulations where N_1 is the increment of the unit normal to the middle surface, one finds in terms of components:

$$\tilde{\mathbf{m}}_o \overline{\frac{\partial V_1}{\partial m_{oo}}}\, \frac{\partial N_1}{\partial m_{oo}} = \mathcal{M}^{\alpha\beta}\Big\{\mathcal{V}_{1\|\gamma}\, g_{\beta\rho}\Big[\mathcal{N}^{\rho}_{1|\alpha} + \mathcal{V}^{\rho}_{o\|x\alpha}\mathcal{N}^{x}_{1} + \mathcal{V}^{\rho}_{o\|x}\mathcal{N}^{x}_{1,\alpha}\Big] + $$
$$+ \mathcal{V}^{3}_{1\|\gamma}\Big[g_{\alpha\beta}\, b^{\beta}_{x}\mathcal{N}^{x}_{1} + \mathcal{V}^{3}_{o\|x\alpha}\mathcal{N}^{x}_{1} + \mathcal{V}^{3}_{o\|x}\mathcal{N}^{x}_{1,\alpha}\Big]\Big\},$$

with

$$\mathcal{V}^{\beta}_{o\|\gamma} = \mathcal{V}^{\beta}_{o|\gamma} - b^{\beta}_{\gamma}\mathcal{V}^{3}_{o}\,; \quad \mathcal{V}^{3}_{o\|\gamma} = g_{\gamma\alpha} b^{\alpha}_{\beta}\mathcal{V}^{\beta}_{o} + \mathcal{V}^{3}_{o,\gamma}\,; \quad \mathcal{V}^{\beta}_{o|\gamma} = \Gamma^{\beta}_{\gamma\alpha}\mathcal{V}^{\alpha}_{o} + \mathcal{V}^{\beta}_{o,\gamma}$$

$b_{\alpha\beta} = -\overline{S_{oo\alpha}}\, \dfrac{\partial N_{oo}}{\partial m_{oo}}\, S_{oo\beta}$; $g_{\alpha\beta} = \overline{S_{oo\alpha}} \cdot S_{oo\beta}$; S_{oo} natural basis of $\Sigma_{m_{oo}}$

$\alpha, \beta, \gamma, x, \rho = 1,2.$

But we stress the fact that the theoretical formulae can be written in an automatic way without any difficulty.

In general the numerical analysis in mixed principles brings into play many more unknowns than in the primal method, and it seems always judicious to apply some computational strategies like

*) Ritzian corrected methods [69,70], possibly through modal synthesis [71] ;

* postponed analysis in critical zones after application of the primal method ;

* coupling of mixed principles with the primal one by partitioning, reserving the former to critical zones, for the buckling problem which remains a global phenomenon even if there exists local weaknesses [72,73].

Important problems are still open and not only from a mathematical point of view. One has evoked in particular those of unknown imperfection data, or that of local effects or also buckling of inelastic shells comprising numerous stiffners or holes, which prompted recent delicate investigations. In spite of a great research activity, the analysis of inelastic shells remains, as we saw, a difficult task.

References

1. Aron H. Das Gleichgewicht und die Bewegung einer unendlicher dünnen beliebig gekrümmten elastichen Schale. J. Reine und Angew. Math. 1874, vol. 78, p. 136-174.

2. Love A.E. The small free vibrations and deformations of a thin elastic shell. Phil. Trans. Roy. Soc. London 1888, Ser. A, vol. 179, p. 491-546.

3. Koiter W.T. A consistent first approximation in the general theory of thin elastic shells. Proc. IUTAM Symp. Theory of thin Shells. Delft 1959. North-Holland Amsterdam 1960, p. 12-33.

4. Budiansky B., Sanders J.L. On the "best" first order linear shell theory. Prog. in Appl. Mech. Prager Anniversary Volume. Vol. 192 Mac Millan N.Y. 1963, p. 129-140.

5. Koiter W.T. On the non-linear theory of thin elastic shells. Proc. Phys. Sciences Mechanics, Series B, V. 69, no 1, 1965.

6. Mushtari K.M., Galimov K.Z. Non-linear theory of elastic shells (In Russian 1957). NASA TT-F 62. U.S. Dept of Commerce, Off. of Technological Services, Washington 1961.

7. Sanders J.L. Non-linear theories for thin shells. Quart. Appl. Math. 1963, Vol. 21, p. 21-36.

8. Novozhilov V.V. The theory of thin shells. P. Noordhoff. 1959.

9. Valid R. An intrinsic formulation for the non-linear theory of shells and some approximations. Trends in Computerized Structural Analysis and Synthesis. Oct 30-Nov 1, 1978 Washington D.C. Comp. and Struct. Vol. 10, 1979, p. 143-194. T.P. ONERA 1979-2.

10. Simmonds J.G., Danielson D.A. Non-linear shell theory with finite rotation and stress function vectors. J. Appl. Mech. Trans. ASME. Ser. E, 1972, vol. 39, No 4, p. 1085-1090.

11. Pietraszkiewicz W. Finite rotations and Lagrangian description in the non-linear theory of shells. Warsawa-Poznen 1979 Polish. Scient. Publ.

12. Schmidt R. A current trend in shell theory : constrained geometrically non-linear Kirchhoff-Love theories bases on polar decomposition of strains and rotations. Proc. Symp. on Advances and Trends in Structures and Dynamics. Arlington, Virginia U.S.A. 22-25 Oct. 1984. Pergamon Press (In print).

13. Valid R. The principle of complementary energy in the non-linear theory of shells. XVth IUTAM Congress Toronto, 17-23 Aug. 1980. T.P. ONERA 1980-16. La Rech. Aérosp. No 1981-1.

14. Zienkiewicz O.C. The Finite Element Method. Third ed. Mc GRAW Hill. 1977.

15. Washizu K. Variational Methods in Elasticity and Plasticity, Pergamon Press, 1975.

16. Reissner E. On a variational theorem in elasticity. J. of Math and Phys. Vol. 29, 1950.

17. Pian T.H.H. Derivation of element stiffness by assumed stress distribution. AIAA J. 2, No 7, 1333-1336, 1964.

18. Valid R. Mechanics of Continuous Media and Analysis of Structures. North-Holland 1981.

19. Valid R. Le principe des travaux virtuels et les principes variationnels associés. Pub. ONERA 1982-1. Trad. ESA-TT-873.

20. Brezzi F. On the existence uniqueness and approximation of saddle-point problems arising from Lagrange multipliers. RAIRO, R2, Août 1974.

21. Thomas J.M. Méthode d'éléments finis hybrides duaux RAIRO, Analyse Numérique, vol. 10, n° 12, 1976.

22. Le Tallec P. Compatibility and existence results in discrete finite incompressible elasticity. Comp. Meth. Appl. Mech. and Engn., Vol. 27, No 2, July 1981, pp. 239-259.

23. Boland P.L., Pian T.H.H. Large deflection analysis of thin elastic structures by the assumed stress hybrid element method. Comp. and Structures, VoL. 7, p. 1-12, Feb. 1977.

24. Schmidt R. Lecture GAMM Annual Conf. Würzburg. April 21-24, 1981, ZAMM 62, T 165-T 167, (1982).

25. Valid R. De la théorie non-linéaire des coques. Principes mixtes et critères de flambage. Conf. Ecole CEA-INRIA-EDF. Rocquencourt 19-14 Janv. 1983. T.P. ONERA n° 1983-149.

26. Zubov L.M. The stationary principle of complementary work in non-linear theory of elasticity. Prikb. Math. Mekh. 34 (1970) p. 241-245. J. Appl. Math. Mech. 34 (1970) p. 228-232.

27. Koiter W.T. On the complementary energy theorem in non-linear elasticity theory. Trends in Applied of Pure Math. in Mech. G. Fichera ed. Pitman Publ. Conf. Univ. of Lecce (Italy) 1975

28. Fraeijs de Veubeke B. A new variational principle for finite elastic displacements Int. J. Engn. Sc. Vol. 10 (1972) p. 745-763.

29. Stumpf H. The derivation of dual extremum and complementary stationary principles in geometrically non-linear shell theory. Ing. Arch. 48, (1979) p. 221-237.

30. Atluri S.N. Alternate stress and conjugate strain measures and mixed variational formulations involving rigid rotations for computational analyses of finitely deformed solids with applications to plates and shells. I Theory. Comp. and Struct. Vol. 18, No 1, p. 93-116.

31. Koiter W.T. On the stability of elastic equilibrium. Thesis, Delft Univ. H.J. Paris Amsterdam 1945 ; Engl. transl. (a) NASA TT-F 10 883 (1967), (b) AFF DL-TR-70-25 (1970).

32. Koiter W.T. On the thermodynamic background of elastic stability theory. Rep. No 360, Dept. Mech. Ing. Techn. Univ. Delft, 1967. Problems of Hydrodynamics and Continuum Mechanics. L.I. Sedov Anniversary Vol. SIAM, Philadelphia (1969) p. 423-433.

33. Koiter W.T. General equations of elastic stability for thin shells. Proc. Symp. in Honour of Lloyd H. Donnel. Ed. by D. Muster (Un. of Houston), April 1966.

34. Sewell M.J. A general theory of equilibrium path through critical points, I, II Proc. Roy. Soc. A 306, 201-223, 225-238.

35. Thomson J.M.T. A general theory for the equilibrium and stability of discrete conservative systems. Zamp Vol. 20, 1969.

36. Thomson J.M.T., Hunt G.W. A general theory of Elastic Stability. J. Wiley. N.Y. (1973).

37. Budiansky B. Theory of buckling and post-buckling behavior of elastic structures. Adv. in Appl. Mech. Vol. 14, Ed. by Chia-Shun Yih. 1974.

38. Valid R. Déformations non-linéaires et flambage statique. Publ. ONERA n° 1977-2.

39. Krätzig W.B. Stability and collapse load computation for elastic shells. Ruhr Univ. Bochum - Hamburg 1984.

40. Vand der Neut A. Mode interaction with stiffened panels. Proc. IUTAM Symp. on Buckling of Structures. Harvard Univ. June 17-21, 1974, Springer 1974.

41. Potier-Ferry M. Amplitude modulation, phase modulation and localization of buckling patterns. IUTAM Symp. on Collapse : The Buckling of Structures London 1982. Cambridge Univ. Press, 1983.

42. Arbocz J. Shell stability analysis. Theory and practice. IUTAM Symp. on Collapse : The Buckling of Structures in Theory and Practice, London 1982. Cambridge Univ. Press, 1983.

43. Elishakoff I. How to introduce the imperfection sensitivity concept into design. IUTAM Symp. : The Buckling of Structures in Theory and Practice, London 1982. Cambridge Univ. Press, 1983.

44. Singer J. Vibrations and buckling of imperfect stiffened panels. Recent developments. IUTAM Symp. on Collapse : The Buckling of Structures in Theory and Practice, London 1982, Cambridge Univ. Press., 1983.

45. Olszak W. Sawczuk A. Inelastic Behaviour in Shells. P. Noordhoff, 1967.

46. Marcal P.V. A comparative study of numerical methods of elastoplastic analysis. Proc. AIAA/ASME 8th Structures, Structural Dynamics and Material Conf. March 1967.

47. Marcal P.V. Finite element analysis with material non-linearities. Theory and Practice. MARC CDC Background papers.

48. Harris H.G., Pifko A. Elasto-plastic buckling of stiffened rectangular plates. Proc. Symp. on Appl. of Finite Element Method in Civil Engineering Nov. 13-14, 1969 (Nashville, Tennessee U.S.A.).

49. Whang B. Elastoplastic orthotropic plates and shells. Proc. Symp. in Appl. of Finite Element method in Civil Engineering Nov. 13-14, 1969 (Nashville, Tennessee, U.S.A.)

50. Wempner G. Discrete approximation of elastic-plastic bodies by variational methods. Int. Conf. on Variational Methods in Engineering. Sept 25-29, 1972 (Southampton).

51. Bushnel D. Bosor 5. Program for buckling of elastic-plastic complex cylindrical shells of revolution including large deflections. Comp. and Struc. Vol. 6, 1976, p. 221-239.

52. Nguyen Quoc Son. Loi de comportement élasto-plastique des plaques et des coques minces. Symp. franco-polonais : Problèmes non-linéaires de Mécanique. Cracovie 1977.

53. Spilker R.L., Pian T.H.H. Hybrid stress models for elasto-plastic analysis by the initial stress approach. Int. J. Num. Meth. Engn. 14, p. 359-378, 1979.

54. Parisch H. Large displacements of shells including materials non-linearities. Comp. Meth. Appl. Mech. and Engn., Vol. 27, No 2, July 1981, p. 183-214.

55. Owen D.R.J., Figueras J.A. Elasto-plastic analysis of anisotropic plates and shells by the semi-loof element. Int. J. Num. Meth. Engn. Vol. 19, p. 521-539, (1983).

56. Blanchard D., Paumier J.C. Une justification de modèles de plaques visco-plastiques. R.A.I.R.O. Analyse Numérique, Vol. 18, No 4, 1984, p. 377-406.

57. Hill R. A general theory of uniqueness and stability in elastic-plastic solids. J. Mech. and Phys. of Solids, 1958, Vol. 6, p. 236-249.

58. Sewell M.J. A survey of plastic buckling. In Stability, ed. by H. Leipholz, 1972, Ch. V, p. 85-197.

59. Hutchinson J.W. Post-bifurcation behavior in the plastic range. J. Mech. Phys. of Solids, 1973, Vol. 21, p. 163-190.

60. Hutchinson J.W. Plastic buckling. Adv. in Appl. Mech. ed. by Chia-Shun Yih, Vol. 14, 1974 Ac. Press.

61. Nguyen Quoc Son, Radenkovic D. Stability of equilibrium of elastic-plastic solids. IUTAM Symp. : Appl. of Math. of Functional Analysis to Problems in Mech. Marseille 1975.

62. Nguyen Quoc Son, Radenkovic D. Stabilité et bifurcation en plasticité. Séminaire sur le Flambement des Structures, 20-24 mai 1980, ed. SEBTP, Coll. FNB, 1981.

63. Valid R. Principes variationnels non-linéaires, flambage et postflambage des coques élastiques et inélastiques en formulations mixtes (To appear).

64. Valid R. The structural stability criterion for mixed principles. In Hybrid and Mixed Finite Element Methods, ed. by. S.N. Atluri, R.H. Gallagher, and O.C. Zienkiewicz, 1983 (J. Wiley).

65. Ilyushin A.A. Plasticity (in Russian 1948), Eyrolles Paris, 1956.

66. Crisfield M.A. On an approximate yield criterion for thin steel shells. Dept. of the Environment, TRRL Report 658, Crowthorne 1974. (Transport and Road Research Laboratory).

67. Eidsheim O.M. and Larsen P.K. Non-linear analysis of elasto-plastic shells by hybrid stress finite elements. Comp. Meth. in Appl. Mech. and Engn. 34 (1982), p. 989-1018.

68. Naghdi P.M. Foundations of Elastic Shell Theory. In Progress in Solid Mechanics, I.N. Sneddon and R. Hill ed., North-Holland, 1963.

69. Almroth B.O., Stern P., Brogan F.A. Automatic choice of global shape functions in structural analysis. AIAA Journal, Vol. 16, No 5, (May 1978).

70. Noor A.K. Recent advances in reduction methods for the non-linear problems. Comp. and Struct. Vol. 13, 1981, p. 31-44.

71. Valid R. Une méthode de calcul des structures au flambage par sous-structuration et synthèse modale. C.R.A.Sc. Paris, t. 294 (1982). Série II p. 299-302.

72. Ohayon R. et Valid R. Principes variationnels symétriques couplés de type primal-dual en élastodynamique linéaire. C.R. Acad. Sciences, Paris, t. 297, série II, 28 Nov. 1983.

73. Ohayon R. et Valid R. Principes variationnels couplés primal-dual en élastodynamique linéaire : cas des coques minces. C.R. Acad. Sciences. Paris T. 298, série II, No 9, 1984.

Limit State of Steel Cylindrical Structures Under Earth-quake Loadings

H. AKIYAMA
Faculty of Engineering
University of Tokyo
Tokyo, Japan.

T. YUHARA
Nagasaki Technical Institute
Mitsubishi Heavy Industries, Ltd.
Nagasaki, Japan.

S. SHIMIZU
Kobe Shipyards and Engine Works, Mitsubishi Heavy Industries Ltd., Kobe Japan.

T. TAKAHASHI
Fast Breeder Reactor Project
Power Reactor & Nuclear Fuel Development Corporation, Tokyo, Japan.

Summary

Limit state induced by buckling of steel cylindrical structure under earthquake loadings is investigated from standpoint of energy concept. A number of the buckling tests of steel cylindrical shell structures have been made, which showed that they have the stable load-displacement relation and the adequate deformation capacities beyond the buckling.

The authors have proposed that energy input imparted by strong earthquakes to buckled structures and the deformation capacity in the post-buckling are suitable indices for seismic resistance of the steel cylindrical shell structures because the buckling does not cause the structure to immediately collapse in the case of such a repeated loading as earthquake motions.

The purpose of this study is to investigate the energy input to buckled steel cylindrical structures with the increase of intensity of earthquake motions. A series of the nonlinear dynamic analysis were performed under various types of the earthquake records by using the hysteresis loop including buckling, which was derived from the buckling tests.

The limit state could be defined as the state in which deformation and energy input to buckled structures increase divergently when the intensity of the earthquake excitation exceeds a certain value. The result obtained in this paper is intended to adopt the limit state in the post-buckling region to evaluate the margin of safety against the buckling resistance of steel cylindrical structures under strong earthquake loadings.

Introduction

The most important and primary question in the earthquake resistant design is:

What is the loading effect of the earthquake?

or

What is the difference between the gravity loading and the earthquake loading?

The simplest answer to these is:

The gravity loading is time-independent and the load itself

Inelastic Behaviour of Plates and Shells
IUTAM Symposium Rio de Janeiro 1985
Editors: L. Bevilacqua, R. Feijóo and R. Valid

absolutely proportional to the mass of structure.

Contrarily, the earthquake loading is time-dependent and is exerted during a very short time, say almost within one minute.

Thus, when these effects are expressed in terms of energy, they can be described as follows:

The gravity loading can supply infinitely large amount of energy to the structure, if the structure loses its static equilibrium. On the other hand, the energy which is applied to a structure during an earthquake is finite.

The energy input exerted by earthquakes has been found out to be a very stable amount which is mainly dependent on the total mass and the fundamental natural period of the structure. Therefore, it can be concluded that the loading effect of the earthquake should be measured by its energy input to the structure.

Fig. 1 demonstrates the loading effect of the gravity loading. The solid line indicates the load-deformation curve of the structure under the vertical load. The level of the gravity load is shown by the lines parallel to the abscissa. Under the level-A, the structure is stable and can resist to the load.

Under the level-C, the structure becomes unstable and cannot resist to the load. The limit-state for the gravity load corresponds to the loading level-B. Beyond the level-B, no equilibrium exists. Thus, when the buckling phenomena limit the load carrying capacity, to investigate the post-buckling behavior is meaningless.

Fig. 2 demonstrates the loading effect of the earthquake loading. The solid line indicates the load-deformation curve of the structure subjected to the horizontal load. The level of the seismic loading is shown by the lines parallel to the ordinate. Below the level-A, the structure remains elastic.

Under the level-B, the structure develops its maximum strength. Beyond the level-B, while the strength of structure decreases, the structure is still stable in the sense of dynamic equilibrium. Under the level-C, the structure loses its horizontal resistance and collapses ultimately.

The energy required to make the structure collapse is written as

$$E = \int_0^{\delta_u} Q \, d\delta \qquad (1)$$

where δ_u = deformation at the collapse point.

Therefore, the criterion for the aseismic design can be basically described by

$$E_R > E \qquad (2)$$

where E : energy input due to the earthquake
E_R: energy absorption capacity of the structure.

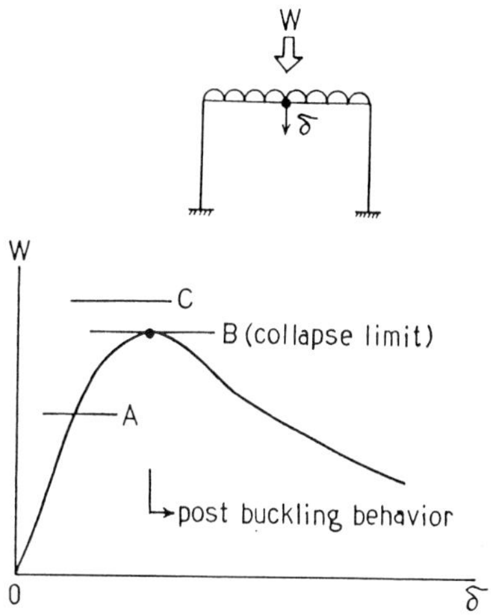

Fig. 1. Gravity Loading

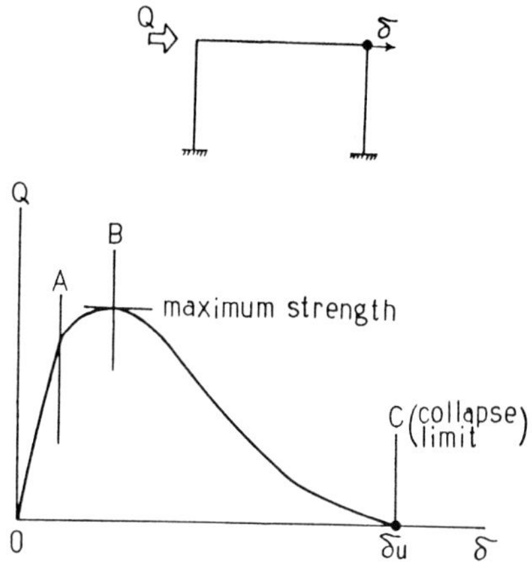

Fig. 2. Seismic Loading

Research on the buckling of the steel containment vessel of the fast breeder eactor "Monju" were carried out in 1983 and 1984 to evaluate the seismic safety against the buckling of the steel cylindrical shells with ring-stiffeners [3].

In this research programme, a number of static buckling tests of large, fabricated steel models have been included.

In this paper, the buckling resistance under strong earthquake motions will be tried to evaluate the results of the buckling tests by energy input concept.

The hysteresis under repeated transverse loads and deformation capacity are obtained from buckling tests of large, fabricated steel cylinder models under transverse loadings, simulating the overturning moment with transverse shear during the horizontal earthquake loadings.

Non-linear dynamic analysis of single-degree-of-freedom system is carried out, non-linear spring constant of which is the restoring force characteristics of regarding type. That is modelized from the experimental hysteresis loop under repeated loading.

By changing the stepwise the maximum acceleration of various seismic motion, the limit of the structure is investigated to consider the maximum displacement as the deformation capacity and to compare the energy input imparted by earthquakes with energy absorption capacity of the structure [4].

Buckling Test

Free standing, ring-stiffened steel containment vessel of FBR "Monju" as shown in Fig. 3(a), has the radius-to-thickness ratio R/t of 651.3 and ring-spacing-to-radius ratio ℓ/R of 0.27 except for 0.14 in the lowest part. Two steel cylindrical models, about 1/19 of full size, were fabricated by rolling and welding. These models are designated by BP-1 and BP-2 of which ring-stiffeners were disposed equivalently to the lower part of the actual containment vessel, as shown in Fig. 3(b).

The mechanical properties of the steel plating of each model are given in Table 1 by the tensile tests on the specimens cut from the same sheet of materials (JIS SAPH-45).

The initial imperfections of the fabricated steel models were measured precisely to obtain the local out-of-roundness and local out-of-straightness of the surface of the models. For example out-of-straightness is given in Table 2, which is the local shell imperfection between two ring-stiffeners.

Static buckling tests of transverse-loaded, ring-stiffened cylinders were carried out, as shown in Fig. 4, simulating the characteristic load distribution at the earthquake as shown in Fig. 5.

Under the constant axial compressive loads equivalent to dead weight and vertical earthquake load, the transverse load V was

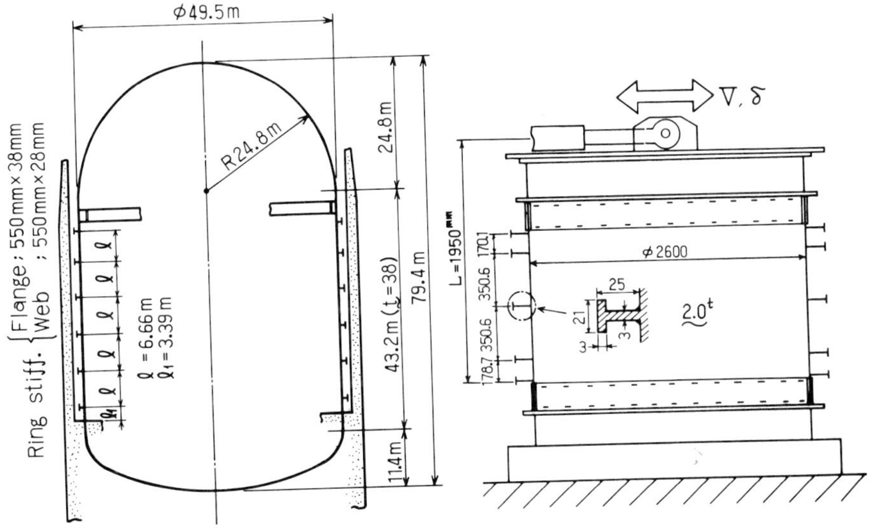

(a) Containment Vessel (b) Model

Fig. 3. Steel Containment Vessel and Cylindrical Model.

Table 1 Mechanical Properties

Model	Young's Modulus E (kg/mm^2)	Yield Stress σ_y (kg/mm^2)	Tensile Stress σ_B (kg/mm^2)	Elongation δ (%)
BP-1	20400	31.6	48.1	33.8
BP-2	20000	33.4	49.1	27.4

° Cross head speed 0.2 mm/min.

Table 2 Initial Imperfection (out-of-straightness)

(mm)

	θ	0°	45°	90°	135°	180°	225°	270°	315°
BP-1	C	1.15	1.17	1.10	1.10	0.95	1.20	1.15	0.98
	D	0.66	0.90	0.68	0.60	0.90	0.83	0.85	0.65
BP-2	C	0.90	1.15	0.65	0.78	0.60	0.57	1.06	0.52
	D	0.42	0.80	0.20	0.57	0.43	0.42	1.00	0.40

° Value is lateral displacement of shell plate between two ring-stiffeners. D is for lowest bay with $\ell/R=0.14$ and C is the bay above D with $\ell/R=0.27$.

Table 3 Results of Buckling Tests

Model	Buckling Load V_{cr}(ton)	Buckling Stress		Buckling Coefficient	
		Max. Bending Stress at Buckled Bay σ_b (kg/mm^2)	Max. Shearing Stress τ (kg/mm^2)	$C_b = \frac{\sigma_a^* + \sigma_b}{E \times (t/R)}$	$C_v = \frac{\tau}{E \times (t/R)}$
BP-1	<u>68.9</u>	11.4	8.4	0.39	0.26
BP-2	<u>69.1</u>	11.4	8.4	0.40	0.27

* σ_a; Axial stress 0.8 kg/mm^2 due to constant dead weight equal to 13 ton.

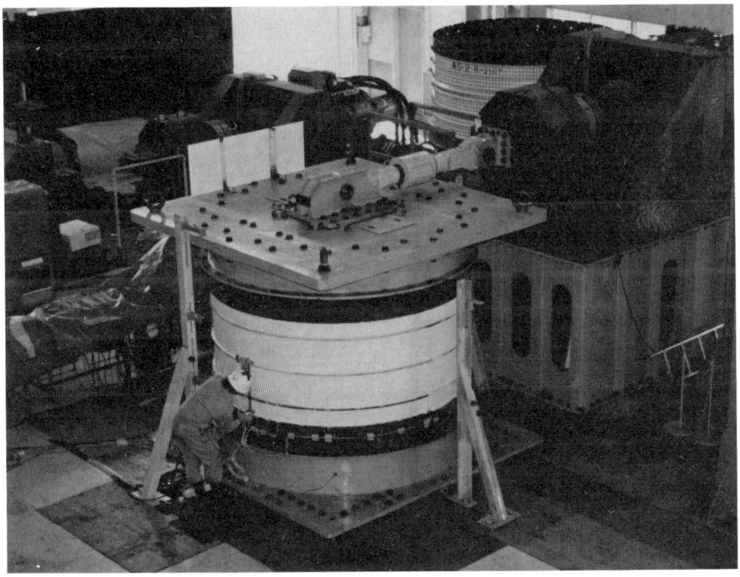

Fig.4 Test Set-up

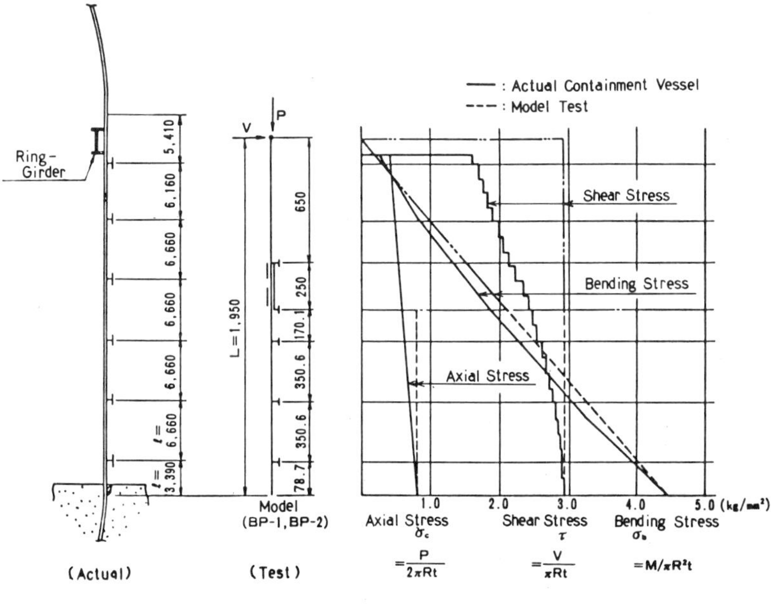

Fig.5 Load Condition under and Earthquake

applied to the top of the model at a distance L equal to 1.5R from the bottom. As shown in Fig. 6, a load V produces the maximum-bending-to-shear stress ratio σ_b/τ equal to 1.5 at the bottom of the model, which was predicted by a modal analysis of the containment vessel for a horizontal earthquake loading. Under constant dead weight load of 13 tonf, horizontal load was applied in forward direction until the buckling occurred by the hydraulic actuator (the maximum capacity of 125 tonf).

The results of the two tests are shown in Table 3.

The dominant mode of buckling is the shell buckling between two ring-stiffeners in the lower part due to transverse shear and bending loads, as shown in Fig. 7.

Because the moment of inertia of the ring-stiffener is adequate, the mode of buckling was limited to local one between two ring-stiffeners as shown in Fig. 7. The adjacent ring-stiffeners to this bay maintained the original shapes with no distortion.

The buckling load could be estimated by the extrapolation curves from the previous data [5][6], which were obtained by the buckling tests of the metal, unstiffened cylindrical models, as shown in Fig. 8.

Extrapolation from unstiffened, short cylinders to the ring-stiffened cylinders for the interframe buckling could be given by

$$\left(\frac{\sigma_{b,cr}}{\sigma_{bo}}\right)^2 + \left(\frac{\tau_{cr}}{\tau_s}\right)^2 = 1 \qquad (1)$$

Where $\sigma_{b,cr}$ in the maximum bending stress at buckling, and τ_{cr} is the maximum shear at 90° away from $\sigma_{b,cr}$.

σ_{bo} is the buckling stress due to pure bending, expressed by

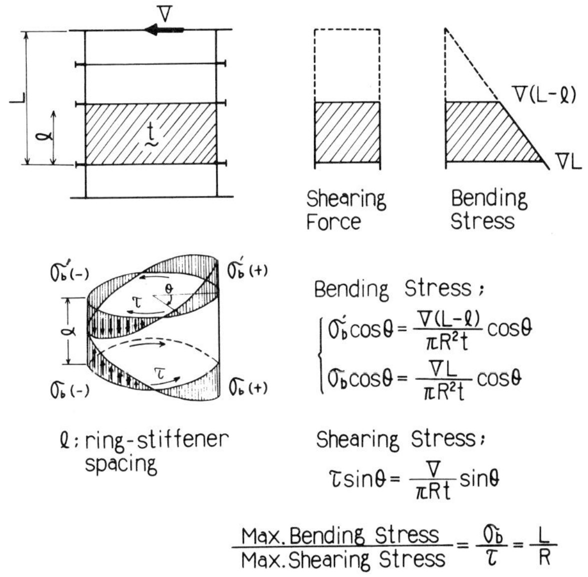

Fig.6 Maximum Bending-to-Shear Stress Ratio

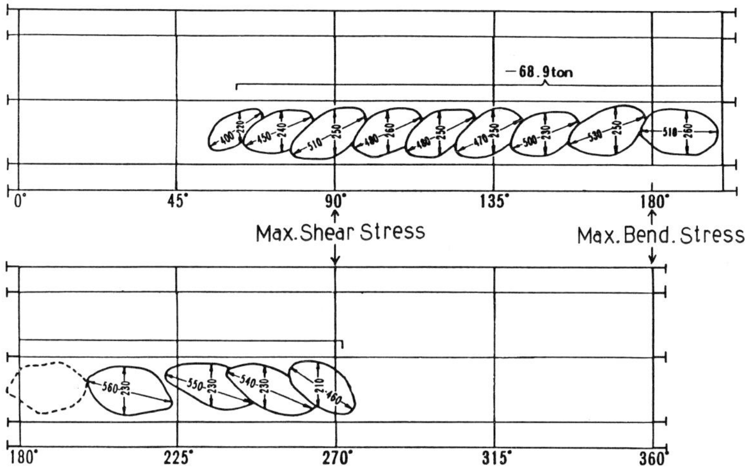

Fig.7 Buckled Pattern just after Initial Buckling of BP-1 Model

$$\sigma_{bo} = a \times \left(\frac{\ell}{\sqrt{RT}} \right)^{n} E \frac{t}{R} \tag{2}$$

and τ_s is the buckling stress due to pure shear with a sinusoidal distribution, expressed by

$$\tau_s = b \times \tau_t \tag{3}$$

Where τ_t is buckling stress due to pure torsion [8]

$$\tau_t = 4.83 \times (1/(\ell/\sqrt{Rt})^2) \times \sqrt{1+0.0239(\ell/\sqrt{Rt})^3} \times E(t/R) \tag{4}$$

From the NACA data of R/t=682, ℓ/R=0.5 and ℓ/R=1.0 shown in Fig. 8, a=0.64, b=1.3 and n=-0.14 were obtained as extrapolation curves.

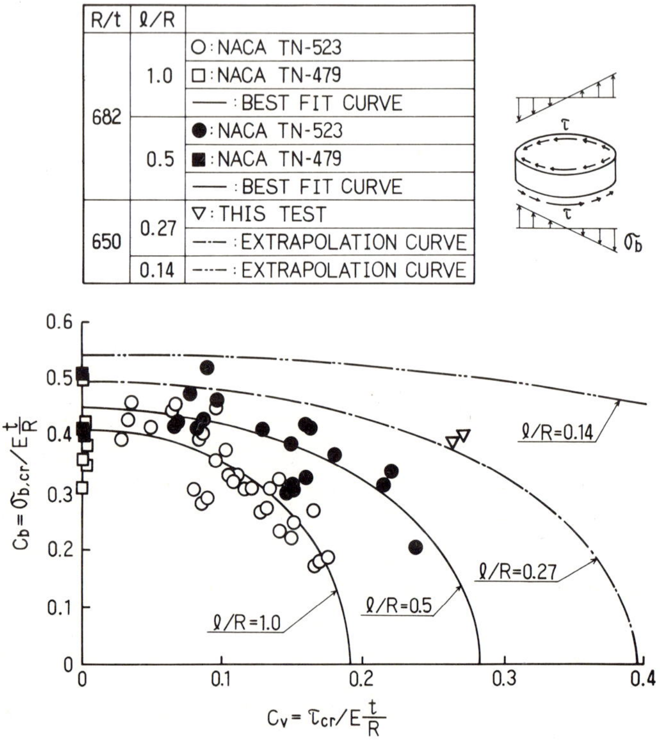

Fig. 8. Interaction Buckling with Bending Stress and Shear.

The results of these tests have the connection with the interaction-buckling results for shear and bending compression of the previous data in short, unstiffened metal cylinders.

The buckling loads in BP-1 and BP-2 scattered within a narrow range and reproduction of the bucklings under the same condition was quite satisfactory in these large, fabricated steel cylinders.

The ring-stiffener does increase only the buckling strength of the cylinders under shear and bending loadings, but also prevented the spread of the buckling deformation to the overall.

Post-Buckling Behaviour

After buckling due to bending moment with transverse shear, the load was applied continuously, and reversedly.

It was proved that the steel cylinders have the high capacity of energy absorption even after buckling with stable hysteresis of regular characteristics, typically as shown in Fig. 9. The deformation of the buckled cylinder was concentrated to the initially buckled bay, and a large deformation capacity of the whole structural system was shown. The final shape of the tested model was shown in Fig. 10.

The different loading-processes were given to each model in the post-buckling region.

Fig. 11 shows the load-displacement relationship between the transverse load V and horizontal displacement δ in the BP-1 model.

The initial buckling occurred over half the circumference of the shell plating, between two ring-stiffeners, second as counts from the bottom. Therefore, reverse transverse load was applied to develop buckling in the remaining half the circumference. Furthermore, repeated transverse load was applied under displacement control. The stable load-displacement hysteresis were obtained and deformation developed mainly in the buckled section

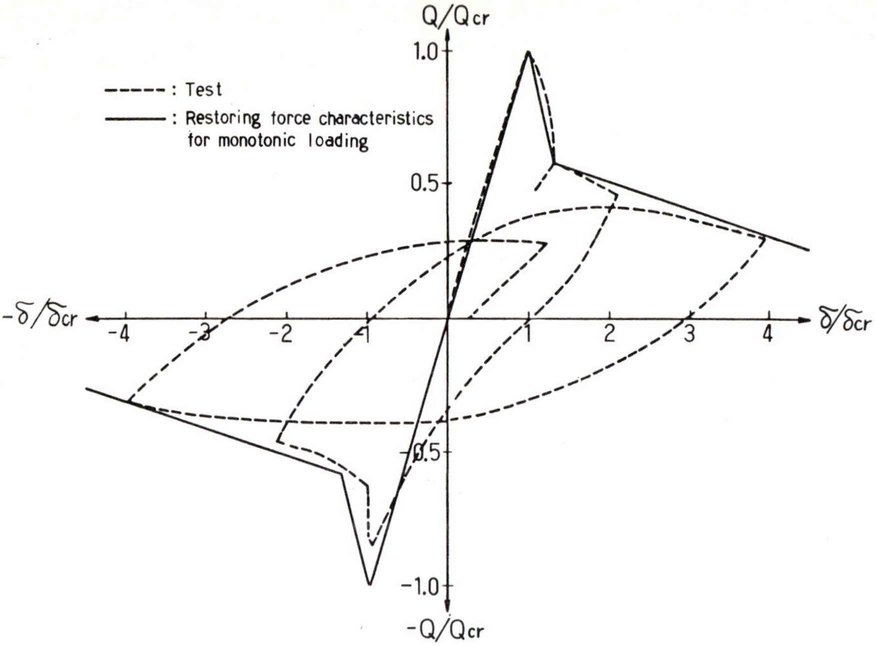

Fig.9 Typical Restoring Force Characteristics in Test

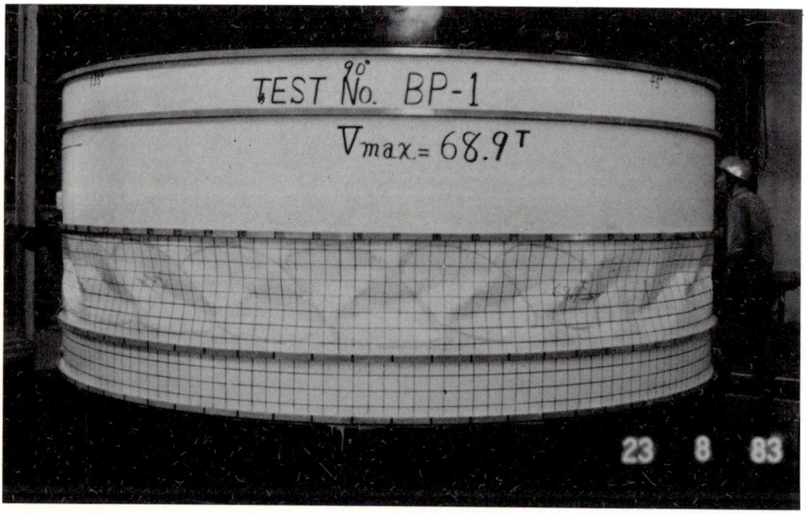

Fig.10 Buckled Shell between Ring-Stiffeners

between two ring-stiffeners. Buckling mode was diamond pattern between two ring-stiffeners which were not buckled and maintained the original shape.

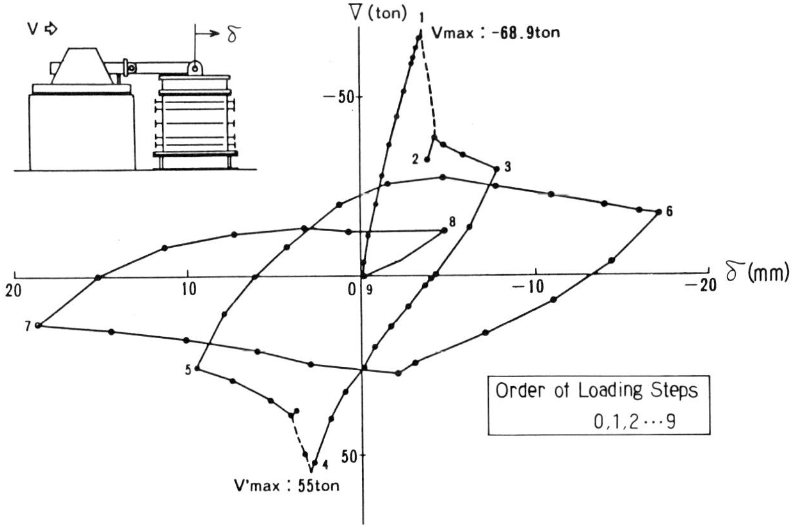

Fig. 11 Load-Displacement Relationship at the Top of Model BP-1

In BP-2 models shown in Fig. 12 a transverse load was applied repeatedly only in one direction after initial buckling, and thereafter a reverse one was applied to buckle the opposite side. In this case, the stable hysteresis with regular characteristics could be obtained.

Table 4 shows the total energy input E to the model applied by a transverse load measured in the tests. The quantities E were 12.5 to 14.5 times as large as the elastic strain W_e until buckling occurred (i.e. $W_e = V_{cr} \times \delta_{cr}/2$, where V_{cr} is the buckling load and δ_{cr} is elastic deformation at buckling).

Nevertheless the model neither becomes fissured nor collapsed, and behaved stably. From this fact, it is considered that the model was highly capable of energy absorption after buckling due to reciprocal transverse loads.

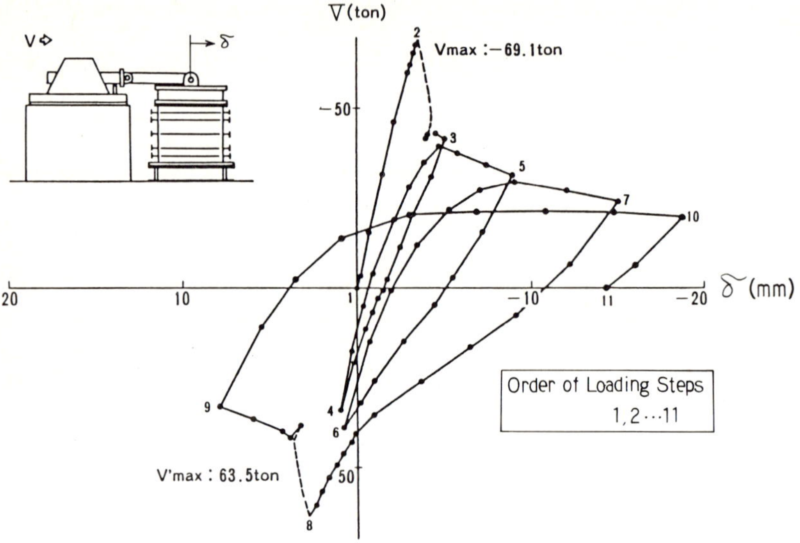

Fig. 12. Load-Displacement Relationship at the Top of Model BP-1.

Table 4. Total input energy into model before and after buckling.

Model	Energy input per half cycle ΔE_i (ton-mm)	Total energy input E (ton-mm)	Elastic energy W_e*1 (ton-mm)	Ratio E/W_e
BP-1	250, 352, 456, 538, 196	1972	124.0	14.5
BP-2	140, 20, 192, 56, 262, 500, 392	1562	120.9	12.9

*1 $W_e = \frac{1}{2} \times V_{cr} \times \delta_{cr}$

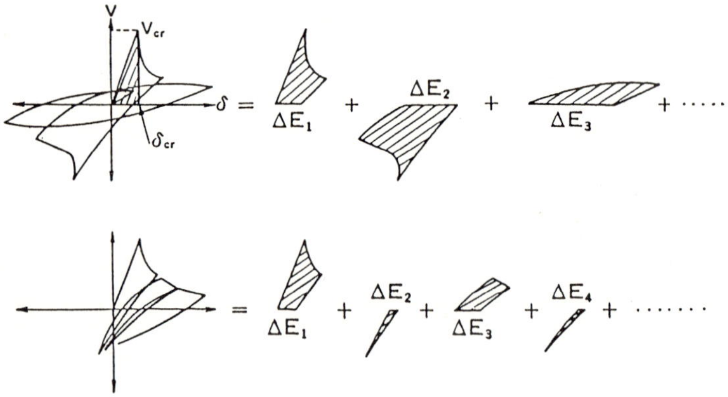

Hysteretic Rules

The results of the load-displacement relationships showed that a simplified rule for hysteresis and restoring force characteristics could be produced, as follows.

The converted force Q is used, which represents the transverse shear load concentrated to the center of gravity of an actual containment vessel and displacement δ is corresponding to this point.

1) Restoring force characteristics under monotonic loading

The Q-δ relation under positive loading and that under negative loading are doomed to be equal to each other. The restoring force characteristics under monotonic loading are shown in comparison with the experimental results in Fig. 9.

2) Hysteretic rules under repeated loading

Summarizing these test results, the post-buckling behavior of the cylindrical shells has been reduced to a hysteresis rule which governs the restoring-force characteristics of this type of structure. This hysteresis rule is of primary concern for us to evaluate the energy absorption capacity of the structure influenced by buckling.

To describe the hysteresis rule, some definitions are introduced.

- °The loading path and the unloading path are defined as follows:
  - a) The loading path is identified by the condition;

    $$Q\, d\delta > 0 \qquad (5)$$

  - b) The unloading path is identified by the condition;

    $$Q\, d\delta < 0 \qquad (6)$$

  where Q: load or stress
  δ: deformation
- °The skeleton curve is identified by the Q-δ curve under the monotonic loading.
- °The unloading point is defined by the point which rests

on the skeleton curve and terminates the loading path.

°The initial unloading path is defined by the point on which buckling occurs.

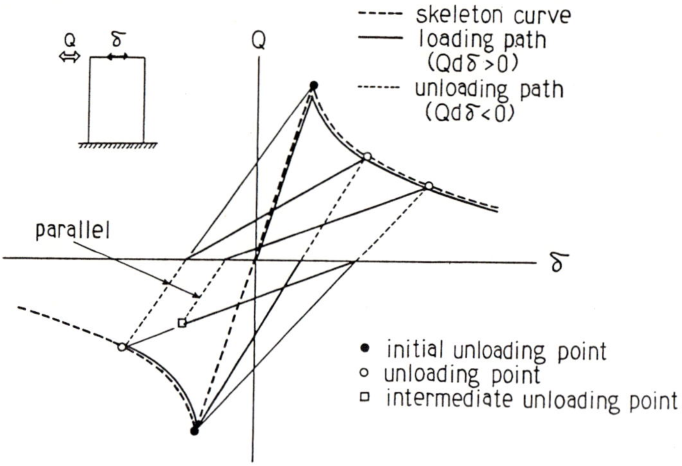

Fig. 13. Hysteresis Rule.

°The intermediate unloading point is defined by the point which terminates the loading path but does not reach the skeleton curve.

Under the assumption that the initial unloading points under positive and negative loading domains have been already experienced, the hysteresis rule is described as follows.

a) The loading path points to the preceding unloading point in the same loading domain. After the preceding unloading point is reached, the loading path follows the skeleton curve.
b) The loading path from the unloading point points to the initial unloading point in the reverse loading domain. The unloading path from the intermediate unloading point has the same slope as that of the preceding unloading path from the unloading point in the same loading domain.

Fig. 13 illustrates a typical pattern of restoring-force characteristics according to the above-mentioned hysteretic rule. In

Figs. 14 and 15, restoring-force characteristics are shown and compared with the predicted curve.

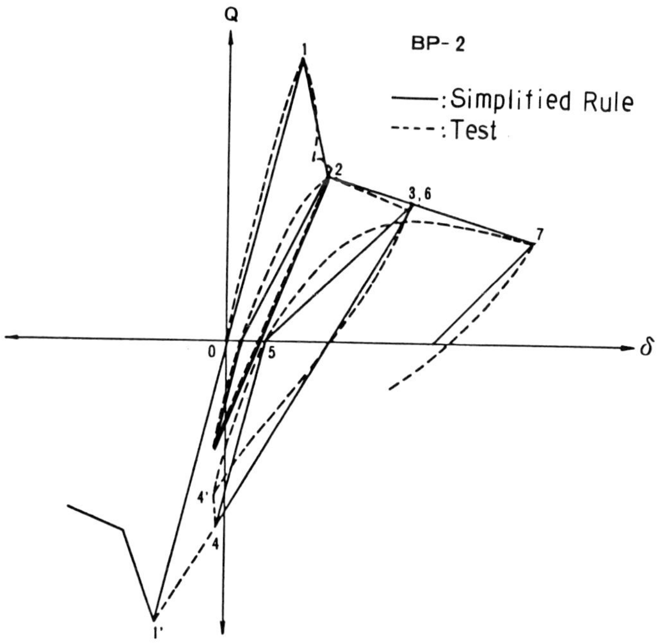

Fig. 14. Comparison of Simplified Load-Displacement Relation with Test Results(One Side Buckling).

It is considered that these rules give conservative estimates of areas enclosed by hysteresis loops.

The proposed hysteresis rule has been ascertained to be applicable to general cylindrical shells with a large variety of geometry [7].

Energy Input Imparted by Earthquake Motions

Buckling of shells limits the load carrying capacity of steel cylindrical shell, which is subjected to the overturning moment with transverse shear load during the horizontal earthquake motions.

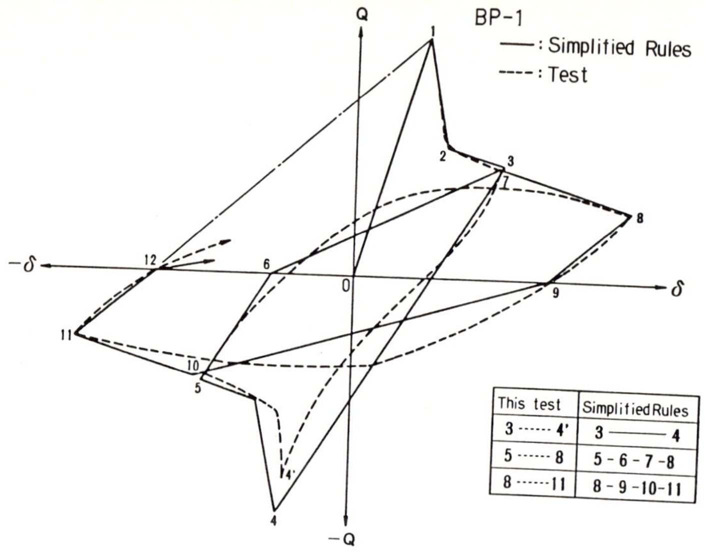

Fig. 15. Comparison of Simplified Load-Displacement Relation with Test Results (Both Sides Buckling).

The energy absorption and the maximum deformation in the post-buckling behavior of the steel containment vessel under strong earthquake excitations are investigated to evaluate the buckling resistance by the energy input concept [2].

The motion at the center of the gravity of the containment vessel can be expressed as for the fundamental mode of the deformation in the structural system as follows;

$$M\ddot{\delta} + C\dot{\delta} + F(\delta) = F_e \tag{7}$$

where M ; mass of the structure concentrated to the center of gravity

$C\dot{\delta}$; damping force

$F(\delta)$; restoring force

F_e ; seismic force $(= -M\ddot{Z}_o)$

Z_o ; horizontal ground motion

δ ; displacement at the center of gravity relative to the ground.

Multiplied by $d\dot{\delta} = \delta dt$ on both sides, and integrated over the entire duration of an earthquake, Eq. 7 is reduced to

$$M \int_{o}^{to} \ddot{\delta}\ \dot{\delta} dt + C \int_{o}^{to} \dot{\delta}^2 dt + \int_{o}^{to} F(\dot{\delta})\ \delta dt = \int_{o}^{to} F_e \dot{\delta} dt \qquad (8)$$

The first term of the left hand side expresses the kinetic energy. Second term expresses the energy consumed by the damping mechanism W_h. Third term expresses the strain energy deposited in the spring system, which consist of cumulative plastic strain energy W_p, and elastic strain energy at the instant when the earthquake motion fades away. The kinetic energy and the elastic strain constitute the elastic vibrational energy, W_e.

The right hand side of Eq. 8 expresses the total amount of input energy exerted by an earthquake.

The total energy input E exerted by an earthquake causes the structure to oscillate and is transformed into the elastic vibrational energy, W_e, the energy absorption caused by inelastic deformation W_p and the energy absorbed by damping W_h. Thus following equation holds

$$W_e + W_p + W_h = E \qquad (9)$$

The elastic vibrational energy has a range of

$$0 \leqq W_e < \frac{Q_{cr} \cdot \delta_{cr}}{2}$$

where Q_{cr}; buckling force

δ_{cr}; displacement at buckling

W_p is the accumulated effect of the post-buckling deformation of the spring-system.

The quantity $E_D = W_e + W_p$ means the energy input attributable to the damage, which is evaluated in comparison with the energy absorption capacity of the structure beyond the buckling and with the maximum deformation in post-buckling region.

In this study E_D is converted to the equivalent velocity V_D as follows:

$$V_D = \sqrt{2E_D/M} \tag{10}$$

The index related to energy absorption is denoted by nondimensional quantity V_D/V_E as follows.

$$V_D/V_E = \sqrt{(W_p + W_e)/W_e} = \sqrt{2E_D/(Q_{cr} \cdot \delta_{cr})} \tag{11}$$

Where the V_E is converted velocity related to the upper limit of the elastic strain energy W_e. W_e may be defined as $(Q_{cr} \cdot \delta_{cr})/2$, where Q_{cr} is the buckling load and δ_{cr} is elastic displacement at buckling.

Table 5. Maximum acceleration of ground motions.

Ground motion (Original wave)	Damping h (% of critical)	Maximum acceleration* (cm/s^2)	Maximum Velocity* (cm/s)
El-centro	0	1998	187
	1	3098	184
	2	4007	186
Sendai	0	1400	181
	1	2930	157
Tokachi-oki	0	2172	175
Artificial	0	2281	200

*Peak acceleration of ground motion was amplified to this acceleration which causes the buckling to structure.

Non-linear dynamic response was performed for the single-degree-of-freedom system, of which restoring force characteristics are modelled from the above mentioned. By integrating the response the energy input to a structure and cumulated inelastic strained energy were obtained.

Analysis and Results

The four earthquake records are selected as input ground motions, of which accelerations α_{max} are scaled and changed stepwise from the initial value α_{cr} upwards. Seismic waves with peak acceleration equal to α_{cr} cause the maximum displacement δ_{cr} related to buckling at the fundamental natural period of the structure with no damping.

The investigation focuses on the total energy input, inelastic strain energy and the maximum displacement δ_{max}, in comparison with the intensity of the earthquake, as characterised by the peak acceleration ratio $\gamma = \alpha_{max}/\alpha_{cr}$.

Parametric analyses were carried out for the steel containment vessel subjected to scaled earthquake motions, acceleration ratio γ of which ranges from 1.0 to 2.0.

The steel containment vessel was represented by one-degree-of-freedom system with the fundamental natural frequency ω of 6.4Hz and with the aforementioned restoring force characteristics which was scaled to the actual containment vessel. The buckling force Q_{cr} is equal to 2.5×10^7kg and corresponding displacement δ_{cr} is 5.1 cm. Input ground motions used are shown in Table 4 with the peak acceleration which causes the displacement equal to δ_{cr}. Fig. 16 shows the energy spectrum in elastic responses by these earthquake records with the peak accelerations scaled to 100 cm/s^2.

Fig. 17 shows the relation between the total energy input V_D/V_E and the intensity of earthquakes in the case of no damping, where total energy input to a structure is equal to total energy absorption. V_D/V_E begins to diverge when the intensity of the

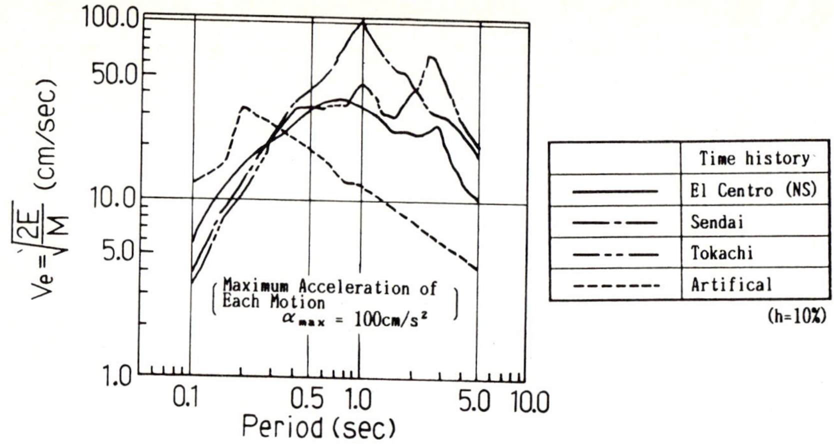

Fig. 16. Four Earthquake Waves.

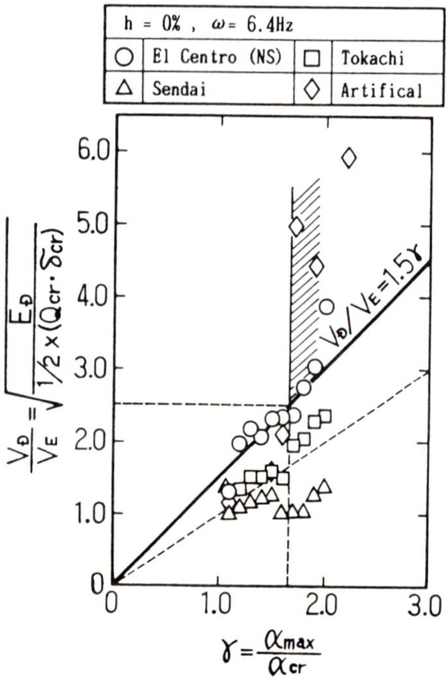

Fig. 17. Energy input to Structure E_D(h=0.0%).

earthquake reaches to some value beyond γ equal to 1.0. The portion beyond this value is shadowed in Fig. 17. Limit state can be defined as the state where V_D/V_E begins to diverge.

Fig. 18 shows the maximum displacement during the earthquake for the intensity of the earthquakes in the case of the no damping. In the same manner, such relations including the effects of damping of 1% and 2% of critical are shown in Fig. 19 and 20. Fig. 21 shows the relation between the maximum displacement and the inelastic strain energy absorptions by the post-buckling behaviors during the earthquakes by which the limit state is not reached.

The following results were obtained as to the effect of the severity of the earthquake on the total energy absorption and maximum displacement in the post-buckling region.

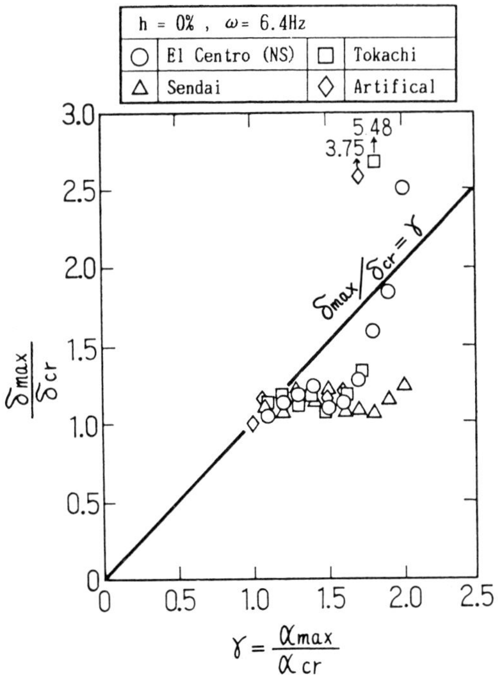

Fig. 18. Maximum Displacement.

(1) Limit state

The limit state where the energy absorption E_D and maximum displacement δ_{max} commence to increase divergently is reached when the severity of the earthquake γ reaches to a certain level as follows.

$\gamma = 1.67$ with no damping

($\gamma = 2.25$ with a damping of 1% and 2% of critical)

(2) Energy absorption from the initiation of buckling to the limit state.

Despite high irregularity of ground motions, the energy input to a buckled structure is a comparatively stable quantity.

The energy absorption V_D/V_E increases with a constant rate when the intensity of the earthquake γ increases, as shown in Fig. 17 and 19.

The relation between them can be shown as follows:

$$V_D/V_E = 1.5\gamma \qquad (12)$$

The constant coefficient equal to 1.5 in Eq. 12 shows the upper limit of the increasing effect of energy input to a buckled structure below the limit state.

Such an increase of energy input due to plastification is a significant feature of the high frequency structures while the energy input in the lower frequency structures such as buildings is generally stable amount[2].

(3) Maximum displacement.

When the earthquakes do not make the structure reach the limit state, the maximum displacement increases in proportion to the intensity of the earthquake γ. The maximum displacement is also of importance from the point of view of the functional requirement of a nuclear containment vessel. Before the limit state

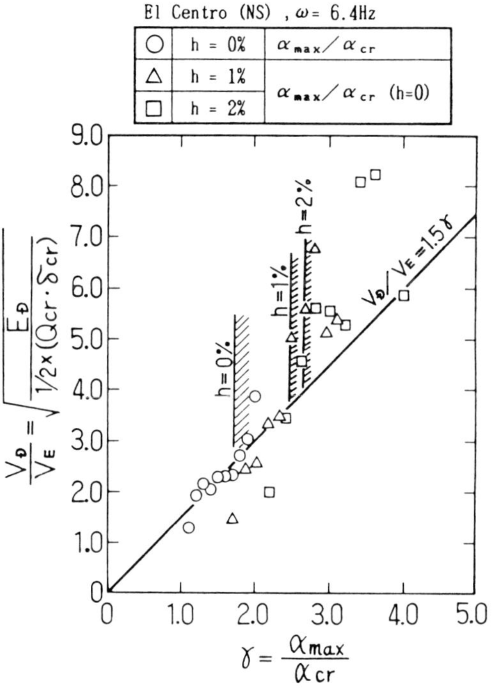

Fig. 19. Energy Input and Damping Effects.

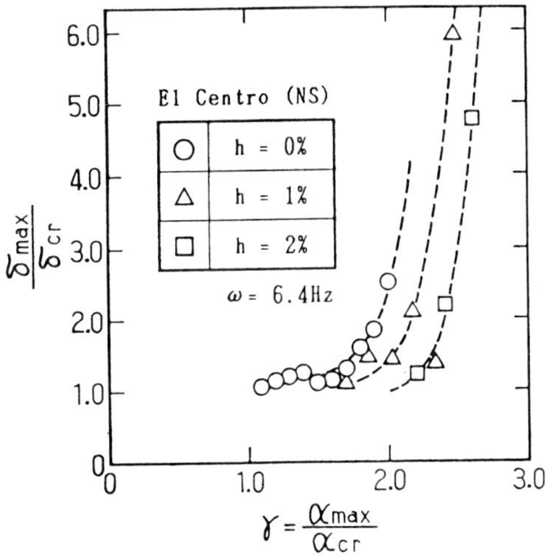

Fig. 20. Damping Effects on Maximum Displacement δ_{max}.

is not reached, it is of the same order as the elastic deformation at buckling and does not show a rapid rate of increase as shown in Fig. 18. Within such an extent, it is considered that the maximum displacement does not make an integrity of a containment vessel lose.

Below the limit state the maximum displacement ratio δ_{max}/δ_{cr} is proportional to the total inelastic strain energy imparted to a structure W_p normalized by its elastic strain W_e.

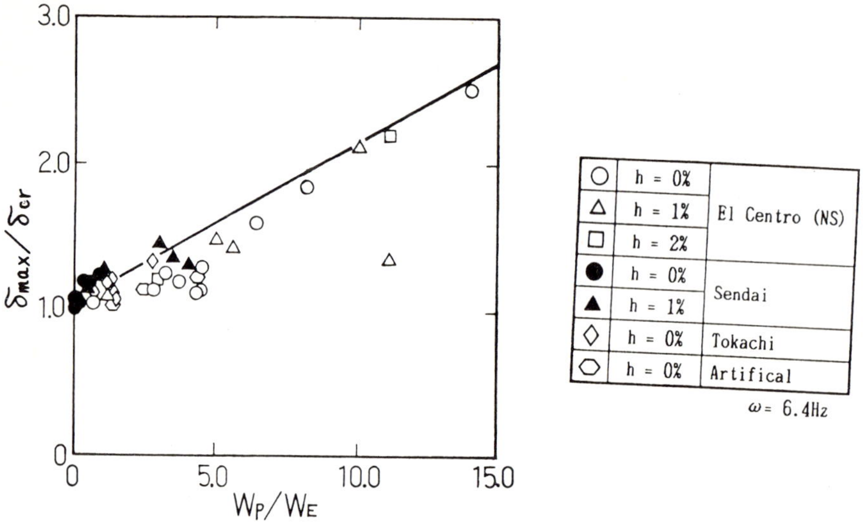

Fig. 21. Maximum Displacement δ_{max} and Inelastic Strain Energy W_p.

Conclusion

Occurrence of buckling due to seismic loading does not mean the catastrophic failure or immediate collapse, even in the case of the elastic buckling region of cylindrical shells of large R/t ratio dealt in this paper.

If the limit state is not reached during excitations, the total energy input to structure, which dissipated by post-buckling behavior, is limited to a certain extent. The maximum displacement remains about the same order of the initial critical displacement.

The intensity of the earthquake, by which the limit state can be reached, is 1.67 times as strong as the earthquake that can cause the buckling in the case of the steel containment vessel with no damping.

The margin of safety for the buckling resistance against a seismic loading should be set up for this limit state so far as the maximum displacement does not lose the integrity of steel containment vessel (in many cases it can be satisfied because the maximum displacement is not so excessive below the limit state).

At least, it may be inadequate and too conservative to apply the lower bounds of previous, scattered buckling data as a single criterion to prevent the buckling mode of failure.

Acknowledgement

Authors wish to thank Prof.Y. Asada of the University of Tokyo, Prof. H. Hangai of the University of Tokyo, Dr. S. Miyazono of the Japan Atomic Energy Research Institute for their valuable comments, who were the members of the committee (Chairman H. Akiyama) that evaluated the structural integrity of seismic buckling in the steel containment vessel of Prototype Fast Breeder Reactor. The buckling tests quoted in this paper are a part of the research and experimental work [3], carried out by Mitsubishi Heavy Industries, Ltd., sponsored by the Power Reactor & Nuclear Fuel Development Corporation (PNC).

References

1. Housner, G.W., "Limit design of structures to resist earthquakes", Proc. of 1st WCEE, 1956.
2. Akiyama, H., "Earthquake resistant limit-state design of buildings", University of Tokyo Press, April 1985.
3. Hagiwara, K., Yuhara, T., Imamura, N., Aoki, I., Matsumoto,

M., Shimizu, S., Yamada, Y., and Hayashi, K., "Buckling Tests on Monju Reactor Containment Vessel", (Confidential Research Report-PNC SJ 222 84-01, preparing for Power Reactor & Nuclear Fuel Development Corporation) Mitsubishi Heavy Indsutries, Ltd., March 1984.

4. Akiyama, H., Shimizu, S., Yuhara, T., and Morishita, M., "Ultimate limit state of metal containment shell buckling under seismic loadings", 8TH-SMIRT Conference, Brussels, August 1985.

5. Lundquist, E.E., "Strength tests of thin walled duralumin cylinders in pure bending", NACA-TN-No. 479, 1933.

6. Lundquist, E.E., "Strength tests of thin walled duralumin cylinders in combined transverse shear and bending", NACA-TN-No. 523, 1935.

7. Akiyama, H., et al., "Buckling Tests on Steel Cylindrical Shells Subjected to Combined Shear and Bending", Trans. of Architectural Institute of Japan (to be published).

8. Timoshenko, S.P., Gere, J.M., "Theory of Elastic Stability", McGraw-Hill, 1961.

Buckling of Spherical Domes Made of Microconcrete and Creep Buckling of Such Domes Under Long-term Loading

D. VANDEPITTE and G. LAGAE

Laboratorium voor Modelonderzoek
State University at Ghent

Summary

The major part of an investigation that began about 15 years ago consisted in testing 75 spherical model domes made of microconcrete. Some were subjected to rapidly increasing uniform radial pressure until they buckled, and some were subjected to a radial pressure held constant until they buckled (two of the latter specimens have not yet failed, ten years after the constant load was applied). Non-linear calculations reflecting the effects of the rapid loading and of the loading of long duration have been carried out and their results have been compared with the data obtained experimentally. The numerical results do not support GERARD's and KOLLAR's creep buckling hypothesis, while the set of experimental results of the long-term loading tests supports the hypothesis quite well in a statistical sense.

Test specimens

All the test specimens were nominally identical. Their geometry (fig. 1) was such that buckling occurred before the microconcrete

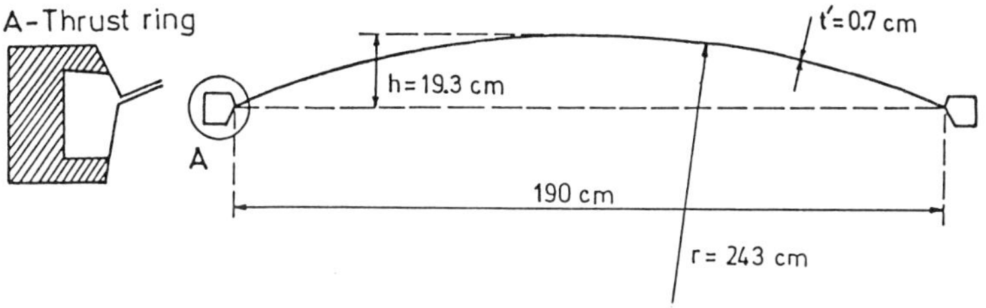

Fig. 1 - Geometry of test domes

would be crushed by excessive compression when a liquid applied inward radial pressure to the spherical domes. The microconcrete of the caps, the very rigid steel mould in which these were cast and the casting method were described in detail in reference [1]. It took us nearly a year of fruitless attempts before we succeeded in producing sound microconcrete shells of 0.7 cm

Inelastic Behaviour of Plates and Shells
IUTAM Symposium Rio de Janeiro 1985
Editors: L. Bevilacqua, R. Feijóo and R. Valid

thickness and with a circumference of 1.9 m diameter.

The edge of each model dome was cast in a groove in a steel ring (detail A in figure 1) designed to withstand the thrust exerted by the loaded shell. The thrust ring of more than half of the specimens was prestressed by means of 4 1/2" circumferential steel strands of the kind used for prestressing concrete structures and tensioned around the ring. The purpose of the prestressing was to compensate for the elastic extension of the steel ring due to the thrust and for the contraction of the concrete due to the liquid pressure, so that a membrane state of stress was achieved in the cap (in the creep tests) or so that a membrane state was achieved approximately (in the rapid loading tests).

The modulus of elasticity and the creep strains of the micro-concrete were measured on thin bars ($2 \times 1 \times 10$ cm³) which were cast on the same day and from the same batch as the corresponding model dome. The bars were loaded so that the uniaxial compressive stress equalled the biaxial compressive stress in the cap as calculated with membrane theory. For most of the model domes subjected to permanent loading, the changes of length of unloaded bars, otherwise kept under the same conditions as the loaded bars and due to changes of temperature and moisture, were also recorded.

Buckling formulae and computer codes for spheres and for spherical caps

Actual buckling pressures p_u can be expressed in terms of the critical uniform radial pressure p_{cl} given for a perfect elastic sphere by ZOELLY's classical formula [2] :

$$p_{cl} = \frac{2E}{\sqrt{3(1-\nu^2)}} \left(\frac{t'}{r}\right)^2 \qquad (1)$$

where r and t' are the radius and the thickness of the sphere, respectively.

When POISSON's coefficient ν is assumed to be 0.2 , a reasonable value for uncracked concrete, equation *(1)* becomes

$$p_{cl} = \gamma E \left(\frac{t'}{r}\right)^2 , \quad \text{with} \quad \gamma = 1.18 \qquad (2)$$

Several authors have tried to account for low experimental buckling pressures by non-linear theories, which also led to formula *(2)*, but with lower values of γ. von KARMAN and TSIEN found $\gamma = 0.365$ and TSIEN [3] found $\gamma = 0.34$.

The buckling pressure p_1 for a spherical cap depends on its geometry, which, according to BUDIANSKY [4], is represented by the parameter

$$\lambda = 2\sqrt[4]{3(1-\nu^2)}\sqrt{\frac{h}{t'}} \qquad (3)$$

where t' is the thickness of the cap and h is its height (fig. 1). In figure 2, the ratio p_1/p_{cl} found by BUDIANSKY on the assumption that the buckling mode is axisymmetric is given as a function of λ.

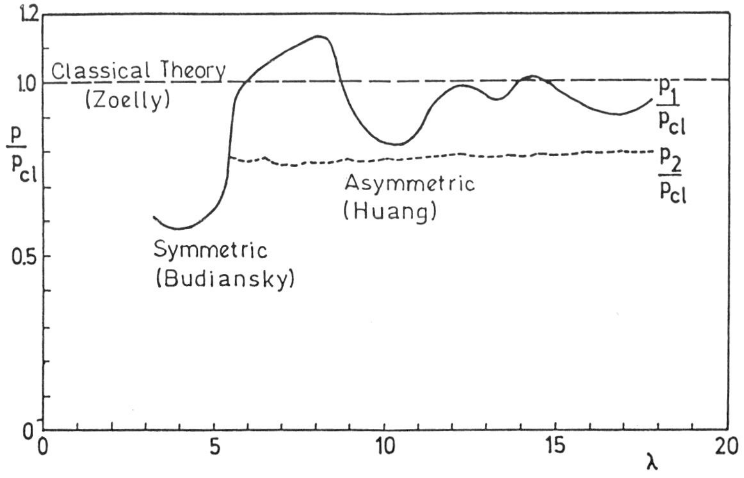

Fig. 2 - Theoretical buckling pressures

Other authors have considered unsymmetrical buckling modes, characterized by a number, n, of complete buckling waves along a parallel circle, when studying bifurcation of the equilibrium of the cap. HUANG [5] has obtained buckling pressures p_2 which are also represented in terms of p_{cl} in figure 2. For perfect calottes with $\lambda > 5.5$ HUANG's curve is located below BUDIANSKY's and buckling should be expected to be unsymmetrical.

Nominally, all our test domes had the same λ, but in fact, there were slight variations in the real geometry and λ actually ranged between *13.321* and *14.266* , with an average slightly below *14*.

The computer program BOSOR 4 [6] allows the buckling load of elastic shells of revolution to be determined, taking into account possible unsymmetrical buckling modes and non-linear relations between displacements, strains and curvatures, as well as the real boundary conditions. Thus the increase in diameter of the steel ring under the influence of the thrust, the rotational restraint of the edge of the calotte provided by the thrust ring, and the effect of prestressing a thrust ring could be considered in the calculations. p_b denotes the elastic buckling pressure resulting from BOSOR 4 computations. BOSOR 4 does not account for non-linear material behavior.

Rapid tests on domes with unprestressed thrust rings

In the "rapid" tests the liquid pressure was increased gradually until failure of the dome ensued. Each test was completed within a few hours.

Table I contains the main results of the tests on 23 caps whose thrust ring was not prestressed. $\bar{t}'$ is the average thickness and $\bar{r}$ is the average radius of curvature of the model shell considered. f_{21} is the strength of the concrete, measured on 8×8×8 cm³ cubes

Table I - Rapid tests - Ring not prestressed

Nr	$\bar{t}'$ (mm)	$\bar{r}$ (mm)	f_{21} (N/mm²)	p_u (N/cm²)	E_o (N/mm²)	γ	$\frac{p_u}{p_{cl}}$	$\frac{p_u}{p_1}$	$\frac{p_u}{p_2}$	$\frac{p_u}{p_b}$
K4	7.12	2431	60	14.2	30 356	0.545	0.465	0.484	0.592	0.703
K6	7.07	2438	46	11.8	25 667	0.546	0.465	0.482	0.593	0.703
K7	6.91	2428	59	11.3	22 799	0.611	0.520	0.533	0.665	0.787
K9	6.98	2450	66	13.7	29 013	0.583	0.497	0.509	0.635	0.751
K33	6.91	2469	51	11.9	33 652	0.450	0.384	0.391	0.489	0.580
K36	6.90	2479	76	13.2	31 527	0.542	0.462	0.469	0.589	0.698
K38	6.86	2524	65	11.3	29 269	0.521	0.444	0.449	0.564	0.672
KN12	7.07	2538	64	11.6	26 126	0.568	0.484	0.492	0.617	0.732
KN15	6.94	2554	66	11.8	26 537	0.604	0.512	0.518	0.650	0.774
KN19	6.99	2537	71	12.9	34 230	0.498	0.425	0.430	0.541	0.642
KN24	7.10	2560	65	12.6	32 716	0.499	0.425	0.430	0.541	0.643
KN26	7.06	2546	59	10.2	26 766	0.496	0.422	0.428	0.538	0.638
KN27	7.09	2557	62	11.9	30 523	0.506	0.431	0.437	0.549	0.652
KN28	7.01	2540	61	11.8	30 033	0.515	0.438	0.444	0.558	0.663
KN29	6.76	2535	58	11.8	31 449	0.526	0.448	0.452	0.568	0.678
KN30	6.98	2515	65	12.6	32 958	0.496	0.423	0.428	0.538	0.639
KN31	7.04	-	58	13.5	33 825	0.518	0.441	0.448	0.562	0.667
KN32	6.93	-	55	10.4	26 904	0.516	0.440	0.445	0.560	0.665
KN33	7.02	-	62	12.1	26 888	0.584	0.498	0.504	0.634	0.753
KN34	6.94	-	63	13.1	28 962	0.604	0.515	0.521	0.655	0.779
KN35	6.83	-	61	14.9	34 208	0.600	0.511	0.516	0.647	0.772
KN36	6.98	-	61	15.5	34 760	0.587	0.500	0.507	0.637	0.756
KN37	6.89	-	59	11.8	33 326	0.477	0.407	0.411	0.517	0.615
$\bar{x}$	6.97	2506	62	12.5	30 108	0.539	0.459	0.466	0.584	0.694
δ(%)	1.3	1.9	9.8	10.7	11.1	8.5	8.5	8.6	8.5	8.5

that were 21 days old. p_u is the experimental buckling pressure. $E_o = \sigma/\varepsilon$ is the modulus of elasticity derived from the strain ε of $(2\times1\times10\ cm^3)$ microconcrete bars on the day of the test under a compressive stress σ equal to the membrane stress produced in the test dome by the failure load. Hence E_o is a secans modulus.

Buckling strength of the model domes

p_u varies between 10.2 and $15.5\ N/cm^2$, with an average of $12.5\ N/cm^2$ and a coefficient of variation δ of $10.7\,\%$, in spite of the fact that all test specimens were made with the same aggregates and with the same kind of cement (but the properties of cement exhibit seasonal fluctuations). E_o varies even more than f_{21} and p_u.

The numbers γ in Table I were calculated from the equation

$$p_u = \gamma E_o \left(\frac{\bar{t}'}{\bar{r}}\right)^2 \qquad (4)$$

The correlation between the buckling load of a shell and $\bar{t}'/\bar{r}$ seems to be better than that between the buckling load and $(t'/r)_{min}$ [7]. In our tests failure occurred mostly in an area where t' was smaller than $\bar{t}'$, but not systematically where t' was minimal. γ is 0.539 on the average, but ranges from 0.450 to 0.611.

In the last four columns of Table I, p_u is compared with p_{cl}, p_1, p_2 and p_b. Although much care has been taken to keep imperfections to a minimum, p_u is always considerably lower than ZOELLY's classical buckling pressure p_{cl}. The average value of p_u/p_{cl} is 0.459. Nevertheless, γ appreciably exceeds von KARMAN's and TSIEN's value 0.365 for every single test specimen.

p_u is also lower than p_1, p_2 and p_b, but is closer to p_2 than to p_1 and closer to p_b than to p_2. Average ratios are $p_u/p_1 = 0.466$, $p_u/p_2 = 0.584$ and $p_u/p_b = 0.694$, and δ in each case is about $8.5\,\%$. It is not surprising that BOSOR 4 provides a better approximation than either BUDIANSKY's or HUANG's theory, for the BOSOR calculation reflects the boundary conditions more faithfully.

Yet p_u is still $30.6\,\%$ lower than p_b, on an average. This is presumably due to a number of factors :

1) The inherent heterogeneity of the material microconcrete.
2) The non-linear behavior of the material.
3) The fact that we measured E_o on bars compressed in one direction, while the material of the model domes, in their membrane state of stress, was subjected to equal compression in all

directions parallel to a tangent plane.

4) The radial displacements of the thrust ring were somewhat impeded by the steel dish which, together with the model dome, enclosed the water exerting the test pressure on the dome [1]. Hence the boundary conditions assumed in the calculations were not quite correct.

5) The imperfections of the test shells. All theoretical results, including the BOSOR 4 values p_b , hold true for shells that are perfectly spherical in the unloaded condition.

In view of the notorious imperfection sensitivity of spherical shells subjected to radial pressure, the last factor is probably the most influential one.

An indication to the same effect is given by the results of two comparative calculations, both assuming the overall geometry shown in figure 1, a rigidly clamped edge along the perimeter of the dome, $E = 29\,430\ N/mm^2$ and $\nu = 0.18$. One of the shells is perfectly spherical, with a circular meridian of radius $r = 245\ cm$.

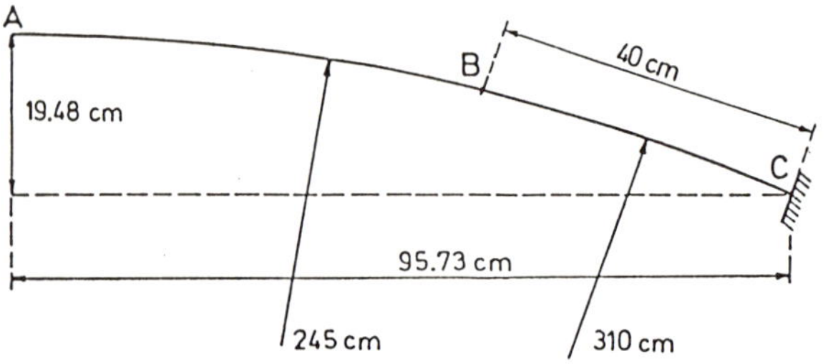

Fig.3 - Meridian of imperfect dome

Each half meridian *ABC* (fig. 3) of the other shell is obtained by replacing the *40 cm* long stretch *BC* of a perfectly circular arc of radius *245 cm* by a circular arc of radius *310 cm* adjoining the clamped edge *C* and thus contains an angular discontinuity of *0.98* degrees at the junction of the two arcs in *B* . *310 cm* was, roughly, the average of the highest radii of curvature locally recorded in the test caps and we assumed that value to hold throughout the part where buckling was likely to occur (as will be seen in the next section). The buckling pressure obtained by means of the BOSOR 4 program was $21.7\ N/cm^2$ for the perfectly spherical dome and $17.7\ N/cm^2$ for the dome comprising a flatter area along its circumference. Hence, according to these computations, the presence of this flatter annular zone would account for a *19%* decrease in buckling strength.

Buckling mode

HUANG's theory and the BOSOR 4 program predict an unsymmetrical buckling failure, with a certain number of dents and bulges in the circumferential direction.

All our model domes failed in the same manner : an almost circular, sometimes slightly elliptic part of the shell wall, broken into a number of sectors or pieces of another shape, was punched out towards the center of the sphere [12]. This may be interpreted as follows : along the perimeter of one of the buckling dents, perhaps of the deepest dimple, the brittle unreinforced concrete cracked at the convex side of the dome and was crushed by excessive local bending at the concave side, while it was crushed along a number of radiuses of the dent at the convex side, and the punched hole was a frozen materialization of the buckling dent. After each collapse the fragments of the punched out disk could be pieced together. The disk was always nearly circular : the ratio of its greatest to its smallest dimension seldom exceeded *1.15*. Its size was also remarkably constant and the average of the mean of both diameters was *36 cm*, which is equal to $2.76\sqrt{rt'}$.

The BOSOR 4 calculation predicted oblong buckling waves with their greatest dimension in the meridional and their smallest dimension in the circumferential direction. On an average, the actual size of the punched out disks was only *61 %* of the size the BOSOR program led to expect in the meridional direction, and it was *139 %* of the theoretical size in the circumferential direction.

According to the calculation the buckling waves should appear close to the edge of the dome. In most cases a hole was indeed punched into the cap in the vicinity of its edge. There were very few cases in which failure occurred in the neighborhood of the crown of the shell, perhaps because the shell wall had a local weakness there.

Rapid tests on domes with prestressed thrust rings

Table II, in which the symbols have the same meaning as in Table I, is a collection of data concerning 11 tests in which the radial pressure on the dome and the circumferential prestressing force around the ring were increased together in four stages, up to $8\ N/cm^2$ and to *400 kN*, respectively. Thereafter, the prestressing force was kept constant, while the liquid pressure was augmented

Table II - Rapid tests - Ring prestressed

Nr	$\bar{t}'$ (mm)	$\bar{r}$ (mm)	f_{21} (N/mm^2)	p_u (N/cm^2)	E_o (N/mm^2)	γ	$\frac{p_u}{p_{cl}}$	$\frac{p_u}{p_1}$	$\frac{p_u}{p_2}$	$\frac{p_u}{p_b}$
K11	7.04	2418	70	14.0	23 300	0.710	0.605	0.628	0.771	0.771
K12	6.97	2431	54	12.8	30 409	0.510	0.435	0.447	0.554	0.554
K13	7.18	2419	67	14.2	27 124	0.595	0.507	0.530	0.646	0.646
K14	6.95	2425	72	15.7	25 852	0.739	0.630	0.648	0.803	0.803
K15	7.01	2402	78	14.7	28 729	0.601	0.512	0.533	0.653	0.653
K16	7.14	2339	60	13.2	24 786	0.573	0.488	0.512	0.624	0.623
K17	7.13	2435	66	16.2	26 432	0.714	0.608	0.634	0.775	0.776
K19	7.10	-	72	15.2	24 016	0.750	0.639	0.662	0.813	0.814
K27	6.90	2416	71	14.4	28 092	0.629	0.536	0.550	0.685	0.683
K28	6.85	2418	65	11.8	28 712	0.511	0.435	0.445	0.556	0.555
K34	6.93	2484	50	11.1	28 481	0.504	0.430	0.436	0.547	0.547
$\bar{x}$	7.02	2418.7	66	13.9	26 903	0.621	0.530	0.548	0.675	0.674
δ (%)	1.6	1.5	12.9	11.3	8.3	15.2	15.1	15.5	15.1	15.1

until failure resulted. Since the buckling pressure in one case was as high as $16.2\ N/cm^2$, the prestressing force was insufficient to produce a membrane state of stress in the dome and bending did arise during the last loading stages. Nevertheless, the carrying capacity was higher on the whole (though not in every single test) than for the domes whose thrust ring was not prestressed : $\bar{p}_u = 13.9\ N/cm^2$ instead of $\bar{p}_u = 12.5\ N/cm^2$. $\bar{\gamma}$ is also higher, of course : 0.621 instead of 0.539, and much higher than von KARMAN's and TSIEN's value 0.365 for complete spheres. γ varies more from specimen to specimen than in the series of tests with unprestressed rings.

BUDIANSKY and HUANG did not assume a membrane state of stress in their theories. Yet it is not inappropriate to list the values of p_u/p_1 and p_u/p_2 in Table II, because such a state did not exist just prior to failure.

Although p_u is generally higher than for the domes with unprestressed thrust rings, the ratio p_u/p_b is somewhat lower : the average value of the ratio is only 0.674 instead of 0.694, because the beneficial effect of the prestress is taken into account in the BOSOR calculation and leads to a $19\,\%$ higher value of p_b.

The failure mode was the same as in the tests with unprestressed rings, but the punched holes were slightly smaller.

Creep buckling calculations

The BOSOR 5 version of BUSHNELL's computer code [8 – 9] has the capability of accounting for the effect of creep deformations on the buckling load. It is assumed in the BOSOR 5 program that the effective creep strain $\bar{\varepsilon}_c$ varies with time in accordance with an equation of the type

$$\bar{\varepsilon}_c = \left(\frac{\bar{\sigma}}{\sigma_o}\right)^{\alpha} t^{\beta} \qquad (5)$$

where σ_o , α and β are constants,

t is the time elapsed after application of the load,

$\bar{\sigma} = (\sigma_1^2 + \sigma_2^2 - \sigma_1\sigma_2)^{1/2}$ represents the effective stress, σ_1 and σ_2 being the meridional and the circumferential stress in the shell, respectively. For a spherical dome in a membrane state of stress we have $\sigma_1 = \sigma_2$ and $\bar{\sigma} = \sigma_1 = \sigma_2$.

If the shell material is loaded into the plastic range during a load increment, the total strains ε_1 and ε_2 are composed of the total strains at the beginning of the load step, plus the new elastic strains

$$\frac{1}{E}(\Delta\sigma_1 - \nu.\Delta\sigma_2) \quad \text{and} \quad \frac{1}{E}(\Delta\sigma_2 - \nu.\Delta\sigma_1)$$

due to the current load step, plus the new plastic and creep increments

$$\Delta\bar{\varepsilon}_p \cdot \frac{\partial\bar{\sigma}}{\partial\sigma_i} \quad \text{and} \quad \Delta\bar{\varepsilon}_c \cdot \frac{\partial\bar{\sigma}}{\partial\sigma_i} \qquad (i = 1, 2)$$

due to the current load step. The strain increments are thus

$$\Delta\varepsilon_1 = \frac{1}{E}(\Delta\sigma_1 - \nu.\Delta\sigma_2) + (\Delta\bar{\varepsilon}_p + \Delta\bar{\varepsilon}_c)\frac{\partial\bar{\sigma}}{\partial\sigma_1} \qquad (6)$$

$$\Delta\varepsilon_2 = \frac{1}{E}(\Delta\sigma_2 - \nu.\Delta\sigma_1) + (\Delta\bar{\varepsilon}_p + \Delta\bar{\varepsilon}_c)\frac{\partial\bar{\sigma}}{\partial\sigma_2} \qquad (7)$$

where

$$\frac{\partial\bar{\sigma}}{\partial\sigma_1} = (\sigma_1 - \frac{\sigma_2}{2})/\bar{\sigma} \quad \text{and} \quad \frac{\partial\bar{\sigma}}{\partial\sigma_2} = \left(\sigma_2 - \frac{\sigma_1}{2}\right)/\bar{\sigma} \qquad (8)$$

The system of equations *(6)* and *(7)* contains 3 unknowns : $\Delta\sigma_1$, $\Delta\sigma_2$ and $\Delta\bar{\varepsilon}_p$. $\Delta\bar{\varepsilon}_c$ is a known function of $\bar{\sigma}$ and t, and $\bar{\sigma}$ is a known function of $\sigma_{io} + \Delta\sigma_i$, in which σ_{io} , the stress at the beginning of the load step, is known. The third equation, needed in general for solving the system, is the given stress-strain curve, by which an effective stress increment is related to an effective plastic strain increment.

When employing the BOSOR 5 computer program, we have assumed linearly elastic material behavior, except for creep, and we have dropped the terms having the index p in the above equations. The acting stress seldom exceeded $f_{21}/4$.

The creep strains we measured on the $2\times1\times10\ cm^3$ microconcrete bars exhibited considerable scatter, as creep strains always do. Moreover, the compressive stress applied to the bars varied only within the rather narrow interval $7 - 17.5\ N/mm^2$. The rather bewildering array of recorded creep strains could not easily be compressed into a law of the type defined by equation (5). The constants that seemed to produce the best data fitting formula of that type are $\sigma_o = 98\,100\ N/mm^2$, $\alpha = 1$, $\beta = 0.4$, and the corresponding creep equation is

$$\bar{\varepsilon}_c = \frac{\bar{\sigma}}{98\,100}\, t^{0.4} \qquad (9)$$

with $\bar{\sigma}$ expressed in N/mm^2 and t in days. Equation (9) holds fairly well for $0 < t < 125$ days, but its validity is somewhat doubtful for longer loading periods, because not many bars were loaded during a longer span of time.

We performed creep buckling calculations, using equation (9), for the imperfect dome whose meridian is defined in figure 3, assuming that its support precludes any linear displacement and any rotation along its circumference and again letting $E = 29\,430\ N/mm^2$ and $\nu = 0.18$. Besides the rapid loading case mentioned before, three levels of permanent pressure were considered :

$p = 14.71\ N/cm^2$ $\qquad p = 11.77\ N/cm^2$ $\qquad$ and $\qquad p = 8.83\ N/cm^2$.

The calculated maximum immediate deflection of the shell under the radial pressure $p = 14.71\ N/cm^2$ is $w = 2.83\ mm$. When the pressure is kept constant, the maximum deflection amounts to $w = 4.86\ mm$ after $t = 10$ days and creep buckling occurs after t_u days, with $10 < t_u < 12$, the number of full buckling waves in the circumferential direction being $n = 8$. The critical duration of the loading, t_u, varies little, however, with n.

When p is taken equal to $11.77\ N/cm^2$, the deflection increases from the initial value $w = 2.14\ mm$ to $w = 7.06\ mm$ after $t = 110$ days and $110 < t_u < 115$. n is again 8.

When the creep buckling calculation is carried out for the constant load $p = 8.83\ N/cm^2$, w is found to be $1.54\ mm$ at $t = 0$ and $13.75\ mm$ at $t = 1150$ days. Further : $1150 < t_u < 1200$ and $n = 3$. It should be noted that not as many iterative computations were performed for each of the numerous time increments required for the loading level $p = 8.83\ N/cm^2$ as may have been desirable.

Together with the weak sensitivity of the critical duration of the loading to n, this may explain why $n = 3$ is found for this loading case, whereas $n = 8$ was found for higher pressure levels and correspondingly shorter loading periods.

GERARD's and KOLLAR's creep buckling hypothesis

GERARD [10] suggested that a critical strain be derived from known solutions of shell buckling problems without creep and surmised that a shell should collapse under constant load when the total strain ε_t, including the creep strain, equals that critical strain.

KOLLAR's proposition [11] is equivalent : according to him, the pressure causing creep buckling can be obtained by replacing YOUNG's modulus in the formula for the elastic critical pressure by σ/ε_t, where σ is the constant stress and ε_t is the total strain at the time, t, of collapse.

A possible formulation of GERARD's and KOLLAR's assumption is

$$\frac{p}{\bar{E}_t} = \frac{p}{\frac{\bar{\sigma}}{\bar{\varepsilon}_t}} = \text{constant, independent from } p \qquad (10)$$

The value of $p/\bar{E}_t$, for example for $p = 11.77\ N/cm^2$, can be obtained as follows. The stress computation gives the effective membrane stress $\bar{\sigma}$ in the middle of the $40\ cm$ wide, flatter edge zone of the imperfect shell : $\bar{\sigma} = 23.54\ N/mm^2$. The corresponding effective elastic strain is $\frac{2(1+\nu)\bar{\sigma}}{3E} = \frac{2(1+0.18)23.54}{3 \times 29\,430} = 629\times10^{-6}$, and the corresponding creep strain at $t = 110$ days is $\frac{23.54}{98100}\,110^{0.4} = 1573\times10^{-6}$. Hence $\bar{\varepsilon}_t = 2202\times10^{-6}$, $\bar{E}_t = \bar{\sigma}/\bar{\varepsilon}_t = 10690\ N/mm^2$ and $p/\bar{E}_t = 0.1177/10690 = 11.01\times10^{-6}$.

Similar calculations for $p = 14.71\ N/cm^2$ and $p = 8.83\ N/cm^2$ yield $p/\bar{E}_t = 7.84\times10^{-6}$ and $p/\bar{E}_t = 17.44\times10^{-6}$, respectively. For $t_u = 0$ (rapid loading) we had $p_u \doteq 17.66\ N/cm^2$ and consequently $p_u/\bar{E} = \frac{2(1+\nu)p_u}{3E} = \frac{2(1+0.18)0.1766}{3 \times 29430} = 4.72\times10^{-6}$.

Hence, we obtain for $p/\bar{E}_t$ the numbers

4.72×10^{-6}, 7.84×10^{-6}, 11.01×10^{-6} and 17.44×10^{-6}, instead of a constant,

when p amounts to

17.66, 14.71, 11.77 and 8.83 N/cm^2, respectively.

The calculations described above manifestly do not corroborate GERARD's and KOLLAR's hypothesis.

Creep tests on domes with unprestressed thrust rings

We have previously reported [12] upon the major part of the experimental results of our series of creep tests, which "appears to have been the first experimental study of the effect of creep on buckling of concrete shells" [13]. The present report provides more complete data.

Data concerning 18 tests of long duration on model domes whose thrust rings were not prestressed are collected in Table III.

Table III - Tests of long duration - Ring not prestressed

Nr	$\bar{t}'$ (mm)	$\bar{r}$ (mm)	f_{21} (N/mm^2)	t_u (days)	p (N/cm^2)	E_o (N/mm^2)	$\frac{p}{p_o}$	E'_t (N/mm^2)	$\frac{p}{p'_t}$	E_t (N/mm^2)	$\frac{p}{p_t}$	n
KN4	6.89	2538	60	26.5	5.9	31 618	0.466	14 468	1.019	16 071	0.917	9
KN5	6.96	2538	62	574	4.9	34 111	0.353	6 833	1.762	8 093	1.488	
KN6	6.97	2566	57	80	4.9	27 810	0.441	8 670	1.416	8 906	1.377	9
KN7	6.95	2540	64	49	5.9	32 454	0.447	10 781	1.345	11 095	1.307	8
KN8	6.87	2536	66	1733	3.9	30 445	0.324	8 229	1.199	8 457	1.166	
KN9	6.79	2526	64	-	3.9	28 634	0.350	still carried p after 2065 days				
KN10	6.82	2523	66	33	5.9	30 152	0.493	17 000	0.875	-	-	9
KN11	6.96	2530	72	14	6.9	33 882	0.494	15 932	1.051	17 509	0.956	9
KN13	7.00	2557	61	316	4.9	31 303	0.386	7 647	1.577	13 015	0.928	9
KN14	6.84	2516	63	35	7.8	34 769	0.563	14 842	1.320	15 899	1.232	9
KN16	6.96	2562	67	0.5	8.8	34 422	0.641	28 878	0.765	30 023	0.735	
KN17	7.00	2553	60	4	8.8	30 059	0.721	22 354	0.970	24 500	0.884	
KN18	6.94	2580	60	4.5h	8.8	29 749	0.733	27 732	0.812	29 047	0.775	
KN20	6.93	2526	69	8.5	7.8	28 768	0.669	20 092	0.957	21 120	0.911	
KN21	6.96	2537	68	13.5	7.8	35 547	0.541	23 167	0.831	23 983	0.802	10
KN22	6.90	2530	65	42.5	6.9	30 916	0.551	11 586	1.471	12 626	1.349	10
KN23	7.08	2501	61	63.5	6.9	36 985	0.427	19 058	0.829	21 543	0.734	9
KN25	6.96	2522	64	45	7.8	35 403	0.537	17 993	1.057	17 993	1.057	9
$\bar{x}$	6.93	2538	64			32 057			1.133		1.039	
δ(%)	1.02	0.77	5.9			8.5			26.5		23.9	

$\bar{t}'$, $\bar{r}$ and f_{21} have the same meaning as in Tables I and II. The constant pressure p was applied 21 days after each shell was cast. t_u is the time that elapsed between the application of the load and the failure of the dome. t_u ranges between 4.5 hours and 1733 days.

E_o is the modulus of elasticity at $t = 0$, i.e. : measured on $2 \times 1 \times 10\ cm^3$ bars at the time the load was applied to the dome. p_o in the column under the heading p/p_o was obtained from the equation $p_o = \bar{\gamma} E_o (\bar{t}'/\bar{r})^2$, where $\bar{\gamma} = 0.539$ is the average value of γ derived from rapid tests on domes with unprestressed thrust rings (see Table I). We do not know the pressure that would have caused the model domes in Table III to buckle in a rapid test and p_o is the best estimate of that pressure that is available to us.

It cannot be presumed to be a very accurate estimate, since the lowest γ is *17%* lower than $\bar{\gamma}$ and the highest γ is *13%* higher than $\bar{\gamma}$. The p/p_o column in Table III shows that the permanently applied pressure ranged from *32.4%* to *73.3%* of the most probable value of p_o.

In Table III $E_t' = \sigma/\varepsilon_t'$, where σ is equal to the permanent membrane stress in the dome in question and also to the constant compressive stress in the corresponding bars, and where ε_t' is the total strain of these bars on the day the shell failed. p_t' in the column under the heading p/p_t' is given by $p_t' = \bar{\gamma} E_t' (\frac{\bar{t}'}{\bar{r}})^2$. One can see from that column that the ratio p/p_t' fluctuates round unity and has an average value of *1.133* and a coefficient of variation $\delta = 26.5\%$.

ε_t' and E_t' contain the effect of temperature and moisture changes on the length of the bars. It is logical to subtract this effect, as measured on identical unloaded bars, from ε_t'. From the remaining sum ε_t of the elastic strain and the creep strain a higher modulus of elasticity $E_t = \sigma/\varepsilon_t$ and a higher pressure $p_t = \bar{\gamma} E_t (\frac{\bar{t}'}{\bar{r}})^2$ were calculated. In the last column but one of Table III, p is compared with p_t. The average value of the ratio p/p_t for 16 tests is *1.039*. This tallies very well with GERARD's and KOLLAR's proposition, since the ratio would be *1* if the proposition were true. The standard deviation is, however, quite high : it is *0.239×1.039*. The lowest ratio p/p_t is *0.734* and the highest is *1.488*.

It is not astounding that the numerical values of p/p_t are scattered so widely, since we are dealing with a type of failure involving simultaneously instability of shells and creep of concrete. Each of these two phenomena by itself is notoriously precarious and the combination of both is bound to be highly capricious.

The failure zone of the model caps was subjected to direct compression and bending, whereas creep strains were measured on bars loaded in uniaxial compression. It is conceivable, though by no means certain, that the values of p/p_t would have been less widely scattered if the creep strains had been measured on specimens subjected to biaxial compression or to simultaneous compression and bending.

One model dome, KN8, finally buckled after 4 years and 9 months under a load which was only *32.4%* of our best estimate of its carrying capacity under rapid loading.

As to model cap KN9, after it had carried a constant radial pressure $p = 3.9\ N/cm^2$ for 2065 days, the pressure was increased to $p_u = 6.9\ N/cm^2$, which caused buckling of the specimen. On the day of this (swift) increase of the load, E_t' and E_t, determined from measurements on small bars made at the same time and from the same batch of microconcrete as dome KN9, amounted to $E_t' = 6882\ N/mm^2$ and $E_t = 8307\ N/mm^2$. The corresponding value $p_t = 3.24\ N/cm^2$ is lower than p and the test dome should already have buckled before if GERARD's assumption held strictly true for every single shell.

Deformation of the domes under long-term loading

Theoretically the deformation of a perfectly spherical dome is axisymmetric until an asymmetric pattern of dimples and bulges suddenly appears under the bifurcation load. Our test shells were not perfectly spherical, but their shape was nearly axisymmetric. Nevertheless, their deformation was already asymmetric soon after the loading was applied. In the tests of long duration we have measured the deformation of the domes after certain intervals and in most cases, but not in all, a number of full waves could be detected in the circumferential direction. The last column of Table III shows that this number, n, is almost always equal to 9, once to 8 and twice to 10. This squares with theory, for HUANG's diagram predicts 9 full waves when $\lambda = 14$ and the BOSOR 4 calculation also predicts 9 waves.

The wave pattern was never regular and the amplitude of the radial displacement was different from wave to wave. In most creep tests the curvature due to the deformation was notably greater in one wave than everywhere else. There a disk would eventually be punched out and in many cases we could thus predict where buckling would occur. The critical area became almost flat and in 13 cases, especially under low radial pressures, the sign of the actual curvature of the shell even changed locally, sometimes several weeks before failure.

Creep tests on domes with prestressed thrust rings

Table IV, which is made up in the same way as Table III, contains data regarding 23 domes which were permanently subjected to a constant radial pressure p and whose thrust ring was (or still is) prestressed. Two caps have now withstood $p = 4.9\ N/cm^2$ for 10 years. The other 21 domes failed after being pressurized for

Table IV - Tests of long duration - Ring prestressed

Nr	$\bar{t}'$ (mm)	$\bar{r}$ (mm)	f_{21} (N/mm^2)	t_u (days)	p (N/cm^2)	E_o (N/mm^2)	$\frac{p}{p_o}$	E'_t (N/mm^2)	$\frac{p}{p'_t}$	E_t (N/mm^2)	$\frac{p}{p_t}$	n
K18	6.97	2407	65	1	11.8	30 095	0.755	22 717	1.0	-	-	
K22	7.16	2428	69	132	7.8	27 273	0.535	11 971	1.220	-	-	
K24	6.99	-	70	10	8.8	29 340	0.580	19 151	0.888	-	-	
K25	7.08	2402	64	79	7.8	22 239	0.657	18 242	0.801	-	-	
K26	6.92	2428	75	70	7.8	28 824	0.542	13 623	1.148	-	-	
K30	6.87	2462	58	3	9.8	28 678	0.711	19 312	1.056	-	-	
K31	6.91	-	67	8	9.8	35 851	0.548	20 833	0.942	-	-	
K32	6.85	2441	64	98	5.9	35 089	0.345	12 578	0.962	-	-	
K35	6.93	2462	-	2675	5.9	33 667	0.357	7 328	1.640	9 523	1.262	
K37	6.97	2480	72	167	6.9	28 367	0.496	10 552	1.333	13 059	1.077	9
K39	7.03	2493	71	-	4.9	37 007	0.270	has carried p since 01-7-75				
K40	6.99	2470	71	-	4.9	35 127	0.283	has carried p since 15-7-75				
K41	7.07	2498	59	45	6.9	31 263	0.444	10 232	1.356	11 382	1.219	9
K42	6.86	2527	64	27	7.8	33 659	0.512	13 657	1.262	14 907	1.156	9
K43	7.26	2477	69	34	7.8	33 219	0.445	13 367	1.106	15 855	0.932	9
K44	6.90	2493	70	6	8.8	29 445	0.633	20 444	0.912	20 254	0.921	
K45	6.91	2552	69	31	8.8	36 915	0.528	15 484	1.258	18 135	1.075	9
K46	6.80	2556	60	66	5.9	33 516	0.402	10 804	1.245	13 527	0.995	
K48	6.99	2554	62	21.5	6.9	34 736	0.427	13 584	1.092	15 863	0.935	9
K49	6.92	2558	66	5	8.8	35 442	0.551	26 234	0.744	27 071	0.721	
K50	6.87	2511	67	40	8.8	33 616	0.568	15 793	1.208	17 324	1.102	9
K53	6.89	2532	65	0.3	8.8	29 729	0.649	-	-	-	-	
K54	6.91	2555	68	0.3	9.8	31 318	0.713	24 401	0.916	25 011	0.893	
$\bar{x}$	6.96	2490	67			31 804			1.104		1.024	
δ(%)	1.5	2.0	6.6			11.5			19.7		15.0	

periods t_u ranging between 0.3 and 2675 days.

p_o in Table IV has also been calculated with the formula $p_o = \bar{\gamma} E_o (\bar{t}'/\bar{r})^2$, but with a higher $\bar{\gamma}$ than the one used in Table III, namely with the average value $\bar{\gamma} = 0.621$ determined from rapid tests on caps with prestressed rings (see Table II). The test pressure p ranged from $0.270 p_o$ to $0.755 p_o$.

The average value of the ratio p/p'_t is 1.104, with a coefficient of variation $\delta = 19.7\,\%$. E_t and p_t are not known for all the domes in this series, because we did not measure the length variations of unloaded microconcrete bars at the beginning of our test program. The average of the 12 known values of p/p_t is 1.024, the coefficient of variation being $\delta = 15\,\%$. The scatter is much less than for the creep tests on calottes with unprestressed thrust rings.

In all cases where a number, n, of circumferential waves could be counted, we found $n = 9$.

Figure 4 and figure 5 show lines of equal spatial curvature for the test domes K39 and K40 respectively, in the deflected shape

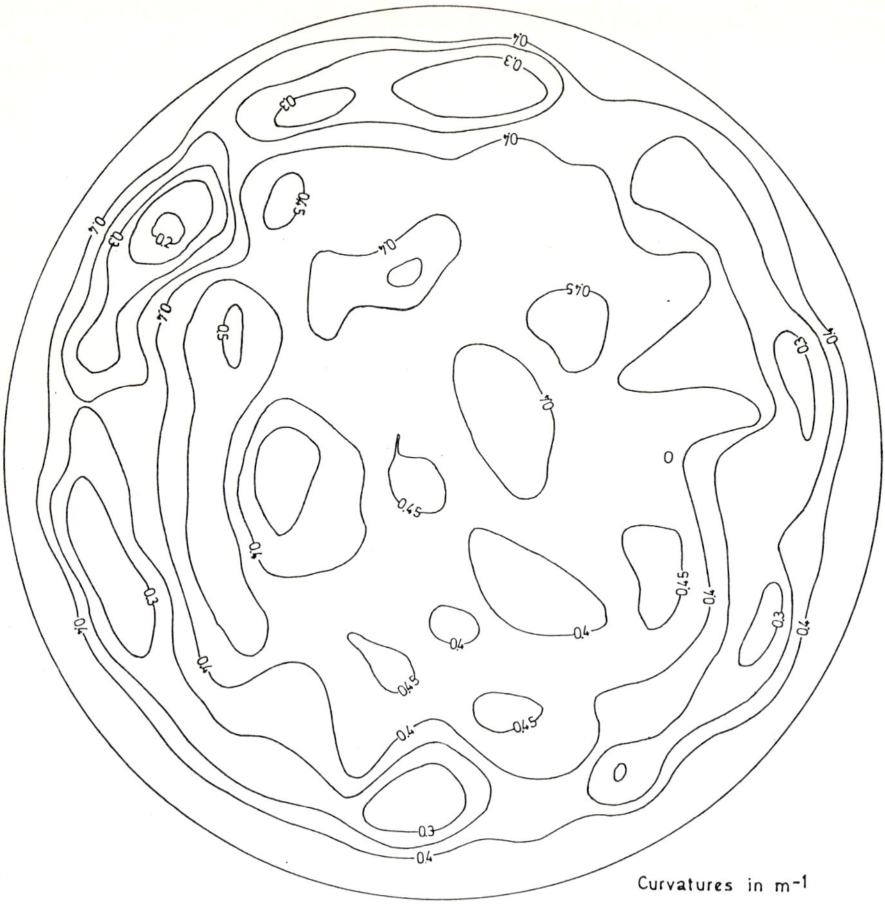

Fig. 4 - Model dome K39 - t = 3543 days - Lines of equal spatial curvature

brought about by a constant load of almost 10 years duration. Two adjacent curves correspond with a $0.05\,m^{-1}$ curvature differential. The actual curvature of dome K40 is zero along the perimeter of the small shaded area in figure 5 and it is just slightly convex towards the center of the nominal sphere within the shaded area. If K40 ever buckles under $p = 4.9\ N/cm^2$ or if we caused it to buckle by increasing the pressure, the shaded area would no doubt be part of the disk that would be punched out.

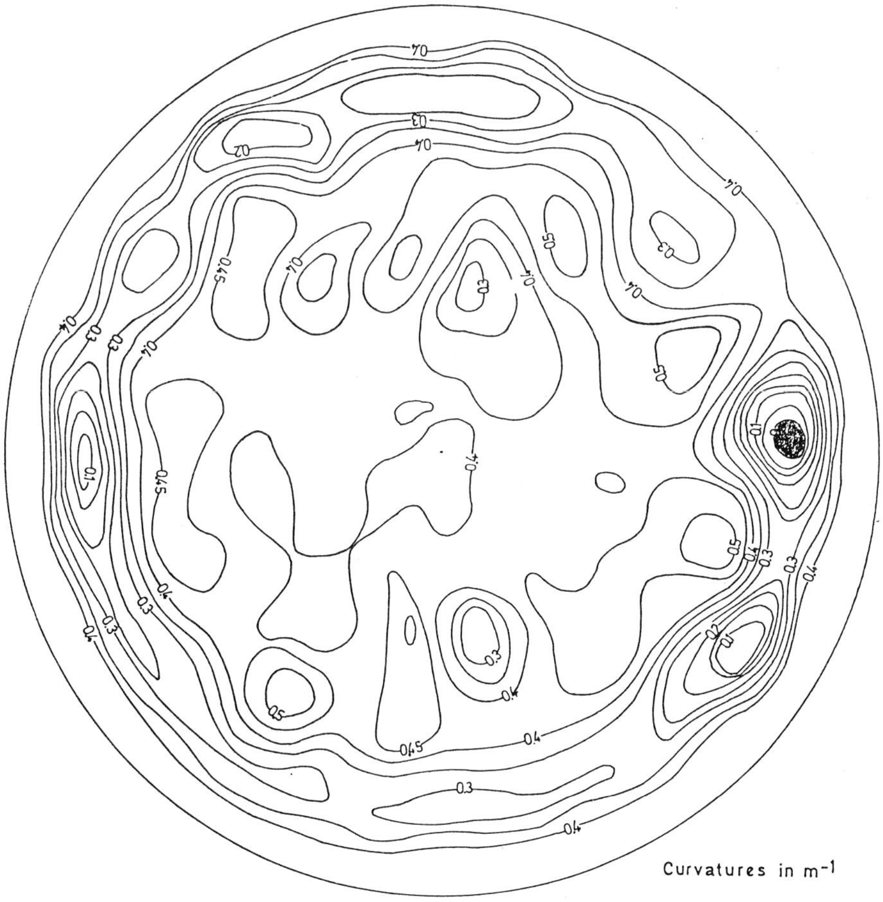

Fig. 5 - Model dome K40 - t = 3529 days
Lines of equal curvature

Creep buckling failure mode

Long-term radial pressure has invariably caused the model domes to fail in the same way as rapid loading, namely by punching out a circular or slightly elliptical disk, normally near the edge of the shell. The disks were somewhat larger than in the case of rapidly increasing load. The mean diameter of the disks was, on an average, about equal to $2.94\sqrt{rt'}$.

Conclusions

1) In "rapid" tests, carefully made spherical unreinforced micro-concrete model domes buckled under a uniform radial pressure

p_u which was much lower than the critical pressure p_{cl} given by the ZOELLY formula *(1)*, but higher than the pressure given by von KARMAN and TSIEN, and which, on an average, was *30.6 %* lower or, when a state of stress approaching a membrane state was achieved by prestressing the thrust rings, *33.6 %* lower than the buckling pressure predicted by a BOSOR 4 calculation for a perfect dome. For individual model domes the experimental buckling pressure could be as low as *54.7 %* of the buckling pressure calculated with the BOSOR 4 program.

2) Achieving a state of stress which approached a membrane state increased the buckling load for a given geometry of the spherical cap : the average value of γ in formula *(4)* rose from *0.539* to *0.621*, i.e. by about *15 %*, when the domes were tested to failure within a few hours. The average value of γ would be lower for more imperfect shells.

3) The test shells invariably failed by the punching out of a circular or slightly oval disk, usually quite near the edge of the cap. In rapid tests, the mean diameter of the disk was about $2.76\sqrt{rt'}$ when the thrust ring was not prestressed and somewhat smaller when the thrust ring was prestressed. In tests of long duration the diameter of the disk was about $2.94\sqrt{rt'}$.

4) Under long-term loading, the carrying capacity of the domes decreased drastically, owing to creep of the microconcrete. One dome failed after having been subjected for 4 years and 9 months to a constant radial pressure p which was only *32.4 %* of the best estimate that we can make of its carrying capacity p_o under short-time loading. Another test cap, which was kept more nearly in a membrane state of stress, buckled after having carried for 7 years and 4 months a constant pressure amounting to only *35.7 %* of p_o.

5) Given the geometry of the dome, its carrying capacity under long-term loading increases when it is kept in a membrane state of stress.

6) Two model domes, in which the state of stress approaches a membrane state, have now withstood for 10 years a pressure $p = 0.270p_o$ and a pressure $p = 0.283p_o$, respectively. This suggests, but does not prove, that spherical concrete shells may be able to withstand permanently a pressure amounting approximately to $0.3p_o$.

7) Due to the imperfection of the model domes, their deformation under permanent uniform radial pressure was not axisymmetric,

even long before they buckled. If it were feasible to regularly measure the curvature of a full-size spherical dome in a number of points, it might perhaps be possible to foresee a creep buckling failure.

8) A BOSOR 5 calculation, carried out on the assumption that the creep strain varies with time in accordance with equation *(9)*, enables an analyst to estimate the period t_u during which a spherical concrete dome might be expected to withstand a uniform radial pressure of given intensity. The critical time span t_u thus estimated gives, at best, only an idea of the order of magnitude of the real critical time span.

9) In a statistical sense, GERARD's and KOLLAR's creep buckling hypothesis for shells, which, to the best of the authors' knowledge, is unproven so far, is corroborated quite well by the tests described in this paper. Neglecting the scatter, it may be said that a spherical concrete dome fails under the permanent radial pressure given by the formula

$$p = p_t = \gamma E_t \left(\frac{t'}{r}\right)^2 = \gamma \frac{\sigma}{\varepsilon_t} \left(\frac{t'}{r}\right)^2 \qquad (11)$$

which, again in a statistical sense, is about *3%* conservative. Substitution of equation *(11)* into the expression $\sigma = \frac{rp}{2t'}$ for the compressive membrane stress in the dome yields the equation $\varepsilon_t = \frac{\gamma t'}{2r}$. The conclusion is that the dome buckles when the total strain ε_t , produced by the membrane stresses and increasing on account of creep, equals a critical value given by

$$\varepsilon_t = \frac{\gamma t'}{2r} \qquad (12)$$

whether the limit is reached within a short time or after a long span of time. In equation *(12)* γ is equal to either *0.539* or *0.621*, as explained above, for the excellent shells reported upon in this paper. The limit value *(12)* of ε_t does not depend on the acting pressure. ε_t is the total strain produced by the membrane stresses, but it does not include climatic length variations.

10) As may be expected when shell instability and creep of concrete are combined, reality may deviate considerably from the statistical truth embodied in statement 9) and the limit value of ε_t may diminish to *0.72* times that given by equation *(12)*.

11) In view of the scatter characterizing the limit value of ε_t and since the corresponding time intervals are scattered even more widely, it does not seem possible to predict with any

degree of accuracy the span of time t_u that will elapse before creep buckling of a concrete shell occurs. The column headed t_u in Table IV provides an illustration : three nominally identical spherical domes, subjected to a permanent radial pressure $p = 5.9\ N/cm^2$, buckled after 66 days, 98 days and 2675 days, respectively.

12) BOSOR 5 calculations, in combination with creep equation *(9)*, do not support GERARD's and KOLLAR's creep buckling assumption. The experimental evidence, that is referred to in 9) and that bears out the assumption, is more conclusive. The contradiction between theory and experiment may be due to the use of a creep strain equation of type *(5)* in the computer program. A creep strain equation of the more commonly employed type $\bar{\varepsilon}_c = \bar{\varepsilon}_{c\infty}(1 - e^{-\theta t})$, where $\bar{\varepsilon}_{c\infty}$ is the final value of the creep strain (at $t = \infty$) and θ is a constant, might lead to results that would perhaps tally better with GERARD's and KOLLAR's proposition.

13) Not discussed in this paper is the fact that the rapid tests described above show that a Southwell plot, in which changes of curvature are used instead of absolute displacements, is able to predict the buckling pressure for a spherical concrete dome with good accuracy, but only if its thrust ring is not prestressed [14].

References

1. Vandepitte, D.; Rathé, J. : An experimental investigation of the buckling load of spherical concrete shells, subjected to uniform radial pressure. RILEM International Symposium, Buenos Aires. September 1971. p. 427-442.

2. Zoelly, R. : Über ein Knickungsproblem an der Kugelschale. Dissertation. Zürich. 1915.

3. Tsien, H.S. : A theory for the buckling of thin shells. Journal of the Aeronautical Sciences. August 1942. p. 373-384.

4. Budiansky, B. : Buckling of clamped shallow spherical shells. IUTAM Symposium on The Theory of Thin Elastic Shells, Delft. 1959. p. 64-94.

5. Huang, N.C. : Unsymmetrical buckling of thin shallow spherical shells. Journal of Applied Mechanics. September 1964. p. 447-457.

6. Bushnell, D. : Stress, stability and vibration of complex, branched shells of revolution. Computers and Structures. March 1974. p. 399-435.

7. Litle, W.A. : Reliability of shell buckling predictions. Research Monograph n° 25. M.I.T. Press, Cambridge. 1964.

8. Bushnell, D. : BOSOR 5 - A computer program for buckling of elastic-plastic complex shells of revolution including large deflections and creep. Lockheed Missiles & Space company D407168. December 1974.

9. Bushnell, D. : A strategy for the solution of problems involving large deflections, plasticity and creep. International Journal for Numerical Methods in Engineering. Vol. 11, 1977, 683-708.

10. Gerard, G. : A creep buckling hypothesis - A critical strain approach to creep buckling of plates and shells. Journal of the Aeronautical Sciences. September 1956.

11. Kollar, L. : Remarks on the safety factor against shell buckling and on the influence of creep on the critical load. Symposium Pipes and Tanks, Weimar. May 1968. 16 pages.

12. Vandepitte, D. ; Rathé, J. ; Weymeis, G. : Experimental investigation into the buckling and creep buckling of shallow spherical caps subjected to uniform radial pressure. World Congress on Shells and Spatial Structures, Madrid. September 1979. Volume 1. p. 1.1-1.15. International Association for Shell and Spatial Structures.

13. Medwadowski, S. : General Report for Theme 1 : Stability of shells and spatial structures. Bulletin of the International Association for Shell and Spatial Structures. December 1979 - April 1980. n. 71/72.

14. Vandepitte, D. ; Rathé, J. ; Weymeis, G. : Experimental verification of the validity of the Southwell plot for radially loaded spherical caps of microconcrete. Euromech Colloquium, Delft. March-April 1980. 14 pages.

Dynamic Stability of Voscoelastic Shallow Shells

J. BRILLA

Institute of Applied Mathematics and
Computing Technique
Comenius University, Bratislava

Summary

The quasistatic and dynamic stability and post-buckling behaviour of viscoelastic shallow shells are studied on the basis of the generalized Kármán-Donnell-Vlasov equations. An analysis is made for a thin-walled cylindrical panel subject to a compression and a periodical loading.

Introduction

When dealing with buckling and postbuckling analysis of viscoelastic shells similarly as in the case of elastic shells we derive governing equations applying linear constitutive equations and geometrically non-linear theory. The first information about critical loads can be obtained from the analysis of linearized problems, too. When considering infinitesimal perturbations we may feel justified in neglecting non-linear terms. Then we can easily to determine critical loads with infinite critical time and instant critical loads. However, in the case of instability with respect to infinitisimal perturbations we arrive at an apparent contradiction. We assume infinitesimal perturbations and find out that they grow without bounds [1] . Therefore it is necessary to deal with non-linear analysis of stability problems.

When studying non-linear problems we may find out that in contradiction to linear problems postbuckling deformations tend to finite values instead of growing to infinity. Therefore the solution of a postbuckling viscoelastic problem cannot be solved by successive aproximation that require only the solution of linear problems.

Inelastic Behaviour of Plates and Shells
IUTAM Symposium Rio de Janeiro 1985
Editors: L. Bevilacqua, R. Feijóo and R. Valid

We shall show some characteristic features of buckling analysis of quasistatic problems and deal with dynamic stability of viscoelastic shallow shells.

Basic equations

When deriving the governing equations of large deflection theory of viscoelastic shallow shells similarly as in the case of elastic shells we apply physically linear constitutive equations and the tensor of large deflections. Then the governing equations of viscoelastic shallow shells assume the form

$$\frac{h^3}{12}\int_0^t G_{ijkl}(t-\tau)D_\tau w_{,ijkl}(\tau)\,d\tau = q - N_{ij}(w_{,ij} - b_{ij}) + h\, e_{ik}\, e_{jl}\,(w_{,ij} - b_{ij})\, F_{,kl},$$

$$\int_0^t K_{ijkl}(t-\tau)\, D_\tau F_{,ijkl}(\tau)\,d\tau = - e_{ik}\, e_{jl}\left(\frac{1}{2} w_{,ij} - b_{ij}\right) w_{,kl}, \tag{1}$$

where w is the transverse displacement, positive in the outward direction, F the Airy stress function, q the external loading, positive in the outward direction, $N_{ij} = - h\,\sigma_{ij}$ the boundary loading, b_{ij} the curvature tensor, h the shell thickness, e_{ij} the alternating tensor, $D_\tau = \partial/\partial\tau$, $G_{ijkl}(t)$ the tensor of relaxation functions,

$$K_{ijkl}(t) = e_{im}\, e_{jn}\, e_{kr}\, e_{ls}\, J_{mnrs}(t), \tag{2}$$

where $J_{mnrs}(t)$ is the tensor of crep functions.

We assume that the domain of definition Ω is a rectangular panel or a closed cylinder. We shall consider the following boundary conditions

$$w = \frac{\partial w}{\partial n} = 0 \qquad \text{on } \partial\Omega, \tag{3}$$

or

$$w = M_{nn} = 0 \qquad \text{on } \partial\Omega \tag{4}$$

and

$$F_{ss} = - F_{sn} = 0 \qquad \text{on } \partial\Omega , \tag{5}$$

or

$$\frac{\partial F}{\partial n} = \frac{\partial^3 F}{\partial n^3} = 0 \qquad \text{on } \partial\Omega , \tag{6}$$

where n is the direction of the outward normal to $\partial\Omega$ and s is the direction of the tangent to $\partial\Omega$.

Relaxation and creep functions are not independent. When we denote the coresponding tensor operator

$$G = \int_0^t G_{ijkl}(t-\tau) \frac{\partial}{\partial \tau}(\, . \,) \, d\tau \tag{7}$$

and similarly J. Then it holds

$$G^{-1} = J = \int_0^t J_{ijkl}(t-\tau) \frac{\partial}{\partial \tau}(\, . \,) \, d\tau \tag{8}$$

or in the form of Laplace transform

$$p \tilde{J}_{ijkl}(p) = \left[p \tilde{G}_{ijkl}(p)\right]^{-1} . \tag{9}$$

As this inversion can be too complicated very often it is more convenient to consider the governing equations for a viscoelastic shallow shells in terms of displacements. They assume the form

$$\frac{h^3}{12} \int_0^t G_{ijkl}(t-\tau) \, D_\tau w_{,ijkl}(\tau) \, d\tau = q - N_{ij}(w_{,ij} - b_{ij})$$

$$- h \, (w_{,ij} - b_{ij}) \int_0^t G_{ijkl}(t-\tau) \, D_\tau (u_{k,l}(\tau) +$$

$$+ \frac{1}{2} w_{,k}(\tau) \, w_{,l}(\tau) + b_{kl} \, w(\tau)) \, d\tau , \tag{10}$$

$$\int_0^t G_{ijkl}(t-\tau)\, D_\tau (u_{k,lj}(\tau) + w_{,k}(\tau)\, w_{,lj}(\tau) + + b_{kl}\, w_{,j}(\tau))\, d\tau = 0. \tag{10}$$

When we have to solve the classical time dependent stability problems with respect to initial perturbations we have to analyse the influence of initial perturbations;. As Boltzmann equations have been derived for homogeneous initial conditions, it is necessary to express governing equations in a differential form or to derive and add to the governing equations (1,10) additional terms for initial values. These terms are different for different relaxation and creep functions.

In the case of a standard material with constitutive equations

$$(1 + \alpha D_t)\, \sigma_{ij} = (E_{ijkl} + \Omega_{ijkl}\, D_t)\, \varepsilon_{kl}. \tag{11}$$

the governing equations are

$$\frac{h^3}{12} (E_{ijkl} + \Omega_{ijkl}\, D_t)\, w_{,ijkl} = (1 + \alpha D_t)\, q - - N_{ij}\, (w_{,ij} - b_{ij}) + h\, e_{ik}\, e_{jl}\, (w_{,ij} - b_{ij})\, F_{,kl},$$

$$e_{im}\, e_{jn}\, e_{kr}\, e_{ls}\, H_{ijkl}\, (D_t)\, F_{,mnrs} = = - e_{ik}\, e_{jl}\, \Delta(D_t)(\tfrac{1}{2}\, w_{,ij} - b_{ij})\, w_{,kl} \tag{12}$$

where $H_{ijkl}(D_t)$ is the adjoint matrix and $\nabla(D_t)$ the determinant of $E_{ijkl} + \Omega_{ijkl}\, D_t$, considering common contracted indices for pairs of indices ij and kl. In the integral form we get

$$\frac{h^3}{12} \left(\int_0^t G_{ijkl}(t-\tau)\, D_\tau w_{,ijkl}(\tau)\, d\tau - \frac{1}{\alpha}\, \Omega_{ijkl}\, w^{o}_{,ijkl}\, e^{-t/\alpha} \right)$$

$$= q - N_{ij}\, (w_{,ij} - b_{ij}) + h\, e_{ik}\, e_{jl} (w_{,ij} - b_{ij})\, F_{,kl}$$

$$+ N^0_{ij} (w^0_{,ij} - b_{ij}) e^{-t/\alpha} , \qquad (13)$$

$$\int_0^t K_{ijkl}(t-\tau) D_\tau F_{,ijkl}(\tau) d\tau = - e_{ik} e_{jl} (\tfrac{1}{2} w_{,ij} -$$

$$- b_{ij}) w_{,kl} + e_{ik} e_{jl} (\tfrac{1}{2} w^0_{,ij} - b_{ij}) w^0_{,kl} e^{-t/\beta} , \qquad (14)$$

where

$$G_{ijkl}(t) = E_{ijkl} H(t) - (E_{ijkl} - \tfrac{1}{\alpha}\Omega_{ijkl}) e^{-t/\alpha} ; \qquad (15)$$

$$K_{ijkl} = e_{im} e_{jn} e_{kr} e_{ls} \left[p^2 G \right]^{-1}_{mnrs} \qquad (16)$$

and

$$\beta = [\Omega]^{-1}_{ijkl} E_{ijkl} . \qquad (17)$$

As derivative D_τ in governing equations have to be taken in the sense of distributions it is better to rewrite them in the form

$$\frac{h^3}{12} (G_{ijkl}(0) w_{,ijkl} + \int_0^t D_t G_{ijkl}(t-\tau) w_{,ijkl}(\tau) d\tau$$

$$- \tfrac{1}{\alpha} \Omega_{ijkl} w^0_{,ijkl} e^{-t/\alpha}) = q - N_{ij} (w_{,ij} - b_{ij}) \qquad (18)$$

$$+ h e_{ik} e_{jl} (w_{,ij} - b_{ij}) F_{,kl} + N^0_{ij} (w^0_{,ij} - b_{ij}) e^{-t/\alpha}$$

$$K_{ijkl}(0) F_{,ijkl} + \int_0^t D_t K_{ijkl}(t-\tau) F_{,ijkl}(\tau) d\tau =$$

$$= - e_{ik} e_{jl} (\tfrac{1}{2} w_{,ij} - b_{ij}) w_{,kl} +$$

$$+ e_{ik}\, e_{jl}\, (\tfrac{1}{2}\, w^{0}_{,ij} - b_{ij})\, w^{0}_{,kl}\, e^{-t/\beta} \,. \qquad (18)$$

Quasistatic stability

When dealing with analysis of the quasistatic stability and postbuckling behaviour of viscoelastic shallow shells we restrict ourselves to a cylindrical rectangular panel or a closed cylinder. Then $b_{11} = 0$, $b_{12} = 0$ and $b_{22} = 1/R$, where R is the radius of the cylinder or of the panel. We put $q = 0$ and $N_{11} = p(t)$, $N_{12} = N_{22} = 0$.

Then applying formally to (1) Laplace transform and then Tauber's theorem [2] we arrive at

$$\begin{aligned} \frac{h^3}{12}\, G_{ijkl}(0)\, w_{,ijkl}(0) &= -\, p(0)\, w_{,11}(0) \\ &\quad + h\, e_{ik}\, e_{jl}\, w_{,ij}(0)\, F_{,kl}(0) - F_{,11}(0)/R \\ K_{ijkl}(0)\, F_{,ijkl}(0) &= -\, \frac{1}{2}\, e_{ik}\, e_{jl}\, w_{,ij}(0)\, w_{,kl}(0) \\ &\quad + w_{,11}(0)/R \end{aligned} \qquad (19)$$

for $t = 0$ and

$$\begin{aligned} \frac{h^3}{12}\, G_{ijkl}(\infty)\, w_{,ijkl}(\infty) &= -\, p(\infty)\, w_{,11}(\infty) \\ &\quad + h\, e_{ik}\, e_{jl}\, w_{,ij}(\infty)\, F_{,kl}(\infty) - F_{,11}(\infty)/R \,, \\ K_{ijkl}(\infty)\, F_{,ijkl}(\infty) &= -\, \frac{1}{2}\, e_{ik}\, e_{jl}\, w_{,ij}(\infty)\, w_{,kl}(\infty) \\ &\quad + w_{,11}(\infty)/R \end{aligned} \qquad (20)$$

for $t = \infty$.

For the limit values of the tensors $G(t)$ and $J(t)$ we have the following relations

$$J(0) = G^{-1}(0) \quad , \quad J(\infty) = G^{-1}(\infty) \tag{21}$$

and

$$K_{ijkl} = e_{im}\, e_{jn}\, e_{kr}\, e_{ls}\, J_{mnrs} \ . \tag{22}$$

Equations (19,20) corespond to Kármán-Donnell-Vlasov the equations for cylindrical shells with material constants given by limit values (21) of relaxation and creep functions.

Thus the buckling load and post-buckling deformations of a viscoelastic cylindrical shell at $t = 0$ are equal to that load and deformations of an elastic shell with the instant moduli. We denote the corresponding critical load p_{icr}.

Similarly buckling load and post-buckling deformations for $t = \infty$ are the same as the buckling loads and post-buckling deformations of an elastic shallow shell with the longtime moduli. We denote the critical load for $t = \infty$ by p_{lcr}. Then from $G_{ijkl}(0) > G_{ijkl}(\infty)$ it follows $p_{icr} > p_{lcr}$.

A detailed analysis similar to those for isotropic viscoelastic plates [3] shows: For $p < p_{lcr}$ there exists only the trivial unperturbed solution $w = 0$ and initial perturbations decrease to zero. For $p_{lcr} < p < p_{icr}$ there exists only a trivial unperturbed solution and initial perturbations grow to the post-buckling solution which for $t = \infty$ assumes values determined by the equation (20) . For $p = p_{icr}$ there exists a non-trivial unperturbed solution increasing continuously from zero at $t = 0$ to a finite value determined by the equation (20) . For $p > p_{icr}$ the postcritical solution starts by a jump at $t = 0$, given by (19) .

Dynamic stability

In the following we shall deal with analysis of dynamic stability of viscoelastic cylindrical shells and assume $q = -\rho D_t^2 w$ and $N_{11} = p(t) = p_1 + p_2 \cos \Theta t$. In order to get qualitative results we restrict ourselves to analysis of linearized problems. For the sake of simplicity we consider a shell of standard solid with homogeneous relaxation spectrum $\Omega_{ijkl} = \beta E_{ijkl}$. Then the linearized problem leads to analysis of the equations

$$\frac{h^3}{12} (1 + \beta D_t) E_{ijkl} w_{,ijkl} = (1 + \alpha D_t)(- \rho D_t^2 w - p(t) w_{,11} - h F_{,11}/R) ,$$

$$(1 + \alpha D_t) K_{ijkl} F_{,ijkl} = (1 + \beta D_t) w_{,11}/R \qquad (23)$$

where

$$K_{ijkl} = e_{im} e_{jn} e_{kr} e_{ls} C_{mnrs} , \quad C = E^{-1} . \qquad (24)$$

We consider the following initial conditions

$$w(0) = w_o , \quad D_t w(0) = D_t^2 w(0) = 0 , \quad F(0) = 0 \qquad (25)$$

Then eliminating F we arrive at

$$(1 + \beta D_t)(\frac{h^3}{12} K_{mnrs} E_{ijkl} w_{,ijklmnrs} + h w_{,1111}/R^2)$$
$$= (1 + \alpha D_t)(- \rho D_t^2 K_{ijkl} w_{,ijkl} - p(t) K_{ijkl} w_{,11ijkl}) \qquad (26)$$

Now we consider orthotropic cylindrical shells with simply supported boundaries. Then the solution can be sought in the form

$$w = \sum_{m=1}^{\infty} \sum_{n=1}^{\infty} w_{mn}(t) \sin \frac{m\pi x}{a} \sin \frac{n\pi y}{b} \qquad (27)$$

in the case of a cylindrical panel, where a,b are dimensions of the panel and in the form

$$w = \sum_{m=1}^{\infty} \sum_{n=0}^{\infty} w_{mn}(t) \sin \frac{m\pi x}{L} \cos \frac{n\pi y}{R} \tag{28}$$

in the case of a closed cylinder, where L is the length of the cylinder and R its radius .

Denoting $m = n_1$, $n = n_2$, $x = x_1$, $y = y_2$, $a = a_1$, $b = a_2$, $L = a_1$, $R = a_2$ for both cases we arrive at the following differential equation for $w_{n_1 n_2}(t)$

$$\{A - p(t) + \alpha\, p'(t)\ B\}\, w_{mn}(t) + \{\beta\, A - \alpha\, B\, p(t)\}\, w'_{mn}(t) + C\, w''_{mn}(t) + \alpha\, C\, w'''_{mn}(t) = 0, \tag{29}$$

where

$$A = \frac{h^3}{12} K_{mnrs} E_{ijkl} \frac{n_i\, n_j\, n_k\, n_l\, n_m\, n_n\, n_r\, n_s}{a_i\, a_j\, a_k\, a_l\, a_m\, a_n\, a_r\, a_s} + h \frac{n_1^4}{a_1^4} \frac{1}{R^2},$$

$$B = K_{ijkl} \frac{n_1^2\, n_i\, n_j\, n_k\, n_l}{a_1^2\, a_i\, a_j\, a_k\, a_l}, \tag{30}$$

$$C = \rho\, B \frac{a_1^2}{n_1^2}.$$

Then in the quasistatic case when

$$\rho = 0 , \quad p_2 = 0 \tag{31}$$

when looking for the solution in the form $w_{mn} = w^o_{mn} \exp(\omega\, t)$ we get

$$w_{mn} = w^o_{mn} \exp\left(- \frac{A_{mn} - p\, B_{mn}}{\beta\, A_{mn} - \alpha\, p\, B_{mn}}\, t\right) \tag{32}$$

As $\beta > \alpha$ the solution decreases to zero for $p < A_{mn}/B_{mn}$ and increases to infinity for $A_{mn}/B_{mn} < p < \beta\, A_{mn}/\alpha\, B_{mn}$ and becomes instantly unstable for $p = \beta\, A_{mn}/\alpha\, B_{mn}$. Therefore we call

$$p_{lcr} = A_{mn}/B_{mn} \tag{33}$$

the longtime critical values and

$$p_{icr} = \frac{\beta}{\alpha}\, p_{mnlcr} \tag{34}$$

the instant critical values. These values correspond to critical values of equations (20) and (19) .

In the dynamic case with a constant pressure p_1 we arrive at the characteristic equation

$$(A - p_1\, B) + (\beta\, A - \alpha\, p_1\, B)\, \omega + C\, \omega^2 + \alpha\, C\, \omega^3 = 0 \tag{35}$$

Properties of the solution depend on the roots of this equation. For $A - p_1\, B < 0$ at least one root is positive and thus for $p_1 > p_{lcr} = A/B$ the solution is unstable. From the analysis of relations between roots and coefficients of (35) it is obvious that $p_{lcr} = A_{mn}/B_{mn}$ and thus the longtime values are similar, as in the case of quasistatic stability problems, critical values with respect to initial perturbations.

Now we shall deal with the dynamical stability of the viscoelastic cylindrical shell subjected to periodic axial pressure $p(t) = p_1 + p_2 \cos \Theta t$.

We shall look for the solution of the corresponding differential equation for w_{mn} in the form

$$w(t) = \sum_{k=-\infty}^{\infty} w_k\, e^{ik\Theta t/2} , \tag{36}$$

where for the sake of simplicity we leave out indices m,n.

Then inserting (36) into (29) and comparing coefficients at exp (ik t/2) we arrive at

$$\{A - p_1 B + i (\beta A - \alpha p_1 B) k \Theta/2 - C k^2 \Theta^2/4$$

$$- i \alpha C k^3 \Theta^3/8\} w_k - \frac{1}{2} p_2 B (1 + \tag{37}$$

$$+ \alpha i k \Theta/2)(w_{k-2} + w_{k+2}) = 0 .$$

In fact we have two independent systems of equations. For odd indices we have a system with the period $T_1 = 2T = 4\pi/\Theta$ and for even indices we get a system with the period $T = 2\pi/\Theta$. These systems have solutions only when their determinants are equal to zero. Values of Θ which fulfil the determinental equation $\Delta(\Theta) = 0$ are critical frequencies. An analysis of critical frequencies has to be done numerically as also approximate analysis is connected with the analysis of equations of sixth degree.

When we restrict our analysis to shells of Voigt solids $\alpha = 0$ and we arrive at

$$(A_1 - p_1 B + i \beta A k \Theta/2 - C k^2 \Theta^2/4) w_k$$

$$- p_2 B (w_{k-2} + w_{k+2})/2 = 0 , \tag{38}$$

which leads to similar analysis as in the case of elastic shells with damping [4] .

Conclusions :

We have analysed qualitative properties of quasistatic stability of viscoelastic cylindrical shells and have shown that a shell is unstable to initial perturbations for an axial compression greater than the longtime critical load p_{1cr} which is less than the instant critical load corresponding to the critical load for an elastic shell.

Similar qualitative results have been shown in analysis of the dynamic stability of the cylindrical shell subject to a constant axial pressure.

In the case of a periodic load the analysis leads to problems what it is necessary to analyse numerically.

References

1. Brilla, J.: Stability of viscoelastic shallow shells in Koiter, W.T.; Mikhailov, G.K.: Theory of Shells. Amsterdam, New York, Oxford: North-Holland 1980.

2. Carslaw, H.S.; Jaeger, J.C.: Operational methods in applied mathematics, Oxford: University Press 1948.

3. Brilla, I.: Bifurcation theory of the time-dependent von Karman equations. Aplikace matematiky 29(1984) 3-13.

4. Bolotin, V.V.: Dinamicheskaya ustojchivosť uprugikh sistem. Moskva: Gostekhizdat 1956.

Optimum Design of Structures Made of Elastic-Plastic Materials

N.V. BANICHUK

Institute for Problems in Mechanics
USSR Academy of Sciences, Moscow.

Summary

The shape optimum design for elastic-plastic structures loaded by fixed and variable repeated forces is considered. Formulated variational problems with unknown boundaries and partial derivatives are investigated by the methods of distributed parameter control theory. The necessary optimality conditions for unknown boundaries are obtained taking into account load carrying capacity and shakedown constraints. The derived sensitivity analysis formulas are applied to the different problems of elastic-plastic design.

Introduction

Strength and weight requirements are the most important factors in the theory of optimum design. Various failure modes are considered in modern studies. But because of the complexity of the collapse mechanism, it is difficult to take the real strength condition into account in the solutions. This probably explains why the theory of optimum design is far from being complete. Contemporary investigations in this field are concerned with the research of the new statement of the problems, taking into account the different types of nonlinear behaviour of the material, consideration of the complex structural elements (plates, shells, three-dimensional solid bodies). Considerable efforts are being applied to the development of design sensitivity analysis and effective optimization methods for minimization of the structural weight or another cost functionals under strength constraints. Related problems of the strength maximization under some inequality constraints are also considered. Research made in this field can be classified into two basic groups.

In the first group belong those studies in which behaviour of the structures is supposed to be pure elastic (see, for example

Inelastic Behaviour of Plates and Shells
IUTAM Symposium Rio de Janeiro 1985
Editors: L. Bevilacqua, R. Feijóo and R. Valid

[1-3]). Typical problems consist of minimizing the stress concentration for the structural elements with rapidly varying geometry and weight minimization for shell structures under constraint on admissible stress intensity level. Besides rigorous solutions, some results were obtained with the application of intuitively selected criteria such as equal stress or equal strength criteria. Note refs. [4-6] which contain the discussion of the designs based on the concept of equal strength.

In the second group belong the studies based on representations about load carrying capacity and behaviour of the structure immediately before the collapse. The problems of weight minimization for given critical loads [7-16] were solved here in the frame of the theory of elastic-plastic behaviour of the materials. Some problems with shakedown requirement have been considered in [17-19] for the case of variable repeated loads. Note also that numerous optimal design problems for plastic and elastic-plastic structural elements were discussed in [20-26]. In the developments described herein, it is well established that if plastic properties are taken into account, then it is possible to achieve an additional meaningful economy of materials. Here we mean that the weight of plastic design is compared with the weight of pure elastic design for the same stress intensities.

Shape Optimal Design of Elastic-Plastic Structures under Carrying Capacity Constraint.

Let us consider the equilibrium state of a solid body, occupying a domain Ω. The body is under action of volume forces q_i and external loads T_i applied to the part Γ_σ of the body surface Γ. The rigid clamping conditions are supposed to be satisfied for another part Γ_u of the body surface ($\Gamma_\sigma + \Gamma_u = \Gamma$). This signifies that the displacement vector is equal to zero on this part of the surface: $(u)_{\Gamma_u} = 0$. Behaviour of the material is assumed to be ideal elastic-plastic. An admissible stress state is characterized by means of the inequality $g(\sigma_{ij}, k) \leqslant 0$. Here k - plasticity constant, σ_{ij} - stress tensor components, g - given function. The yield state is attained if the sign of equality holds for this relationship. The behaviour of the material is pure elastic if the stresses satisfy the discussed condition with rigorous inequality sign. The family of convex closed surfaces, which contain the origin of coordinates in stress space and correspond to different values of k, is given by equation $g(\sigma_{ij},k) = 0$. These surfaces tend to the origin when $k \to 0$.

We assume that plasticity regions are raised for applied loads. However, we consider that the spreading of plasticity zones is such that the carrying capacity of the body is not exhausted. This is the only constraint we take into account. Exhaustion of carrying capacity means unbounded increasing of strains for constant loads.

Analysis of stresses and strains can be fulfilled for elastic-plastic body by means of developed numerical methods. Performing the corresponding calculations is possible only with the help of high speed computers. An especially large volume of calculations are necessary for the optimum design of elastic-plastic bodies. These calculations are needed to compare different admissible design variants. This is the reason why, up to now, there are essential difficulties in solving optimal structural design problems for elastic-plastic bodies using full elastic-plastic analysis.

Essential simplification is achieved for problems of elastic-plastic analysis, when the load carrying capacity is evaluated and no additional information concerning stress-strain state and stiffness (rigidity) characteristics of the structure is required. Fundamental theorems of limit equilibrium state can be used in this case [27,28]. According to the statical theorem of the limit equilibrium theory, the body carries the applied loads, if a statically admissible stress field σ_{ij} is possible satisfying the safety requirements and the equilibrium conditions

$$\sigma_{ij,j} + q_i = 0 \qquad (1)$$

$$(\sigma_{ij} n_j)_{\Gamma_\sigma} = T_i \qquad (2)$$

and such that

$$g(\sigma_{ij},k) < 0 \tag{3}$$

where n_j represents the unit vector along the external normal to the surface of the body. A subscript after the comma denotes the differentiation with respect to the corresponding coordinate.

Taking into account the statical theorem of limit analysis, we can formulate the next optimization problems: it is required to minimize the volume of the material

$$\int_{\Omega} d\Omega \; - \; \min_{\Gamma_V} \tag{4}$$

subject to constraints (1)-(3) on load carrying capacity. Part of the boundary $\Gamma_V \subset \Gamma_\sigma$ of the body is taken as a design variable.

The set of σ_{ij}, satisfying (1)-(3) (safety stresses), is non-closed. This is the reason why the statement of the optimal design problem in the form (1)-(4) is incorrect. For regularization of the problem let us consider next an approximate transformation. Consider two yield surfaces corresponding to the original parameter k and a varied parameter k_f, where $k > k_f$ and $k = k_f + \varepsilon$, $\varepsilon > 0$ - small number. Considering the classical geometrical representation of the yield surfaces, it is not difficult to note that the rigorous inequality (3) is fulfilled "automatically" if statically admissible stress fields satisfy relationships (1),(2) and condition

$$g(\sigma_{ij},k_f) \leqslant 0 \tag{5}$$

Thus, these fields are at the same time safety statically admissible for original yield surface. The solution of the optimal design problem can be obtained with any desired accuracy by defining a sufficiently small value ε and taking into account (5) instead of (3) in basic relations (1)-(4).

Application of this approach, based on the introduction of the modified yield surface, is not the reason for additional mistakes if small enough, value ε is taken. Note here that there is some roughness in the experimental evaluation of plasticity constant k for real materials and the obtained values characterised by some dispersion.

To investigate the formulated problem, let us apply an approach based on sensitivity analysis. Introducing the slack variable μ, we transform the inequality (5) into

$$g(\sigma_{ij},k_f) + \mu^2 = 0 \tag{6}$$

Augmented Lagrange functional has the following form:

$$J = \int_\Omega d\Omega + \int_\Omega \psi_i(\sigma_{ij,j}+q_i)d\Omega + \int_\Omega \lambda(g+\mu^2)d\Omega \tag{7}$$

where ψ_i (i=1,2,3) and λ are adjoint variables. Next the expression for δJ may be sought by varying the surface Γ_v and taking into account the boundary conditions

$$\delta J = \int_\Omega (\lambda\frac{\partial g}{\partial\sigma_{ij}} - \psi_{i,j})(\delta\sigma_{ij} - \sigma_{ij,k}\delta x_k)d\Omega +$$

$$+ \int_\Omega 2\lambda\mu d\Omega + \int_\Gamma \psi_i n_j(\delta\sigma_{ij} - \sigma_{ij}\ \delta x_k)d\Gamma +$$

$$+ \int_{\Gamma_v} \{1 + \lambda(g + \mu^2)\}n_k\delta x_k\ d\Gamma \tag{8}$$

In the derivation of the relation between δJ and variation of surface Γ_v, we use the property of the symmetry of the stress tensor ($\sigma_{ij}=\sigma_{ji}$) and define the adjoint variables ψ_i as a function satisfying the partial differential equations and boundary conditions

$$\frac{1}{2}(\psi_{i,j}+\psi_{j,i}) = \frac{\partial g}{\partial\sigma_{ij}} \tag{9}$$

$$(\psi_i)_{\Gamma_u} = 0 \tag{10}$$

Let us suppose that λ and μ are subjected to the relation

$$\lambda\mu = 0 \quad x \in \Omega \tag{11}$$

The two last integrals in the right hand of (8) are equal to zero, requiring the functions ψ_i, λ, μ to obey (9)-(11). The loads T_i and q_i are supposed to be given as functions of spatial coordinates $T_i = T_i(x_k)$, $q_i = q_i(x_k)$. For shape variations of the body we assume also that variations of the boundary conditions have the form $\delta(\sigma_{ij}n_j) = n_j\delta\sigma_{ij} + \sigma_{ij}\delta n_j = T_{i,j}\delta x_j$. Variation δn_j is expressed in terms of derivatives of 1 and coefficients of the first quadratic form of varied surface. The term including δn_j is transformed with the help of Gaussian equations. The derived expression depends explicitly on the mean curvature H. We have as a result the next expression for δJ

$$\delta J = \int_{\Gamma_V} \{1 + \lambda g - 2H\psi_i T_i - (\psi_i \sigma_{ij})_{,j} + (\psi_i T_i)_{,k} n_k\} \delta l d\Gamma \tag{12}$$

Formula (12) determines the sensitivity of the optimized functional to the variations of design variables (shape variations sensitivity) and gives us the possibility to develop the numerical methods of successive optimization [2]. Using the expression (12) and the stationarity principle $\delta J=0$ we can write the necessary optimality condition:

$$1+\lambda g-2H\psi_i T_i-(\psi_i \sigma_{ij})_{,j}+(\psi_i T_i)_{,k} n_k = 0 \text{ on } \Gamma_V \tag{13}$$

Consider the particular case when the volume forces are absent ($q_i=0$) and free boundary ($T_i=0$) is varied. We suppose also that $g=g_o(\sigma_{ij})-k_f^2$, where $g_o(\sigma_{ij})$ is a second order homogeneous function of stress tensor components. The optimality condition is simplified

$$\lambda\,[2k_f^2 + g] = 1 \quad \text{on } \Gamma_V \tag{14}$$

In this case the optimality condition doesn't depend on adjoint variables ψ_i. Thus the solution of the optimum design problem may be found on the base of equations (1),(2),(6),(11) and the expression for δJ

$$\delta J = \int_{\Gamma_v} \{1 - \lambda[2k_f^2+g]\}\delta l d\Gamma \tag{15}$$

In general, the case determination of the optimal solution is based on a full system of relations (1),(2),(6),(9)-(12).

It is not difficult to note that the analogy between the equations (1),(2),(9),(10) and the equations of rigid-plastic model. For this analogy there is a correspondence of adjoint variables ψ_i to displacement velocities v_i, values $\frac{1}{2}(\psi_{i,j}+\psi_{j,i})$ to strain velocities, and relation (9) to associate flow rule of plasticity theory.

Using the analogy we introduce the notation:

$$\dot{\varepsilon}_{ij} = \frac{1}{2}(\psi_{i,j} + \psi_{j,i})$$

Multiplying (9) by σ_{ij} and taking the property of homogeneity of $g_o(\sigma_{ij})$ into account, we obtain

$$\sigma_{ij}\,\dot{\varepsilon}_{ij} = 2\lambda g_o \tag{16}$$

With the help of the equality (16) we can transform the optimality condition (14). Consider two cases: $g \neq 0$ and $g = 0$. If for the varied part of the boundary $g \neq 0 (g_0 < k^2)$ then $\mu \neq 0$. This inequality and condition (11) give us $\lambda = 0$. It means that the optimality condition (14) is not fulfilled if the given assumption ($g \neq 0$) takes place. Suppose that $g = 0 (g_0 = k_f^2)$ in (14). From this condition and on the basis of (6),(11) we have $\lambda = \sigma_{ij}\,\dot{\varepsilon}_{ij}/2k_f^2$. Hence the optimality condition can be written in the form

$$\sigma_{ij}\dot{\varepsilon}_{ij} = 1 \quad \text{on } \Gamma_v \tag{17}$$

Thus, we attain the classical optimality condition, obtained in the paper [9]. Condition (17) means that the rate of energy dissipation is constant along the unknown part of the boundary

Γ_v. Note that stress intensity is also constant

$$g_0 = k_f^2$$

along Γ_v. Consequently, if for solving the problem the improved succession of shapes converge, then the limit optimal boundary Γ_v will be uniformly stressed.

A numerical solution of the optimization problem can be obtained with an application of the successive optimization method [2] or some other iteration method. Computations of the next two types must be performed when the successive optimization method is used. Computations of the first type are connected with the solving of the rigid-plastic analysis problem. As a result, the values $\sigma_{ij},\psi_i,\lambda,\mu$ are determined for a given shape of the body. Computation of the second type furnish the improved variations of the boundary shape. Calculated values $\sigma_{ij},\psi_i,\lambda,\mu$ and the derived formulas of sensitivity analysis are used for this purpose. Many different methods of calculus of variations and the mathematical programming theory can be used for effective realization of sensitivity analysis. Note that various methods were used in the frame of successive optimization algorithm, but the gradient method appeared to be the most efficient from the practical point of view [1,2].

The approach based on the successive optimization method has been developed for solving minimum weight design problems with constraints on load carrying capacity. The results of computations are presented in Figs. 1-3 for plane elastic-plastic elements. The behaviour of these elements is characterized by two-dimensional equilibrium equations and Mises yield condition.

In Fig. 1 an improved supporting element is shown. We take into account the symmetry of external loads boundary conditions and the geometry of the element with respect to the axis x. Therefore only half of the element is presented in Fig. 1 and also in Fig. 2. Compressive loads are uniformly distributed and applied to the part of the boundary CD. The unknown part of the boundary BC is free from loads. The line BC is considered as a design variable and improved by iterations. There is contact

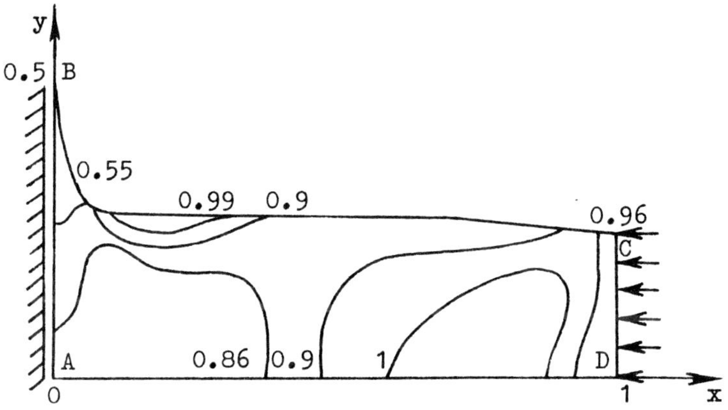

Fig. 1. Plastic design for compressed element.

between the element under compression and the perfectly rigid and smooth surface AB. We suppose that the friction on AB is absent and the displacements in the x-direction are equal to zero.

The results obtained were compared to the corresponding pure elastic minimum weight design. For this purpose the minimum weight design problem was solved under the stress intensity constraint $g_0(\sigma_{ij}) \leqslant 1$. The solution was obtained by the application of the algorithm of elastic optimization proposed in [29]. Computations were performed for the same geometrical and loading parameters (datas) with the application of closed system of equations of the elasticity theory. An approximate solution is presented in Fig. 2. Comparison of numerical results, corresponding to elastic designs, show that taking into account the stress intensity constraint in elastic property of material we can achieve significant reduction of weight. Additional weight decreasing of plastic design with respect to elastic design is more than 6.6%.

Equal stress intensity lines are presented by solid curves in

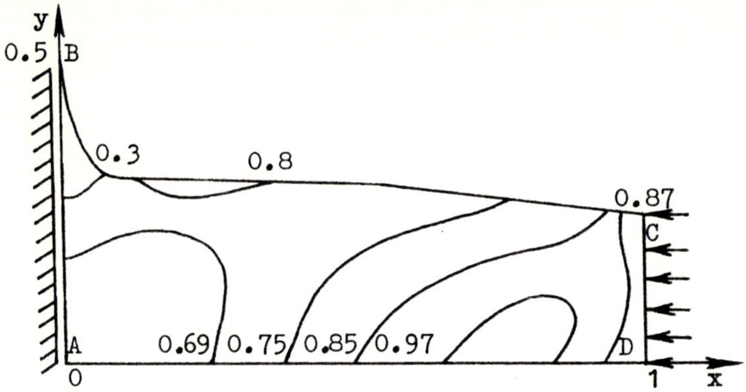

Fig. 2. Elastic design for compressed element.

Figs. 1 and 2. We see that for plastic design the stress fields become smoother.

As another example, we shall consider the plane structural element shown in Fig. 3a,b. The loads are uniformly distributed along the boundary AB and oriented in x-direction. For the part of the boundary AD we have a rigid clamping condition. Free of the loads part of the boundary BC has been varied in designing process. Beginning with the initial shape presented in Fig. 3a, we obtain, after computation, the improved boundary shape shown in Fig. 3b. In the present example as in the previous one, the solid curves in Fig. 3a,b signify the equal stress intensity lines.

Shape Optimum Design Under Shakedown Constraint

In the case of invariable loads it is sufficient for structural safety to take the load carrying capacity constraint into account. But in the general case of varying loads, this constraint is a necessary but not a sufficient condition and the collapse of the structure can occur. The reason for this is the following: the cycles of the rates of plastic deformations can arise when repeated variable loads are less than the critical value.

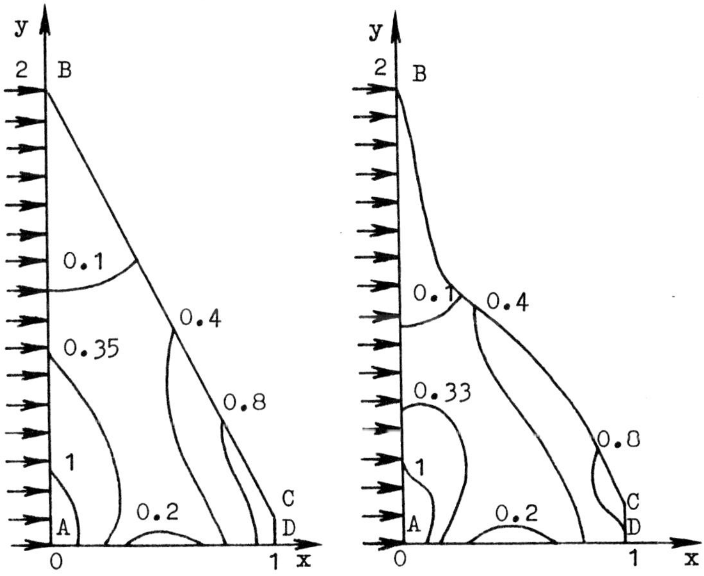

Fig. 3. Initial (a) and improved (b) shapes

This is the so called collapse by cyclic plastic deformations which furnish the increasing of general deformation. There is no collapse and the shakedown takes place if during the loading the body accumulates residual stresses such that for the next part of the loading programme the deformations become pure elastic. Below we use the main statical theorem (Melan's theorem) of the shakedown theory [27]. If it is possible to find the independent time distribution of the residual stresses σ^r_{ij}, such that their sum with elastic stresses σ^e_{ij} gives safety stress state $(g(\sigma^e_{ij} + \sigma^r_{ij}, k) < 0)$ for any point of the body and for all combinations of loads, lying in given limits, then the shakedown will be reached by the structure. For the future variations of the loads in given limits the behaviour of the structure will be perfectly elastic.

Let us formulate the optimization problem with shakedown constraint. It is necessary to determine the part of the body surface Γ_v, minimizing the functional (4) such that the stresses σ^e_{ij}, σ^r_{ij} satisfy the system of equations

$$\sigma^{e}_{ij,j} + q_i = 0, \quad \sigma^{e}_{ij} = \frac{1}{2} C_{ijkl}(u_{k,l} + u^{e}_{l,k}) \tag{18}$$

$$(\sigma^{e}_{ij} n_j)_{\Gamma_\sigma} = T_i, \quad (u^{e}_i)_{\Gamma_u} = 0$$

$$\sigma^{r}_{ij,j} = 0, \; \dot{\sigma}^{r}_{ij} = 0, \; (\sigma^{r}_{ij} n_j)_{\Gamma_\sigma} = 0 \tag{19}$$

and the inequality

$$g(\sigma^{e}_{ij} + \sigma^{r}_{ij}, k_f) \leqslant 0 \tag{20}$$

for all loads T_i, q_i. We assume that all loads are included into the loading programme $T_i = T_i(x,t)$, $q_i = q_i(x,t)$, $t \in [0,t_k]$. Here t_k is given and the dot above the symbol denotes the differentiation with respect to t(the remaining symbols are usual). As was done previously, the constant k in plasticity condition (20) is replaced by the constant k_f.

We express the minimized volume of the structure in the following form

$$\frac{1}{t_k} \int_0^{t_k} \int_\Omega d\Omega dt \tag{21}$$

Taking into account (21) and (18)-(20) we construct the augmented Lagrange functional

$$J = \int_0^{t_k} \int_\Omega f^L d\Omega dt, \qquad f^L = \frac{1}{t_k}[1 + (g+\mu^2)]+$$

$$\psi^{e}_i(\sigma^{e}_{ij,j}+q_i) + \chi^{e}_{ij}(\sigma^{e}_{ij}-C_{ijkl}u^{e}_{k,l})+\psi^{r}_i\sigma^{r}_{ij,j} \tag{22}$$

By means of ψ^{e}_i, ψ^{r}_i, χ^{e}_{ij}, λ we denote the adjoint variables. To obtain the expression for δJ vary the surface Γ_v and use the boundary conditions for state variables. Also we assume that the variations of σ^{r}_{ij} do not depend on t. The adjoint variables are defined with the help of the next system of equations

$$\frac{1}{2}(\psi^e_{i,j} + \psi^e_{j,i}) = \chi^e_{ij} + \lambda\frac{\partial g}{\partial \sigma^e_{ij}} \quad \text{in } \Omega \tag{23}$$

$$(\psi^e_i)_{\Gamma_u} = 0 \tag{24}$$

$$(C_{ijkl}\,\chi^e_{kl})_{,j} = 0 \text{ in } \Omega \tag{25}$$

$$(C_{klij}\,\chi^e_{kl} n_j)_{\Gamma_\sigma} = 0 \tag{26}$$

$$\frac{1}{2}(\psi^r_{i,j} + \psi^r_{j,i}) = \frac{1}{t_k}\int_0^{t_k} \lambda \frac{\partial g}{\partial \sigma^r_{ij}}\, dt \text{ in } \Omega \tag{27}$$

$$(\psi^r_i)_{\Gamma_u} = 0 \tag{28}$$

$$\lambda.\mu = 0 \text{ in } \Omega \tag{29}$$

If the relations (23)-(29) are satisfied then the next formula for δJ is valid

$$\delta J = \int_{\Gamma_v} \{\frac{1}{t_k}\int_0^{t_k} [1 + \lambda g - 2H\psi^e_i T_i - (\psi^e_i \sigma^e_{ij})_{,j} - (\psi^r_i \sigma^r_{ij})_{,j} + (\psi^e_i T_i)_{,k} n_k] dt\} \delta l d\Gamma \tag{30}$$

Here H is the mean curvature of surface Γ_v, and δl denotes the variation of the unit normal.

The main formula of sensitivity analysis (30) connects the variation of the functional with the shape variation of the body. To construct the improved boundary shape it is necessary to obtain the information concerning elastic stresses σ^e_{ij} and adjoint variables $\psi^e_i, \psi^r_i, \chi^e_{ij}, \lambda$ for $t \in [0,t_k]$.

Straightforward consideration gives us the necessary optimality condition

$$(\psi_i^r \sigma_{ij}^r)_{,j} - 1 = \frac{1}{t_k} \int_0^{t_k} [\lambda g - 2H\psi_i^e T_i -$$

$$- (\psi_i^e \sigma_{ij}^e)_{,j} + (\psi_i^e T_i)_{,k} n_k]dt \tag{31}$$

In the case, when the volume forces and surface loads on Γ_v are absent ($q_i = 0$ in Ω, $T_i = 0$ on Γ_v) and the yield condition is given as $g(\sigma_{ij}, k_f) = g_0(\sigma_{ij}) - k_f^2 \leqslant 0$, where $g_0(\sigma_{ij})$ is the second order homogeneous function, optimality condition is simplified

$$(\psi_i^r \sigma_{ij}^r)_{,j} - 1 = \frac{1}{t_k} \int^{t_k} [\lambda g - (\psi_i^e \sigma_{ij}^e)_{,j}]dt \tag{32}$$

Optimal Plastic Anisotropy

Various optimization problems may be formulated within the framework of the theory of anisotropic bodies; one of them will be discussed here. The problem is concerned with optimum distribution of plastic constants (moduli) B_{ijkl}, which are contained in the formula $g = B_{ijkl}\sigma_{ij}\sigma_{kl} - 1$ for yield function. For definiteness let us assume that the critical loading parameter will be maximized. Suppose the body is to consist of identical infinitesimal crystals arbitrarily oriented with respect to each other. The fact that the crystals are identical but arbitrarily oriented means that the positions of the axes of plastic symmetry with respect to a fixed Cartesian reference frame change with the position within the body, but the values of the plastic moduli measured along the axes of plastic symmetry remain unchanged. Let us denote the orientation of the axes of anisotropy at each point $x = \{x_1, x_2, x_3\}$ of the medium with respect to a fixed Cartesian coordinate system (x_1, x_2, x_3) by the angles $\alpha_1(x)$, $\alpha_2(x)$, $\alpha_3(x)$ representing the components of a vectorial function $\alpha(x)$, that is $\alpha(x) = \{\alpha_1(x), \alpha_2(x), \alpha_3(x)\}$. Let α_j $(j = 1,2,3)$ denote the angle made by the plastic symmetry axis x_j' and the fixed coordinate axis x_j. The problem of determination of the optimum orientation of the axes anisotropy, i.e. finding the vector function $\alpha(x)$ from the condition of maximum critical loading parameter is now reduced to

$$p_* = \max_{\alpha} p(\alpha) \tag{33}$$

$$\nabla . \sigma + q = 0 \tag{34}$$

$$(\sigma . n)_{\Gamma_\sigma} = T \tag{35}$$

$$g(\sigma) = \sigma \cdot\cdot B \cdot\cdot \sigma - 1 \leqslant 0 \tag{36}$$

where $q = pq^0$, $T = pT^0$ (q^0, T^0 - given vector-functions of space coordinates), σ - second order stress tensor, B - fourth order tensor of plasticity constants. The dot between the symbols signifies the scalar product and two dots denote the double scalar product.

The solution of the problems (33)-(36) and the determination of the optimum distribution of plastic moduli makes it possible to find the most suitable directions of the reinforcement, and to evaluate the quality of the structures traditionally used in practice. Even in cases in which the optimum structural anisotropy proves to be difficult to realize in practice, solutions of the optimization problems may be used to determine the limiting possibilities and the quasi-optimum reinforcement schemes.

Let us derive optimality conditions for the problems (33)-(36). To this end we denote the tensor of plasticity constants in principal axes by means of b and write the relation between tensor b and tensor B in space coordinates

$$B = 0 \cdot (0 \cdot b \cdot 0) \cdot 0, \quad 0^{*} \cdot 0 = E \tag{37}$$

where 0 - orthogonal rotational tensor of the second order. Variation of B corresponding to the variation $\delta 0$ is written in the form

$$\delta B = 4B \cdot 0^{*} \cdot \delta 0 \tag{38}$$

We can express the optimized functional (33) in the form

$$p = \frac{1}{\mathrm{mes}\Omega} \int_{\Omega} p d\Omega \quad (\nabla p = 0) \tag{39}$$

Taking into account (33)-(37), (39) we construct the augmented Lagrange functional

$$J = \frac{1}{\mathrm{mes}\Omega} \int_{\Omega} p d\Omega + \int_{\Omega} \psi \cdot (\nabla \cdot \sigma + q) d\Omega + + \int_{\Omega} \lambda (g + \mu^2) d\Omega + \int_{\Omega} \eta \cdot \cdot (0^* \cdot 0 - E) d\Omega \tag{40}$$

To obtain the expression for δJ we vary the tensor of plasticity constants in accordance with formula (38) and use boundary conditions (35). We also suppose that the variations of p in (40) do not depend on spatial coordinates ($\nabla p = 0$) and that the adjoint variables and Lagrange multiplier λ satisfy the equations

$$\frac{1}{2}[\nabla\psi + (\nabla\psi)^*] = \lambda \frac{\partial g}{\partial \sigma} = 2\lambda B \cdot \cdot \sigma \tag{41}$$

$$\lambda\mu = 0 \qquad (g + \mu^2 = 0) \tag{42}$$

Taking into account the equations (41), (42) and the principle $\delta J = 0$, we obtain the optimality condition

$$\sigma \cdot \cdot B \cdot \sigma = \sigma \cdot B \cdot \cdot \sigma \tag{43}$$

Using indexing notation we rewrite the criteria (43) $\sigma_{ij}\sigma_{kl}B_{ijkp} = \sigma_{ij}\sigma_{kp}B_{ijkl}$. The derived criteria (43) shows that the plastic anisotropy axes have a special orientation and for optimal body tensor $\sigma \cdot \cdot B \cdot \sigma$ is a symmetric one.

Note that the paper was mainly concerned with the weight and critical loading parameter functionals. The shape of the structure and the orientation of anisotropy axes were considered as the unknown design variables. No additional constraints were imposed on the distribution of design variables. However, from mathematical and physical points of view in some cases we must restrict the variations of design functions to obtain more regular solutions. First of all, we mean the correctness of the mechanical model of the structure and the existence of the optimum design. These aspects were discussed in [30,31].

Acknowledgement

The author wishes to thank V.I. German, who performed the numerical calculations, and S.Iu. Ivanova, V.V. Kobelev for useful discussions and valuable comments.

References

1. Haug, E.J.; Arora, J.: Applied optimal design. Mechanical and structural Systems. New York; John Wiley 1979.

2. Banichuk, N.V.: Problems and methods of optimal structural design. New York: Plenum Press 1983.

3. Troitskii, V.A.; Petuchov, L.V.: Optimization of shape of elastic body (in Russian). Moscow: Nauka 1982.

4. Ishlinskii, A.Yu.: A beam having cross sections of equal strength. Scientific Notes. Moscow State University. 39 (1940) 87-90.

5. Save, M.: Some aspects of minimum-weight design. Engineering plasticity (eds. J. Heyman, F.A. Leckie). Cambridge University Press (1968) 611-626.

6. Banichuk, N.V.: Optimality conditions and analytical methods of shape optimization. Proc. Advanced Study Institute Optimization of Distributed Parameters Structures. Alphen aan den Rijn : Sijthoff-Noordhoff (1981) 973-1004.

7. Banichuk, N.V.; Kobelev, V.V.: Optimal nonequal-strength shape of beams cross-sections. MTT (Mech. of Solids) 18 (1983) 162-167.

8. Hopkins, H.G.; Prager, W.: Limits of economy of material in plates. J. Appl. Mech., 22 (1955) 372-374.

9. Drucker, D.C.; Shield, R.T.: Design for minimum weight. Proc. 9th Int. Congr. Appl. Mech., Brussels, Vol. 5 (1956) 212-222.

10. Sawczuk, A.: Some problems of load carrying capacities of orthotropic and nonhomogeneous plates. Zakblad Mech. Osrod. Cragl. Polsk. Acad. Nauk. YIII, 4 Warsaw (1956).

11. Drucker, D.C.; Shield, R.T.: Bounds on minimum weight design. Quart. Appl. Math., 15 (1957) 269-281.

12. Prager, W.; Taylor, J.E.: Problems of optimal structural design. J. Appl. Mech. 35 (1968) 102-106.

13. Save, M.A.: A Unified formulation of the theory of optimal plastic design with convex cost function. J. Struct. Mech. 1 (1972) 267-276.

14. Zavelani, A.; Maier, G.; Binda, L.: Shape optimization of plastic structures. Proc. IUTAM Symposium Opt. in Struct. Design. Berlin: Springer-Verlag (1975) 541-554.

15. Polizzotto, C.: Optimal plastic design for multiple sets of loads. Meccanica. 9(1974) 206-213.

16. Rozvany, G.I.N.; Olhoff, N.; Cheng, K.T.; Taylor, J.E.: On the solid plate paradox in structural optimization. J. Struct. Mech. 10 (1982).

17. Konig., J.S.: On optimum shakedown design. Proc. IUTAM Symposium Opt. in Struct. Design. Berlin: Springer-Verlag (1975) 405-414.

18. Polizzotto, C.; Mazzarella, C.; Panzeca, T.: Optimum design for workhardening adaptation. Computer Meth. Appl. Mech. Eng. 12 (1977) 129-144.

19. Mazzarella, C.; Polizzotto, C.: Optimum design of rigid-workhardening structures with constraints on deformation. Engineering Structures. 2(1980) 138-146.

20. Prager, W.: Introduction to structural optimization. Wien: Springer-Verlag 1974.

21. Rozvany, G.I.N.: Optimal design of flexural systems. Oxford Pergamon Press 1976.

22. Ciras, A.A.: Mathematical models of analysis and optimization of elastic-plastic systems (in Russian). Vilnius Mokslas Publishers 1982.

23. Lepik, U.: Optimal design of inelastic structures under dynamic loading (in Russian). Tallin, Valgus 1982.

24. Save, M.A.; Massonnet, C.E.: Plastic analysis and design of plates, shells and disks. Amsterdam, London: North-Holland Publishing Company 1972.

25. Cinquini, C.; Sacchi, G.: Problems of optimal design for elastic and plastic structures. J. Mecanique Appl. 4 (1980) 1-29.

26. Haftka, R.T.; Prasad, B.: Optimum structural design with plate bending elements - A survey. AIAA Journal 19 (1981) 517-522.

27. Koiter, W.T.: General theorems for elastic-plastic solids. Progress in Solid Mechanics (eds. Sneddon, I.N. and Hill, R.) Amsterdam: North-Holland Publishing Company 1960.

28. Hodge, P.G.: Plastic analysis of structures. New York: McGraw-Hill 1959.

29. Banichuk, N.V.; Belsky, V.G:; Kobelev, V.V.: Optimisation for the theory elasticity problems with unknown boundaries. MTT (Mech. of Solids) 19 (1984).

30. Niordson, F.I.: Optimal design of elastic plates with a constraint on the slope of the thickness function. Internat. J. Solids Struct. 19 (1983) 141-151.

31. Banichuk, N.V.; Kartvelishvili, V.M.: Design of plates with allowance for constraints on the variability of the thickness. MTT (Mech. of Solids) 18 (1983)130-136.

Mathematical Models for the Analysis and Optimization of Elasto-plastic Plates and Shells Under Complete Plastic Failure

A. CYRAS and A. DANIUNAS

Vilnius Civil Engineering Institute
Vilnius, Lithuanian SSR, USSR.

Summary

Mathematical models for the analysis and optimization of elastic perfectly plastic structures under different types of loading are derived on the basis of the dual extremum principle for the minimum elastic potential of residual forces and maximum complementary work of residual displacements. The application of these types of mathematical models to the design of plates and shells is illustrated by examples.

1. Formulation of the Problem.

The mathematical models for the optimization of elastic perfectly plastic structures based on the method of limit equilibrium do not impose any constraints on the displacements. Nevertheless, the displacements corresponding to the stage of plastic failure frequently are considerable and these must be taken into account in the design problem, otherwise the optimization of the structure has no sense. Hence, it follows that for the optimum design of elastoplastic structures, it is necessary to consider the constraints on the displacements. Subsequently, there arises a problem of elastoplastic design of structure for which the stresses and strains are not taken to the limit state, but to the state corresponding to incomplete plastic failure. It is known that this state is defined by relations which are quite different from those used in the method of limit equilibrium. In this paper the relations of this type are considered, and the application of these to the formulation of an optimization problem are presented.

Consider an elastic perfectly plastic discrete system subjected to an external loading, which does not exceed its limit value,

Inelastic Behaviour of Plates and Shells
IUTAM Symposium Rio de Janeiro 1985
Editors: L. Bevilacqua, R. Feijóo and R. Valid

though it causes plastic strains. The problem of the analysis of such a structure consists of the determination of its stresses and strains, with the physical characteristics, both elastic (stiffness) and plastic (limit forces), being prescribed. In the optimization problem the plasticity characteristics are the values to be searched for, and to determine these it is necessary to satisfy the condition of the quality of the structure.

2. Problem of Analysis

To determine the actual stresses and strains of an elastic perfectly plastic structure we use an extremum principle for the minimum elastic potential of residual forces and for the maximum complementary work of residual displacements [1], [2].

In the case of monotonically increasing loading, the following dual pair of mathematical programming problems correspond to this principle:

Static Formulation

$$\frac{1}{2}\{S_r\}^T [D]\{S_r\} \rightarrow \min,$$

$$\{f(\{S_e\} + \{S_r\})\} < \{S_o\},$$

$$[A]\{S_r\} = \{0\}. \tag{1a}$$

Kinematic formulation

$$(-\frac{1}{2}\{S_r\}^T[D]\{S_r\} - \{\lambda\}^T[\tilde{f}(\{S_e\} + \{S_r\})]^T\{\lambda\}^T(\{S_o\} -$$

$$- \{f(\{S_e\} + \{S_r\})\}) \rightarrow \max,$$

$$[D]\{S_r\} + [\tilde{f}(\{S_e\} + \{S_r\})]\{\lambda\} - [A]^T\{U_r\} = \{0\},$$

$$\{\lambda\} > \{0\} \tag{1b}$$

Where $\{S_r\}$ is the vector of the residual forces; $\{S_e\}$ is the vector of the forces of the elastic solution; $\{S_o\}$ is the vector of the limit forces; $\{f(\ )\}$ is the yield vector-function, and $[f(\ )]$ is its gradient; $[A]$ is the algebraic operator of the equilibrium equations; $[D]$

is the flexibility matrix of the structure; $\{U_r\}$ is the vector of the residual displacements; $\{\lambda\}$ is the vector of the multipliers. We assume that the force vector of the elastic solution $\{S_e\}$ is known, and this is determined through $\{S_e\} = [\alpha]\{F\}$, where $[\alpha]$ is the influence matrix of the forces, and $\{F\}$ is the vector of the external loading.

By solving problem (1) we obtain the unknowns $\{S_r^*\}, \{\lambda^*\}$ and $\{U_r^*\}$ The values of the actual forces and the displacements are obtained through:

$$\{S\} = \{S_e\} + \{S_r^*\}, \quad \{U\} = \{U_e\} + \{U_r^*\} \tag{2}$$

where the values of the displacements of the elastic solution $\{U_e\}$ are assumed to be known, and these are determined through the relation $\{U_e\} = [\beta]\{F\}$, where $[\beta]$ is the influence matrix of the displacements.

In cyclic loading when the external loading is defined by the variation bounds $\{F^+\}$ and $\{F^-\}$, we have the following dual pair of problems [2]:

Static formulation

$$\frac{1}{2}\{S_r\}^T[D]\{S_r\} \to \min,$$

$$\left.\begin{array}{l}\{f(\{S_{ei}^+\} + \{S_r\})\} \leqslant \{S_o\},\\ \{f(\{S_{ei}^-\} + \{S_r\})\} \leqslant \{S_o\}\end{array}\right\} \quad i \in J,$$

$$[A]\{S_r\} = \{0\}. \tag{3a}$$

Kinematic formulation

$$\begin{aligned}&(-\frac{1}{2}\{S_r\}^T[D]\{S_r\} - \sum_i\{\lambda_i^+\}^T[\tilde{f}(\{S_{ei}^+\}+\{S_r\})]^T\{S_r\} -\\ &-\sum_i\{\lambda_i^-\}^T[\tilde{f}(\{S_{ei}^-\}+\{S_r\})]^T\{S_r\} - \sum_i\{\lambda_i^+\}^T(\{S_o\} - \{f(\{S_{ei}^+\}+\{S_r\})\}) -\\ &-\sum_i\{\lambda_i^-\}^T(\{S_o\} - \{f(\{S_{ei}^-\}+\{S_r\})\})) \to \max,\end{aligned}$$

$$[D]\{S_r\}+\sum_i [\tilde{f}(\{S_{ei}^+\}+\{S_r\})]\{\lambda_i^+\}+\sum_i [\tilde{f}(\{S_{ei}^-\}+\{S_r\})]\{\lambda_i^-\}-$$

$$-[A]^T\{U_r\} = \{0\},$$

$$\{\lambda_i^+\} \geqslant \{0\},\ \{\lambda_i^-\} \geqslant \{0\},\ i\varepsilon J, \tag{3b}$$

where the notation is the same as in problem (1). The extremum forces of the elastic solution corresponding to the i-th vertex of the polyhedron of the elastic forces [2] which is symmetric about its centre, are determined on the basis of the following relations:

$$\left.\begin{aligned} \{S_{ei}^+\} &= [\alpha_i^+]\{F^+\}+[\alpha_i^-]\{F^-\}, \\ \{S_{ei}^-\} &= [\alpha_i^-]\{F^+\}+[\alpha_i^+]\{F^-\} \end{aligned}\right\} \tag{4}$$

where the matrices $[\alpha_i^+]$, $[\alpha_i^-]$ are formed from the influence matrix $[\alpha]$. The vectors $\{\alpha_{ij}^+\}$ and $\{\alpha_{ij}^-\}$ formed the matrices $[\alpha^+]$ and $[\alpha_i^-]$ are chosen according to the following rule: $\{\alpha_{ij}^+\} = \{\alpha_{ij}\}$ if $\{\alpha_{ij}^-\} = \{0\}$, and $\{\alpha_{ij}^-\} = \{\alpha_{ij}\}$ if $\{\alpha_{ij}^+\} = \{0\}$. It is not difficult to see that $\{\alpha_{ij}^+\}+\{\alpha_{ij}^-\}=\{\alpha_{ij}\}$.

By solving the mathematical programming problem (3) we obtain the values of the unknowns $\{S_r^*\},\{\lambda^{+*}\},\{\lambda^{-*}\}$ and $\{U_r^*\}$. The variation bounds of the forces are determined on the basis of the relation:

$$\{S^{+*}\} = \{S_{ei}^{+*}\}+\{S_r^*\},\{S^{-*}\} = \{S_{ei}^{-*}\}+\{S_r^*\}, \tag{5}$$

where $\{S_{ei}^{+*}\}$ and $\{S_{ei}^{-*}\}$ are taken corresponding to the multipliers $\{\lambda_i^{+*}\}$ and $\{\lambda_i^{-*}\}$.

Thus, the mathematical models (1) and (3) enable the actual stresses and strains of the structure to be determined i.e. the problem of analysis to be solved.

3. Optimization Problem

The optimization problem may be formulated as follows: the external loading and the configuration of the elastoplastic structure being prescribed, it is necessary to determine the forces and displacements satisfying the conditions of strength and stiffness, as well as the limit force vector $\{S_o\}$ corresponding to the optimality criterion. To be in line with the above statements we assume that the optimality criterion is as follows:

$$\phi(\{S_o\}) \to \min, \tag{6}$$

where $\phi(\ )$ is a scalar function. The form of the function may be varied, and by suitably choosing it we can obtain for an actual structure a definite economical interpretation of the optimality criterion [3]. Firstly, the conditions determining the actual stresses and strains must be considered as the constraints in the problem of optimization of the elastoplastic structure. These conditions are the dual pairs (1) and (3). According to the mathematical programming theory, the dual pair (1) and (3) are equivalent to the generalized Lagrange problem consisting of all conditions of the relevant dual pair and of Kuhn-Tuckers conditions. Secondly, the constraints for the elastoplastic structure must be the constraints on the displacements, the directions and position of these being prescribed. The general form of these constraints may be prescribed as follows:

$$\{U_n^-\} \leqslant [L]\{U\} \leqslant \{U_n^+\} \tag{7}$$

where $\{U_n^-\}$ and $\{U_n^+\}$ are the specified values of the displacements, and the elements of the matrix $[L]$ are chosen according to the following logical rule:

$L_{j\ell} = 1$, if the constraint is imposed on the displacement at the j-th section in the direction ℓ;

$L_{j\ell} = 0$, otherwise.

Besides these necessary conditions-constraints there may exist some others. For example, according to the design codes the values of the limit force S_{oj} in some cases must have equal values at some design sections. Then the number of the components of the unknown vector $\{S_o\}$ is fewer than the number of inequality of the yield conditions. In this case it is convenient to introduce in the calculation the so-called configuration matrix the structure of which is determined according to the following logical rule:

G_{kj} = 1, if the k-th component of the limit force is included in the j-th yield condition;

G_{kj} = 0, otherwise.

Some other technological limitations are possible with respect to both limit forces and the displacements and strains.

Thus, the optimization problem for monotonically increasing loading, when the above statements are taken into account, will have the form:

$$\phi(\{S_o\}) \rightarrow \min,$$

$$\{\psi\} = [G]\{S_o\} - \{f(\{S_e\}+\{S_r\})\} \geqslant \{0\},$$

$$[A]\{S_r\} = \{0\},$$

$$[D]\{S_r\} + [\tilde{f}(\{S_e\}+\{S_r\})]\{\lambda\} - [A]^T\{U_r\} = \{0\},$$

$$\{\lambda\}^T\{\psi\} = 0, \ \{\lambda\} \geqslant \{0\}, \ \{S_o\} \geqslant \{0\},$$

$$\{U_n^-\} \leqslant [L](\{U_e\}+\{U_r\}) \leqslant \{U_n^+\}. \qquad (8)$$

We obtain a non-linear multiextremum mathematical programming problem, in which unknowns are $\{S_o^*\}$, $\{S_r^*\}$, $\{\lambda^*\}$ and $\{U_r^*\}$. The non-linearity is caused by the form of the function $\phi(\ )$, and the multiextremum character - by the Kuhn-Tucker's condition $\{\lambda\}^T\{\psi\} = 0$.

Now we shall discuss some aspects of the application and the truth of the mathematical model (8).

The formulation of the problem can be based on two different initial conditions: 1) the limit force $\{S_o\}$ is not related to the elastic characteristics of the design element; 2) the limit force $\{S_o\}$ is related to the elastic parameters of the structure. In the former case the solution of the problem is obtained by the one step procedure. The latter requires relationships between the components $\{S_o\}$ and the stiffnesses of the elements to be available, and the design is carried out by the iteration technique. It should be noted, that for the structures in which the specific dimensions of the section have a fixed value, e.g. the thickness of reinforced concrete plates and shells, the stiffness of the design element may be considered as not related to the limit force. In this case, the solution may be obtained by the one step procedure.

The extremum principle which has been assumed as the basis of the design, leads to a unique distribution of the residual forces, but not always to a unique distribution of the residual displacements. To ensure the uniqueness of the residual displacements we need to observe the holonomic law, i.e. the plastic strain in loading should not decrease and unloading is not permitted. Unloading phenomen is rare in monotonically increasing loading of real structures. Besides, when unloading occurs, the residual strains are less than those obtained by solving problem (1). Thus, the result obtained through mathematical model (8) will always have some margin.

Likewise, we can derive a mathematical model of the optimization problem for cyclic loading. In this case, however, certain difficulties arise. The fact is, that for dissipative structures the strains are related to the way of loading, while in cyclic loading the way of loading itself is not known. Though we have reliable methods for the determination of the limits of displacements, e.g. see [4], nevertheless, these methods can be applied only when the distribution of the limit forces $\{S_o\}$ are known. But in case of optimization the limit forces $\{S_o\}$ are the unknown values to be obtained. Hence, the solution of the problem in this case is of a complex iteration character.

4. Numerical Example of the Problems of Analysis and Optimization

We consider a thin flexible simply supported plate which obeys the assumptions of the technical analysis theory. The dimensions of the plate in the directions of the coordinates X_1, and

X_2 are equal to L. The material of the plate is isotropic, the section is a sandwich type. The limit moment M_o is constant through the plate. The Poisson's ratio $\nu = 0.3$. The plate is subjected to a monotomically increasing uniformly distributed loading of the intensity $F = 22.6\eta M_o/L^2$. We calculate the plate in three levels of loading $\eta = (0.93, 0.95, 0.97)$. Up to the stage $\eta = 0.69$ the behaviour of the plate is elastic, but at $\eta = 1.0$ it reaches the limit state. Thus, the plate is in an elastoplastic state when $0.69 \leqslant \eta \leqslant 1.0$.

The design is based on the finite elements technique, the stress distribution is prescribed. The finite element discretization of the equilibrium equations is accomplished by the Bubnow-Galiorkin method. This provides the duality of adjoint algebraic operators of equilibrium and geometric equations. Plastic behaviour of plate is defined by the linearized Tresca yield conditions, which are derived for the vicinity of the nodal point of the element. Due to the axial symmetry only one quadrant of the plate is considered with the finite element mesh 6 x 6. The elements are assumed to have four nodes.

The problem is solved by using the algorithm of linear complementarity [5]. In Fig. 1, the curves are plotted representing the variation of the deflection values (to within the constant multiplier M_oL^2/D, where D is the stiffness of the section) on the axis of symmetry corresponding to various levels of loading. The development plastic strains in relation to the value of loading is shown in Fig. 2.

To illustrate the solution of an optimization problem we undertake the design of an elastoplastic simply supported circular plate of a constant thickness. The plate is subjected to a uniformly distributed loading F (Fig. 3a). The Poisson's ration is $\nu = 0.3$, the cylindrical stiffness is D = const. The plate is divided into four design elements of equal width (Fig. 3a).

It is necessary to find an optimum distribution of limit bending moments, when the constraint on the deflection in the centre of the plate is being prescribed by $U^+_{1n} = 0.0673\ FR^4/D$. The problem is solved on the basis of mathematical model (8), which for the linear yield conditions of Tresca has the form:

$$\{\Lambda\}^T\{M_o\} \rightarrow \min,$$

$$\{\psi\} = [G]\{M_o\} - [\Phi](\{M_e\} + \{M_r\}) \geqslant \{0\},$$

$$[A]\{M_r\} = \{0\},$$

$$[D]\{M_r\} + [\Phi]^T\{\lambda\} - [A]^T\{U_r\} = \{0\},$$

$$\{\lambda\}^T\{\psi\} = 0,\ \{\lambda\} \geqslant \{0\},\ \{M_o\} \geqslant \{0\},$$

$$\{U_n^-\} \leqslant [L](\{U_e\} + \{U_r\}) \leqslant \{U_n^+\}. \qquad (9)$$

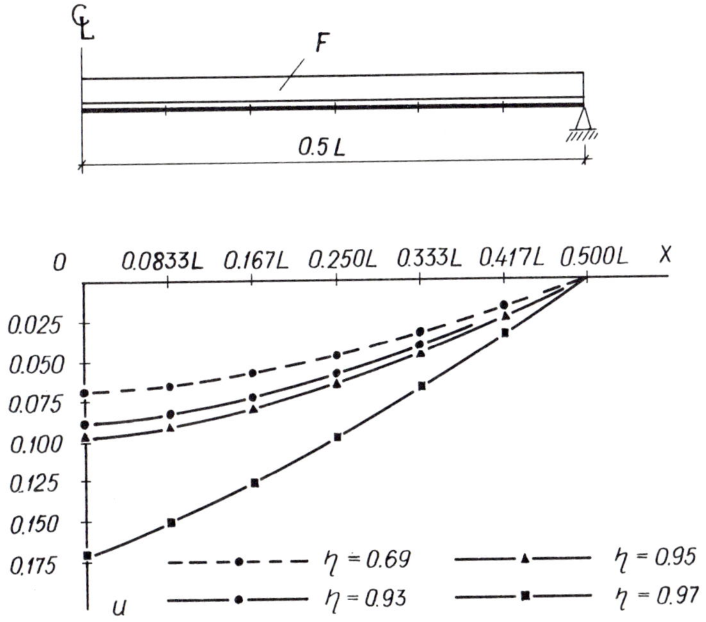

Fig. 1. The deflections on a centre-line of the simply supported square plate at the different load levels.

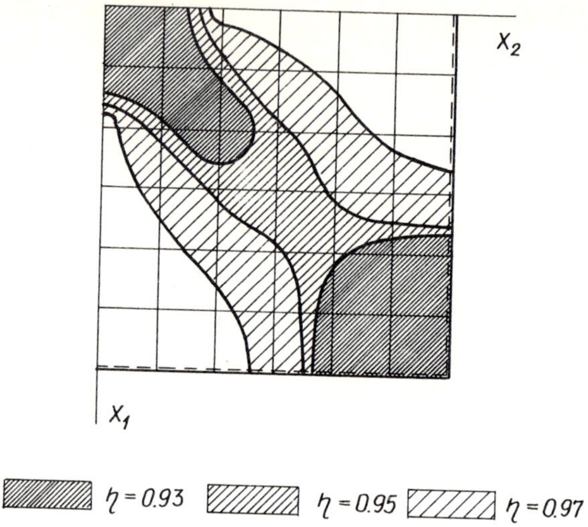

Fig. 2. The plastic zones in the simply supported square plate at the different load levels.

The cost function for the plate has the form:

$$\{\lambda\}^T\{M_o\} = 0.25M_{01} + 0.75M_{02} \rightarrow \min,$$

By the branch-and-bound method [6], [7] we obtain: $M_{01} = 0.214FR^2$, $M_{02} = 0.158\ FR^2$, the cost function $\{\lambda\}^T\{M_o\} = 0.179$. The values of the actual bending moments (radial M_R and circular M_ϕ to within the multiplier FR^2) are shown in Fig. 3b,c. The values of the actual deflections (to within the multiplier FR^4/D) are shown in Fig. 3d. The positioning of plastic circular lines is given as well.

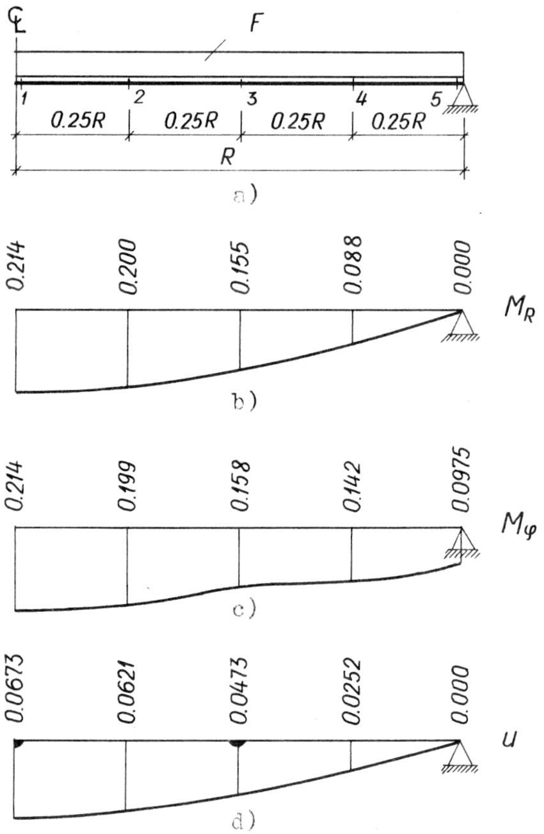

Fig. 3. The distribution of bending moments and deflections in the simply supported optimal circular plate.

References

1. Maier, G.: A matrix structural theory of piecewiselinear elastoplasticity with interacting yield planes. Meccanica 5 (1970) 54-56.

2. Cyras, A.: Mathematical models for the analysis and optimization of elastic-plastic systems. John Wiley, 1983.

3. Cyras, A.: Optimization theory in limit analysis of a solid deformable body. Vilnius, Mintis, 1971 (in Russian).

4. Cyras, A., Atkociunas, J.: Mathematical model for the analysis of elastic-plastic structures under repeat-variable loading. Mechanics Research Communications II(5) (1984) 353-360.

5. Bazzaraa, M.S., Shetty, C.M.: Nonlinear programming. Theory and algorithms, John Wiley, 1979.

6. Lawler, E.L., Wood, D.E.: Branch-and-bound methods a survey. Operations Research, 16(1966) 699-719.

7. Mitten, L.G.: Branch-and-bound methods: General formulation and properties. Operations Research 18(1970) 24-34.

Optimal Plastic Design of Plates, Shells and Shellgrids

G.I.N. ROZVANY

Dept. of Civil Engrg.
Monash University
Clayton, Victoria
Australia;

FB 10
Universität Essen
4300 Essen 1
Fed. Rep. of Germany

ONG T.G.

Dept. of Civil Engrg.
Monash University
Clayton, Victoria
Australia

Summary

In this paper, recent work on the optimal plastic design of plates, shells and shellgrids is reviewed critically and certain important conclusions are arrived at. Of particular significance is the finding that the minimum weight design of solid plates and shells with a maximum thickness constraint contains a theoretically infinite number of rib-like formations. At relatively low load intensities, the layout of such ribs is furnished by the classical optimal grillage theory but at higher load levels a more advanced formulation is necessary. The latter has also been extended from optimal plastic design to optimal elastic design with stress, compliance and deflection constraints and this extended theory has been applied to plates. Moreover, it is shown that ribs in the solution can be suppressed by introducing additional geometrical constraints (termed "Niordson-constraints") or segmentation.

The above developments are based on a general theory of optimal layouts which was developed by Professor W. Prager (Brown University) and the first author in the late seventies. A further application of this theory concerns grid-shells (arch-grids) and membrane shells for which a large number of closed-form solutions are now available.

Introduction

Recent research into optimal plastic plate design has unusually far-reaching implications, revealing some entirely unexpected features of solid, minimum-weight structures. The above breakthrough has been made possible by two theoretical developments: *Static-kinematic optimality criteria*, first introduced by Prager and Shield [1] and later considerably extended by the first author's research group [2,3]; and *the theory of optimal structural*

Inelastic Behaviour of Plates and Shells
IUTAM Symposium Rio de Janeiro 1985
Editors: L. Bevilacqua, R. Feijóo and R. Valid

l a y o u t s , developed in the late seventies by Prager (Brown Univ.) and the first author |4-9|.

It may appear surprising that the search for the most efficient thickness distribution of plates and shells involves layout optimization, which is usually aimed at the selection of the best configuration for isolated members. Undoubtedly, if the optimal thickness functions turned out to be (at least piece-wise) continuous or even smooth, then traditional methods for distributed parameter problems could prove adequate. However, as most optimal solutions for the considered problems contain a theoretically infinite number of discontinuities, the layout (as well as the density) of the infinitesimally spaced "ribs" must be optimized. A similar finding was obtained in the context of plastic torsion problems by Strang and Kohn |10| who established that unconstrained shape optimization usually results in an infinite number of internal boundaries ("holes") which require layout optimization.

After reviewing briefly the concepts of static-kinematic optimality criteria and optimal layout theory and their applications to sandwich plates and grillages, the optimal design of solid plates is discussed in detail. Finally, a brief selective review of shell and shellgrid optimization is given.

Static-Kinematic Optimality Criteria and Optimal Layout Theory

Using Prager's terminology, the "basic variables" of structural mechanics are "generalised" stresses (local stresses or stress-resultants) $\underline{Q}$, strains $\underline{q}$, loads $\underline{p}$ and displacements $\underline{u}$. In optimal plastic design, the specific cost (cost per unit length, area or volume) ψ can be expressed in terms of the generalised stresses $\psi=\psi(\underline{Q})$ and the total cost Φ is then minimized subject to statical admissibility (^{S}):

$$\min_{\underline{Q}^S} \Phi = \int_D \psi\,(\underline{Q})\,d\underline{x} \tag{1}$$

where D is the structural domain and $\underline{x}$ is an element of D. The Prager-Shield condition [1] converts the problem in (1) into an optimal strain-stress relation

$$\text{on } D, \qquad \underline{q}^k = \underline{G}\left[\psi\ (\underline{Q}^s)\right] \tag{2}$$

where ^{k} denotes kinematic admissibility and $\underline{G}$ is the generalised gradient [2,3], which has the following meaning: If $\psi(\underline{Q})$ is differentiable at a $\underline{Q}$ value then $\underline{G} = \underline{\text{grad}} = (\partial/\partial Q_1,\ldots,\partial/\partial Q_n)$. At slope discontinuities of $\psi(\underline{Q})$, $\underline{G}[\psi(\underline{Q})]$ consists of any convex combination of the limiting gradient values for the considered $\underline{Q}$-value. Finally if $\psi(\underline{Q})$ is discontinuous at a stress value $\underline{Q}$ then $G[\psi(\underline{Q})]$ contains an impulse (Dirac distribution) at $\underline{Q}$. The above extended form of the Prager-Shield condition was proposed by the first author [2,3] and constitutes a necessary and sufficient condition for optimality when $\psi(Q)$ is convex and the equilibrium equations linear and a necessary one for non-convex problems, provided that the solution exists.

Static-kinematic optimality criteria convert, in effect, a problem of optimization into a problem of structural analysis.

A large number of generalisations of the Prager-Shield condition have been obtained [2,3]. For example, the effect of self-weight is automatically taken into consideration if (2) is replaced by [11]

$$\text{on } D, \qquad \underline{q}^k = (1+u^k)\ \underline{G}\left[\psi\ (Q^s)\right] \tag{3}$$

where u is the fictitious vertical deflection associated with the strains $\underline{q}$.

If the structural domain D is divided into segments $D_i (i=1,2,\ldots,n)$ such that on each segment i the design value $\bar{\psi}$ of the specific cost must be a multiple of a prescribed "shape function" $\gamma_i(\underline{x})$

$$\bar{\psi} = \Lambda_i\ \gamma_i(\underline{x}) \quad , \quad \bar{\psi} \geqslant \psi\ (\underline{Q}) \tag{4}$$

where Λ_i is an unspecified constant then (2) is replaced by [2,12]

$$\text{on } D_i, \qquad \underline{q}^k = \lambda_i(\underline{x})\ \underline{G}\ [\psi\ (\underline{Q}^s)],$$

$$\int_{D_i} \gamma_i(\underline{x})\ dx = \int_{D_i} \gamma_i(\underline{x})\ \lambda_i(\underline{x})\ d\underline{x} \qquad (5)$$

$$\text{at } \underline{x}, \qquad \lambda_i > 0 \quad \text{only if } \overline{\psi}\ (\underline{x}) = \psi\ [\underline{Q}\ (\underline{x})] \qquad (6)$$

By (6) the non-negative Lagrangian functions $\lambda_i(x)$ are usually nonzero at isolated points (lines) only and hence they often consist of impulses (Dirac distributions).

More recently, the above optimality criteria were extended to e l a s t i c p r o b l e m s [13-15] and were supplemented by new duality theorems [16-18].

It is important to mention that the Prager-Shield condition can also be derived from theorems of Save [19], Mroz [20] or Masur [21].

The optimal structural layout theory is based partly on static-kinematic optimality criteria and partly on the concept of s t r u c t u r a l u n i v e r s e. The latter consists of all potential (or feasible or "candidate") members. Since a static-kinematic optimality condition gives a strain requirement (usually inequality) also for vanishing stresses (i.e. non-optimal members), its fulfilment for the entire structural universe constitutes a necessary and sufficient condition of layout optimality for convex specific cost functions. This means that by imbedding the layout problem into a structural universe, its convexity can be preserved which would not be possible in the case of other formulations.

Application of Static-Kinematic Optimality Criteria: Sandwich Plates

A "sandwich plate" is a rather theoretical concept which is used mostly for illustrating principles of plastic analysis and optimal design. It consists of two cover plates whose thickness is so small that the lever arm between the middle surfaces of the cover plates can be assumed to be constant. The shear is transmitted by an infill (in the space in between the cover plates) whose cost (or weight) is not taken into consideration.

It follows that the specific cost (weight of cover plates per unit plate area) is a linear function of the "yield moment",

$$\psi = k\ M_Y \tag{7}$$

where k is a given constant.

For sandwich plates obeying the Tresca yield condition, for example, the yield moment can be expressed as

$$M_Y = \frac{1}{2}\ (|M_1| + |M_2| + |M_1 - M_2|) \tag{8}$$

and hence the specific cost becomes

$$\psi = k\ (|M_1| + |M_2| + |M_1 - M_2|)/2 \tag{9}$$

In this case, the "generalized stresses" Q_1 and Q_2 become the principal moments M_1 and M_2. The above specific cost function is represented graphically by cost contours for $\psi = 1/2$ and $\psi = 1{\cdot}0$ in Fig. 1a. The gradient vectors $G\left[\psi(M_1, M_2)\right]$ for the sides "a" and "f" in Fig. 1a are (k,O) and (O,k), see vectors normal to "a" and "f" in Fig. 1a, and for the corner "A" the gradient becomes $\nu(k,0)+(1-\nu)\ (0,k)$ with $0 \leqslant \nu \leqslant 1$, see the set of vectors at A. The optimal principal curvature vectors $(q_1, q_2) = (\kappa_1, \kappa_2)$ for various sides and corners in Fig. 1a are shown in Fig. 1b. This means that an optimal solution for a sandwich plate with any loading and boundary conditions can be established if we find a statically admissible moment field $(M_1^s,\ M_2^s)$and a kinematically admissible curvature field $(\kappa_1^k,\ \kappa_2^k)$ such that the "strain-stress" relation between the above two is given by Fig. 1.

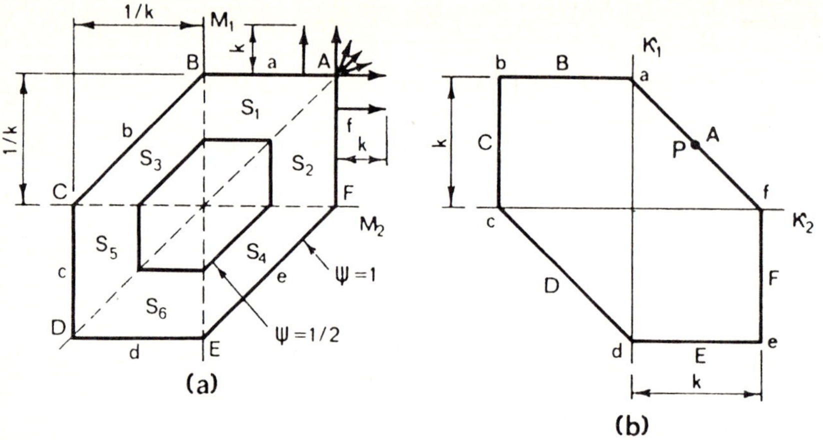

Fig. 1. Specific cost function and cost gradients for Tresca sandwich plates.

Considering a circular, clamped, uniformly loaded Tresca-plate with a diameter 2R [2, pp. 48-51], for example, the optimal moment and curvature fields become

(for $r < 2R/3$)

$$M_r = M_\Theta = pR^2/9 - pr^2/4, \quad u = kR^2/6 - kr^2/4$$

(for $> 2R/3$)

$$M_\Theta \equiv 0, \quad M_r = 4pR^3/81r - pr^2/6, \quad u = k\ (R-r)^2/2 \qquad (10)$$

where M_r and M_Θ are radial and circumferential moments and p=const. is the uniform load intensity. The corresponding curvatures ($\kappa_\Theta=-u'/r$, $\kappa_r=-u''$) then become (for $r<2R/3$) $\kappa_r=\kappa_\Theta=k/2$ (Point P on side A in Fig. 1b) and (for $r>2R/3$) $\kappa_r=-k$, $0 \leqslant \kappa_\Theta \leqslant k$ (side C in Fig. 1b.). Side A in Fig. 1b corresponds to $M_1=M_2=M_r=M_\Theta \geqslant 0$ in Fig. 1a and side C represents $M_r \leqslant 0$, $M_\Theta=0$. The above optimality requirements are clearly satisfied by the moment fields in (10). Moreover, the static/kinematic boundary and continuity conditions $M_\Theta=M_r$ at $r=0$, $M_r=0$ at $r=2R/3$, u and u' continuous across $r=2R/3$, $u'=0$ at $r=R$ are also satisfied by the above solution. The same result was obtained by Onat, Schumann and Shield [22] in 1957 by another method (uniform energy dissipation). Other solutions for sandwich plates were derived by Megarefs in 1966-68 [23] by a purely statical method and in 1967 by Marçal [24] who also proposed an early version of the Prager-Shield condition. The optimal stress fields for sandwich plates are relatively well-behaved functions except that a moment-impulse (in M_Θ) may occur at free edges [23]. An excellent review of early work on sandwich plates is given by Save and Massonet [35].

An Application of Optimal Layout Theory: Least-Weight Grillages

Although Michell's 1904 classical optimality criteria [25] for least-weight trusses can also be readily derived from optimal layout theory [6], the theory of least-weight grillages is particularly important for the following reasons:

(i) As Prager has pointed out [26], grillages constitute the first class of truly two-dimensional optimization problems for which closed form analytical solutions are available for almost all possible boundary and loading conditions.

(ii) Optimal grillages (beam layouts) are more practical than Michell-structures [25] because the latter is subject to instability which is ignored in the formulation.

(iii) The optimal rib layout in least-weight plates has been found similar to the layout of minimum weight grillages.

The grillage theory is based on the following two assumptions:

(a) The beams have a given depth and variable width and hence their specific cost function becomes

$$\psi = k|M| \qquad (11)$$

where M is the beam bending moment and k is a given constant;

(b) The total beam volume is small compared to the feasible

volume and hence the effect of beam intersections on the total cost is neglected.

Considering the specific cost function in (11), the Prager-Shield condition in (2) furnishes the following optimality criteria:

$$\kappa = k \text{ sgn } M \qquad (\text{for } M \neq 0) \tag{12}$$

$$|\kappa| \leq k \qquad (\text{for } M = 0) \tag{13}$$

where $\kappa = -u'' = d^2u/dz^2$ is the curvature of the "associated" or "Pragerian" beam deflection $u(z)$ and z is the distance measured along a particular beam. Considering the displacement surface $u(x,y)$ at a point, the maximum and minimum curvatures occur in the principal directions, at right angles. Since (12) and (13) require the maximum absolute value of the curvatures to be k, the condition $|\kappa| \leq k$ for a nonzero cross-section can only be fulfilled in one or both principal directions (unless the curvature is the same in all directions). It follows that only the following types of regions are admissible in loaded areas of optimal solutions:

$$R^+: \kappa_1 = k, \quad |\kappa_2| \leq k \qquad M_1 \geq 0, \quad M_2 = 0$$

$$R^-: |\kappa_1| \leq k, \quad \kappa_2 = -k \qquad M_1 = 0, \quad M_2 \leq 0$$

$$S^+: \kappa_1 = \kappa_2 = k \qquad M_1 \geq 0, \quad M_2 \geq 0$$

$$S^-: \kappa_1 = \kappa_2 = -k \qquad M_1 \leq 0, \quad M_2 \leq 0$$

$$T: \kappa_1 = -\kappa_2 = k \qquad M_1 \geq 0, \quad M_2 \leq 0 \tag{14}$$

The least-weight grillage problem has therefore been transformed into the following geometrical problem: the area of the grillage must be covered with the optimal regions given in (14) such that (i) the deflection $u(x,y)$ and its slope are continuous along region boundaries and (ii) the kinematic boundary conditions are fulfilled by $u(x,y)$. Optimal beams must then be placed in the directions of principal curvatures with $|\kappa| = k$ and the sign of beam moments must match those of the principal curvatures. A simple example of an optimal solution for a simply supported square grillage is given in Fig. 2.

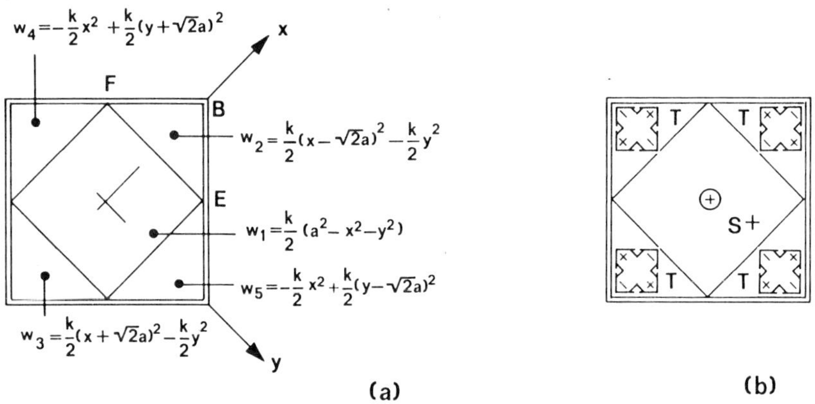

Fig. 2. Example of an optimal grillage layout.

Extensive reviews of the grillage theory were given by Prager and the first author [2,6,26,27,28]. Analytical solutions are now available for simply supported and clamped boundaries, free edges, beam supported edges, non-uniform depth, allowance for cost of supports, bending and shear dependent cost [29], upper constraint on the beam cross-sectional area per unit width [60], partial discretization [68], and allowance for self-weight [7,30,31,32]. The latest development concerns optimal grillage layouts for prescribed elastic deflection [33]. It was shown earlier [3,34] that plastically designed least-weight grillages are also valid for elastic grillages with prescribed maximum stress, compliance or natural frequency.

Another unique feature of the optimal grillage theory is the availability of a computer algorithm [36,37] for generating a n a l y t i c a l l y and plotting least-weight layouts for a broad class of loading and boundary conditions.

Optimal Plastic Design of Solid Plates

It was established already in the late sixties (e.g. by Kozlowski and Mroz [38]) and shown more rigorously recently [39] that the weight of a solid plate can be reduced to an arbitrarily small value by employing a system of sufficiently high and thin ribs. A finite value for the structural weight can be ensured, however, by introducing an upper constraint on the plate thickness. For elastic plates with such constraints, Olhoff and Cheng [40] obtained numerical solutions which show the development of stiffener-like formations of maximum thickness. It was then pointed out by the late Professor Prager in 1980 that the layout of the stiffeners in the Olhoff-Cheng solutions is similar to the optimal layout of grillages for the same boundary and load conditions, obtained about a decade earlier by the first author [41] and later confirmed by Prager [4,5]. Following up Prager's intuitive remarks with a rigorous analysis, the first author, Olhoff, Cheng, Taylor and Wang [39,42] derived exact analytical minimum weight solutions for plastically designed solid plates with a constraint on the maximum thickness. These solutions contain regions with a dense system of ribs (of theoretically infinitesimal spacing) and thus indicate that a number of earlier papers on least-weight plates, based on a "smooth" thickness variation or a small number of finite elements, represent erroneous solutions. The above improved results were obtained by establishing a specific cost function $\psi(M_1, M_2)$ through local optimization of the rib/plate configuration for given values of the principal moments M_1, M_2 and then employing the Prager-Shield condition [1] together with the optimal layout theory [2-6]. All least-weight solutions have been found to consist of the following types of regions:

(a) Ribs in one principal direction only, infinitesimal plate thickness in between ribs;

(b) solid plate of non-maximum thickness with $M_1 = M_2$;

(c) solid plate of maximum thickness.

As an example, Fig. 3 shows the optimal weight Φ_1 of plastically designed circular simply supported solid plates at various levels of non-dimensional load ν. For a comparison, Φ_2 shows the minimum weight of piece-wise smooth solutions and Φ_3 that of solutions of constant thickness. At very low levels of the non-dimensional load ν, the solution always tends to the optimal g r i l l a g e l a y o u t with type (a) regions only.

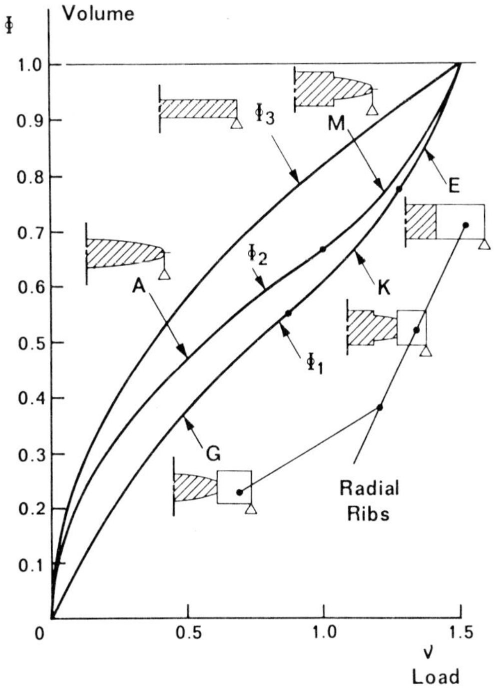

Fig. 3. The total weight of various solutions for plastic circular solid plates.

Optimal Plastic Design of Perforated Plates

A perforated plate may only have two thicknesses: a prescribed maximum thickness or zero thickness. The latter occurs over "perforations" whose in-plane dimensions are assumed to be small so that the load over areas of zero thickness can be transmitted to the adjacent plate segments by some secondary systems of infinitesimal volume. In minimizing the total material volume (or weight) of a perforated plate, it is assumed that the plate material obeys the Tresca yield conditions. In perforated regions, the optimal microstructure can be shown to consist of ribs running in the directions of the principal moments (M_1, M_2). As stresses of the same sign do not influence the yield value of the major stress in the Tresca yield condition, a saving can be achieved at the rib intersections but such a saving is not possible if the stresses are of opposite sign [39]. Consequently the non-dimensionalised specific cost function becomes (Fig. 4)

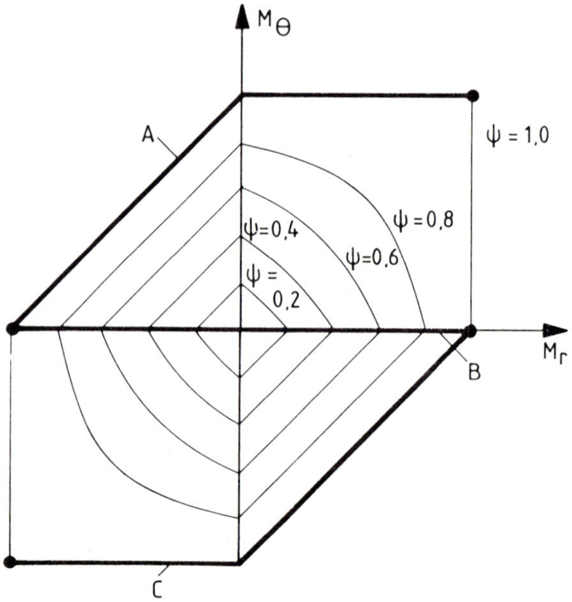

Fig. 4. Specific cost function for plastic perforated plates.

(for $M_1 M_2 > 0$) $$\psi = |M_1| + |M_2| - M_1 M_2$$

(for $M_1 M_2 < 0$) $\phi = |M_1| + |M_2|$

with $|M_1| + |M_2| + |M_1 - M_2| \leqslant 1$ (15)

Making use of the Prager-Shield condition (1), the second author has shown that for plastic axisymmetric perforated plates the least-weight solution may only consist of two types of regions:

(a) unperforated regions (stress regimes A and C in Fig. 4) and

(b) ribs in the radial directions ($M_\Theta=0$).

Introducing the non-dimensional notation $r=\bar{r}/\bar{R}$, $p=\bar{p}\bar{R}^2/\bar{M}$, $M_i=\bar{M}_i/\bar{M}$ ($i=\Theta$, r), where $\bar{r}$ is the radial coordinate, $\bar{R}$ is the plate radius and $\bar{M}$ is the maximum feasible moment capacity, the optimal solution turns out to be the following :

$(0 \leqslant r < g)$ $M_\Theta=1$, $M_r=1-pr^2/6$ (Region "a")

$(g < r \leqslant 1)$ $M_\Theta=0$, $M_r=p(1-r^3)/6r$ (Region "b")

$g=p/6$ (16)

To demonstrate the validity of the above conclusion, the volume of various intuitively selected designs is compared in Fig. 5 where: Design A consists of circumferential ribs only (the shear transmission is assumed to be costless), Design B of radial and circumferential ribs of equal width ($M_\Theta=M_r$) throughout, Design C with $M_r=M_\Theta$ in an inner region and $M_r=0$ (only circumferential ribs) in the outer region and Design D (optimal design) with $M_\Theta=M_r$ in an inner region, and $M_\Theta=0$ (only radial ribs) in the outer region.

The second author has also obtained optimal solutions for clamped circular perforated plates and found that, depending on the load level p, the optimal solution may consist of one or two unperforated regions and one region with only radial ribs.

In the near future, the above conclusions will be extended to n o n - a x i s y m m e t r i c p l a t e s. It is expected that the solution will consist of unperforated regions, governed by the usual yield-line analysis for Tresca-plates and r e - g i o n s w i t h r i b s i n o n e d i r e c t i o n, for which the Prager-Shield condition requires a constant curvature in the directions of the ribs. For plastic perforated plates, therefore, a combination of y i e l d - l i n e a n a l y s i s and o p t i m a l g r i l l a g e t h e o r y will give the optimal solution. A similar result was obtained for constrained Michell frames by Strang and Kohn [43].

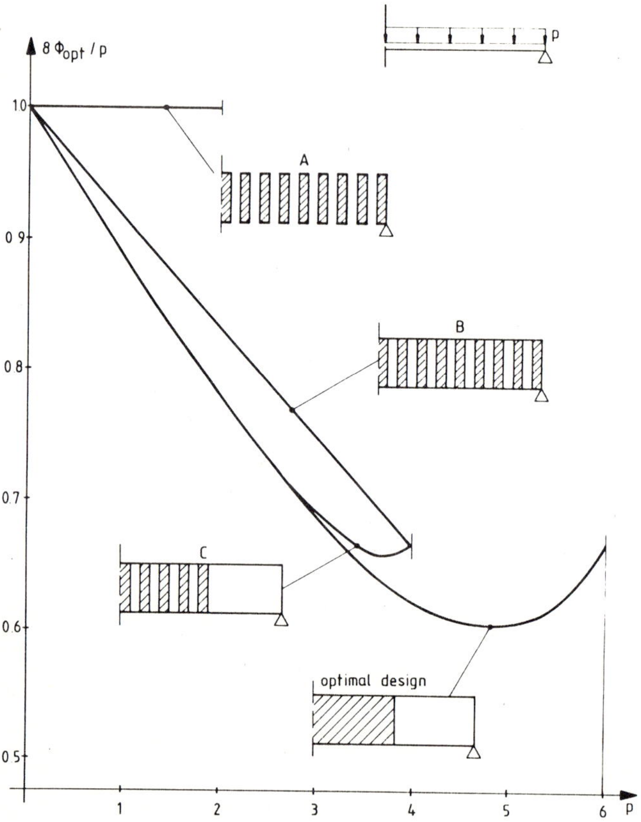

Fig. 5. Comparison of various solutions for plastic circular perforated plates.

Classical and Advanced Layout Theories

"Classical" layout theory [4-9], a generalisation of Michell's theorem [25], which has been used for the optimization of grillages [4,26-34], grid-shells and cable-networks [44-48], is based on two fundamental features: (a) at any point of the structural domain potential members may run in any number of directions (Fig. 6a), but (b) the effect of member intersections on both the cost and strength (or stiffness) is neglected. The above assumptions are realistic in the case of l o w - d e n s i t y systems in which the material volume/feasible volume ratio is relatively low. In fact, detailed investigations [39, 42,49] have shown that the optimal solution for various optimization problems reduces to the one given by the classical layout theory, if the above ratio tends to zero and the effect of intersections is taken into consideration (i.e. the microstructure is also optimized).

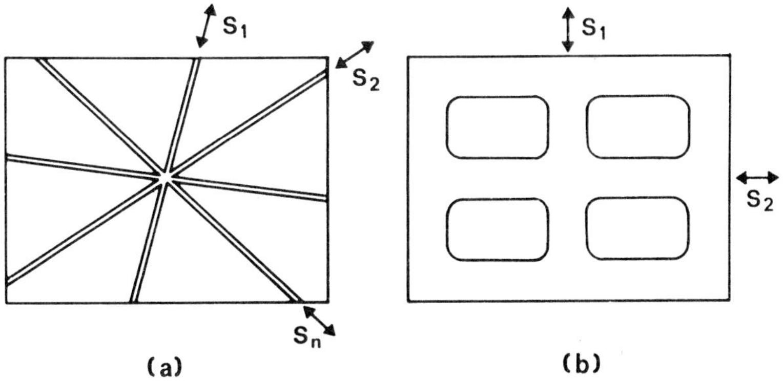

Fig. 6. Classical and advanced layout theories.

The development of "advanced" layout theory was prompted by the discoveries that optimized continua develop an infinite number of internal boundaries [10] and that least-weight plates contain dense systems of ribs [39,40,42]. As a result of these findings, much recent research was devoted to the optimization of the *microstructure* of perforated and composite structures [50-56]. Considering an elastic (or plastic) continuum, the microstructure is first optimized locally by minimizing the material volume ψ per unit area (or volume) for given stiffnesses S_1 and S_2 (or generalised stresses Q_1 and Q_2) in two (or three) principal directions (see Fig. 6b). It follows that the specific cost function $\psi\ (S_1, S_2)$ is a non-separable function of both (or all three) stiffnesses (or generalised stresses) whereas in classical layout theory it was the sum of functions of each stiffness value $\psi = \psi_1(S_1)+\ldots+\psi_n(S_n)$. Advanced layout theory results in substantial extra savings if a high proportion of the feasible space is occupied by structural material. Once the locally optimized microstructure and the corresponding specific cost function ψ are established, *optimal plastic design methods* (e.g. the Prager-Shield condition [1]) can be extended for the optimization of *elastic continua*. This will be demonstrated by considering the optimization of elastic perforated plates.

Optimal Elastic Design of Perforated Plates

In recent papers by Lurie, Cherkaev and Fedorov [50-54] as well as by Strang and Kohn [55,56], least-weight solutions contain regions with two sets of intersecting ribs (strips of material): one such set has a first order infinitesimal spacing [of $O(\delta)$ with $\delta \to 0$] and the other set a second order infinitesimal spacing [of $O(\delta^2)$]. The implications of these results on plate optimization were investigated recently by the authors Olhoff, Bendsøe, Szeto and Sandler [49,57] who have arrived at the following conclusions:

(a) Given a horizontal system of intersecting first and second order ribs whose depth is significantly smaller than its span, it is reasonable to assume that *under a distributed vertical load* the normal stresses in the ribs are proportional to the distance from the middle surface. As a consequence of St. Vénant's principle (and a detailed finite element analysis), any horizontal slice of a rib of second order infinitesimal width is subjected on its interior to stresses in the direction of the rib middle plane only. The same conclusion can be obtained for ribbed systems *in plane stress*. Hence second order ribs do not contribute to the stiffness in the direction normal to their plane. In this respect, the above formulation differs from recent mathematical studies of this problem [50-51].

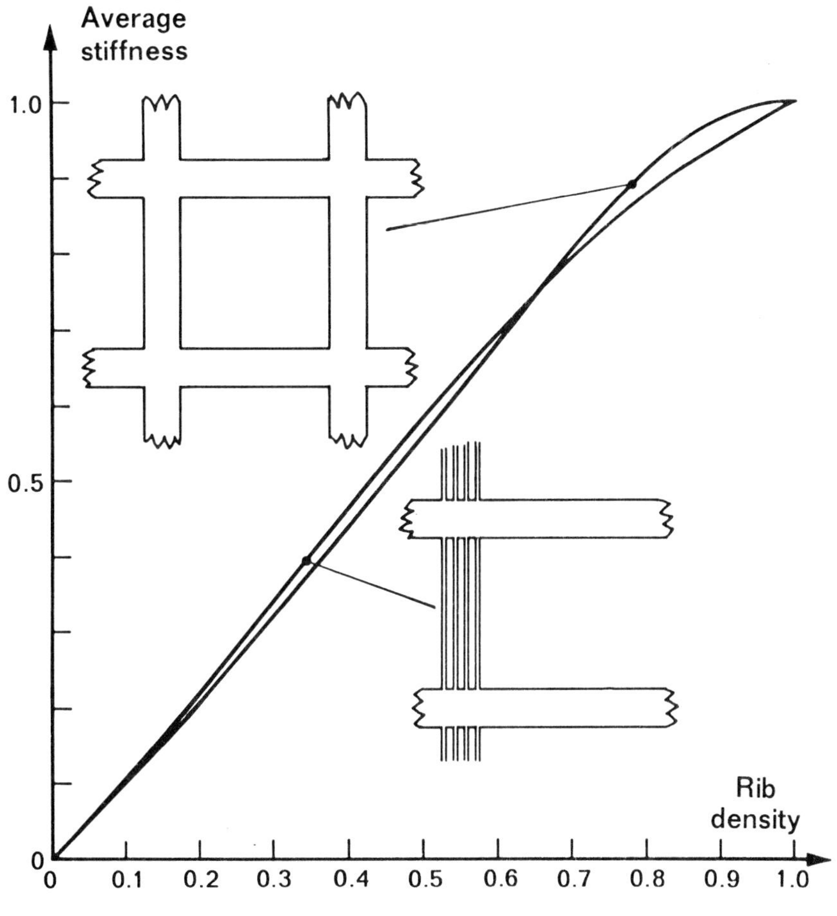

Fig. 7. Comparison of first/second and first/first order microstructures at various rib-densities.

(b) In minimizing the weight of a perforated plate for a given compliance, it was shown on the basis of an intuitive argument that at low rib-densities the material consumption is smaller for a first/ second order ribbed system than for a first/first order system. However, the same conclusion did not seem plausible at high rib-densities. The above arguments were followed up by detailed finite element and finite difference analyses which confirmed the above findings. Fig. 7 indicates the average stiffness for various rib-densities for plane first/second and first/first order ribbed systems having a Poisson's ratio of zero value ($\nu=0$). The savings are more significant if the value of Poisson's ratio is nonzero ($\nu \neq 0$).

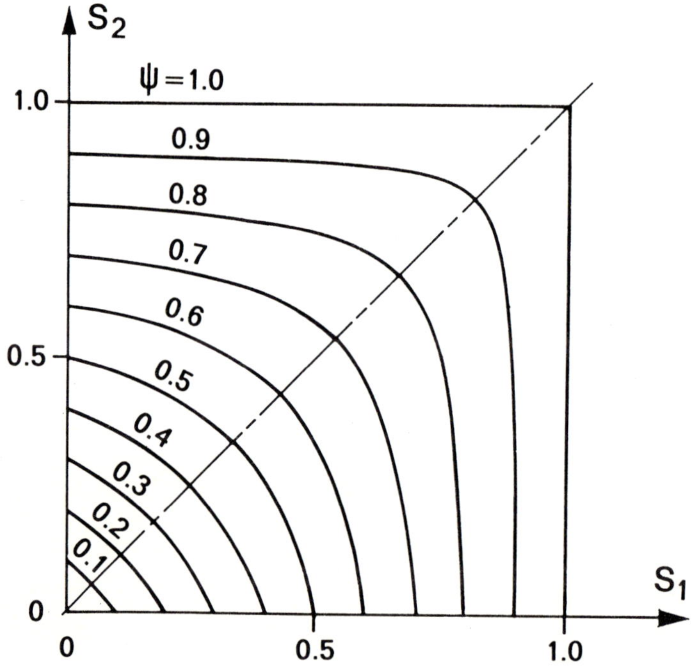

Fig. 8. Specific cost function for elastic perforated plates.

(c) Assuming a first/second order micro-model at a l l rib-densities, a specific cost function for perforated plates in bending or plane stress was derived. For $\nu=0$, the latter gives a relationship between the stiffnesses (s_1, s_2) in the principal directions and the material volume ψ per unit area of the middle surface (Fig. 8):

$$\psi = (s_1 - 2s_1s_2 + s_2) / (1 - s_1s_2) \qquad (17)$$

(d) The proposed specific cost function was then used for examining the design of circular, uniformly transversely loaded elastic perforated plates of given compliance.

(e) It was found that the optimal design for the above problem reduces to that for grillages [2] if the average rib density approaches zero (i.e. at very low load levels).

(f) Static/kinematic optimality criteria (similar to those used in optimal plastic design) were derived by variational analysis, using the proposed microstructure.

(g) The variational formulation indicated that for transversely loaded axially symmetric plates only two types of regions may occur in loaded segments of the optimal solution:

(i) unperforated regions;

(ii) regions consisting of radial ribs only.

(h) On the basis of the above findings, optimal solutions were derived for simply supported and clamped circular plates with uniformly distributed full and partial loading as well as a central point load and for simply supported circular plates with edge loading.

(j) The above results were confirmed by optimizing a number of intuitively selected designs with respect to certain design parameters and also by independent numerical solutions. In Fig. 9, for example, the volume Φ of various partially optimized intuitive designs is compared as a function of the compliance C. As predicted by variational analysis, Design D is optimal at all C-values.

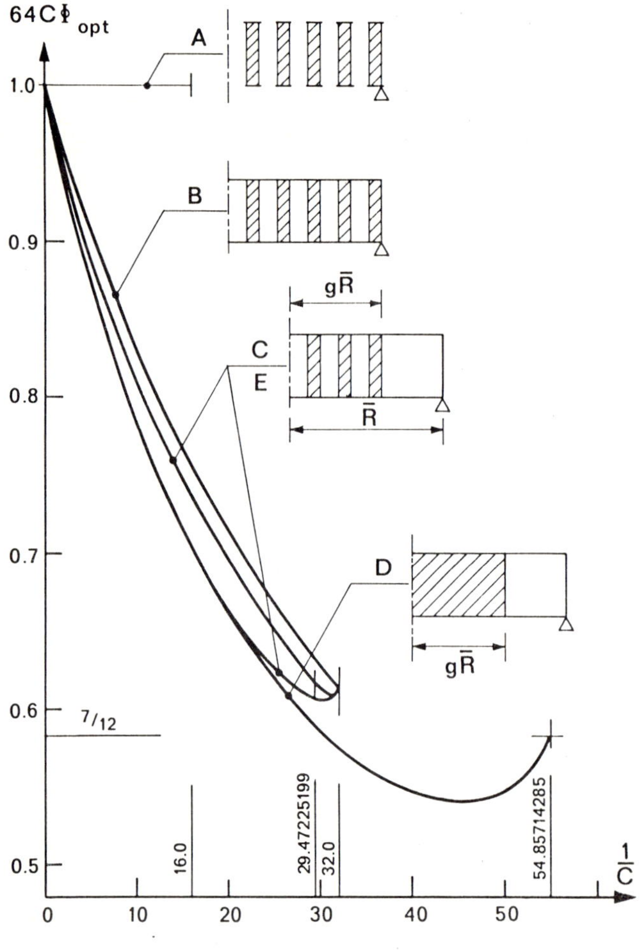

Fig. 9. Comparison of various solutions for elastic circular perforated plates.

(k) More recently, the above results were extended to plates with a non-zero Poisson's ratio ($\nu \neq 0$). Whereas for $\nu=0$ the specific compliance c was given by $c = M_1{}^2/S_1 + M_2{}^2/S_2$ where M_1 and M_2 are the principal moments and S_1 and S_2 principal stiffnesses, for $\nu \neq 0$ the specific compliance can be expressed as $c = M_1{}^2/S_1 + M_2{}^2/S_2 - 2\nu M_1 M_2$ where S_1 and S_2 are parameters depending on the rib-densities (d_1, d_2) in the principal directions. Then the specific cost function will be the same as in (17) but the compliance constraint will be different. Optimal solutions have been determined for the various axisymmetric loading and boundary conditions with $\nu \neq 0$.

(l) It can be concluded from the above investigation that for the class of plates considered, the introduction of first/second order microstructure has not resulted in an improved economy.

(m) The above investigation has also been extended to (i) allowance for the effect of shear and (ii) composite plates consisting of two elastic materials. At a given point of the middle surface, either one or the other material must occupy the entire depth of the plate.

Design Constraints for Suppressing Ribs in Optimal Plate Solutions

Both rib-like formations and vanishing cross-sections can be prevented in least-weight plate solutions by introducing

(a) segmentation, see Eqs. (4)-(6), or
(b) an upper limit on the slope of the plate thickness (="taper"), as proposed originally by Niordson [61].

A general theory of optimal segmented structures is now available for both plastic [2,12] and elastic [15] design. A theory for incorporating "Niordson-constraints" in plastic design has been developed by the first author [59] and later applied to axisymmetric plates [62].

Optimal Design of Long-Span Membrane Shells, Archgrids (Gridshells) and Cable Networks

The optimal plastic design of long-span membrane shells under self-weight and other loads was discussed by Ziegler [63], Issler [64], Prager and Rozvany [65], Nakamura, Dow and Rozvany [66] and Dow, Nakamura and Rozvany [67].

The optimal design of archgrids (shellgrids) and cable-networks was suggested by W. Prager. These surface structures consist of a dense system of intersecting arches or cables in which all members are subjected to axial compression or tension (without bending). Before Prof. Prager's unfortunate death in 1980, he and the first author [44] derived only least-weight archgrids and cable networks in which the direction of arches or cables was prescribed. The effect of selfweight on the optimization of the same type of systems was considered later [45]. The optimization of the *layout* of archgrids and cable networks was investigated by the first author and Wang [9,46-48] who termed the resulting systems *Prager-structures*. The latter have the following remarkable characteristics:

(a) The same optimal solution can be obtained for two different classes of problems, viz.

(i) a surface structure, in which the middle surface of the system as well as the layout of the members within the surface are to be optimized in such a way that all members are in pure tension or compression; or

(ii) a space-structure, in which either all members are in pure tension or all members in pure compression and the elevation of the external loads is also optimized. The latter class of problem is essentially a Michell-frame [25] in which the permissible stress for either compression or tension tends to zero and the external loads are movable along their line of action. It was shown rigorously [9,46] on the basis of the optimal layout theory [4-9] that the solution to this latter class of problem (modified Michell space-frames) reduces to a surface-structure (as under i).

(b) Closed form analytical solutions are now available [46-48] for any axisymmetric and "quasi-axisymmetric" [9,46] boundary and load conditions.

(c) The minimum total weight for Prager-structures is proportional to the product of loads and their optimal elevations [7,9,47].

(d) All the above conclusions have been extended, on the basis of Eq. (3) herein, to Prager-structures subjected to external loads plus self-weight.

(e) It has been shown that Prager-structures are optimal not only for plastic design but also for elastic design with stress, compliance or natural frequency constraints.

Concluding Remarks

It will be seen that the concepts of static/kinematic optimality criteria and optimal layout theory (structural universe) are highly suitable for the treatment of optimal plastic design of plates, shells and shellgrids. Closed form analytical solutions are now available, and can even be generated on the computer, for broad classes of design problems. More recently, the same results have been extended to elastic structures. An unexpected feature of least-weight plate solutions with an upper constraint on the thickness is the appearance of infinitesimally spaced rib-like formations, indicating that a large number of past research papers were based on erroneous solutions.

Acknowledgements

The authors wish to express their gratitude to Ms. Kalinowski, Kramer and Damm and Messrs. Koch and Spengemann for their help in preparing the manuscript of this paper.

References

1. Prager, W.: Shield, R.T.: A general theory of optimal plastic design. J. Appl. Mech. 34 (1967) 184-186.

2. Rozvany, G.I.N.: Optimal design of flexural systems. Oxford: Pergamon Press 1976. Russian version: Moscow, Stroiizdat, 1980.

3. Rozvany, G.I.N.: Variatonal methods and optimality criteria, in Haug, E.J.; Cea, J.(eds.): Optimization of distributed parameter structures, Proc. NATO ASI. Alphen aan der Rijn: Sijthoff and Noordhoff 1981, pp. 82-111.

4. Prager, W.; Rozvany, G.I.N.: Optimization of structural geometry, in: Bednarek, A.R.; Cesari, L. (eds.): Dynamical systems. New York: Academic Press 1977, pp. 265-294.

5. Prager, W.: Introduction to structural optimization. Vienna: Springer-Verlag 1974.

6. Rozvany, G.I.N.: Optimality criteria for grids, shells and arches, in: Haug, E.J.; Cea, J. (eds.): Optimization of distributed parameter structures, Proc. NATO ASI. Alphen aan der Rijn: Sijthoff and Noordhoff 1981, pp. 112-151.

7. Rozvany, G.I.N.; Wang, C.M.: Extensions of Prager's layout theory, in Eschenauer, H.; Olhoff, N. (eds.): Optimization methods in structural design, Proc. Euromech Colloquium, Siegen. Mannheim: Wissenschaftsverlag 1983, pp. 103-110.

8. Rozvany, G.I.N.: A general theory of optimal structural layouts. Proc. Int. Symp. on Optimal Structural Design. Tucson, Arizona: University of Arizona 1981, pp. 34.1-34.10.

9. Rozvany, G.I.N.: Structural layout theory: the present state of knowledge. Chapter 7 in: Atrek, E; Gallagher, R. H.; Ragsdell, K.M.; and Zienkievicz, O.C. (eds.): New directions in structural design. Chichester, England: Wiley & Sons 1984, pp. 167-196.

10. Kohn, R.V.; Strang, G.: Optimal design for torsional rigidity, in: Atlari, S.N.; Gallagher, R.H.; Zienkiewicz, O.C. (eds.): Hybrid and mixed finite element methods. Chichester, England: Wiley & Sons 1983.

11. Rozvany, G.I.N.: Optimal plastic design: allowance for selfweight. J. Engrg. Mech. ASCE 103 (1977) 1165-1170.

12. Rozvany, G.I.N.: Optimal plastic design for partially pre-assigned stress distribution. J. Optimiz. Theory Appl. 11 (1973) 421-436.

13. Rozvany, G.I.N.: Plastic versus elastic optimal strength design. J. Engrg. Mech. ASCE 103 (1977) 210-214.

14. Rozvany, G.I.N.: Optimal elastic design for stress constraints. Computers and Structures. 8 (1978) 455-463.

15. Rozvany, G.I.N.; Ong, T.G.; Karihaloo, B.L.: A general theory of optimal elastic design for structures with segmentation. J. Appl. Mech., accepted.

16. Rozvany, G.I.N.; Hill, R.: Optimal plastic design: superposition principles and bounds on the minimum cost. Comp. Meth. Appl. Mech. Engrg. 13 (1978) 151-174.

17. Rozvany, G.I.N.: The dual of Foulkes and Prager-Shield criteria. J. Engrg. Mech. ASCE. 110 (1984) 1778-1785.

18. Rozvany, G.I.N.: Generalisations of Heyman's and Foulkes' theorems using dual formulation. Int. J. Mech. Sci., in press (1985).

19. Save, M.A.: A unified formulation of the theory of optimal plastic design with convex cost function. J. Struct. Mech. 1 (1972) 267-276.

20. Mroz. Z.: Limit analysis of plastic structures subject to boundary variations. Arch. Mech. Stos. 15 (1963) 63-76.

21. Masur, E.F.: Optimal stiffness and strength of elastic structures. J. Engrg. Mech. ASCE 96 (1970) 621-640.

22. Onat, E.T.; Schumann, W.; Shield, R.T.: Design of circular plates for minimum weight. Zeit. Ang. Math. Phys. 8 (1957) 485-499.

23. Megarefs, G.J.: Minimal design of axisymmetric sandwich plates I and II. J. Engrg. Mech. ASCE 93 (1967) 245-269. 94 (1968) 177-198.

24. Marçal, P.V.: Optimal plastic design of circular plates. Int. J. Solids Struct. 3 (1967) 427-443.

25. Michell, A.G.M.: The limits of economy of material in frame structures. Phil. Mag. 8 (1904) 589-597.

26. Prager, W.; Rozvany, G.I.N.: Optimal layout of grillages. J. Struct. Mech. 5 (1977) 1-18 (also: 14th IUTAM Congress, Delft, 1976, paper No. 310).

27. Rozvany, G.I.N.; Hill, R.: General theory of optimal load transmission by flexure. Advances in Appl. Mech. 16 (1976) 183-308.

28. Rozvany, G.I.N.: New trends in structural optimization. Proc. 6th Australasian Conf. Mech. Struct. Mater., Christchurch, 1977.

29. Rozvany, G.I.N.: Optimal beam layouts: allowance for cost of shear. Comp. Meth. Appl. Mech. Engrg. 19 (1979) 49-58.

30. Rozvany, G.I.N.; Wang, C.M.: Optimal layout theory: allowance for selfweight. J. Engrg. Mech. ASCE 109 (1983) 648-653.

31. Rozvany, G.I.N.; Yep, K.M.; Sandler, R.: Optimal design of long-span truss grids. Proc. 3rd Int. Conf. on Space Structures. Guilford, England.

32. Rozvany, G.I.N.; Sandler, R.; Yep, K.M.: Optimal layout of long-span truss-grids I and II. Int. J. of Solids Struc., in press.

33. Rozvany, G.I.N.; Ong, T.G.: A general theory of optimal layouts for elastic structures, submitted.

34. Olhoff, N.; Rozvany, G.I.N.: Optimal grillage layout for given natural frequency. J. Engrg. Mech. ASCE. 108 (1982) 971-974.

35. Save, M.A.; Massonet, C.E.: Plastic analysis and design of plates shells and disks. Amsterdam: North Holland 1972.

36. Rozvany, G.I.N.; Hill, R.: A computer algorithm for deriving analytically and plotting optimal structural layout. Proc. NASA Symp.: Future trends in computerised structural design, Washington, 1978. Oxford: Pergamon, 1978. Also Computers and Struct. 10 (1979) 295-300.

37. Hill, R.; Rozvany, G.I.N.: Prager's layout theory: a nonnumeric computer method for generating optimal structural configurations and weight-influence surfaces. Comp. Meth. Appl. Mech. Engrg. 49 (1985) 131-148.

38. Kozlowski, W.; Mroz, Z.: Optimal design of solid plates. Int. J. Solids Struct. 5 (1969) 781-794.

39. Rozvany, G.I.N.; Olhoff, N.; Cheng, K.-T.; Taylor, J.: On the solid plate paradox in structural optimization. J. Struct. Mech. 10 (1982) 1-32.

40. Cheng, K.-T.; Olhoff, N.: An investigation concerning optimal design of solid elastic plates. Int. J. Solids Struct. 17 (1981) 305-323.

41. Rozvany, G.I.N.: Grillages of maximum strength and maximum stiffness. Int. J. Mech. Sci. 14 (1972) 651-666.

42. Wang, C.-M.; Rozvany, G.I.N.; Olhoff, N.: Optimal plastic design of axisymmetric solid plates with a maximum thickness constraint. Computers Struct. 18 (1984) 653-665.

43. Strang, G.; Kohn, R.: Hencky-Prandtl nets and constrained Michell trusses. Comp. Meth. Appl. Mech. Engrg. 36 (1983) 207-222.

44. Rozvany, G.I.N.; Prager, W.: A new class of optimization problems: optimal archgrids. Comp. Meth. Appl. Mech. Engrg. 19 (1979) 127-150.

45. Rozvany. G.I.N.; Nakamura, H.; Kuhnell, B.T.: Optimal archgrids: allowance for self-weight. Comp. Meth. Appl. Mech. Engrg. (1980) 287-304.

46. Rozvany, G.I.N.; Wang,C.M.; Dow, M.: Prager-structures: archgrids and cable networks of optimal layout. Comp. Meth. Appl. Mech. Engrg. 31 (1982) 91-114.

47. Rozvany, G.I.N.; Wang, C.M.: On plane Prager-structures I-II. Int. J. Mech. Sci. 25 (1983) 519-527, 529-541

48. Wang, C.M.; Alwis, W.A.M.; Rozvany, G.I.N.: Minimum-weight design of fully stressed archgrids. Proc. 3rd. Int. Conf. Span Structures, Guilford, England, 1984.

49. Rozvany, G.I.N.; Ong, T.G.; Olhoff, N.; Bendsøe, M.P.; Szeto, W.T.: Least-weight design of perforated elastic plates. DCAMM-report. No. 306, 1985.

50. Lurie, K.A.; Fedorov. A.V.; Cherkaev, A.V.: Regularization of optimal design problems for bars and plates. I,II. J. Optimiz. Theory Appl. 37 (1982) 499-521, 523-543.

51. Lurie, K.A.; Fedorov, A.V.; Cherkaev, A.V.: On the existence of solutions to some problems of optimal design for bars and plates. J. Optimiz. Theory Appl. 42 (1984) 247-281.

52. Lurie, K.A.; Cherkaev, A.V.: G-closure of a set of anisotropically conducting media in the two-dimensional case.42 (1984) 283-304.

53. Lurie, K.A.; Cherkaev, A.V.: Exact estimates of conductivity of composites formed by two isotropically conducting media taken in prescribed proportions. Proc. Roy. Soc. Edingburgh 99A (1984) 71-78.

54. Lurie, K.A.; Cherkaev, A.V.: Optimal structural design and relaxed controls. Opt. Control Appl. Meth. 4 (1981) 387-392.

55. Strang, G.; Kohn, R.V.: Optimal design and relaxation of variational problems, to be published.

56. Kohn, R.V.; Strang, G.; Structural design optimization, homogenization and relaxation of variational problems, in: Burridge, R.; Papanicolan, G.; Childress, S.: Lecture Notes in Physics, No. 154. Berlin: Springer-Verlag 1982.

57. Rozvany, G.I.N.; Ong, T.G.; Olhoff, N.; Bendsøe, M.P.; Szeto, W.T.; Sandler, R.: Least-weight design of perforated elastic plates I and II. Int. J. Solids Struct., submitted.

58. Rozvany, G.I.N.; Yep, K.M.; Sandler, R.: Recent extensions of Prager's theories of optimal plastic design and optimal layout. Proc. 5th ASCE EMD Specialty Conf. Wyoming 1984.

59. Rozvany, G.I.N.: Prager-Shield optimality criteria with bounded spatial gradients. J. Engrg. Mech. ASCE 110 (1984) 129-137.

60. Rozvany, G.I.N.; Wang, C.M.: Constrained optimal layouts through Prager-Shield criteria. J. Engrg. Mech. ASCE 109 (1983) 648-653.

61. Niordson, F.I.: Some new results regarding optimal design of elastic plates, in: Eschenauer H.; Olhoff, N. (eds.): Proc. Euromech Coll. 164: Optimization methods in structural design. Mannheim: Wissenschaftsverlag, 1983.

62. Yep, K.M.; Sandler, R.; Rozvany, G.I.N.: Least-weight plates with constrained taper, submitted.

63. Ziegler, H.: Kuppeln gleicher Festigkeit. Ing.-Arch. 26 (1958) 378-382.

64. Issler, W.: Membranschalen gleicher Festigkeit. Ing.-Arch. 33 (1964) 330-345.

65. Prager, W.; Rozvany, G.I.N.: Optimal spherical cupolas of uniform strength. Ing.-Arch. 49 (1980) 287-293.

66. Nakamura, H.; Dow, M.; Rozvany, G.I.N.: Optimal spherical cupola of uniform strength: allowance for selfweight. Ing.-Arch. 51 (1981) 159-181.

67. Dow, M.; Nakamura, H.; Rozvany, G.I.N.: Optimal cupolas of uniform strength: spherical M-shells and axisymmetric T-shells. Ing.-Arch. 52 (1982) 335-353.

68. Rozvany, G.I.N.; Prager, W.: Optimal design of partially discretized grillages. J. Mech. Phys. Solids, 24 (1976) 125-136.

Optimization of Cylindrical Shells Under Combined Loading Against Brittle Creep Rupture

M. ŻYCZKOWSKI and M. RYSZ

Politechnika Krakowska (Technical University of Cracow)
31-155 Krakow, ul. Warszawska 24, Poland.

Summary

Optimal structural design of cylindrical shells under overall bending, torsion, tension and internal pressure in creep conditions is considered. The material is assumed to be governed by the Norton-Odqvist nonlinear steady creep law. Minimal weight of the shell is the design objective, radius and wall thickness are design variables, and the constraint refers to brittle creep rupture as described by the Kachanov-Sdoburev hypothesis. Elimination of circumferential bending in the wall results in a circular profile. The condition of uniform creep strength determines the thickness distribution, whereas the optimal radius is determined numerically.

1. Introductory Remarks

Cylindrical shells work very often in creep conditions, e.g. metal shells at elevated temperatures (pipelines, elements of jet engines etc.), or shells made of plastics or concrete at room temperature. If the loading does not conform to rotational symmetry, then circular cylindrical shells of constant thickness are not optimal inasmuch as their weight is concerned. In general, two functions as design variables may then be considered: current radius $r = r(\phi)$ and wall thickness $g = g(\phi)$, where r and ϕ denote polar (or cylindrical) coordinates. Sometimes the areas of possible longitudinal ribs may also serve as design variables. Such optimization problems under elastic stability constraints were discussed by Życzkowski and Krużelecki [1,2,3].

Optimal structural design in creep conditions was initiated at the end of the sixties by Reytman [4], Prager [5], Nemirovsky [6] and Życzkowski [7], and widely developed at the Technical University of Cracow; more recent results were summarized by Życzkowski in [8]. In contradistinction to static problems of

Inelastic Behaviour of Plates and Shells
IUTAM Symposium Rio de Janeiro 1985
Editors: L. Bevilacqua, R. Feijóo and R. Valid

elastic or plastic optimal design, it introduces a new factor, namely the factor of time. Most structures working in creep conditions are designed for a finite life-time, usually determined by creep rupture or creep buckling; however, in some particular problems the stiffness in creep or stress relaxation may also serve as optimization constraints.

Optimal structural design of shells is very well developed; a survey by Krużelecki and Życzkowski [9] reviews over 600 papers. Most papers, however, deal with elastic or plastic shells, whereas optimal design of shells in creep conditions is represented by very few papers only.

The present paper considers optimal structural design of cylindrical shells under fairly general combined loadings which result in longitudinal homogeneity of the stress state: overall bending by the moment M_b, overall torsion by M_t, axial force N, and uniformly distributed normal internal pressure p, Fig. 1. Such a system of loadings may often be encountered e.g. in pipelines

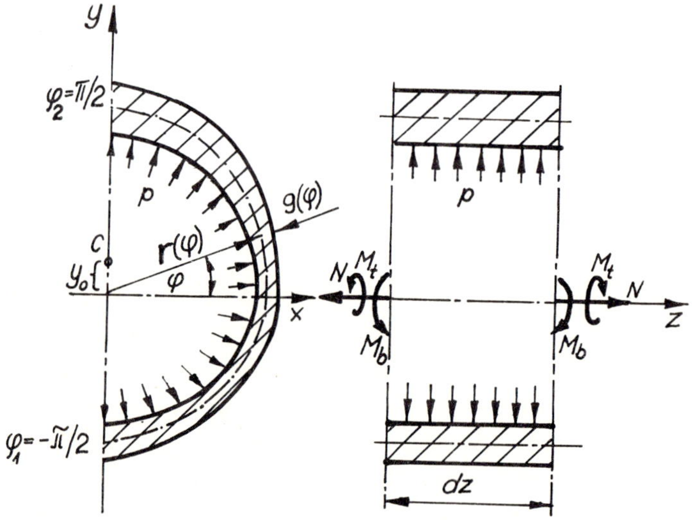

Fig. 1. Cylindrical Shell under Combined Loadings.

The paper constitutes a generalization of previous considerations by the authors, [10,11]: first, torsion is introduced here and axial and circumferential directions are no longer principal; second, the creep rupture hypothesis is more general than that used before, and hence it covers a wider class of materials. Both these generalizations change the problem, mainly complicating it; however, a certain simplification will also appear, since ordering of principal stresses is easier if shearing stresses due to torsion take place and division of the shell into separate zones is no longer necessary.

2. Statement of the Problem

The optimization problem is stated as follows:

(1) Minimal weight of the shell is the design objective. Under the assumptions of a constant bending moment along the axis and of a homogeneous material this design objective reduces to the minimal area of the overall cross section. In most engineering applications the bending moment is variable along the axis; then the most stressed section may be considered as decisive.

(2) As the design variables we consider two functions describing a cylindrical shell (Życzkowski and Gajewski [12]): middle surface of a cylindrical, not necessarily circular shell is described by the function $r = r(\phi)$, and wall thickness - by the function $g=g(\phi)$. Moreover, in the case of a shell reinforced by longitudinal ribs, located at extreme fibres and convenient to carry large bending moment, we should introduce two parameters A_1 and A_2 corresponding to optimal areas of those ribs regarded as concentrated; however, the present paper will not discuss any rib reinforcement, thus leaving prevailing bending beyond consideration.

(3) The optimization constraint refers to creep rupture of the shell under a given system of loadings: M_b, M_t, N and p. In particular, the Kachanov-Sdobyrev hypothesis of brittle creep rupture in its scalar form is adopted [13,14]. According to that hypothesis, a measure of material continuity during the damaging process, ψ, is governed by the evolution equation:

$$\frac{d\psi}{dt} = - R\left(\frac{\sigma_S}{\psi}\right)^{\nu}, \quad 0 \leqslant \psi \leqslant 1 \tag{2.1}$$

where σ_S denotes Sdobyrev's reduced stress,

$$\sigma_S = \delta\sigma_I + (1 - \delta)\sigma_e, \tag{2.2}$$

σ_I and σ_e are the algebraically maximal principal stress, and the effective stress (stress intensity), respectively, finally R, ν and δ are material constants, $0 \leq \delta \leq 1$. Leckie and Hayhurst [15] found that damaging process of some materials is better described by σ_I (e.g. copper), and of others - by σ_e (e.g. steel and aluminium alloys). Hence the combination (2.2) is sufficiently general as to cover a fairly broad class of materials.

If a steady creep is considered and σ_{ij} = const(t), without redistribution of stresses due to elastic effects, geometry changes etc., we may integrate (2.1) in a general form. Making use of the initial condition $\psi(0) = 1$ (perfect continuity, no deterioration) and of the condition of full local deterioration at the point under consideration $\psi(t_R) = 0$, we obtain for creep rupture time t_R the following "local" formula

$$t_R = \frac{1}{(\nu+1)R\sigma_S^{\nu}} . \tag{2.3}$$

It is supposed that in optimal structures designed for a given creep rupture time t_R, this value should be reached, if possible, simultaneously at all points of the body (a structure of uniform creep strength). Denoting the relevant stress in (2.3) by σ_{SR}, introducing a certain safety factor for stresses, j, we obtain the following condition of uniform creep strength

$$\sigma_S = \frac{\sigma_{SR}}{j} = \frac{1}{j\sqrt[\nu]{(\nu+1)Rt_R}} \stackrel{\text{def}}{=} \sigma_o = \text{constr}(r,\sigma). \tag{2.4}$$

It should be noted that the condition of uniform creep strength is, in general, neither a necessary nor a sufficient optimality condition. It may not be necessary either if geometry changes are taken into account (Swisterski, Wroblewski and Życzkowski [16]), or if other constraints are introduced; it may not be sufficient if it does not result in a unique solution. In the case under consideration geometry changes are disregarded and no other constraints are introduced, and hence we regard (2.4) as a necessary condition; on the other hand, it is not sufficient here and further optimization will be performed.

As was mentioned above, other constraints, in particular creep buckling constraints for the shell, will not be considered in the present paper. Such an approach may be justified if the wall thickness is not too small; this case takes place if pressure and axial tension are predominant in comparison to torsion and bending.

(4) As a first step towards optimal design we look for elimination of bending states in the shell, in particular for elimination of circumferential bending. Indeed, any bending of the wall results in transversally nonhomogeneous state of stress and (2.4) cannot be satisfied at any point of the shell. Substituting into the general equilibrium equations of the engineering theory of shells (Wlassow [17], p.201) all the moments and shearing forces equal to zero, we obtain

$$\frac{\partial n_\phi}{\partial \phi} = 0, \qquad - k_2 n_\phi + p = 0, \tag{2.5}$$

and hence

$$n_\phi = \text{const}, \; k_2 = \frac{p}{n_\phi} = \text{const}, \tag{2.6}$$

where $k_2(\phi) = 1/\rho(\phi)$ denotes the circumferential curvature of the shell $n_\phi = \sigma_\phi(\phi) g(\phi)$ is the circumferential membrane force, and the internal pressure p was assumed to be constant (self-weight of the medium inside the shell being disregarded). So, it turns out that the necessary conditions of the membrane state result here in a circular cylindrical shell, though this shape was not assumed a priori. Hence we reduce design variables in the optimization problem under consideration to one function of one variable $g = g(\phi)$ and to one parameter $\rho = r = \text{const}$. Moreover, the design variable $g(\phi)$ and the stress $\sigma_\phi(\phi)$ are interrelated by

$$\sigma_\phi(\phi) g(\phi) = pr, \tag{2.7}$$

resulting from (2.6).

It should be noted that the necessary conditions of membrane state (2.5) are not the sufficient ones. For example, the circumferential bending effects in the wall under similar loadings were studied in the plastic range by Mrowiec and Życzkowski [18, 19]. In the present paper, however, we neglect these effects and assume the membrane state.

(5) Equations of state are assumed as the Norton-Odqvist constitutive equations for an incompressible body:

$$\dot{\varepsilon}_e = K\sigma_e^n , \tag{2.8}$$

$$\dot{e}_{ij} = f\, s_{ij}, \qquad \dot{e}_{ii} = 0, \tag{2.9}$$

where $\dot{e}_{ij}$ and s_{ij} denote deviatoric strain rates and deviatoric stresses respectively, $\dot{\varepsilon}_e$ and σ_e are the effective strain rate and the effective stress as described by the Huber-Mises-Hencky hypothesis, K and n stand for material constants, and in the last equation the summation convention holds. In the case of plane stress under consideration we have

$$\dot{\varepsilon}_e^2 = \frac{4}{3}(\dot{\varepsilon}_z^2 + \dot{\varepsilon}_\phi^2 + \dot{\varepsilon}_z\dot{\varepsilon}_\phi + \frac{1}{4}\dot{\gamma}_{z\phi}^2), \tag{2.10}$$

$$\sigma_e^2 = \sigma_z^2 + \sigma_\phi^2 - \sigma_z\sigma_\phi + 3\tau_{z\phi}^2, \tag{2.11}$$

where the usual engineering notation for stresses and strain rates has been introduced.

(6) Finally, the optimization problem is stated as follows. We minimize the overall cross-sectional area

$$A = 2r \int_{-\pi/2}^{\pi/2} g(\phi)\,d\phi \rightarrow \min \tag{2.12}$$

under the constraint for stresses (2.4), and under the integral constraints

$$N = 2r \int_{-\pi/2}^{\pi/2} \sigma_z(\phi) g(\phi) d\phi = \text{const} \tag{2.13}$$

$$M_t = 2r^2 \int_{-\pi/2}^{\pi/2} \tau_{z\phi}(\phi) g(\phi) d\phi = \text{const}, \tag{2.14}$$

$$M_b = 2r \int_{-\pi/2}^{\pi/2} \sigma_z(\phi)(r \sin\phi - y_o) g(\phi) d\phi = \text{const}, \tag{2.15}$$

where y_o denotes the coordinate of the centre of gravity of the unsymmetric cross-section,

$$y_o = r \int_{-\pi/2}^{\pi/2} g(\phi)\sin\phi d\phi / \int_{-\pi/2}^{\pi/2} g(\phi) d\phi, \tag{2.16}$$

with the equations of state (2.8), (2.9), and the remaining fundamental equations of continuous media (equilibrium, compatibility).

3. Stress and Strain Rate Distribution

The distribution of shearing stresses $\tau_{z\phi}$ follows directly from equilibrium equations. Hydrodynamic analogy yields

$$\tau_{z\phi}(\phi) g(\phi) = \text{const}(\phi) = C. \tag{3.1}$$

Substituting (3.1) into (2.14) we calculate C and express shearing stresses in terms of the twisting moment:

$$\tau_{z\phi} = \frac{M_t}{2\pi r^2 g}; \tag{3.2}$$

now the constraint (2.14) disappears. Further, making use of (2.7) we eliminate the thickness $g(\phi)$ and rewrite (2.12)-(2.16) in the form:

$$A = 2pr^2 \int_{-\pi/2}^{\pi/2} \frac{1}{\sigma_\phi} d\phi \rightarrow \min, \tag{3.3}$$

$$N = 2pr^2 \int_{-\pi/2}^{\pi/2} \frac{\sigma_z}{\sigma_\phi} d\phi = \text{const} \tag{3.4}$$

$$M_b = 2pr^3 \int_{-\pi/2}^{\pi/2} \frac{\sigma_z}{\sigma_\phi} \left(\sin\phi - \frac{\int_{-\pi/2}^{\pi/2} \frac{\sin\phi}{\sigma_\phi} d\phi}{\int_{-\pi/2}^{\pi/2} \frac{1}{\sigma_\phi} d\phi}\right) d\phi = \text{const} \tag{3.5}$$

Moreover, the formula for $\tau_{z\phi}$, (3.2), turns into

$$\tau_{z\phi} = \frac{M_t}{2\pi pr^3} \sigma_\phi \tag{3.6}$$

Compatibility equations make it possible to determine the distribution of axial strain rates $\dot{\varepsilon}_z$. Using Cartesian coordinates we may write

$$\frac{\partial^2 \dot{\varepsilon}_z}{\partial y^2} + \frac{\partial^2 \dot{\varepsilon}_y}{\partial z^2} = \frac{\partial^2 \dot{\gamma}_{yz}}{\partial y \partial z} ; \tag{3.7}$$

in the problem under consideration the strains do not depend on z, hence two derivatives in (3.7) vanish, and $\dot{\varepsilon}_z$ mustbe linear function of y. Returning to cylindrical coordinates we write this function in the form

$$\dot{\varepsilon}_z = \dot{\kappa}\ r\ \sin\phi + \dot{\varepsilon}_o, \tag{3.8}$$

where $\dot{\kappa}$ denotes the curvature rate and $\dot{\varepsilon}_o$ is the rate of elongation of axis of the cylinder (not coinciding here with the centroidal axis, shifted by y_o). Eq. (3.8) resembles Bernoulli's hypothesis, but, in fact, it exceeds that hypothesis: plane sections not necessarily remain plane, warping may occur.

Now, the remaining four unknowns σ_z, σ_ϕ, $\dot{\varepsilon}_\phi$, and $\dot{\gamma}_{z\phi}$ may be found from the Norton-Odqvist equations (2.8), (2.9), and the condition of uniform creep strength (2.4). First we eliminate f from (2.9) performing contraction of the deviators and making use of (2.8):

$$f = \frac{3}{2}K\sigma_e^{n-1}, \quad \dot{e}_{ij} = \frac{3}{2}K\sigma_e^{n-1}s_{ij}. \tag{3.9}$$

Using the last equation as joining $\dot{\varepsilon}_z$ and s_z we obtain

$$\dot{\kappa}\ r\ \sin\phi + \dot{\varepsilon}_o = \frac{K}{2}(\sigma_z^2 + \sigma_\phi^2 - \sigma_z\sigma_\phi + 3m^2\sigma_\phi^2)^{\frac{n-1}{2}}(2\sigma_z - \sigma_\phi), \tag{3.10}$$

where the dimensionless parameter:

$$m = \frac{M_t}{2\pi p r^3} \tag{3.11}$$

joins the loadings and the design variable r. The remaining two independent equations (3.9) determine $\dot{\varepsilon}_\phi$ and $\dot{\gamma}_{z\phi}$, but they will not be used effectively. On the other hand, (2.4) with (2.2) and (3.6) yield

$$\frac{\delta}{2}\left[\sigma_z + \sigma_\phi + \sqrt{(\sigma_z - \sigma_\phi)^2 + 4m^2\sigma_\phi^2}\right] + $$
$$+ (1-\delta)\sqrt{\sigma_z^2 + \sigma_\phi^2 - \sigma_z\sigma_\phi + 3m^2\sigma_\phi^2} = \sigma_o. \tag{3.12}$$

The solution of the system of equations (3.10) and (3.12) with respect to σ_z and σ_ϕ would enable us to determine the stress distribution in terms of the coordinate ϕ and of the parameters

$\dot{\kappa}$, $\dot{\varepsilon}_o$. Then we could use (3.4) and (3.5) to evaluate $\dot{\kappa}$ and $\dot{\varepsilon}_o$ and perform minimization of (3.3) as a function of the design variable r.

4. Change of Variables

An analytical solution of (3.10) and (3.12) with respect to σ_z and σ_ϕ seems impossible. However, the left-hand side of (3.12) is homogeneous of the first degree in stresses and an essential simplification will be obtained by introducing instead of σ_ϕ, σ_z two new, dimensionless unknowns s, ξ by the formulae

$$\sigma_\phi = \sigma_o s \sin\xi, \qquad \sigma_z = \sigma_o s \sin(\xi - \frac{\pi}{3}). \tag{4.1}$$

These formulae resemble the Nadai-Sokolovsky parametrization of the Huber-Mises-Hencky yield condition expressed in principal stresses, but they are used here in some other sense since $\tau_{z\phi}$ is also present and s is not proportional to the stress intensity. In the problem under consideration we have $0 < \xi < \pi$, because σ_ϕ must be positive. Now, in view of the mentioned homogeneity of (3.12), this equation may be solved with respect to $s = s(\xi)$:

$$s = \left\{ \frac{\delta}{2}\left[\sin(\xi - \frac{\pi}{3}) + \sin\xi + \sqrt{\left[\sin(\xi - \frac{\pi}{3}) - \sin\xi\right]^2 + 4m^2\sin^2\xi} \right] + (1-\delta)\sqrt{\frac{3}{4} + 3m^2\sin^2\xi} \right\}^{-1} \tag{4.2}$$

and (3.10) yields

$$\bar{\kappa} r \sin\phi + \bar{\varepsilon}_o = -\frac{\sqrt{3}}{2}\left(\frac{3}{4} + 3m^2\sin^2\xi\right)^{\frac{n-1}{2}} s^n(\xi)\cos\xi, \tag{4.3}$$

where $s(\xi)$ is given by (4.2), and the following dimensionless parameters have been introduced

$$\bar{\kappa} = \frac{\dot{\kappa}}{\sigma_o^n K}, \qquad \bar{\varepsilon}_o = \frac{\dot{\varepsilon}_o}{\sigma_o^n K}. \tag{4.4}$$

So, we have reduced two equations (3.10), (3.12) to only one equation (4.3) with the unknown $\xi = \xi(\phi)$. This equation cannot be solved for ξ, but it can easily be solved with respect to $\phi = \phi(\xi)$, and this inverse function will be used in further calculations:

$$\phi = \arc\sin \frac{f(\xi) - \bar{\varepsilon}_o}{\bar{\kappa}\, r}, \tag{4.5}$$

where $f(\xi)$ stands for the right-hand side of (4.3) with substituted (4.2).

Now, the integrals (3.3)-(3.5) may be rewritten with the integration variable changed into ξ, and so the optimization problem will be presented in an effective form. Moreover, we replace the dimensional design variable r by a dimensionless one, m, substituting, in view of (3.11),

$$r = \sqrt[3]{\frac{M_t}{2\pi p m}} . \tag{4.6}$$

Finally, we look for a minimum of the integral

$$\bar{a} = \bar{a}(\xi_1, \xi_2, m) = \frac{1}{m^{2/3}} \int_{\xi_1}^{\xi_2} \frac{1}{s(\xi)\sin\xi} \frac{d\phi}{d\xi}\, d\xi \rightarrow \min \tag{4.7}$$

under the constraints

$$\bar{n} = \bar{n}(\xi_1, \xi_2, m) = \frac{1}{m^{2/3}} \int_{\xi_1}^{\xi_2} \frac{\sin(\xi - \frac{\pi}{3})}{\sin\xi} \frac{d\phi}{d\xi}\, d\phi = \text{const}$$

(4.8)

$$\bar{m}_b = \bar{m}_b(\xi_1,\xi_2,m) = \frac{1}{m}\int_{\xi_1}^{\xi_2} \frac{\sin(\xi - \frac{\pi}{3})}{\sin\xi} \left[\sin\phi(\xi) - \frac{\int_{\xi_1}^{\xi_2} \frac{\sin\phi(\xi)}{s(\xi)\sin\xi}\frac{d\phi}{d\xi}\,d\xi}{\int_{\xi_1}^{\xi_2} \frac{1}{s(\xi)\sin\xi}\frac{d\phi}{d\xi}\,d\xi} \right] \frac{d\phi}{d\xi}\,d\xi = \text{const}, \tag{4.9}$$

where $s(\xi)$ is given by (4.2), $d\phi/d\xi$ is to be calculated from (4.5) and the dimensionless quantities $\bar{a}$, $\bar{n}$, $\bar{m}_b$ are defined as follows

$$\bar{a} = \frac{\sigma_o}{2p}\left(\frac{2\pi p}{M_t}\right)^{2/3} A, \quad \bar{n} = \frac{1}{2p}\left(\frac{2\pi p}{M_t}\right)^{2/3} N, \ \bar{m} = \frac{\pi}{M_t} M_b. \tag{4.10}$$

The limits of integration ξ_1,ξ_2 may be expressed in terms of $\bar{\varepsilon}_o$ and $\bar{\kappa}$ by solving (4.5) with substituted $\phi = -\pi/2$ and $\phi = \pi/2$, respectively. However, this is not necessary: $\bar{\varepsilon}_o$ and $\bar{\kappa}$ should be determined from (4.8), (4.9) in terms of $\bar{n}$ and $\bar{m}_n$, so we may simply solve (numerically) (4.8), (4.9) with respect to ξ_1 and ξ_2 instead of $\bar{\varepsilon}_o$ and $\bar{\kappa}$. Finally, after ξ_1 and ξ_2 have been evaluated, we look for a minimum of (4.7) as of a function of one variable m (m is hidden in $z(\xi)$ and $\phi(\xi)$ as well).

5. Numerical Examples

The system of two equations (4.8), (4.9) is solved with respect to ξ_1,ξ_2 by Newton's procedure for subsequent values of m and then a numerical minimization of the function $\bar{a} = \bar{a}(m)$ is performed. However, some complications appear when calculating numerically the integrals: they are improper, since $d\phi/d\xi$ increases infinitely for $\phi = -\pi/2$ and $\phi = \pi/2$ i.e. for $\xi = \xi_1$ and $\xi = \xi_2$. Accuracy of most numerical procedures is then poor, but those singularities may be removed by integration per partes of the type

$$\int_{\xi_1}^{\xi_2} u_i \frac{d\phi}{d\xi}\,d\xi = u_i\phi\Big|_{\xi_1}^{\xi_2} - \int_{\xi_1}^{\xi_2} \phi\frac{du_i}{d\xi}\,d\xi, \tag{5.1}$$

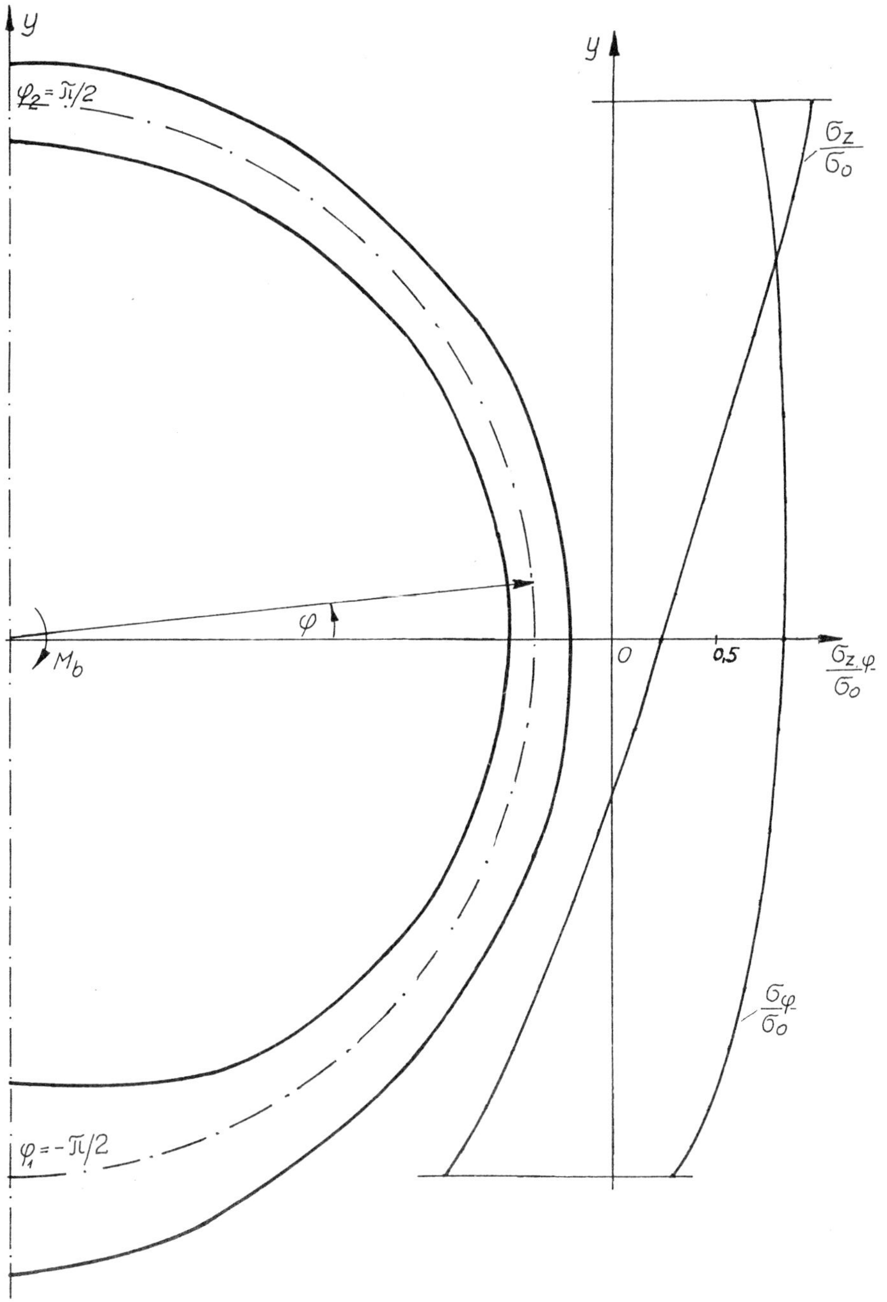

Fig. 2. Optimal shape and stress distribution for $\delta = 0$ (Kachanov-Huber-Mises), n=4, $\bar{m}_b$=5, $\bar{n}$=0.

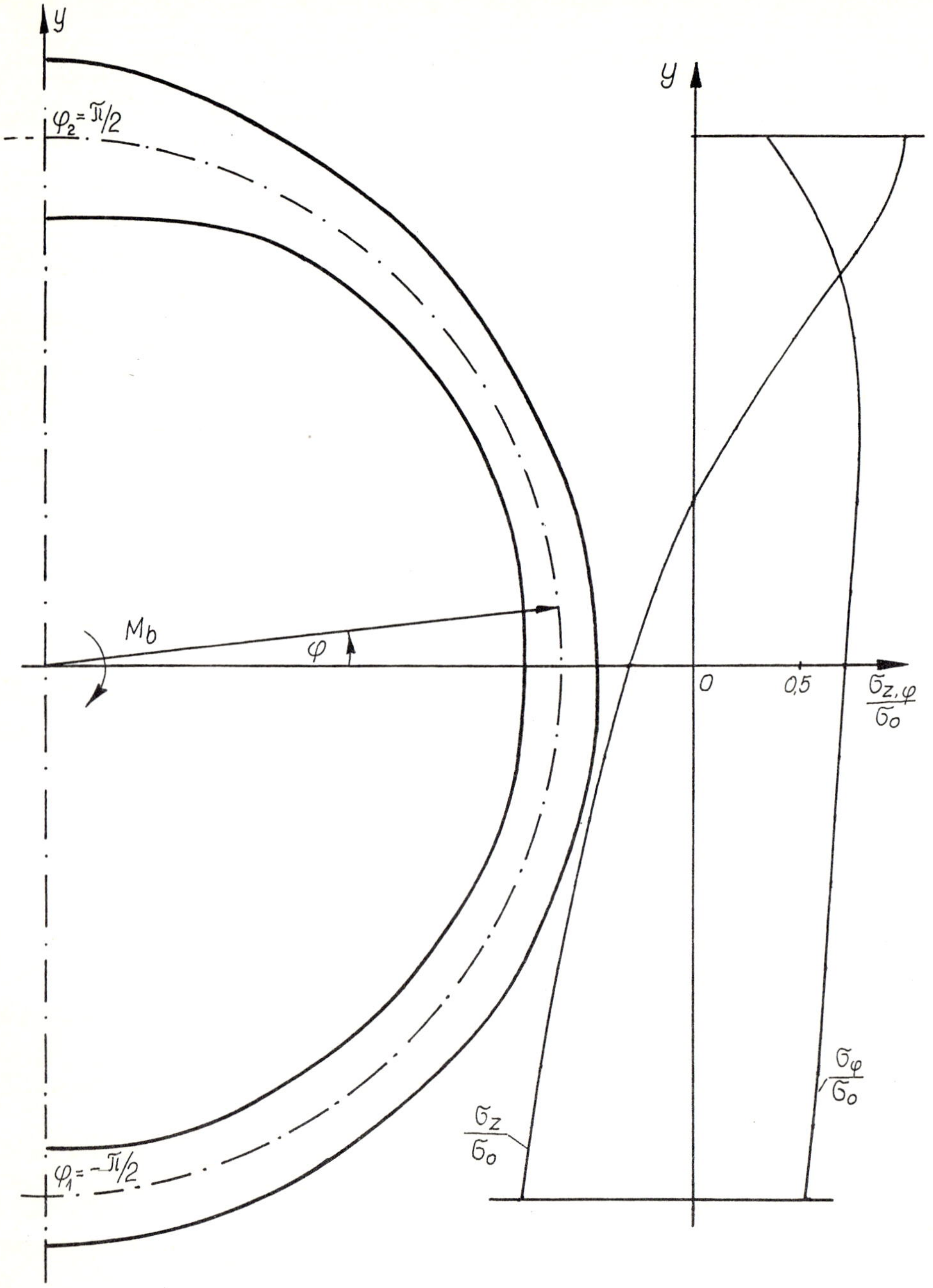

Fig. 3. Optimal shape and stress distribution for δ = 0.5 (Kachanov-Sdobyrev) n=4, $\bar{m}_b$=5, $\bar{n}$=0.

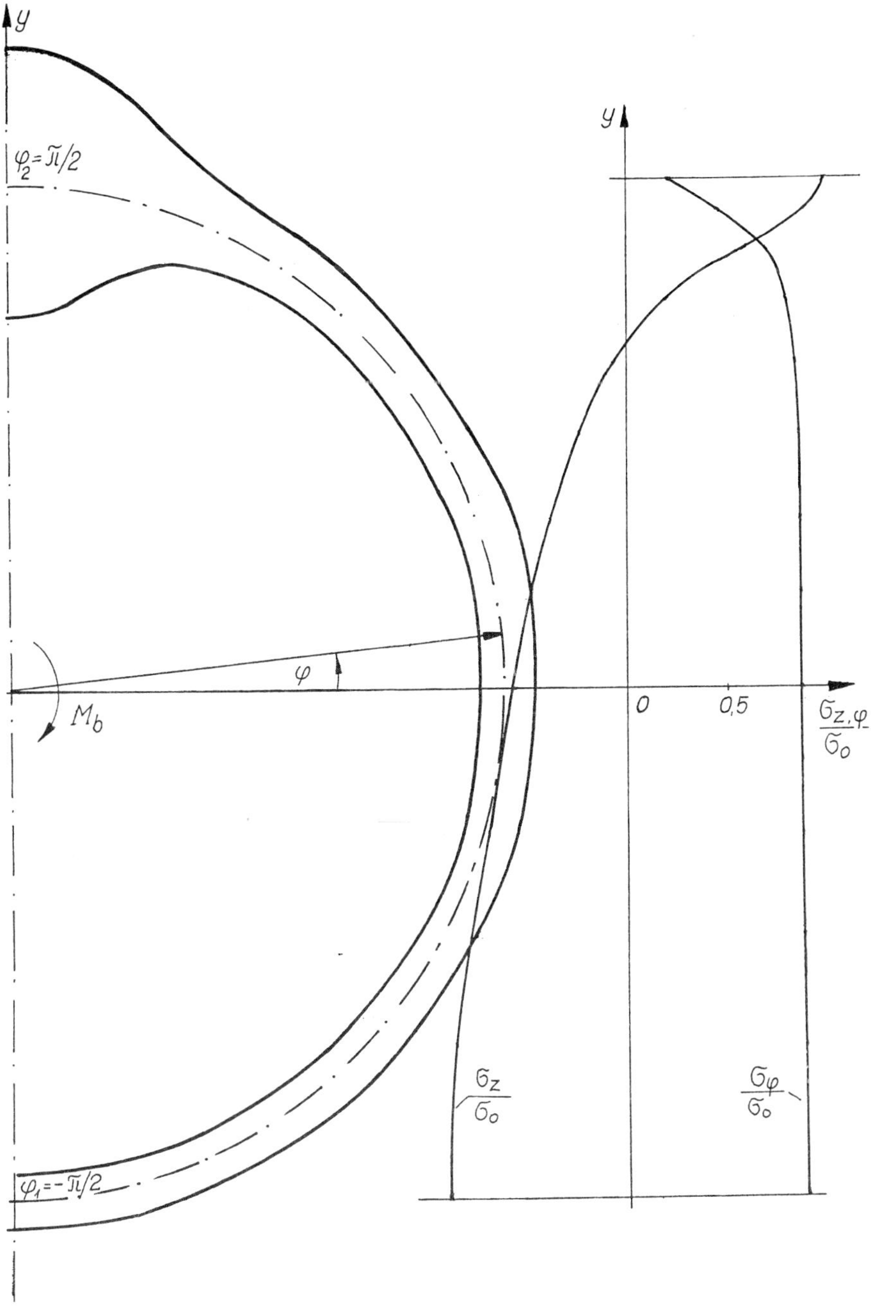

Fig. 4. Optimal shape and stress distribution for $\delta = 1$ (Kachanov-Galileo), n=4, $\bar{m}_b=5$, $\bar{n}=0$.

where u_i stand for the relevant integrands in (4.7)-(4.9).

The resulting optimal shapes and the relevant stress diagrams are shown in Figures 2 (δ = 0, Kachanov-Huber-Mises), 2(δ = 0.5, Kachanov-Sdobyrev), and 4(δ = 1, Kachanov-Galileo).

6. Final Remarks

The shapes shown in Figures 2-4 are optimal, but they may be difficult in manufacturing. Much easier, but less effective optimization may be achieved by applying constant wall thickness reinforced by two longitudinal ribs to carry overall bending of the shell. This problem of purely parametric optimization will be discussed separately.

References

1. Życzkowski, M., Krużelecki, J.: Optimal design of shells with respect to their stability, Proc. IUTAM Symp. Optimization in Structural Design, ed. by A. Sawczuk and Z. Mróz, Springer 1975, 229-247.

2. Krużelecki, J.: Optimization of shells under combined loadings via the concept of uniform stability, Optimization of Distributed Parameter Systems, ed, by E.J. Haug and J.Cea, Sijthoff and Noordhoff 1982, 929-950.

3. Krużelecki, J., Życzkowski, M.: Optimal design of an elastic cylindrical shell under overall bending with torsion. Solid Mech. Arch. 9 (1984), 269-306.

4. Reytman, M.I.: On the theory of optimal design of structures made of plastics with time effects taken into account (in Russian). Mekhanika Polimerov (1967), 2, 357-360.

5. Prager, W.: Optimal structural design for given stiffness in stationary creep. Z. Angew. Math. Physik 19 (1968), 252-256.

6. Nemirovsky, Yu.V.: Optimal design of structures in creep conditions (in Russian). Trudy 3-go Vsesoy. Syezda po Teor. Prikl. Mekh. (1968), 225.

7. Życzkowski, M.: Optimal Structural design in rheology, 12th Int. Congr. Appl. Mech., Stanford 1968; J. Appl. Mech. 38 (1971), 39-46.

8. Życzkowski, M.: Recent results on optimal design in creep conditions. Proc. Euromech. Coll. 164, Optimization Meth-

ods in Structural Design, ed. by H. Eschenauer and N. Olhoff Bibliographisches Institut 1983, 444-449.

9. Krużelecki, J., Życzkowski, M.: Optimal structural design of shells - a survey. Solid Mech. Arch. 10 (1985), 101-170.

10. Życzkowski, M., Rysz, M.: Optimal design of a thin-walled pipeline cross-section in creep conditions. Mechanics of Inelastic Media and Structures, ed. by O. Mahrenholtz and A. Sawczuk, PWN 1982, 329-339.

11. Rysz, M.: Optimal rib-reinforcement of a thin-walled pipeline with respect to creep rupture. J. Pipelines (1985), in print.

12. Życzkowski, M., Gajewski, A.: Optimal structural design under stability constraints. Proc. IUTAM Symp. Collapse - Buckling of Structures, ed. by J.M.T. Thompson and G.W. Hunt, Cambridge Univ. PRess 1983, 299-332.

13. Kachanov, L.M.: On the rupture time in creep conditions (in Russian). Izv. AN SSSR, Otd. Tekhn. Nauk (1958), 8, 26-31; (1960), 5, 88-92.

14. Sdobyrev, V.P.: A long-time strength criterion for certain alloys under combined stresses (in Russian). Izv. AN SSSR, Otd. Tekhn. Nauk (1959), 6, 93-99.

15. Leckie, F.A., Hayhurst, D.R.: Creep rupture of structures. Proc. Roy.Soc. A340 (1974), 323-347.

16. Swisterski, W., Wroblewski, A., Życzkowski, M.: Geometrically non-linear eccentrically compressed columns of uniform creep strength versus optimal columns. Int. J. Non-linear Mech. 18 (1983), 287-296.

17. Wlassow, W.S.: Allgemeine Schalentheorie und ihre Anwendung in der Technik (übersetzt aus dem Russischen). Akademie-Verlag, Berlin 1958.

18. Mrowiec, M.: Limit state of thin pipeline under combined internal pressure and bending moment. Bull. Acad. Pol., Ser. Sci. Techn. 15 (1967), 205-216.

19. Mrowiec, M., Życzkowski, M.: Limit interaction curves for thin-walled pipe-line under internal pressure and bending. Bull. Acad. Pol. Ser. Sci. Techn. 16 (1968), 451-460.

The Collapse of Continuously Welded Stiffened Plates Subjected to Uniaxial Compression Load

K. GHAVAMI

Civil Engineering Department

Pontifícia Universidade Católica do

Rio de Janeiro - Brasil

Summary

This paper outlines the results of several series of tests, which have been carried out at the Structural and Material Laboratory of Pontifícia Universidade Católica do Rio de Janeiro, to determine the collapse load and types of instability failure of continuously welded stiffened plates subjected to uniaxial compression load. The plates were stiffened longitudinally and transversally.

Experimental results are discussed in relation to the influence of stiffeners cross-sections on collapse load and to various currently available methods of analysis.

Introduction

Stiffened plates very frequently form components of structures such as slender ship hulls, box girder bridges, bridge decks, super structures of off-shore oil platforms, buildings and other structures in which high strength-to-weight ratio (S/W) is important. In the case of steel bridges, the need to understand ultimate load behaviours of stiffened plates was emphasized in a tragic way by the collapse of several major box girder bridges in Europe and Australia during the last two decades. Despite a substantial amount of theoretical and experimental research into the ultimate load behaviour of stiffened plates, the collapse load in relation to all possible failure mode, allowing for the complex interactions between plate and stiffeners with the consideration of stiffeners type, residual welding stresses and initial out of plane imperfection is not yet accurately predicted.

As part of a research program directed to provide an improved understanding of the ultimate load behaviour of stiffened

Inelastic Behaviour of Plates and Shells
IUTAM Symposium Rio de Janeiro 1985
Editors: L. Bevilacqua, R. Feijóo and R. Valid

plates hence establishing an accurate design method in the last five years, a modest effort at the Structural and Material Laboratory of Pontifícia Universidade Católica do Rio de Janeiro was devoted to the subject. The results of these investigations have been reported in detail elsewhere (1,2,3,4,5,6,7,8).

The insight gained from the experimental work which is summarized in this paper helped to judge the validity of the assumptions on which the various theories and approximate methods are based and also to assess which method of design can be most efficiently used to answer specific problems.

Description of Test Plates

In this research program simply supported stiffened steel plates of overall length L and width B with three types, of longitudinal stiffeners i.e. rectangular, L and T cross-section, as shown in Fig. 1, with or without transversal stiffener of T cross-section have been studied.

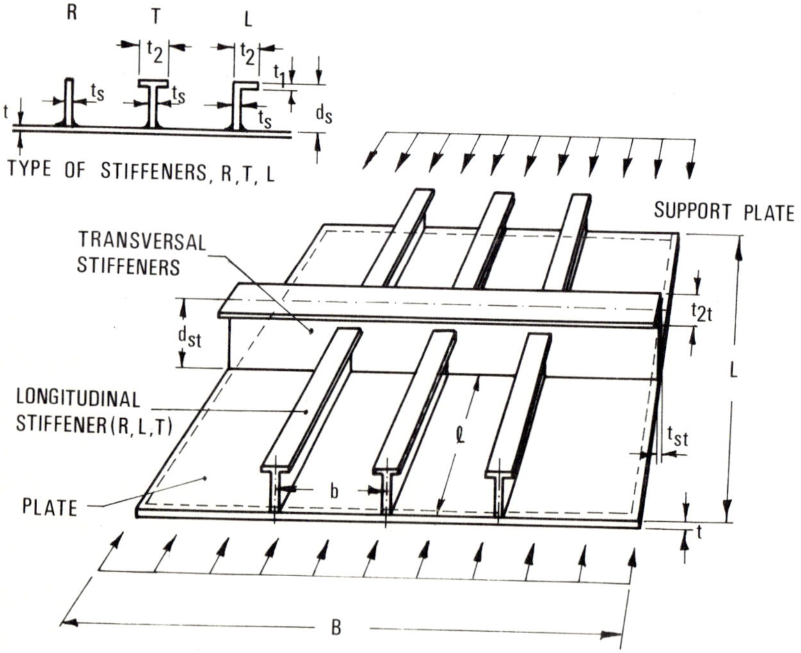

Fig. 1 - A typical plate stiffened longitudinally and transversally.

There were six series of specimens, all in the form of square plates of L x B = 750 x 750 mm .

Series I : 2 isotropic plate (denoted as P).

Series II : Plates with one longitudinal stiffeners of the R, L and T cross-section (denoted as P1R, P1L and P1T).

Series III : Plates with two longitudinal stiffeners of the R, L and T cross-sections (denoted as P2R, P2L and P2T).

Series IV : Plates as in series II with addition of one transversal stiffener (denoted as P1R1T, P1L1T and P1T1T).

Series V : Plates as in series III with addition of one transversal stiffener (denoted as P2R1T, P2L1T and P2T1T).

Series VI : Plates as in series III with addition of two transversal stiffeners (denoted as P2R2T, P2L2T and P2T2T).

The dimensions of the stiffeners and their spacing with the total area of the stiffeners "A_S" are given in Table 1. For the notation refer to Fig. 1. Plate thickness "t" was 4,8 mm and 4,4 mm for the first and the second set respectively. The span of the plates between supports for all series was 650 x 650 mm. The boundary conditions were simply supported. However, the first set had discretized points and the second set continuous supports. All stiffeners were continuously welded to the isotropic plate. The welding was carried out manually by the same welder for all the test series. The throat of the fillet weld was in order of 3,0 mm. Two base plates of 51 x 16 x 750, were welded at the upper and lower parts of the plates. This resulted in applying uniform compression load.

Measurement of Initial Out-of-Plane Deflection

The initial out-of-plane deflection of each plate before and after welding was thoroughly investigated, it was measured relative to the plate ends using a perfectly straight aluminium bar.

Table 1 - Dimensions of Plates Tested

Set	Series	Test Type	Plating ℓ mm	Plating b mm	Longitudinal Stiffeners $t_1=t_s$ mm	Longitudinal Stiffeners t_2 mm	Longitudinal Stiffeners d_1 mm	Longitudinal Stiffeners A_s mm^2	Transversal Stiffeners t_{st} mm	Transversal Stiffeners t_{2t} mm	Transversal Stiffeners d_{st} mm	Transversal Stiffeners A_{st} mm^2
SET 1	Series II	P1R	650	325	7,0		30,0	210	-	-		
		P1L	650	325	6,4	16,4	30,0	256	-		-	-
		P1T	650	325	6,4	26,4	30,0	320	-	-	-	-
	Series III	P2R	650	217	7,0	-	30,0	420	-	-	-	-
		P2L	650	216	6,4	16,4	30,0	512	-	-	-	-
		P2T	650	217	6,4	26,4	30,0	640	-	-	-	-
SET 2	Series IV	P1R1T	325	325	5,1	-	30,0	153	4,7	35,3	41,1	337
		P1L1T	325	325	5,2	14,8	30.2	207	4,8	34,2	40,4	335
		P1T1T	325	325	4,6	25,3	30,0	233	4,9	35,2	40,4	346
	Series V	P2R1T	325	216	5,1	-	30,0	306	4,7	35,7	40,7	337
		P2L1T	325	217	5,1	14,6	30,2	405	4,6	35,9	40,6	331
		P2T1T	325	217	4,7	25,0	28,8	472	4,7	34,8	39,6	328
	Series VI	P2R2T	216	216	5,0	-	30,1	300	4,7	35,7	40,4	336
		P2L2T	217	217	5,1	14,9	30.0	406	4,7	35,5	40,6	336
		P2T2T	216	216	4,6	24,8	29,8	460	4,8	35 5	40,6	342

Measurement of Residual Welding Strains

For the measurement of residual welding stresses on both sides of the plates, a Demec mechanical extensometer with 50 mm gauge length was used. The readings were made before and after welding of the stiffeners. Strain readings were corrected where necessary for temperature effects. Each set of Demec readings was taken at least three times, normally by a different person.

The Testing Rig

A general purpose testing rig as shown in Fig. 2 was designed and constructed in the Structural and Material Laboratory. The maximum capacity of the rig is 1000 kN axial load. The load was applied through two 500 kN jacks to the rigid loading beam which was constrained by sliders and rollers to move to the direction of load, and in the plane of the test plate. In order to distribute the load into the test plate in a uniform manner, avoiding the possibility of premature failure close to the load beams, two bars of 16 x 51 mm cross-sections were fixed at the ends of the test plates. As the boundary conditions were simply supported this type of reinforcement had negligible influence on the overall behaviour of the plate . During the experiment the loads were applied incrementally. At the end of each increment, the strain lateral deflections and shortening of the plate were registered :

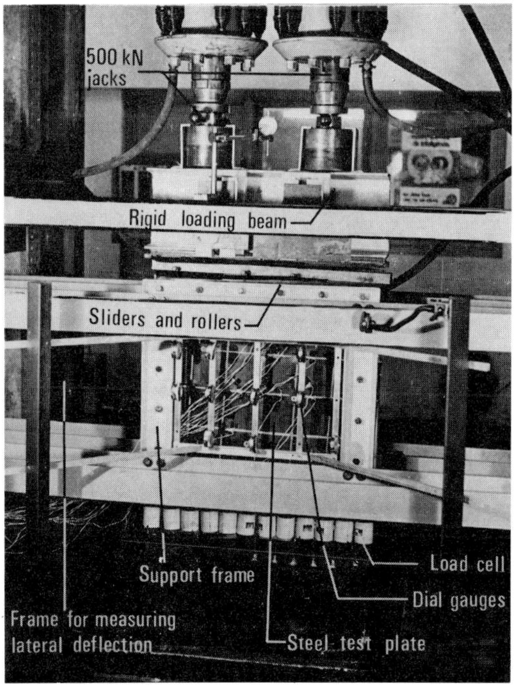

Fig. 2 - Test Plate Positioned in the Testing Rig.

Strain Measurement

For the measurement of the strain in the stiffeners and plating, linear, L and rossete electrical strain gauges of 10 mm and 5 mm gauge length were used. The strain instrumentation in all test plates were almost identical. The linear gauges were fixed on the stiffeners extreme and to the plates opposite side to record both in plane and flexural components of strains. In general about 50 strain gauges were fixed on each test plate . At each load increment complete measurement of strains in stiffeners and plating were made.

Deflection Measurement

Lateral deflections and plate shortening of plates were measured by electrical transducer and mechanical dial gauges of 0,001 mm precision, mounted on an independent deflection frame along the rig as shown in Fig. 2. The deflection frame was designed to allow maximum choice of gauge positions.

Test Results

A summary of material properties of plates and stiffeners with those of maximum initial imperfections "W_o" maximum deflection at collapse "W_{max}", maximum in-plane shortening "U_{max}", slenderness ratio "ℓ/r" and b/t, maximum collapse stress σ_m in relation to the yield stress of plate "σ_{yp}" and the ultimate load "P_u" in relation to squash load "P_{sq}" is given in Table 2.

The maximum collapse stress for each specimen was obtained by dividing the ultimate load to the overall cross-sectional area of the plate.

$$\sigma_m = P_u/(A_p + A_s) \tag{1}$$

and the squash load is calculated as

$$P_{sq} = \sigma_{yp} \cdot A_p + \sigma_{ys} \cdot A_s \tag{2}$$

Discussion of the Results

The effect of the type of longitudinal stiffener on the collapse stress of the plates can be obtained by comparing the results of σ_m/σ_{yp} and P_u/P_{sq} for each series of tests. It was not to our prediction in series II that σ_m/σ_{yp} for the rectangular stiffener was higher than those for L and T. This can be attributed to the form of initial imperfections. In cases of plates with b/t = 45 and b/t = 49 the values of σ_m/σ_{yp} were higher for T and L - type stiffener. In general, in each test series the values of σ_m/σ_{yp} were not significantly different from each other. This may be related to the small percentages of stiffeners area to the overall plates area used in these investigations.

During the tests the redistribution of the deformations in the plates was registered. This indicated that there was an interaction between the local buckling of the plates and buckling of the stiffeners.

Table 2 : Summary of Material Properties, Initial, final deflections and shortening, etc...

			Material properties			Measur. Imperf.	Max. def.	Max. Shortening	Max. Col. Load	Slenderness ratio		
		Test Series	E_p $MP_a \times 10^5$	σ_{yp} MP_a	σ_{ys} MP_a	$\frac{Wo}{t}$ %	$\frac{W_{max}}{t}$ %	$\frac{U_{max}}{L}$ %	$\frac{\sigma_m}{\sigma_{yp}}$ %	b/t	ℓ/r	$\frac{P_u}{P_{sq}}$
SET 1		P	1,81	218	-	61	278	0,38	42,2	135	469	420
SET 1	Series II	P1R	1,81	218	390	69	188	0,31	70,2	67	100	66,9
SET 1	Series II	P1L	1,99	227	270	36	123	0,34	66,5	67	81	65,6
SET 1	Series II	P1T	1,99	227	270	9	33	0,27	60,0	67	71	58,9
SET 1	Series III	P2R	1,95	224	390	25	177	0,41	66,0	45	87	60 3
SET 1	Series III	P2L	2,21	223	270	19	142	0,30	74,0	45	71	71,2
SET 1	Series III	P2T	2,21	223	270	3	128	0,37	74,0	45	63	71,4
SET 2		P	1,78	220	-	20	121	0,40	48,2	147	511	48,2
SET 2	Series IV	P1R1T	1,85	219	326	21	123	0,33	74 0	74	56	72 5
SET 2	Series IV	P1L1T	1,91	225	326	27	27	0,48	71,1	74	43	69,1
SET 2	Series IV	P1T1T	1,75	219	273	33	121	0,33	72,1	74	38	63,6
SET 2	Series V	P2R1T	1,75	219	326	70	52	0,64	88,6	49	46	85,1
SET 2	Series V	P2L1T	1,89	227	326	40	33	0,60	84,6	49	37	80,6
SET 2	Series V	P2T1T	1,78	220	273	23	30	0,51	89,1	49	33	86,6
SET 2	Series VI	P2R2T	1,91	225	326	32	18	0,63	86,2	49	31	82,6
SET 2	Series VI	P2L2T	1,89	227	326	28	53	0,56	97,4	49	22	93,0
SET 2	Series VI	P2T2T	2,09	218	273	32	16	0,51	103,2	49	24	84,6

The maximum shortening of the plates were highest for test series V and VI which was in the range of 0,51 to 0,63 percent. For the other series the shortening changed between 0,31 and 0,48 percent of plate's length.

Comparison with the Predicted values

The predicted values of the maximum mean stress obtained from various methods and codes compared with "σ_m" of the tests are discussed in the following (see Table 3 and Fig. 3) .

Table 3 - Comparison of the Predicted Plate Strength by Various Method with the Mean Collapse Stress of Tests.

Set	Series	Method / Test Type	Collapse stress σ_m Mpa	Murray Test (%)	Carlsen Test (%)	Winter Test (%)	Merrison Test (%)	AISC Test (%)	ASSC Test (%)	Allen Test (%)
SET	Series	P	92	-	62,0	77,0	73,0	-	52,2	-
SET 1	Series II	P1R	153	63,0	64,0	82,0	46,4	99.0	80,4	76,0
		P1L	151	88,0	74,0	99,0	49,0	104,0	96,0	37,0
		P1T	136	98,0	102,0	109,0	56,6	116,0	106,6	104,0
	Series III	P2R	147	99	99,0	130,6	49,6	132,0	128,6	119,0
		P2L	165	106	86,0	119,4	46,7	117,0	115,7	115,0
		P2T	165	116	102,8	119,4	49,6	117,0	115,7	121,0
SET 2	Series IV	P	106	-	-	69,0	76,0	88,7	-	-
		P1R1T	162	85,8	52,5	80,8	56,7	88,3	72,7	136,4
		P1L1T	160	92,0	69,0	85,6	61,2	92,5	78,1	151,2
		P1T1T	158	89,2	81,6	81,0	60,1	91,1	70,2	155,8
	Series V	P2R1T	194	91,0	66,0	89,7	54,1	93,3	89,1	120,1
		P2L1T	192	96,8	84,1	93,7	60,9	96,4	93,2	127,1
		P2T1T	196	93,0	88,0	83,7	61,2	93,4	82,6	178,6
	Series VI	P2R2T	194	96,0	72,0	93,8	61,3	95,4	93,3	121,6
		P2L2T	221	85,5	78,7	82,3	51,6	84,1	80,0	105,7
		P2T2T	225	82,0	77,0	80,0	55,5	80,0	79,1	105,3

In Murray's method [9] the collapse load and the direction of the collapse of steel plates with open section type stiffeners are calculated based on Perry-Robertson formula with an expression for the imperfection term η given by

$$\eta = y' \, (\delta_o + \Delta') / r'^2 \qquad (3)$$

where y' is taken as to appropriate distance to the extreme fibre depending upon whether $(\delta_o + \acute{\Delta})$ is positive or negative and r' is the radius of gyration of stiffener and associated plate which is calculated for effective cross-section. It can be seen in Table 3 that the predicted values are very close to the test results, although this method does not take into account the residual welding stress.

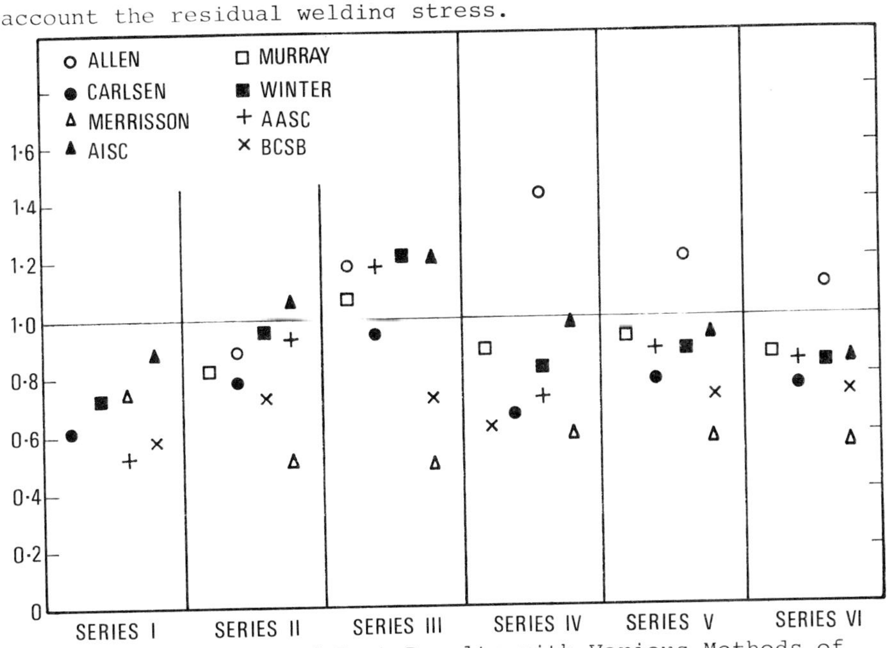

Fig. 3 - Comparison of Test Results with Various Methods of Plate Analysis.

Carlsen's method [10] is also based on the Perry-Robertson concept but with calibrated expression for effective width of the plate. This method accounts for interaction between adjacent stiffener spans. It considers the influence of imperfection, residual-welding stress and the failure modes. Possibly because of an oversimplified consideration of imperfection, the predicted results are relatively conservative.

Winter's method [11] is very simple to use. It is based on the effective width concept. It takes into consideration the influence of imperfection and residual welding stress as a reduction factor for ultimate load. It is noted that the results for plates with low ℓ/r i.e. set two tests were very satisfactory but for the case of P2R it overestimates about 30 percent.

In the Merrison Rule [12] the strength of the plate is calcu -

lated based on nominal plate thickness and σ_{yp}. It considers geometrical imperfections "W_o" residual welding stress "σ_r" and failure mode. It is generally regarded as a too complicated method and overestimates both "σ_r" and "W_o" which results in a very conservative prediction. As can be noted from Table 3 and Fig. 3 this fact was true in this investigation. For the first set the predicted values were about 50 percent of experimental results. However, in each test series the results were very consistent and had almost the same variation.

AISC method [13] is like Winter's method, based on effective width b_e which is inversely proportional to b/t and σ_r. For $\ell/r < 56$ the obtained results by this method were satisfactory. For series V the result differs from the experiment only 7 per cent. However, this method overestimated the collapse stress for the first set of tests.

ASSC method [14] is similar to the previous method but uses equations for the verification of the maximum stress of the plate which leads to the lower predicted values in comparison with AISC. The pattern of the results is the same as for AISC.

Allen's method [14] is based on a general interaction formula based on experimental results which consider only initial imperfections. Except for tests P1R and P1L this method overestimated the maximum collapse stress significantly, especially for test series IV and V.

BCSB - Brazilian Code for Steel Bridges [15] is based on effective width concept with the consideration of initial imperfection of plate as 0,145 b σ_y/E and residual welding stress as 0,1 σ_y . As can be noted in Fig. 3, this method is a conservative method but the results are not as low as those obtained from the Merrison rule.

Conclusions

The results of the investigation based on seventeen tests indicated that :

a - The type of stiffeners cross-section had relatively small influence on the ultimate stress, but had influence on the failure mode as in tests P1R, P1L and P1R1T, P1L1T where stiffeners failure occurred.

b - The introduction of transverse stiffeners did not change the maximum stress significantly but they reduced substantially the maximum out of plane deflections.

c - Among the various methods for the prediction of the maximum stress the Murray's method was the most satisfactory and then AISC. Although the Merrison rule considered most of the variables it gave a very conservative result. The Brazilian Code for Steel Bridge is also a conservative method although not as much as the Merrison rule.

d - The variation between methods was indeed very large (see Fig. 3) which recommended still more research for establishing a general method for the prediction of the ultimate load for such structures.

Acknowledgement

The writer would like to acknowledge the contribution and cooperation of the post-graduate students, Ms. A.Conci, Messrs. S.A.S.Rocha, E.B. de Las Casas, J.El Aschkar, in the execution of the research program into the behaviour of stiffened steel plates. Further, I thank my wife, Ursula, for reading the manuscript and producing the drawings and Tereza Milagres for typing the manuscript.

References

1. Rocha S.A.S. Comportamento último de placas enrijecidas. M.Sc. Thesis, PUC/RJ, 1982.

2. Las Casas, E.B. Análise linear de placas com enrijecedores pelo método das diferenças finitas. M.Sc. Thesis, PUC/RJ, 1981.

3. Conci, A. Instabilidade até o colapso de placas de aço enrijecidas em duas direções. M.Sc. Thesis, PUC/RJ, 1983.

4. El Aschkar, J. Análise não linear de placas com enrijecedores pelo método das diferenças finitas. M.Sc. Thesis, PUC/RJ, 1983.

5. Ghavami, K.; Conci, A.; Rocha, S.A.S. Resistência compressão de placas de aço enrijecidas longitudinalmente. Proc. of the VII Congresso Brasileiro de Engenharia Mecânica - COBEM-83, Uberlandia, December 1983.

6. Ghavami, K.; Conci, A.; Rocha, S.A.S. Métodos de cálculo de placas enrijecidas sob carregamento de compressão axial . Proc. of VII Congresso Brasileiro de Engenharia Mecânica - COBEM 83, Uberlandia, December 1983.

7. Conci, A.; Ghavami, K. Comportamento de placas de aço enrijecidas em duas direções. Proc. IV Congresso Latino-Americano sobre Métodos Computacionais para Engenharia, Santiago, Chile, November 1983.

8. Stramandinoli, Jr. A. Análise de placa metálica enrijecida, considerando-se não-linearidades físicas e geométricas,pré-deformações e tensões residuais. M.Sc. Thesis, PUC/RJ, 1982.

9. Murray, N.W. Analysis and design of stiffened plates for collapse load. The Structural Engineer, Vol. 53, nº 3, March 1975.

10. Carlsen, C.A. A parametric study of collapse of stiffened plates in compression. The Structural Engineer, vol. 58 , nº 2, June 1980.

11. Winter, G. Commentary on the 1968 edition of "Light gauge cold-formed steel design manual", American Iron and Steel Institute, 1970.

12. Merrison Committee, Inquiry into the basis of design and method of erection of steel box girder bridges. Report of the Committee - Appendix 1: Interim design and workmanship rules. Her Majesty's Stationery Office, London 1973.

13. AISC, Specification for the design, fabrication and erection of structural steel for buildings. November 1978.

14. ASSC. Guide to stability design criteria for metal structures. Structural Stability Research Council, New York 1976.

15. Allen, D., discussion of "An approximate method for the design of stiffened steel compression panels", by M.R.Horne and R.Narayanan, Proc. Inst. Civi. Engrs., vol. 61, part 2, June, 1976.

16. BCSB - "Brazilian Code for Steel Bridges". Draft Text Edit. Chapmena & Dowling, Oct., 1982.

17. Horne, M.R. and Narayanan, R. Design of axially loaded stiffened plates. Journal of the Structural Division,Proc. ASCE, vol. 103, nº ST11, Nov. 1977.

Limit Analysis and Design of Plates and Shells

M. SAVE and D. LAMBLIN

Faculté Polytechnique de Mons,
Mons, Belgium

M. KOUAM

Ecole Mohammadia d'Ingénieurs,
Rabat, Morocco

Abstract

For the sake of self-consistency, we recall in the introduction the main aspects of limit analysis and of optimal design for assigned limit load, in which we formulate the optimality criterion used in the examples of the following sections. We then deal with two problems with axial symmetry, namely the circular plate and the circular cylinder, both with piecewise constant thickness. Limit loads and minimum-volume designs are obtained, either by a computer aided analytical approach or by mathematical programming. Minimum-volume of reinforcement of reinforced concrete cylinders under radial pressure is also considered. We finally quote briefly some other related problems, stress the complementarity of the approaches used, and recall the necessity of experimental assessment of the range of validity of the solutions.

Introduction

1. Plastic limit analysis

In a structural shell theory we start from the choice of the generalized strains q_i $(i=1,\dots,n)$ that are sufficient for the complete description of the deformation of the shell when the basic Kirchoff-Love assumptions are accepted. The corresponding generalized stresses Q_i must be so defined as to obtain the virtual specific internal work by the relation

$$W = \sum_i Q_i q_i \ , \tag{1}$$

Inelastic Behaviour of Plates and Shells
IUTAM Symposium Rio de Janeiro 1985
Editors: L. Bevilacqua, R. Feijóo and R. Valid

or equivalently, regarding the q_i and Q_i as components of the two n-dimensional vectors $\overline{q}$ and $\overline{Q}$,

$$W = \overline{Q}.\overline{q}, \tag{2}$$

where the central dot denotes scalar product. Relations (1) or (2) must be valid for any independent $\overline{Q}$ or $\overline{q}$. Similarly, denoting by $\dot{\overline{q}}$ the strain rate vector with components $\dot{q}_i = \partial q_i/\partial t$, where t stands for some measure of time, we have the virtual internal power

$$p = \sum_i Q_i \dot{q}_i = \overline{Q}.\dot{\overline{q}} \ . \tag{3}$$

In the limit analysis and design of structures we then postulate perfect plasticity of a shell element. This property is described in superimposed stress space and strain-rate space with coordinates Q_i and $\dot{q}_i$ respectively by the yield locus of the shell element and the flow rule or normality rule. A generic two-dimensional example is shown in fig. 1. The relevant yield loci for a variety of shells can be found in standard texts (see ref. {1} for example). A yield locus can be either derived from the yield condition applied to every internal component

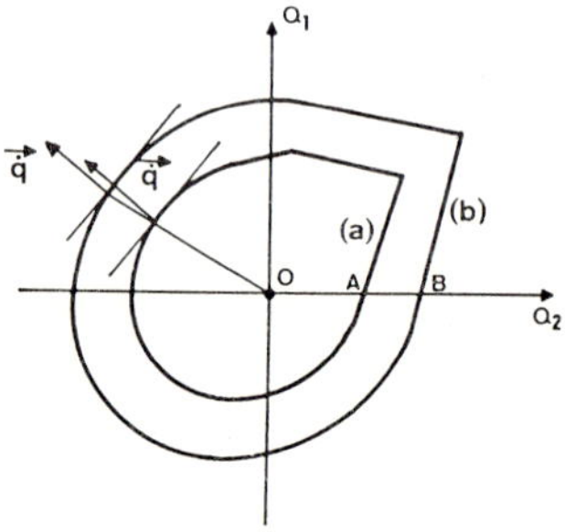

Fig. 1. Two-dimensional yield locus (same structural element, two different yield limits)

sheet of the shell element or, if necessary, considered as a global (maybe experimental) information.
As far as interest is focussed on the collapse state of the structure, and possibly on post-limit behavior, the rigid-perfectly plastic model will be sufficient, as seen in fig. 2. It is the simplest possible model for the analysis and design of plastic structures for the evaluation of the limit load for proportional loading. For purely mechanical loading, the elastic-perfectly plastic model introduces no improvement. Indeed, at incipient collapse, the structure deforms purely plastically under constant load and state of stress. As the undeformed geometry is referred to in describing the limit state, the latter is therefore identical with that of the rigid-perfectly plastic structure. With both models, the limit load

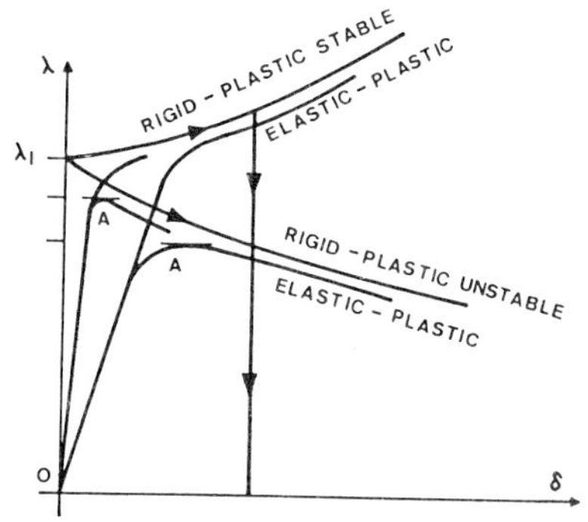

Fig. 2. Typical load vs. deflection curves

will be obtained only if the elastic-plastic deformations for smaller loads do not substantially alter the geometry of the real structure. This can be verified only through a step-by-step analysis of the elastic-plastic deformations, or at least by an evaluation of these deformations just prior to theoretical collapse. In the following we shall assume that such a verification can successfully be made, and that no instability phenomenom of any kind will interfere with the formation of the unrestricted plastic flow mechanism.

On the other hand, the rigid-perfectly plastic model is able to indicate the real physical significance of the limit load. Indeed, a post-yield behavior analysis will give the relation between the load parameter λ and a relevant deflection δ as shown in fig. 2. A rising curve indicates favorable changes of geometry (so-called geometrical work-hardening). The structure remains stable under the limit load and higher loads, but unloading will result in large permanent deflections that most often will render the structure unusable. A downward sloping curve indicates instability of the structure at the limit load, resulting most often in catastrophic, complete collapse (and even fracture), except if important material work-hardening has been neglected. Also, the theoretical limit load is likely not to be reached because prior elastic-plastic deformations will accelerate the onset of the global instability phenomenom (point A in fig. 2).

In rigid-perfectly plastic structures, no deformation occurs before the limit state is reached. Therefore, the limit load also is the "yield point load" of the structure. Denoting by $\overline{F}_\alpha$ the applied forces and by $\overline{\dot{u}}_\alpha$ the velocities of their points of application at collapse, the basic concepts used in limit load evaluation are as follows:

- a stress field is called statically admissible for the given loads $\lambda\overline{F}_\alpha$ if it satisfies all the equilibrium and static boundary conditions for these loads;
- a stress field is called plastically admissible it it does not violate the yield condition at any point;
- a stress field is called licit if it is simultaneously statically and plastically admissible; the load factor corresponding to such a field is called a licit static multiplier and is denoted by λ_s;
- a strain-rate field is called kinematically admissible if it is derived from a field of velocities $\overline{\dot{u}}_\alpha$ that satisfies the kinematic boundary conditions;
- a strain-rate field is called plastically admissible if (i) the velocity field from which it is derived is such that the corresponding exterior rate of dissipation $\Sigma_\alpha \overline{F}.\overline{\dot{u}}$ is nonnegative, and (ii) it belongs to the set on which the interior dissipation function of a typical structural element is

defined;

- a strain-rate field is called licit if it is simultaneously kinematically and plastically admissible; its "power equation", equating the exterior rate of dissipation to the total interior rate of dissipation, yields a load factor λ_k called licit kinematic multiplier.

The basic theorems of limit analysis establish that the limit load factor λ_ℓ at collapse, any licit static multiplier λ_s and any licit kinematic multiplier λ_k satisfy the continued fundamental inequality:

$$\lambda_s \leqslant \lambda_\ell \leqslant \lambda_k \quad . \tag{4}$$

Relation (4) applies to any perfectly-plastic "standard" structure for every element of which:

(i) there exists a convex yield locus, fixed in the stress space;

(ii) the internal dissipation is obtained by application of the normality rule to this yield locus.

For non proportional loading, the basic theorems are generalized as follows {2} :

- if, at any stage of loading, a licit stress field can be found, the structure will not collapse during the loading process;
- if, at some stage of loading, a licit strain-rate field can be found, the structure will not be able to reach this loading stage without exhibiting plastic collapse.

2. Optimal plastic design

Initiated by the pioneering paper of Drucker and Shield in 1956 {3}, optimal limit design has since made enormous progress: optimality criteria and examples of applications successively included: multiple loading {4} and movable loads {5-9}, generalized convex {10,11} and nonconvex {12} cost functions (instead of simply volume or weight), piece-wise varying (or partially pre-assigned) design variables and upper and lower bounds on the design variables {13} (so called technological constraints) and optimal location of discontinuities or supports {14}.

Simultaneously, layout optimization of concrete plate reinforcement, grids and arch-grids developed very rapidly, mostly due to Prager and Rozvany {14-17}. Most applications based on the optimality criteria so derived are analytical, even if computer-aided. Some plate problems have been tackled by a basically numerical approach (by finite elements) using the optimality criterion as the test for design improvement. Most numerical optimal plastic designs however have been treated by mathematical programming methods, resulting in a large number of scientific and technical papers {18-24}.

Consider a structure, the general layout of which is given (axes for systems of beams, midsurface for shells). Denote by ξ an arbitrary point on that layout, and by $d\xi$ the corresponding line or surface element. The yield locus of the structure at a generic point ξ, is as depicted in fig. 1 by curve (a) of a simple two-dimensional case. When ξ varies, the yield surface is allowed to vary its size homothetically, hence retaining its shape and its position with respect to the axes.

We call plastic strength of the structure at point ξ the measure $R(\xi)$ of the length of the ray joining 0 to a point of the yield surface in a fixed but otherwise arbitrary direction (for example the length of OA if the positive Q_2 direction is chosen).

The plastic power of dissipation per unit length or per unit area of the layout is called specific dissipation and is given by

$$D = \sum_i Q_i \dot{q}_i$$

or equivalently by

$$D = \overline{Q} \cdot \overline{\dot{q}}, \qquad (5)$$

where the stress and strain vectors $\overline{Q}$ and $\overline{\dot{q}}$ are associated by the "normality law". When yield surfaces are homothetic, a pro-proportionality relation is obtained between D and R for given $\bar{q}$:

$$D = k.R, \tag{6}$$

where the proportionality factor k depends on $\bar{q}$ only.

By definition, a "design" of the structure will be a function $R(\xi)$, $R(\xi)$ being materialized in any manner (for example, by varying the thickness of the structure with one material, or only varying the strength of the material with constant thickness, or both, etc.), provided relation (6) holds.

Assume next that the specific cost C of the structure (that is, per unit length or area of the layout) is a convex function of R.

We now consider the following problem: design the structure for minimum total cost with the behavioral constraint of supporting assigned loads $\bar{F}_\alpha$. Assuming that bounds R^- and R^+ are set on R by $0 < R^- < R < R^+$, the sufficient optimality condition is (see refs. {11} and {13}) that the collapse mechanism be such that:

$$k\,\bar{q}(\xi) = \begin{cases} \leqslant \left.\dfrac{\partial C}{\partial R}\right|_{R=R(\xi)}, & \text{if } R = R^-, \quad (7a) \\[2ex] = \left.\dfrac{\partial C}{\partial R}\right|_{R=R(\xi)}, & \text{if } R^- < R < R^+, \quad (7b) \\[2ex] \geqslant \left.\dfrac{\partial C}{\partial R}\right|_{R=R(\xi)}, & \text{if } R = R^+ \quad (7c) \end{cases}$$

This condition (7) can be extended to multiple loadings, movable loads and design-dependent forces {11}. Note that it is also a necessary condition for local optimality, independently of convexity considerations.

2. Circular plates with piecewise constant thickness

The generic plate treated is shown in fig. 3: division radii are arbitrary, as well as yield moments $M_i^\circ = k_i M^\circ$. The plate may have a central hole ($b > 0$). When loads p and q exist together, either one is fixed or their ratio is given: only one parameter λ is unknown. Obviously $a > b$. Inner and outer boundary may be either supported or built-in. The yield condition of Tresca is used.

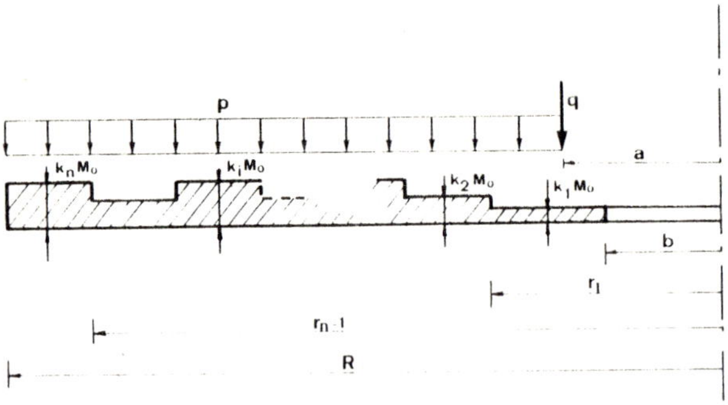

Fig. 3a.

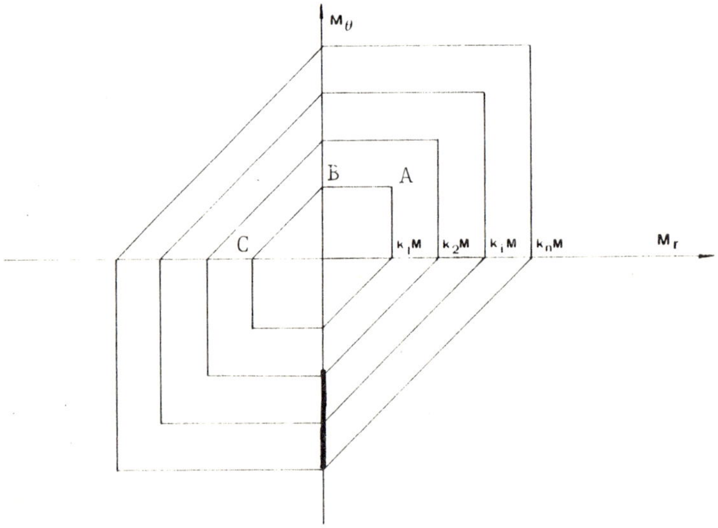

Fig. 3b.

The aim was to write a very efficient "ad hoc" programme for the considered problem. The statical approach was used to achieve complete solutions through the association of corresponding collapse mechanisms, with the reasonable assumption that for both non negative loads p and q the slope of the collapse mechanism does not change sign within the whole plate, or equivalently that only regimes AB and BC (fig. 3b) prevail. Integration of the equilibrium equation with these plastic regimes is straightforward.

The programme assumes a starting value of the limit load parameter λ equal to the average of the limit load parameters for uniform thickness plates with yield moments $k_{max}\ M^{\circ}$ and $k_{min} M^{\circ}$ respectively.

A licit moment field is constructed from the central ring toward the external boundary. Modification of the assumed load parameter occurs to restore plastic admissibility when this turns out to be violated. When M_r is found to be negative in regime AB for some r, the radius r^* of transition from regime AB to BC is obtained from the condition $M_r(r^*) = 0$ in the proper ring, solving the corresponding transcendental equation with a special subroutine. Remark that, whereas M_r is a continuous function of r, M_θ may exhibit discontinuities. It is also worth noting that the collapse mechanism may exhibit rigid regions.

The procedure described above can be performed in a purely analytical manner for the simple uniformly loaded plate made of only two rings (fig. 4 and 5): the limit load p can be obtained for all values of r_1 and k for either $k > 1$ or $k < 1$. From these results, minimum-volume designs can be directly studied: optimal k values are first determined for each r_1 and the corresponding volume plotted versus r_1 as shown in fig. 6 where relative minima are clearly seen. Absolute minimum is reached for $r_1 = 0.85$ with $k = 1.73$.

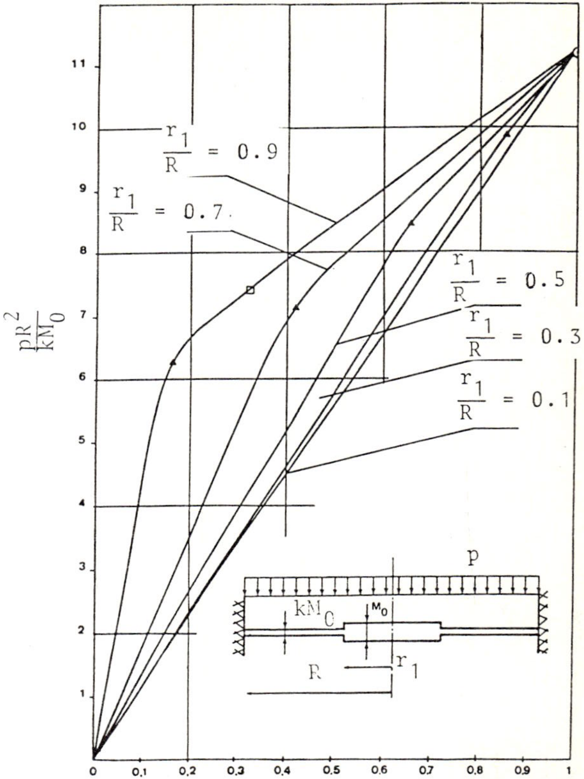

Fig. 4. Limit value of uniform load for a two-ring built-in plate with yield moments M_0 and $kM_0 < M_0$.

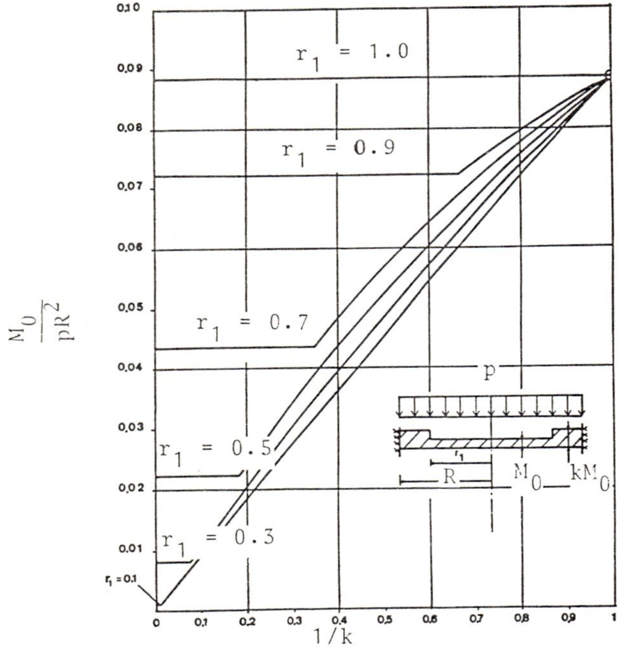

Fig. 5. Limit value of uniform load for a two-ring built-in plate with yield moments M_0 and $kM_0 > M_0$

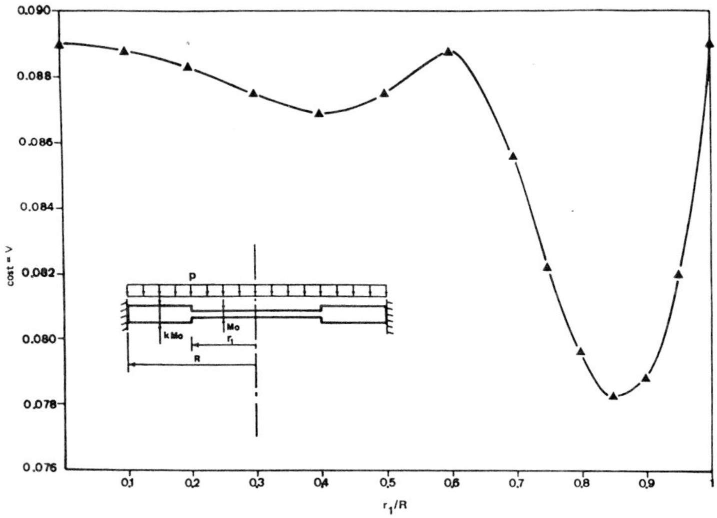

Fig. 6. Minimum values of nondimensional cost $V = \pi M_0 r_1^2 + (R^2 - r_1^2)k / \pi p R^4$, (with respect to k) plotted versus ratio r_1/R

When applied to plates regularly reinforced by equal rings resulting in yield moments as shown in fig. 7 and 8, the method enables comparison with the "smearing-out" technique which replaces the given plate by a so-called equivalent orthotropic plate with yield moments $M_r^\circ = M^\circ$ and $M_\theta^\circ = (k + 1)M^\circ/2$. This substitution corresponds to the case $n = \infty$. The conclusions are: a) for a built-in plate, the "smearing-out" procedure underestimates the limit load up to more than 50 %. To achieve less then 10 % error with $1 \leqslant k \leqslant 3$ it can be used only for n > 16; b) the situation is better for the simply supported plate where overestimation occurs as a rule and where the same accuracy of at least 90 % is already obtained for n even and not smaller than 4. Note that the inaccuracy of the "smearing-out" technique comes from the fact that it does not take the locations of the reinforcing rings into account but only their number and size: for example it makes no difference between the two plates of fig. 9.

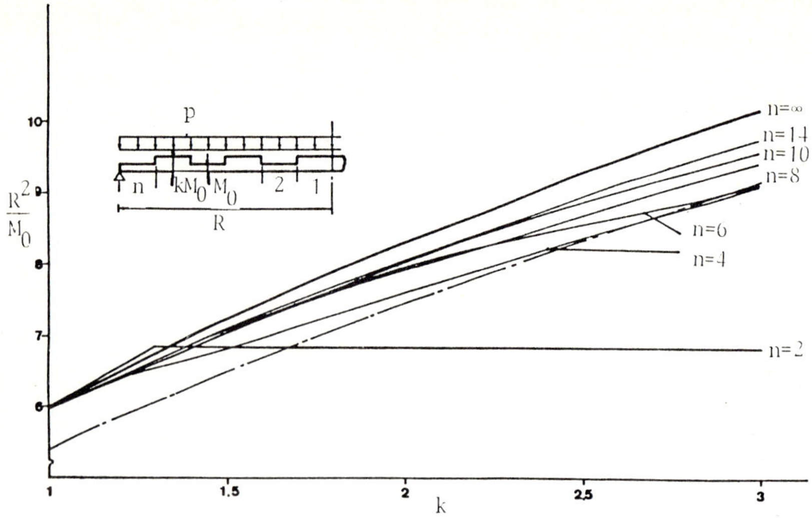

Fig.7. Limit value of uniform load p vs coefficient k of circumferential reinforcement for supported plates with radial yield moment M_0.

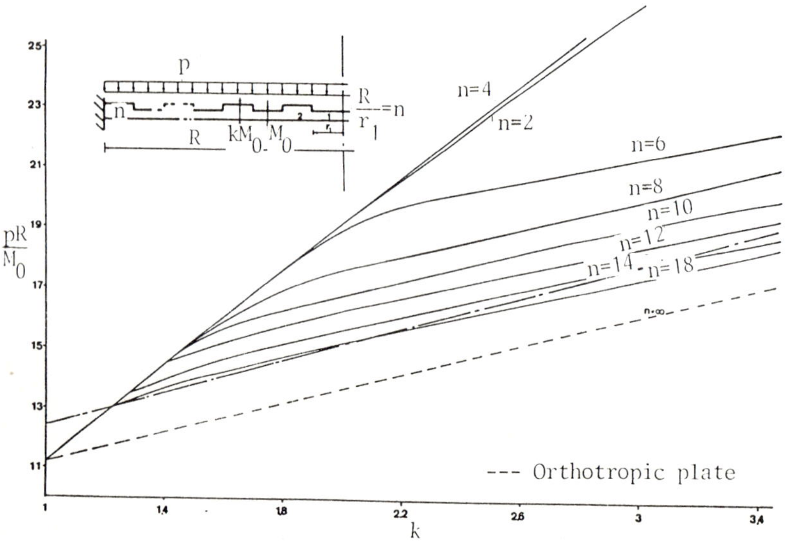

Fig.8. Limit value of uniform load p vs coefficient k of circumferential reinforcement for built-in plates with radial yield moment M_0.

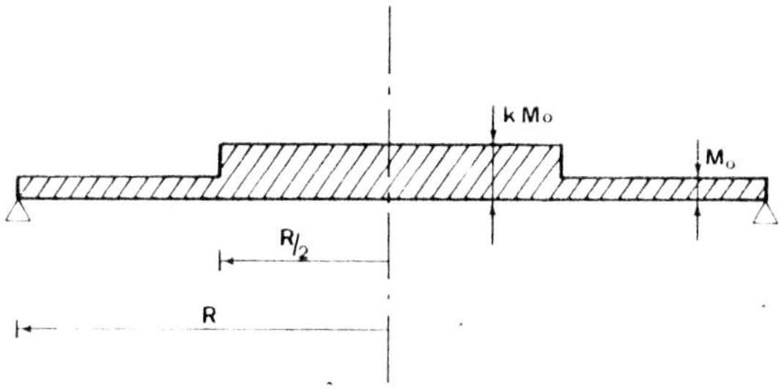

Fig. 9a.

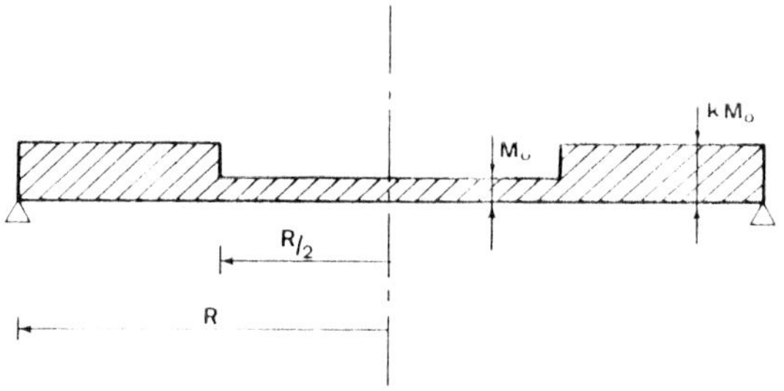

Fig. 9b.

We now turn to minimum-volume design of these plates with piecewise constant thickness, considering successively sandwich plates, in which the volume of the face sheets is to be minimized, and solid plates. For brevity we hereafter restrict ourselves to the discussion of two examples, though a variety of cases have been treated (see ref. {25} and {26}).
The optimality condition (7), when applied to structures with piecewise constant strength must be satisfied in average in each region over which the strength is constant.

The design procedure is as follows:

a) make physically reasonable assumptions on the stress regime (on the yield locus) for each ring, and on the relative values of the yield moments;

b) use the normality law to verify that the optimality condition can be satisfied; otherwise, modify the stress regimes;
c) use the final stress regimes in the equilibrium equation which, upon integration, gives the stress field. The boundary conditions finally enable the design to be found explicitly.

When the uniformly loaded, circular, sandwich plate is built in, the stress profile for any ring is formed of regimes AB or BC or both on the hexagon for that ring. In fig. 10 we show the yield hexagons of all rings superimposed on the same diagram. We assume that M_i^o decreases with r in a central region $0 < i < k$ and then increases, $k < i < n$. In a typical ring of the central region the stress profile goes with increasing r from A to some point between A and B, possibly reaching B. In the outer region it goes from B, or some point between B and C, to point C. In the intermediate ring k where the stress regime changes at radius r^* from AB to BC, three types of regimes can occur, as shown in fig. 10. Either a positive hinge circle occurs at the inner radius of the ring and a negative hinge circle

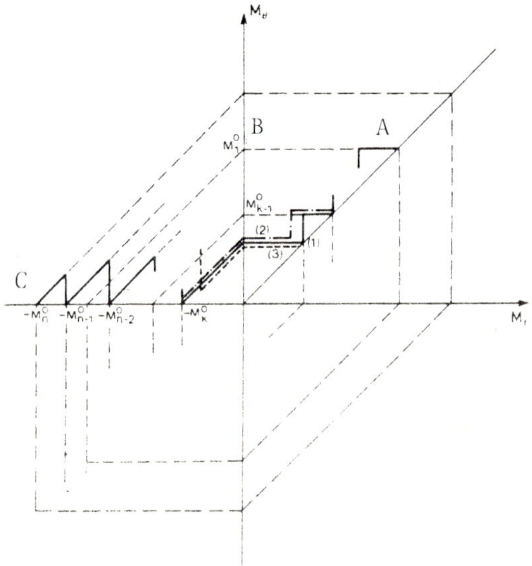

Fig. 10. Stress profile for optimal built-in plates with piecewise-constant yield moment. (——) hinge circle at both ends; (–.–.–)hinge circle in r_k; (---) hinge circle in r_{k+1}.

at the outer radius, or only one of these two hinge circles occurs. For each of these three cases an equation for r^* is obtained by combining equilibrium, continuity, and yield conditions. Since k is unknown, the solution is obtained by successive approximation by a numerical procedure implemented on an IBM 360 computer by Lamblin {25}.

For n = 1 (constant yield moment) we recover $M° = 0.089pR^2$ (Save and Massonnet {1}) with $C_c = 0.089\pi pR^4$. The design with continuously varying thickness and a = R will be obtained by letting n tend to infinity, giving the absolute minimum cost $C_a = 0.058\pi pR^4$. Results for values of i from 1 to 6 are given in Table 1. Optimal values of the dividing radii have been obtained numerically by a grid method. We see that the saving increases very rapidly from i = 2 to i = 5 and then has a tendency to slow down strongly. Caution must be exercised in the numerical search for the optimal radii, because several local minima can exist .

Table 1a.Optimal divison radii

Number of rings	r_1/R	r_2/R	r_3/R	r_4/R	r_5/R	Gain G (%)
1						0
2	0.860					12.10
3	0.816	0.915				15.86
4	0.490	0.801	0.910			19.82
5	0.498	0.774	0.860	0.934		21.94
6	0.411	0.549	0.769	0.858	0.932	23.41
∞						34.00

Table 1b. Optimal plastic yield moments

Number of rings	M_1°/pR^2	M_2°/pR^2	M_3°/pR^2	M_4°/pR^2	M_5°/pR^2	M_6°/pR^2
1	0.08900					
2	0.06565	0.11360				
3	0.05914	0.09150	0.12141			
4	0.08300	0.04300	0.07950	0.11190		
5	0.07945	0.03812	0.06627	0.09187	0.11587	
6	0.08971	0.06156	0.03240	0.06183	0.08769	0.11261

Lamblin and al. {26} give the solutions for various cases of loadings and boundary conditions for circular and annular plates. When an "edge effect" is expected, an upper bound M^+ is set on the yield moment. Obviously, M^+ must be larger than the yield moment M_c° of the plate with constant thickness, otherwise the plate could not support the load. On the other hand, for the condition $M^\circ \leqslant M^+$ to be relevant, the upper bound must be smaller than the maximum yield moment of the corresponding design without bound. Using the superscript $*$ to denote, for comparison, elements pertaining to a design without upper bound, we must have

$$M_c^\circ \leqslant M^+ \leqslant M_{max}^{\circ *} \quad . \tag{8}$$

Consider for example, an annular plate simply supported at the outer edge, free at the inner edge of radius a, and uniformly loaded. Assuming that Eq.(8) is satisfied and $M_{i+1} < M_i$, as in the absence of bounds, we choose the following stress profile (see fig. 11):

a) in the central ring and possibly in some neighboring rings $i = 1,\ldots,j$, where $M^\circ = M^+$, regime AB, from B to some intermediate point;
b) in the neighboring ring j+1, regime AB between two intermediate points;
c) in all other rings, regime AB from A to an intermediate point as in the simply supported circular plate.

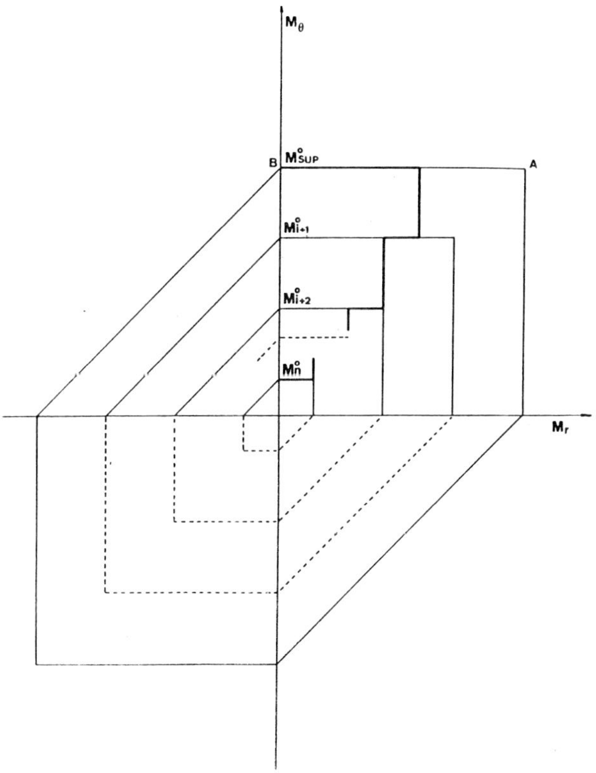

Fig. 11. Stress profiles for optimal supported annular plate.

We then show that this stress profile is compatible with the optimality condition. The design is next obtained as usual from equilibrium, continuity, and yield conditions. Because the number j of rings where $M^\circ = M^+$ is unknown, the solution is obtained numerically by successive approximations starting from j=1 and going up to the point where all conditions are satisfied. An example of results in the case of a plate made of four rings can be found in fig. 12.

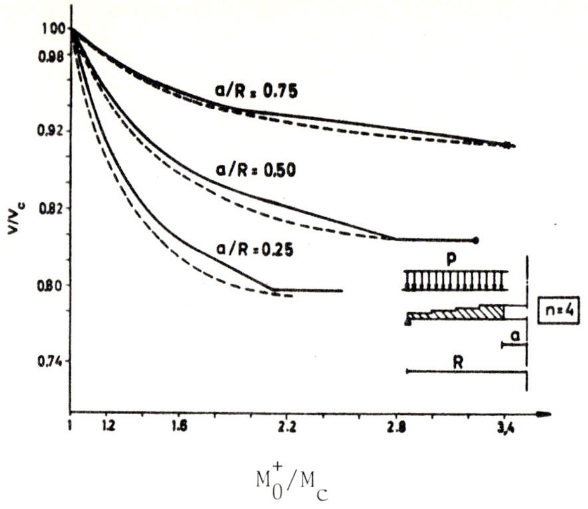

Fig. 12. Optimal supported plates with central hole. V is the plate volume, V_c is the volume of uniform thickness plate with the same limit load, M_0^+ is the upper bound on yield moment, and M_c is the yield moment of uniform thickness plate with the same limit load. (——) Equal rings. (---) Optimized rings.

3. Cylindrical shells

Both sandwich and solid cylindrical shells with piecewise constant thickness were studied by Kouam in his thesis {27}. He successively considered limit analysis and minimum-weight limit design with both the "limited interaction" , the hexagonal yield condition of Hodge and the sandwich Mises condition for cylinders without end load. Constant or linearily varying pressure was applied with various end conditions. For the simpler cases with a small number of different thicknesses (2 or 3), analytical computer-aided solutions were obtained following a procedure similar to that described in section 2 for circular plates. Linear and non linear mathematical programming were then used for larger number of rings. Each ring was divided into sub-rings regarded as finite elements. The method was to describe in each element either the radial velocity or the generalized stresses in polynomials of degree three, with suitable boundary conditions, and find the unknown coefficients by the mathematical programming procedure.

Some conclusions are worth pointing out:

a) for limit analysis, from a systematic variation of the number of rings, the number of elements per ring, the shell geometry and the boundary conditions, it was found that: 1) for a shell with uniform thickness, 4 elements are sufficient (the increase in accuracy is smaller than 1.16 % when using 6 elements;) 2) for a simply supported shell made of 2 to 5 rings, 2 elements per ring are enough; 3) for a built-in shell (at one or both ends) at least 3 elements per ring are necessary;

b) practical designs of vertical cylindrical tanks filled with liquid are classically made according to limitation of membrane stress, possibly empirically modified as indicated in the API 650 Code, or following the so-called Zick and Mc Grath procedure. Examples of such tanks made of 6 and 8 rings were analysed very easily and their safety with respect to plastic collapse were found to vary from 1.70 to 1.81, under the assumption of free upper edge and simply supported lower edge;

c) for optimal design, whereas a number of cases of sandwich shells were treated by various methods, only the symply supported uniformly loaded solid shell made of 3 rings was solved using the optimality criterion and writing an ad hoc program to solve the set of equations. Application to experimental models of short steel shells gave satisfactory comparison (less than 10 % difference).

4. Indications on some other problems

Cylindrical shells subjected to radial pressure often exhibit ring stiffeners. The treatment of these ring stiffeners has been as a rule done by the smearing-out technique which, as for stiffened plates, will give bad results in the case of a small number of strong stiffeners. In his thesis {28} and in subsequent papers with co-workers {29-30}, Guerlement has given computer-aided analytical solutions of the limit behavior of (a) simply supported cylindrical shells with one central stiffening ring, with and without axial load, and (b) infinitely long cylindrical shells with regularly spaced stiffening rings, with and without end-loads. The linearized Tresca yield condition is used, and due account is taken of the curvature of the stiffeners that are treated in bi-axial state of stress. Stiffeners of rectangular and cross-section are studied, the latter giving birth, by extension of the flanges, to the case of the double-layer shell.

At Liège University, Belgium, Nguyen Dang Hung and co-workers {32} {33} have developed the numerical approach to limit loads of axisymmetric shells using conical finite elements and non-linear programming.

Information on limit analysis and design of containment vessels can be found in the review paper {34} by the senior author, including indications on theoretical designs for minimum volume of reinforcement of concrete cylindrical tanks.

Finally, it is worth remarking that limit load evaluation can sometime prove useful even for plastic buckling {35} and for estimation of residual safety of cracked structures {36}.

5. Conclusions

It is firmly believed that limit analysis and design are and remain adequate for a number of practical situations where they can be of much value in providing safer and more economical structures, provided the range of applicability of the solutions have been evaluated experimentaly and good computing tools are proposed to the designer, both directions in which efforts should be pursued.

References

1. Save, M.A.; Massonnet, C.E.: Plastic analysis and design of plates, shells and disks (North-Holland, 1972).

2. Drucker, D.C.; Prager, W.; Greenberg, H.J.: Extended limit design theorem d for continuous media, Quart. Appl. Math. 9 (1952) 381-389.

3. Drucker, D.C.; Shield, R.T.: Design for minimum weight, Proc. 9th. Int. Congr. Appl. Mech., Brussels, Vol. 5 (1956) 212-222.

4. Shield, R.T.: Optimum design method for multiple loading. J. Appl. Math. Phys. (Zamp) 14(1963) 38-45.

5. Save, M.A.; Prager, W.: Minimum weight design of beams subjected to fixed and moving loads, J. Mech. Phys. Solids 11 (1963) 38-45.

6. Save, M.A.; Shield, R.T.: Minimum weight design of sandwich shells subjected to fixed and moving loads, Proc. 11th Int. Congr. Appl. Mech., Munich, 1964 (Springer, Berlin, Heidelberg, New York, 1966) pp. 341-349.

7. Lamblin, D.O; Save, M.A.: Minimum volume plastic design of beams for movable loads, Meccanica 6(3)(1971) 157-163.

8. Lamblin, D.O.: Minimum weight plastic design of continuous beams subjected to one single movable load, J. Struct. Mech. 1(1)(1972) 133-157.

9. Sacchi, G., Maier, G.; Save, M.A.: Limit design of frames for movable loads by linear programming, in: Optimization in Structural Design, IUATAM Symp. Warsaw, Poland, 1973, eds. A. Sawczuk and Z. Mroz (Springer, 1975).

10. Prager, W.; Shield, R.T.: The general theory of optimal plastic design, J. Appl. Mech. 34(1967) 184-186.

11. Save, M.A.: An unified formulation of the theory of optimal plastic design with convex cost functions, J. Struct. Mech. 1(2) (1972) 267-276.

12. Rozvany, G.I.N: Non-convex structural optimization problems, Proc. ASCE 99, EM1(1973) pp. 243-248.

13. Prager, W.: Introduction to structural optimization (Springer, Vienna, 1974).

14. Rozvany, G.I.N.: Optimal design of flexural systems (Pergamon Press, Oxford, 1976).

15. Prager, W.; Rozvany, G.I.N.: Optimization of structural geometry, in: Dynamical systems, Proc. Int. Symp., Florida, 1976, ed. Bednarek (Academic Press, New York, 1977) pp. 265-294.

16. Rozvany, G.I.N.; Hill, R.D.: The theory of optimal load transmission by flexure, Advan. Appl. Mech. 16(1976)183-306.

17. Rozvany, G.I.N.; Prager, W.: A new class of structural optimization problems: optimal archgrids, Computer Meth. Appl. Mech. Engrg. 19(1970) 127-150.

18. Munro, J.: Optimal plastic design of frames, in: Engineering Plasticity by Mathematical Programming, ch. 7, Proc. NATO Advanced Study Institute, Univ. of Waterloo, Waterloo, Ontario, Canada, August 2-12, 1977, eds. M.Z. Cohn and G. Maier, pp. 135-171.

19. Maier, G.; Srinivisan, R.; Save, M.A.: On limit design of frames using linear programming, Proc. Int. Symp. Computer-aided Structures. Design, Univ. of Warwick, 1972.

20. Smith, D.L.: Plastic limit analysis and synthesis of structures by linear programming, PhD Thesis, Univ. of London, 1974.

21. Zavelani-Rossi, A.: Optimal design of discretized continua, in: Engineering Plasticity by Mathematical Programming, ch. 10, Proc. NATO Advanced Study Institute, Univ. of Waterloo, Waterloo, Ontario, Canada, August 2-12, 1977, eds. M.Z. Cohn and G. Maier, pp. 223-1238.

22. Zavelani, A.; Maier, G.; Binda, L.: Shape optimization of plastic structures by zero-one programming, IUATAM Symp., Warsaw, Poland, 1973, eds. A. Sawczuk and Z. Mroz (Springer, 1975).

23. Zavelani, A.: A compact linear programming formulation for optimal design in plane stresses, J. Struct. Mech. 2 (4) (1973) 301-324.

24. Maier, G.; Zavelani, A.; Benedetti, D.; A finite element approach to optimal design of plastic structures in plane stress, Int. J. Num. Meth. Engrg. 4(1972) 455-473.

25. Lamblin, D.: Analyse et dimensionnement plastique de coût minimum de plaques circulaires (Thèse de Doctorat en Sciences appliquées, Faculté Polytechnique de Mons, 1975).

26. Lamblin, D.O.; Guerlement, G.; Save, M.A.: Solutions de dimensionnement plastique de volume minimal de plaques circulaires pleines et sandwhiches en présence de contraintes technologiques, Journal de Mécanique théorique et appliquée, Vol. 4, n° 4, 1985 (in press).

27. Kouam, M.: Contribution à l'analyse limite et au dimensionnement optimal des coques cylindriques. Thèse de Doctorat en Sciences appliquées, Faculté Polytechnique de Mons, Belgium, 1983.

28. Guerlement, G.: Contribution à l'analyse limite des coques cylindriques, Doctoral Thesis, Faculté Polytechnique de Mons, 1975.

29. Guerlement, G.; Lamblin, D.O.; Save, M.A.: Limit analysis of a cylindrical shell with reinforcing rings, to be published in Engineering Structures.

30. Cinquini, C.; Guerlement, G.; Lamblin, D.O.: Shell-stiffener interaction, Application to simply supported cylindrical shell under uniform pressure, to be published.

31. Cinquini, C.; Kouam, M.: Optimal plastic design of stiffened shells, Int. J. Sol. Struct., 19, 9, 773-783, 1983.

32. Nguyen Dang Hung: Sur la plasticité et le calcul des états limites par éléments finis. Thèse de docteur spécial, Université de Liège, 1984.

33. Dang Hung, N.; Trapletti, M.; Ransart, D.: Bornes quasi inférieures et bornes supérieures de la pression de ruine des coques de révolution par la méthode des éléments finis et par la programmation non-linéaire, Int. J. Non-Linear Mech. 13 (1978) 79-102.

34. Save, M.: Limit analysis and design of containment vessels. Nuclear Engineering and Design 79 (1984) 343-361.

35. Kemper, M.J.: Conf. on vessels under buckling conditions, I. Mech. E., Dec. 1972.

36. Goodall, I.W.; Griffiths, J.E.: On the limit analysis of a spherical pressure vessel with fully circumferential defects, Int. J. Mech. Sci. 24 (1982) 635-645.

Analysis of the Finite Deflection of Visco-plastic Plates Using the Finite System Method

A. Baltov

Institute of Mechanics and Biomechanics
Bulgarian Academy of Sciences, Sofia, Bulgaria.

Summary

A method for the determination of finite deflections of flexible elasto-visco-plastic plates under dynamic impulsive loading is given. The energetic visco-plastic model is adopted for the description of the material behaviour. Velocities and stress resultants are determined by using the Finite System Method (FSM). The results of the calculations, concerning square and circular plates, are presented as an illustration.

Statement of the Problem

Flexible metal plates, under impulsive loading, are considered. The load intensity is taken to be such, that inelastic strain rates in a range $1\text{-}10^5\ s^{-1}$ develop in the plate. The deflections are of the same order as the plate thickness and the elastic strain is not to be neglected. The impulsive load is taken as a transient regime, so that no shock waves occur in the plate. The aim of the present study is to propose a method for determination of deflections (including residual deflection as well) when the plate undergoes dynamic inelastic deformation. One might come across such a problem in the cases of impulsively loading structures or during dynamic forming processes. The dynamic behaviour of inelastic plates have been studied in numerous papers (see refs. [1], [2], [3] etc.) The specific aim of the present paper is:

(a) to use the energetic visco-plastic model to describe the material behaviour of the plate [4];

(b) to use the Finite System Method [5] to solve the geometrically and physically nonlinear problems under consideration.

(c) to consider the elastic part of the strain together with the visco-plastic one.

Inelastic Behaviour of Plates and Shells
IUTAM Symposium Rio de Janeiro 1985
Editors: L. Bevilacqua, R. Feijóo and R. Valid

Model of the Flexible Plate

Taking into account the range of the deflections, the Von Karman's theory of flexible plates is adopted [6]. A material description of the deformation process is assumed. The curvilinear orthogonal material coordinate system OX^k, $(K = 1,2,3)$ is introduced. X^α, $(\alpha = 1,2)$ are the coordinates of the points in the middle plane of the plate (before deformation) and the axis OX^3 is normal to that plane. The plate is loaded dynamically by $P^3(X^\alpha,t)$. The following material variables are used:

(1) Displacements of the middle plane points $U^K(X^\alpha,t)$ and its velocities $V^k = \dot{U}^K$. The dot denotes time differentiation;

(2) Material strain tensor $\varepsilon_{KL}(X^M,t)$ and its values at points (X^α): $\overset{o}{\varepsilon}_{\alpha\beta}(X^\gamma,t),(\alpha,\beta,\gamma = 1,2)$. The strain rate tensors: $\dot{\overset{o}{\varepsilon}}_{\alpha\beta}$, $\dot{\varepsilon}_{KL}$;

(3) Measure of the change of middle surface curvature of the plate during deformation: $\chi_{\alpha\beta}$ and its velocity $\dot{\chi}_{\alpha\beta}$;

(4) Second Piola-Kirhoff stress tensor $\Sigma^{KL}(X^M,t)$;

(5) Stress resultants in the plate:

$$N^{\alpha\beta}(X^\gamma,t) = \int_{-H}^{H} \Sigma^{\alpha\beta} dX^3, \quad M^{\alpha\beta}(X^\gamma,t) = \int_{-H}^{H} \Sigma^{\alpha\beta} X^3 dX^3 \tag{1}$$

where $2H = \text{const}$ is the plate thickness.

Following Von Karman's approach we introduce:

a) Geometrical equations in the form:

$$\varepsilon_{\alpha\beta} = \overset{o}{\varepsilon}_{\alpha\beta} + X^3 . \chi_{\alpha\beta}, \quad (\alpha,\beta = 1,2) \tag{2}$$

where

$$\chi_{\alpha\beta} = -U^3_{|\alpha\beta}, \quad \overset{o}{\varepsilon}_{\alpha\beta} = \frac{1}{2}(U^\alpha_{|\beta} + U^\beta_{|\alpha}) + \frac{1}{2} U^3_{|\alpha} U^3_{|\beta} \tag{3}$$

The vertical line denotes covariant differentiation with respect to X^α.

Regarding dynamic processes, it is convenient to use the geometrical equations, expressed in terms of velocities:

$$\dot{\varepsilon}_{\alpha\beta} = \dot{\overset{o}{\varepsilon}}_{\alpha\beta} + X^3 . \dot{\chi}_{\alpha\beta} \tag{4}$$

$$\dot{\varepsilon}^{o}_{\alpha\beta} = \frac{1}{2}(V^{\alpha}_{|\beta} + V^{\beta}_{|\alpha}) + U^{3}_{|\alpha}V^{3}_{|\beta} + V^{3}_{|\alpha}U^{3}_{|\beta},$$

$$\chi_{\alpha\beta} = -V^{3}_{|\alpha\beta}. \tag{4}$$

b) Equations of motion:

$$\bar{\rho}\dot{V}^{\alpha} = N^{\alpha\beta}_{|\beta}$$

$$\bar{\rho}\dot{V}^{3} = M^{\alpha\beta}_{|\beta\alpha} + P^{3} - N^{\alpha\beta}\chi_{\beta\alpha}, \tag{5}$$

where $\bar{\rho}$ is density, assigned to a middle plane unit area.

Material Model

Since the plate deflections are finite, but the strains are small, mechano-mathematical models for infinitesimal deformations might be used. We adopt the energetic visco-plastic model of the body material behaviour [4]. The model is based on the following assumptions:

(a) Deformation consists of two parts: elastic and visco-plastic, i.e. $\varepsilon_{KL} = \varepsilon^{(e)}_{KL} + \varepsilon^{(a)}_{KL}$;

(b) The elastic strain tensor is related to the stress tensor by means of the Hooke's law;

(c) Material is inelastically incompressible, i.e. $\varepsilon^{(a)}_{KK} = 0$;

(d) Visco-plastic strain rate depends on the surpassing of an energetic measure over the corresponding energetic barrier. This barrier corresponds to the energetic state of the infinitesimal neighbourhood of a material point, when the strain rate intensity reaches the static value β_s. β_s is a limit rate, over which the material expresses a strain rate sensitivity. The main idea of the model formulation is based on the thermo-fluctuation micro-mechanism of the visco-plastic deformation, according to Boltzman's statistical law [7]. In the case of isotropic elastically deformed material the energetic measure reads:

$$Z_f = \frac{1}{2\mu}(\Sigma_o)^2 + \Sigma_o . \ni^{(a)}_o, \tag{6}$$

where

$$\Sigma_o = \sqrt{\tfrac{1}{2}\, S^{KL} S_{LK}}\,, \quad S^{KL} = \Sigma^{KL} - \tfrac{1}{3}\,\delta^{KL}\Sigma^{MM} \tag{7}$$

μ is the elastic constant and δ^{KL} is the Kronecker symbol. The inelastic deformation condition is assumed to be of Mises type:

$$\Sigma_{eq} = \Sigma_p \tag{8}$$

where

$$\Sigma_p = \Sigma_p(\ni_o^{(a)}, \dot{\ni}_o), \quad \dot{\ni}_o = \sqrt{\tfrac{2}{3}\dot{\ni}_{KL} \cdot \dot{\ni}^{LK}}\,,$$

$$\dot{\ni}_{KL} = \dot{\varepsilon}_{KL} - \tfrac{1}{3}\delta_{KL}\,\dot{\varepsilon}_{MM}, \quad \Sigma_{eq} = \Sigma_o . \mathrm{sign}(\Sigma_p). \tag{9}$$

A dimensionless parameter is introduced:

$$\mathfrak{z} = \frac{Z_f}{Z_{fs}} - 1 = \sqrt{\frac{(\Sigma_o)^2 + 2\mu\Sigma_o . \ni_o^{(a)}}{(\Sigma_s)^2}} - 1, \tag{10}$$

where the energetic barrier Z_{fs} is:

$$Z_{fs} = \frac{1}{2\mu}(\Sigma_s)^2, \quad \Sigma_s = \Sigma_p(\ni_o^{(a)}, \beta_s). \tag{11}$$

Visco-plastic strain rate tensor is expressed as follows:

$$\dot{\varepsilon}_{KL}^{(a)} = \begin{cases} 0, & \Sigma_{eq} < \Sigma_s \\ \Lambda \exp\{D\mathfrak{z}\} S_{KL}/\Sigma_o, & \Sigma_{eq} \geqslant \Sigma_s \end{cases} \tag{12}$$

The material functions $\Lambda(\ni_o^{(a)})$, $D(\ni_o^{(a)}, \dot{\ni}_o)$ and $\Sigma_p(\ni_o^{(a)}, \dot{\ni}_o)$ are to be determined by performing uniaxial dynamic tensile tests [7]. According to computational procedure of FSM, it is convenient to express the constitutive equations in an equivalent form, involving deformation measures:

$$\dot{\varepsilon}_{KL}^{(a)} = \begin{cases} 0, & \Gamma < \ni_s, \\ \Lambda \exp\{D\mathfrak{z}\}(\ni_{KL} - \varepsilon_{KL}^{(a)})/\Gamma, & \Gamma \geqslant \ni_s, \end{cases} \tag{13}$$

where

$$\ni_{KL} = \varepsilon_{KL} - \tfrac{1}{3}\delta_{KL}\,\varepsilon_{MM}, \quad \mathfrak{z} = \sqrt{\frac{(\ni_o)^2 - (\ni_o^{(a)})^2}{(\ni_s)^2}} - 1$$

$$\Gamma = \sqrt{(\ni_o)^2 + (\ni_o^{(a)})^2 - \Delta}, \quad \Delta = \ni_o \cdot \ni_o^{(a)},$$

$$\ni_s = \frac{1}{2\mu} \Sigma_s . \tag{14}$$

The constitutive relations, written in terms of stress resultants-strain in the middle plane are as follows:

$$N^{\alpha\beta} = K^{\alpha\beta\gamma\delta}(\varepsilon^o_{\gamma\delta} - \varepsilon^{o(a)}_{\gamma\delta}),$$
$$M^{\alpha\beta} = D^{\alpha\beta\gamma\delta}(\chi_{\gamma\delta} - \chi^{(a)}_{\gamma\delta}), \tag{15}$$

where $K^{\alpha\beta\gamma\delta}$ and $D^{\alpha\beta\gamma\delta}$ are the tensors of elastic stiffnesses [8]. $\chi^{(a)}_{\alpha\beta}$ is the measure of the inelastic change of the middle surface curvature during deformation:

$$\chi^{(a)}_{\alpha\beta} = \frac{3}{2H^3} \int_{-H}^{H} \Sigma^{(a)}_{\alpha\beta} \; x^3 dx^3 \tag{16}$$

We assume that the residual inelastic deflection $U^{(a)3}$ can be determined by using the equation $\chi^{(a)}_{\alpha\beta} = -U^{(a)3}_{|\alpha\beta}$.

Application of the FSM

The FSM, proposed by the author [5], has been successfully applied to the analysis of the dynamic behaviour of structural elements ([8], [9] etc.). The FSM is also suitable to solve the problems, formulated in the present paper. The main idea of the method consists of treating the body as a cybernetic system, that consists of finite number (after discretization) of interacting subsystems. The FSM is applied according to the algorithm, presented in Fig. 1. The process is discretized in time with step $\Delta\tau$. The middle plane is covered by mesh with steps Δa^1 and Δa^2. The differential operators are substituted by other ones. The iteration procedure, presented in Fig. 1, is realized at each time step in each mesh point. The input parameters at the first stage of the subsystem action are $\varepsilon^o_{\alpha\beta}$ and $\chi_{\alpha\beta}$, while the corresponding output parameters are $N^{\alpha\beta}$ and $M^{\alpha\beta}$. The inelastic deformation measures $\varepsilon^{(a)}_{\alpha\beta}$ and $\chi^{(a)}_{\alpha\beta}$ describe the subsystem state. Equations (13) and (16) give the change of

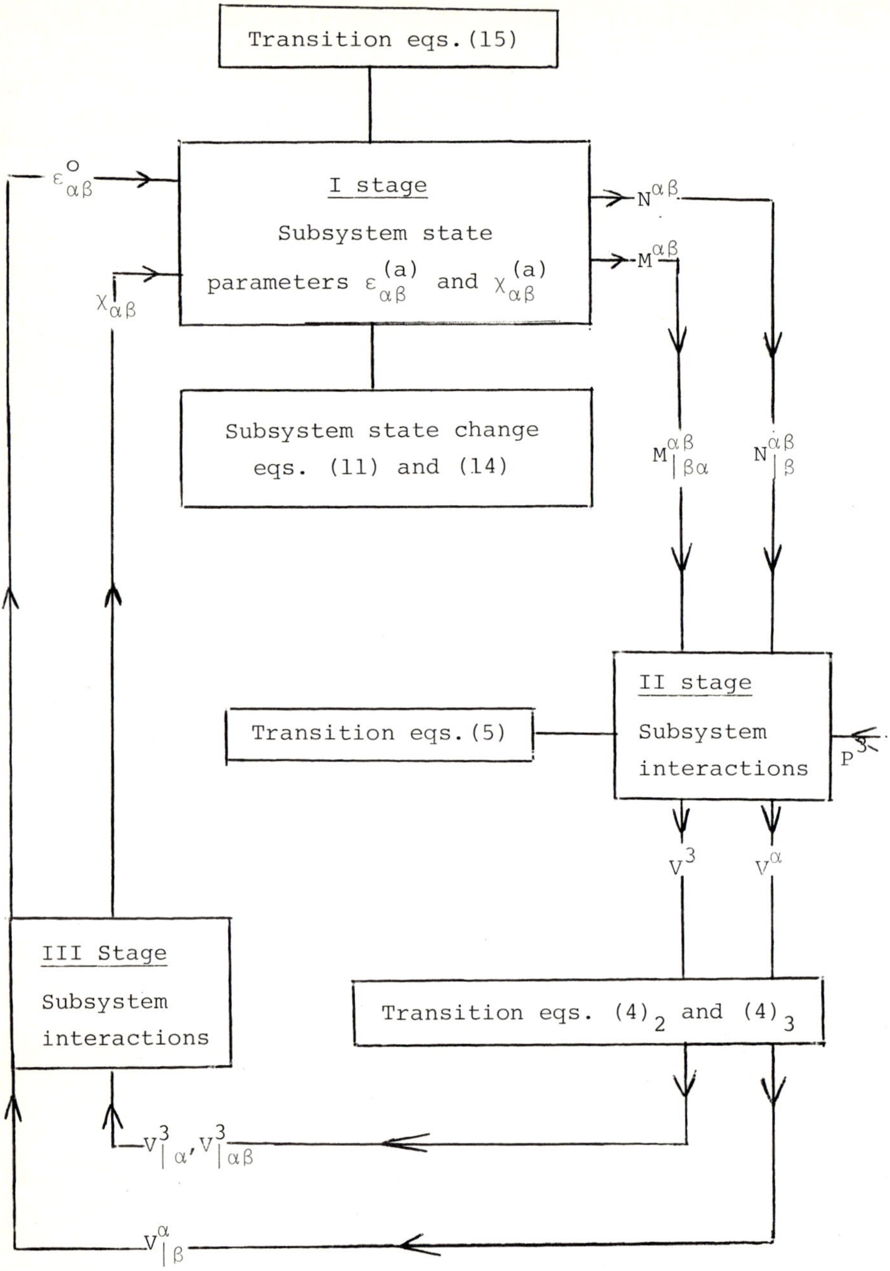
Transition eqs.(15)
I stage
Subsystem state
parameters $\varepsilon_{\alpha\beta}^{(a)}$ and $\chi_{\alpha\beta}^{(a)}$
$\varepsilon_{\alpha\beta}^{o}$
$\chi_{\alpha\beta}$
$N^{\alpha\beta}$
$M^{\alpha\beta}$
Subsystem state change
eqs. (11) and (14)
$M^{\alpha\beta}_{|\beta\alpha}$
$N^{\alpha\beta}_{|\beta}$
II stage
Subsystem
interactions
P^3
Transition eqs.(5)
V^3
V^α
III Stage
Subsystem
interactions
Transition eqs. $(4)_2$ and $(4)_3$
$V^3_{|\alpha}, V^3_{|\alpha\beta}$
$V^\alpha_{|\beta}$

Figure 1

the subsystem state. The output-input transition is described by means of the constitutive equations (15). The second stage is characterized by interaction between the neighbouring subsystems. The input parameters now are $N^{\alpha\beta}_{|\beta}$ and $M^{\alpha\beta}_{|\beta\alpha}$. The corresponding output parameters are V^{α} and V^{3} and the transition equations are given by expressions (5).

The load P^3 stands for an external input parameter. Infinitesimally neighbouring subsystems interact again at the third stage. The input parameters are $V^{\alpha}_{|\beta}$, $V^{3}_{|\alpha}$ and $V^{3}_{|\alpha\beta}$. The output parameters are the input parameters of the first stage, e.g. $\overset{o}{\varepsilon}_{\alpha\beta}$ and $\chi_{\alpha\beta}$. Equations $(4)_2$ and $(4)_3$ are the transition equations. Quantities $U^{3}_{|\alpha}$, appearing in these equations, are determined by the integration in time of $V^{3}_{|\alpha}$. The subsystem action is closed in a cycle, which allows to organize the iteration procedure. Displacements are determined through velocities and the residual deflection - through the inelastic curvature changes. The mesh steps and the time step obey some restrictions to provide the convergence of the iteration process [8]. However, the FSM possesses some additional advantages - the possibility to change the material model easily because of the change only to the corresponding block in the computing programme. This allows for an unification, regarding a set of material models.

Numerical Examples - Discussion.

The method, proposed for the determination of velocities, stress resultants and deflections of flexible elasto-visco-plastic plates, undergoing impulsive loading, is illustrated by solving some characteristic problems:

A. Square plate, loaded by uniformly distributed load:

$$\dot{P}^3(t) = \begin{cases} \frac{1}{2}P[1-\mathrm{Cos}(\frac{\pi}{5\Delta t}t)], & t \in [0,5\Delta\tau], \\ P, & t \in [5\Delta\tau, T]. \end{cases} \qquad (17)$$

Material is aluminium with material characteristics: $E = 7.10^{10}$ N/m^2; $\nu = 0,36$; $\bar{\rho} = 2700\ Ns^2/m^4$; $\Lambda = 0,93354 + 10,71$ $\ni^{(a)}_o - 440,24(\ni^{(a)}_o)^2$; $\ni_s = 0,014 + 5,3\ \ni^{(a)}_o - 0,46.10^3$

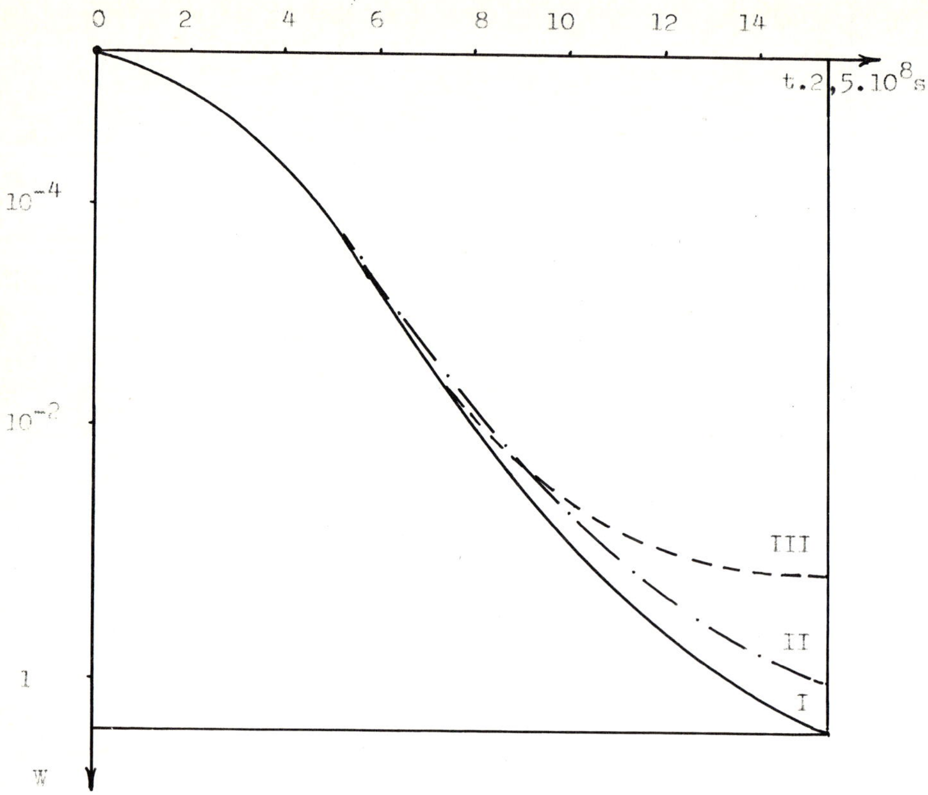

Fig. 2: Maximum deflection at the centre of the plate.

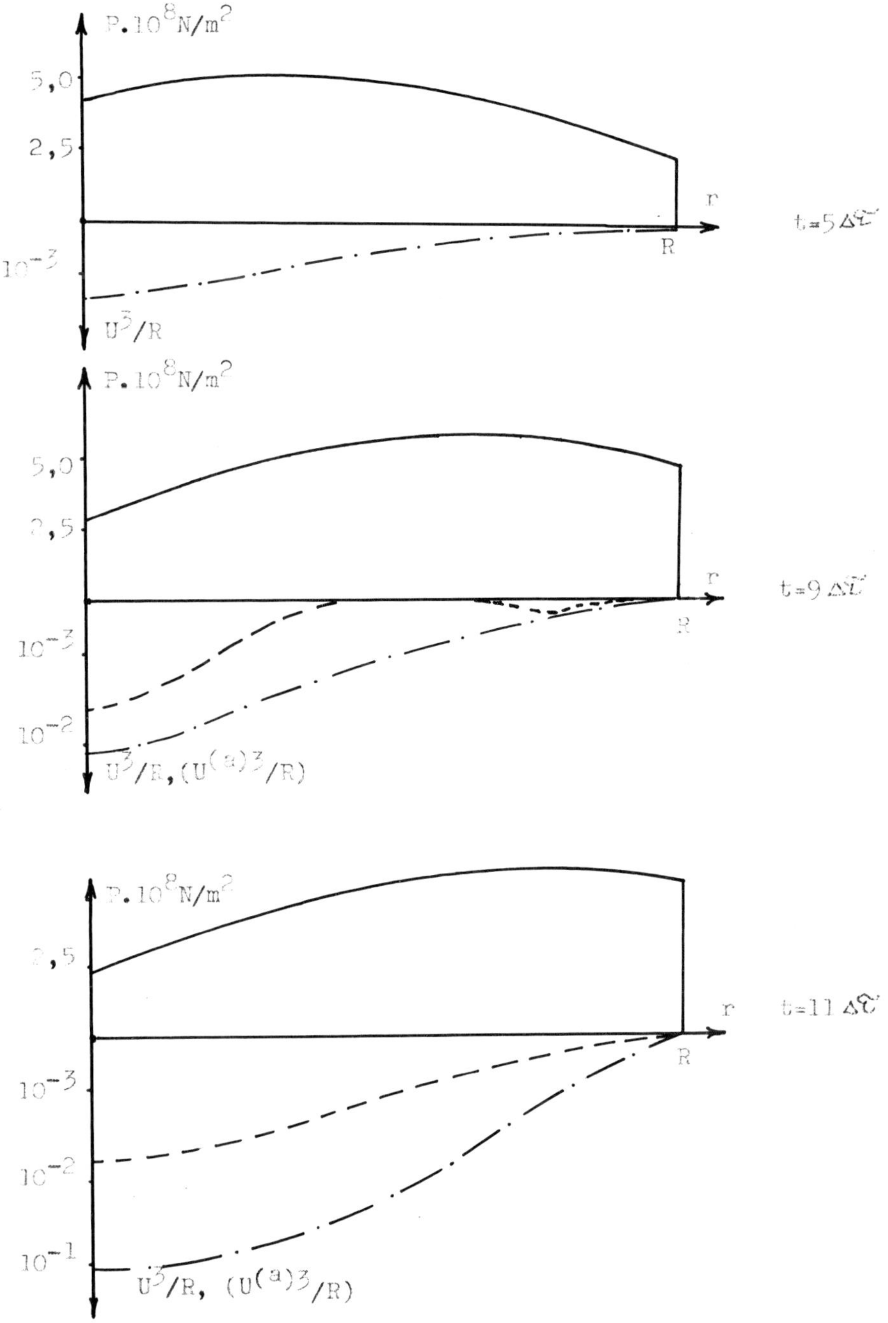

Fig. 3: Total residual deflections and load distribution for different times.

$(\ni_o^{(a)})^2 + 1{,}4.10^4(\ni_o^{(a)})^3$; $D = 8{,}3 - 338\ni_o^{(a)} + 7960(\ni_o^{(a)})^2$.

The plate sizes are: $H = 0{,}002$ m; $a = 0{,}8$ m; Load is equal to $P = 2.10^8$ N/m^2. The discretization is performed with steps: $\Delta\tau = 10^{-5}$s.; $\Delta a = a/5$. The time variation of the maximum deflection $W = 2U^3/a$ in the center of the plate is given in Fig. 2. The three characteristic cases are considered: (I) geometrically non-linear visco-plastic case; (II) geometrically linear visco-plastic case; (III) geometrically nonlinear elastic case. The analysis of the solution shows that: (a) the FSM stands effective for the cases under consideration: (b) the elastic strains are comparable with the visco-plastic strains; (c) the geometrical non-linearity is to be taken into account.

B. Circular plate, loaded by variable loading:

$$P^3(\tau,t) = \begin{cases} \frac{1}{2}P\{1-\mathrm{Cos}[\frac{\pi}{4\Delta\tau}(t-\psi_t)]\}, & t \in [\psi_t, \psi_t+4\Delta\tau], \\ P.\exp\{\frac{\psi_t+4\Delta t-t}{t_a}\}, & t \in [\psi_t+4\Delta\tau, T] \end{cases}$$

$$\psi_t = -\frac{L_o}{V_d} + \frac{1}{V_d}\sqrt{L_o^2+\tau^2}, \quad \tau \in [0,R] \tag{18}$$

Material is aluminium with characteristics given in the previous case A. Load parameters are: $P = 5.10^8$N/m^2; $L_o = 0{,}04$,; $V_d = 0{,}006$ m/μs; $t_a = 10$μs. Discretization steps are: $\Delta\tau = 0{,}1R$; $\Delta\tau = 5.10^{-2}$μs. The plate sizes are: $R = 0{,}264$ m; $H = 0{,}01$ m. The load distribution, as well as the total and residual deflections at the moments: $t = 5\Delta\tau$, $t = 9\Delta\tau$, $t = 11\Delta\tau$, are given in Fig. 3. The residual deflection determination permits the use of the solving method in the cases of dynamical forming of the plate elements.

References

1. Jones, N.: Recent progress in the dynamic plastic behaviour of structures. Shock and Vibr. Digest 10 (1978) 21-33.

2. Symonds, P.S.: Dynamics of non-elastic structures, Moskow 1982 (in Russian).

3. Wierzbicki, T.: Dynamika plyt i powlok in Zastosowania lepkoplastycznosci. Warszawa Osolineum 1971, 229-295.

4. Baltov, A.: Application of an energetic measure in viscoplasticity ZAMM 64(1983) 108-110.

5. Baltov, A.: Application of mathematical system theory dynamical thermoplasticity ZAMM 58(1978) 195-197.

6. Fung, Y.E.: Foundation of solid mechanics, N.Y. 1965.

7. Baltov, A.: On the energetic conception in viscoplasticity Theor. Appl. Mech. 15 (1984) 28-34 (in Bulgarian).

8. Baltov, A.: Application of the Finite System Method in dynamics of structures. Theor. Appl. Mech. 13 (1982) 60-67 (in Bulgarian).

9. Baltov, A.; Minchev, Orl.: Visco.plastic flexible rectangular plates under dynamic loading, in Proceed. X Int. Conf. on Nonlinear Osc. Varna 1984, 536-539.

On a Nonlocal Biaxial Strength Criterion for Concrete and its Application to Ultimate Load Analysis of RC Shells by the FEM

H.A. MANG, J. EBERHARDSTEINER, H. WALTER

Institute of Strength of Materials
Technical University of Vienna, Vienna, Austria

Summary

A change of the mechanical state of a point of a body such as the transition of the material from the elastic to the plastic state or the initiation of a crack usually also depends on the mechanical states of points in the neighborhood of the considered point. This is referred to as a "nonlocal dependence". As a first approximation of this unprecise notion, the dependence of the change of the mechanical state of the point on the spatial gradients of local strains and stresses may be determined. The aim of this paper is to report on a nonlocal biaxial strength criterion for concrete and on its application to ultimate load analysis of reinforced concrete (RC) shells by the finite element method (FEM). It is based on the assumption that the biaxial strength of concrete increases with increasing gradient of the strain energy density in the considered point of the shell. In the numerical study it is shown that the influence of the nonlocal character of the proposed biaxial strength criterion on the global response of two different RC shells is rather small.

Background of Problem

It is well known that a change of the mechanical state of a point of a body usually does not only depend on the considered instant of time but also on prior states of this point. By "change of the mechanical state", e.g., the transition of the material from the elastic to the plastic state or the initiation of a crack is meant. The term "prior states" refers to a temporal neighborhood, e.g., to a finite interval in time before the considered instant.

Lacking a precise definition, the notion "temporal neighborhood" does not represent a useful basis for quantifying the dependence of the change of the mechanical state of a point on prior states of this point. However, as a first approximation of this quantity, the dependence of the change of the mechanical state of this point on the time gradients of mechanical quantities defining the state of the point at the considered instant

Inelastic Behaviour of Plates and Shells
IUTAM Symposium Rio de Janeiro 1985
Editors: L. Bevilacqua, R. Feijóo and R. Valid

of time may be determined. The dependence of the yield limit on the rate of loading and deformation, representing an important item of the theory of viscoplasticity, is an example for such a first approximation.

It seems to be less well known that a change of the mechanical state of a point of a body usually also depends on the mechanical states of points in the neighborhood of the considered point. The phenomenon of the so-called "overelasticity" [1],[2],[3] is an example for this nonlocal dependence. The notion "nonlocal dependence" means that this dependence is not restricted to the considered point. It is characterized by an increase of the yield limit for the case of a nonuniform stress distribution as compared to the yield limit for a uniform stress distribution.

Campus [1] reported on experimental evidence of overelasticity. He tested steel beams of rectangular cross-section, subjected to pure bending. The experiments showed that the stress gradient leads to an increase of the yield limit. Analogous to the role of the time gradients of mechanical quantities in the theory of viscoplasticity, the spatial gradients of these quantities may be used to determine a first approximation of the dependence of a change of the physical state of a point of a body on the spatial neighborhood of this point.

König and Olszak [4] presented a theory of overelasticity, characterized by the following general form of the yield condition:

$$f(\sigma_{ij}, \varepsilon_{ij}, \sigma_{ij,k}, \varepsilon_{ij,k}) = 0 \qquad (1)$$

where $\sigma_{ij,k}$ is the gradient of the stress tensor σ_{ij} and $\varepsilon_{ij,k}$ is the gradient of the strain tensor ε_{ij}. König and Olszak applied their theory to the problem of pure bending of a rectangular beam corresponding to the experiments by Comptu. Unfortunately, experimental investigations concerning overelasticity in the context of multiaxial states of stress do not exist. Nevertheless, the application of a multiaxial theory of overelasticity to a multiaxial stress problem would have been important in its own right.

Analogous to the yield limit of metals, the uniaxial tensile strength of concrete depends on the mechanical states of points in the neighborhood of the considered point. There is much experimental evidence of this nonlocal dependence.

Evaluating a large number of test results from the literature, Hellmann [5] presented the following expression for the modulus of rupture (bending tensile strength) of concrete of beams of 10 cm depth, $f_{btu}^{(10)}$:

$$f_{btu}^{(10)} = c_{bt}(f_{cu}')^{\frac{2}{3}} \tag{2}$$

where f_{cu}' is the cube strength of concrete and $c_{bt} = 0.98$ is a proportionality factor. Hellmann also presented a similar formula for the direct tensile strength of concrete, f_{tu}:

$$f_{tu} = c_t(f_{cu}')^{\frac{2}{3}} \tag{3}$$

where $c_t = 0.52$ is a proportionality factor. Equations (2) and (3) represent coarse approximations of the true situation. From these equations it follows that

$$\frac{f_{btu}^{(10)}}{f_{tu}} = \frac{c_{bt}}{c_t} = 1.88. \tag{4}$$

Thus, analogous to the increase of the yield limit of metals, the stress gradient yields an increase of the uniaxial tensile strength of concrete (Fig. 1(a)). This observation is corroborated by the $f_{btu}/f_{btu}^{(10)}$ - h diagram (Fig. 1(b)) presented by Mayer [6] who has evaluated a large number of test results from the literature. This diagram shows that the modulus of rupture, f_{btu}, increases with decreasing depth of the beam, h. Since it is justified to assume the validity of the Navier stress distribution at incipient cracking of concrete, a decrease of h is associated with an increase of the stress gradient.

Statement of Problem

For ultimate load analysis of plates and shells made of reinforced concrete (RC), knowledge of the biaxial strength of concrete is necessary. Fig. 2(a) shows a frequently employed fracture envelope representing the geometric locus of the biaxial fracture strength $\sigma_{1u}(\alpha_1)$, $\sigma_{2u}(\alpha_1)$, where $\alpha_1 = \sigma_2/\sigma_1$ is kept constant as the uniformly distributed principal stresses σ_1, σ_2 are increased up to σ_{1u}, σ_{2u}. This fracture envelope was obtained by Kupfer [7] using plates of 20 cm length and width and 5 cm thickness as test specimens. Because of the uniform distributions of σ_1 and σ_2, the gradients of stress and strain are zero. Unfortunately, experimental investigations concerning the dependence of σ_{1u} and σ_{2u} on the gradients of stress and strain do not exist.

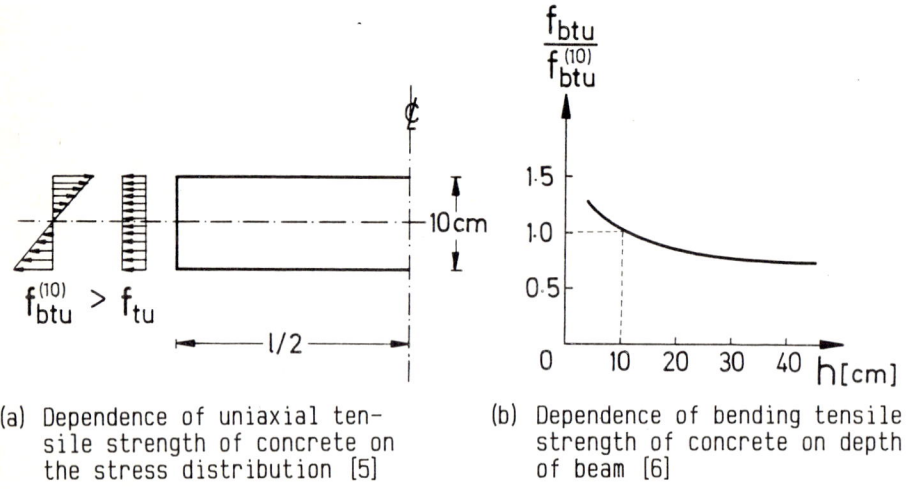

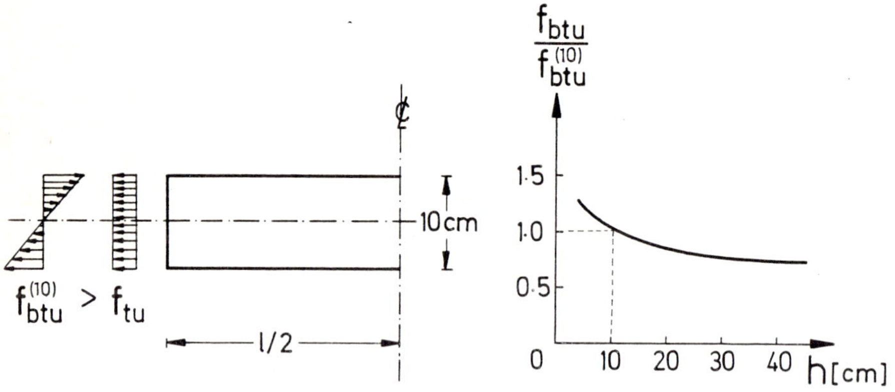

(a) Dependence of uniaxial tensile strength of concrete on the stress distribution [5]

(b) Dependence of bending tensile strength of concrete on depth of beam [6]

Fig.1. Dependence of uniaxial tensile strength of concrete on the stress gradient

Concerning Kupfer's fracture envelope, for $1 \leqq \alpha_2 = 1/\alpha_1 \leqq \infty$ (biaxial-tension domain) and $-\infty \leqq \alpha_2 \leqq \alpha_{2L}$ (part of tension-compression domain) with $\alpha_{2L} = f(f_{cu}/f_{tu})$, where f_{cu} is the prism strength of concrete, it is assumed that failure of concrete occurs by cracking of concrete normal to the direction of σ_1. Cracked concrete retains its capacity to carry internal forces in the direction parallel to the cracks so long as $f_{cu} < \sigma_2 < f_{tu}$. Because of tension stiffening, activated by bond-slip between the reinforcement and the surrounding concrete, the latter also retains part of its strength normal to the cracks. For $\alpha_{2L} < \alpha_2 \leqq 1$ (part of tension-compression domain, biaxial compression domain), it is assumed that concrete fails by crushing, thereby losing all of its strength. The concept of a sudden change of the fracture mode of concrete for a certain ratio of principal stresses ($\alpha_2 = \alpha_{2L}$) represents a coarse simplification of the true physical situation. This concept facilitates the mathematical analysis.

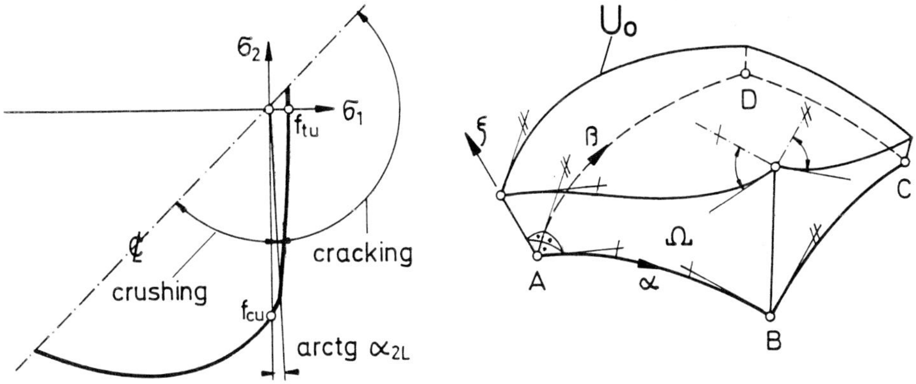

(a) Fracture envelope based on uniform distribution of σ_1 and σ_2 [7]

(b) Strain energy density U_0 in the points of the middle surface Ω of a shell

Fig.2. On the dependence of the biaxial strength of concrete on the gradient of the strain energy density

The aim of the present paper is to report on a nonlocal biaxial strength criterion for concrete and on its application to ultimate load analysis of RC shells by the finite element method (FEM). It is based on the assumption that the biaxial strength of concrete increases with increasing gradient of the strain energy density U_0 in the considered point of the shell. This assumption is consistent with the experimentally observed fact that the uniaxial tensile strength of concrete increases with increasing gradient of stress.

For the limiting case of $\nabla U_0 = \mathbf{0}$, the biaxial strength of concrete is taken from Kupfer's failure envelope. Point A in Fig. 2(b) refers to this limiting case, providing $\partial U_0/\partial\zeta = 0$. However, e.g., at point B in Fig. 2(b) $\nabla U_0 \neq \mathbf{0}$, leading to an increase of σ_{iu}, i=1,2. Thus, the proposed strength criterion for concrete can be formulated mathematically as

$$\sigma_i \geq \sigma^*_{iu} = k\sigma_{iu} \tag{5}$$

where $k \geq 1$ is a function of $|\nabla U_0|$. Conceptually, equation (5) represents a nonlocal strength criterion. Computationally, however, it is a local criterion.

For $-0.1 \leq \alpha_2 \leq 1$ the magnifying factor k is set equal to 1, irrespective of the value for $|\nabla U_0|$. Reasons for disregarding a possible increase of σ_{iu} in this case are

(a) the fact that the process of the structural degradation of thin RC shells is usually controlled by cracking rather than crushing of concrete,

(b) the lack of experimentally corroborated knowledge about the dependence of failure of concrete by crushing on the stress gradient for uniaxial states of stress, not to speak of the dependence of σ_{iu} on $|\nabla U_0|$ for multiaxial states of stress, and

(c) the avoidance of mathematical difficulties resulting from the consideration of the influence of plastic deformations of concrete on k.

By contrast with the situation for $-0.1 \leq \alpha_2 \leq 1$, plastic deformations of concrete outside this interval are rather insignificant. This justifies replacing U_0 by the elastic strain energy density $\hat{U}_0$ for the purpose of computing k.

Determination of this factor is the nucleus of the theoretical part of the present paper. The nonlocal biaxial strength criterion, given by equation (5), is applied within the framework of ultimate load analyses of a built RC natural draught cooling tower shell subjected to dead load and quasi-static wind load and of a RC hypar groined vault subjected to dead load and snow load.

The main purpose of the numerical investigation is to study the influence of the nonlocal character of the proposed strength criterion on the global response of the two shells. Another reason for analyzing the hypar groined vault, representing a relatively brittle shell, is to investigate whether the conventional restriction to a fixed failure envelope (Fig. 2(a)) yields an objective result for the collapse load and, if this should not be the case, whether the proposed nonlocal strength criterion for concrete leads to the objectification of the result for the collapse load. By objectivity, convergence of a sequence of corresponding numerical results to the exact solution in the limit of consistently refined FE discretizations is meant. For a rather ductile shell such as the mentioned natural draught cooling tower the objectivity of the numerical results is not an issue.

Criticism of a fixed failure envelope was raised first by Bažant and Cedolin [8],[9],[10] in the context of FE analysis of blunt crack propagation. Restricting attention to linear elasticity, for a sharp crack, representing the limiting case of a blunt crack band, the stresses at the crack tip tend to infinity. Thus, for this limiting case the conventional fracture criterion

$$\sigma_i \geqq \sigma_{iu}, \qquad (6)$$

characterized by fixed quantities $\sigma_{iu}(\alpha_1)$, is always satisfied, indicating the lack of objectivity of this criterion.

Recently, Bažant [11] investigated the size effect in fracture criteria for concrete. He introduced a so-called "size reduced-strength" representing a characteristic of the entire structure rather than of the material. Although it is doubtful whether, in general, such a global characteristic exists, Bažant's criticism of fixed quantities $\sigma_{iu}(\alpha_1)$ is justified.

Determination of Magnifying Factor k for σ_{iu}

Supposing that the depth h of a fictitious concrete beam with rectangular cross-section (Fig.3) was known which satisfies the condition

$$\frac{f_{btu}(h)}{f_{tu}} = \frac{\sigma^*_{iu}}{\sigma_{iu}} \qquad (7)$$

where the right hand side refers to the considered point of the investigated shell. Then, by combining equations (5) and (7), k could be determined as

$$k = \frac{f_{btu}(h)}{f_{tu}} \; . \qquad (8)$$

In equations (7) and (8), $f_{btu}(h)$ is the modulus of rupture of the beam shown in Fig. 3(a). By contrast with f_{btu}, the direct tensile strength of concrete, f_{tu}, may be considered as a material property. Thus, its dependence on the depth of a beam subjected to axial tension (Fig. 3(c)) may be neglected. The expression on the right hand side of equation (7) requires that

$$\frac{\sigma^*_{2u}}{\sigma^*_{1u}} = \frac{\sigma_{2u}}{\sigma_{1u}} = \alpha_1 \tag{9}$$

which is consistent with Kupfer's tests carried out for proportional loading. In Ref. [12], equation (7) did not serve as the starting point for determination of k. Instead of it, this equation was derived from a condition involving strain energy densities released when concrete fails by cracking.

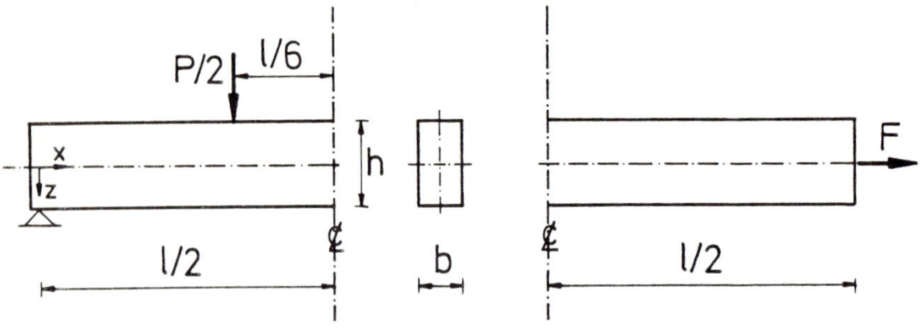

Fig.3. Fictitious concrete beam [12]

Multiplying the numerator and the denominator of equation (8) by $f^{(10)}_{btu}$ and making use of equation (4) yields

$$k = 1.88\,\frac{f_{btu}}{f^{(10)}_{btu}}\ . \tag{10}$$

Thus, the k - h diagram (Fig.4) is obtained by multiplying the ordinates of the $f_{btu}/f^{(10)}_{btu}$ - h diagram shown in Fig.1(b) by 1.88. The dashed parts of the k - h diagram represent extrapolations of the part of the curve which is based on experimental results. As will be shown in the following, the lines h = 0 and k = 1 must be asymptotes of the k - h diagram.

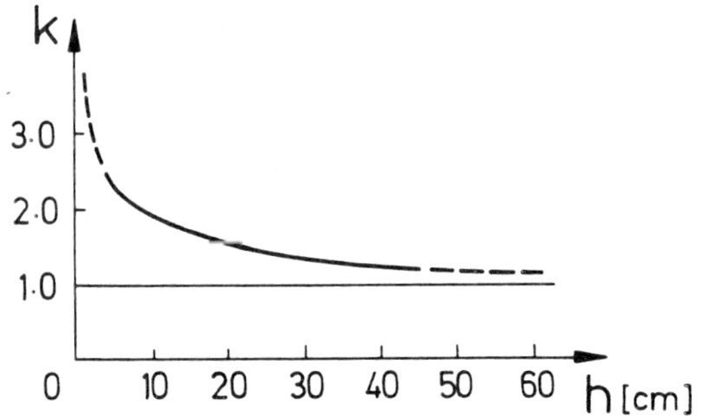

Fig.4. Dependence of the magnifying factor k for σ_{iu} on the depth h of a fictitious concrete beam

The depth h of the fictitious concrete beam is determined from a hypothesis of comparison, given as [12]:

$$\frac{|\nabla \hat{U}_0|}{\hat{U}_0} = \frac{\frac{dU_0^{(b)}}{dz}}{U_0^{(b)}} \tag{11}$$

where the left hand side refers to the considered point of the investigated shell,

$$U_0^{(b)} = \frac{1}{2}\sigma\varepsilon \tag{12}$$

is the strain energy density in a point of the middle third of the tensile fiber of the fictitious concrete beam of Fig.3(a) and

$$\frac{dU_0^{(b)}}{dz} = \frac{1}{2}\left(\frac{d\sigma}{dz}\varepsilon + \sigma\frac{d\varepsilon}{dz}\right) = \frac{2\sigma\varepsilon}{h} \tag{13}$$

is the gradient of $U_0^{(b)}$ in this point. Since Hooke's law holds for the uncracked beam, it appears logical to use $\hat{U}_0$ and $|\nabla\hat{U}_0|$ for the comparison expressed by equation (11).

Substituting the expressions for $U_0^{(b)}$ and $dU_0^{(b)}/dz$ into equation (11) yields

$$h = 4\,\frac{\hat{U}_0}{|\nabla\hat{U}_0|}\,. \qquad (14)$$

Details of computation of $\hat{U}_0$ and $|\nabla\hat{U}_0|$ are given in Ref. [12].

For the limiting case $\hat{U}_0 \to \infty$, $|\nabla\hat{U}_0|$ tends more strongly to ∞ than $\hat{U}_0$. Thus, according to equation (14), $h \to 0$. Since $k \to \infty$ for this limiting case which is related to linear fracture mechanics, the line $h = 0$ must be an asymptote of the $k - h$ diagram. Needless to say, this information is insufficient for a reliable extrapolation of the part of the curve in Fig.4 which is based on experimental results. However, by contrast with FE analysis of blunt crack propagation, this extrapolation is rather insignificant for ultimate load analysis of RC plates and shells by the FEM because, in general, the values taken on by h are outside the range of the mentioned extrapolation. According to equation (14), for the limiting case $|\nabla\hat{U}_0| \to 0$, $h \to \infty$. Since $k = 1$ for $|\nabla\hat{U}_0| = 0$, the line $k = 1$ must be the second asymptote of the $k - h$ diagram. When h takes on a value outside the range of the depth of a beam of length and width of typical test beams, this value should be regarded merely as an abstract analysis quantity related to $\hat{U}_0$ and $|\nabla\hat{U}_0|$ in the considered point of the investigated shell.

Ultimate Load Analysis of RC Plates and Shells by the FEM

The proposed nonlocal biaxial strength criterion was incorporated into a computer program for ultimate load analysis of RC plates and shells by the FEM. Mechanical fundamentals and analysis algorithms of this program are described in detail by Floegl [13].

Geometric nonlinearity is considered within the framework of Koiter's theory of small displacements and moderately large rotations [14] which has proved to be adequate for ultimate load analysis of RC shells. The constitutive model for the intact concrete is a so-called "equivalent uniaxial model" proposed by Liu, Nilson and Slate [15]. It is based on the previously mentioned biaxial tests conducted by Kupfer [7].

Curved, triangular, C^1-conforming, thin-shell elements [16] are used for discretization. They are subdivided into sufficiently many thin layers of concrete such that, approximately, a plane state of stress may be assumed to exist in each layer. The reinforcement is "smeared" to thin layers. The states of strain and stress are determined at seven integration points each in the individual concrete and steel layers.

If the strength criterion signals cracking of concrete at a certain integration point, it is assumed that a band of equidistant parallel cracks normal to σ_1 $(\sigma_1 > \sigma_2)$, distributed over the tributary domain of this point, will open. This is the basis for the stress-controlled "smeared" technique for modelling the cracked concrete.

Numerical Investigation

The numerical investigation consists of ultimate load analysis of a built RC natural draught cooling tower shell subjected to dead load and quasi-static wind load and of a hypar groined vault subjected to dead load and snow load.

RC natural draught cooling tower shell

Fig. 5 shows characteristic dimensions, the thickness profile and the FE mesh of a hyperbolic cooling tower of approximately 93 m height, erected in Voitsberg, Styria, Austria, designated as cooling tower Voitsberg III.

In Fig. 5, r(z) is the radius of the middle surface of the shell at $z \equiv \beta$. The coefficients A,B,...F for determination of the generatrix of the shell,

$$Az^2 + 2Brz + Cr^2 + 2Dz + 2Er + F = 0, \qquad (15)$$

are listed in Ref. [12]. According to Mehl [18], the simplifying assumption of a rigid, hinged base is justified. Material parameters are compiled in Ref. [12]. Because of symmetry of the quasi-static wind load with respect to the luff and the lee meridian, only one half of the shell needs to be analyzed. The distribution of the wind-load function $\bar{p}_S$ was chosen as proposed by Krätzig, Peters and Zerna [19].

Ultimate load analysis of the Voitsberg III cooling tower, reported in more detail in Ref. [12], is based on the load combination $g + \chi\bar{p}_S$ where g is the dead load and χ is a magnifying factor for $\bar{p}_S$. Fig. 6 shows diagrams of displacement components, at two different points of the shell, versus χ. The solid (dashed) curves were obtained by means of the new (conventional) strength criterion. The wind load associated with point A (B) is termed "crack load". It represents a lower bound to the ultimate load. The latter is characterized by the failure of the equilibrium iteration to converge while the displacements increase rapidly. Fig. 6 shows that the influence of the nonlocal character of the proposed strength criterion on the global response of the shell is rather small in spite of relatively large values of k_{MAX} (e.g., $k_{MAX}(\chi = 2.35) = 1.84$, $k_{MAX}(\chi = 2.81) = 2.14$). The values are located on the part of the k - h diagram (Fig. 4) which is based on experimental results.

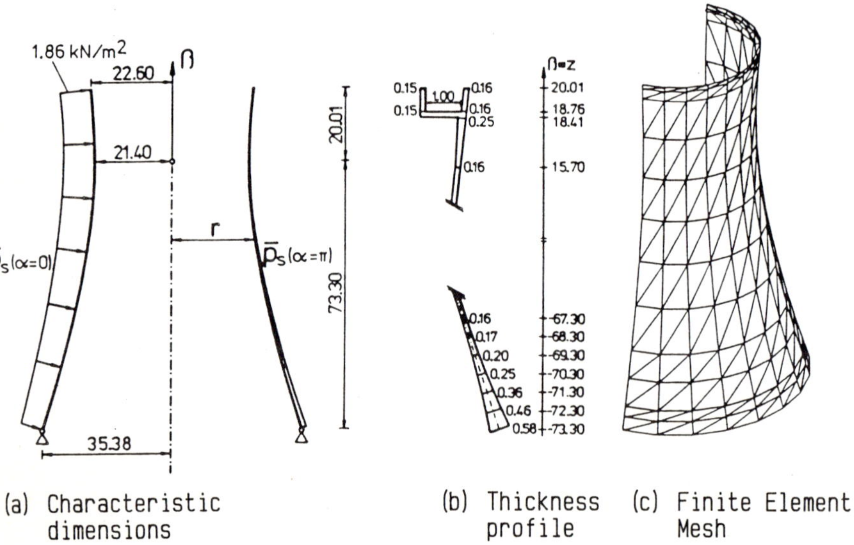

Fig.5. Cooling tower Voitsberg III, Styria, Austria [17].

Fig. 7 shows plots of "smeared cracks" in three different concrete layers, at $\chi = 2.35$, for the proposed nonlocal strength criterion ($k \geq 1$) as well as for the conventional strength criterion ($k = 1$). The direction of the dashes indicates the direction of the cracks. Expectedly, the nonlocal strength criterion leads to a reduction of the amount of cracking. The conclusion that the biaxial tensile strength of concrete does not have much influence on the global response of a rather ductile shell such as the Voitsberg III cooling tower corresponds with an analogous conclusion for RC beams, drawn by Dodds, Darwin and Leibengood [20] on the basis of ultimate load analysis of the beams by the FEM.

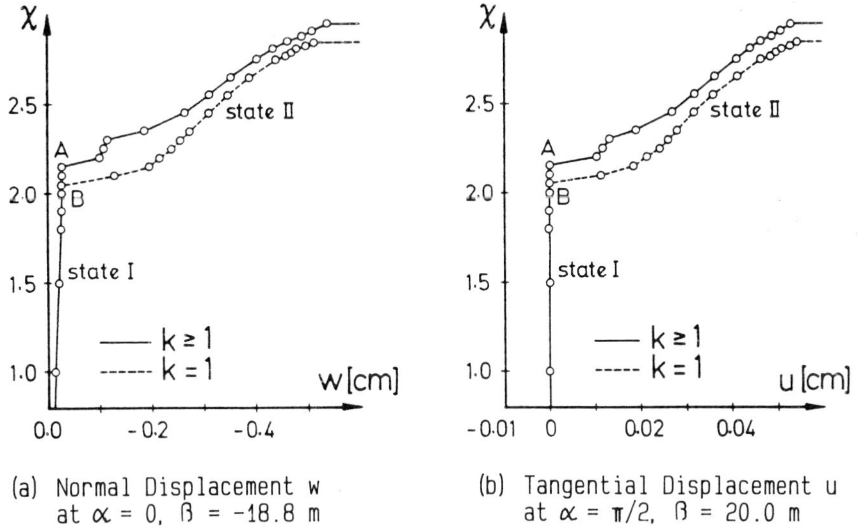

Fig.6. Load-displacement diagrams [12]

RC hypar groined vault

Fig. 8(a) shows a hypar groined vault previously analyzed by Schaper [21]. The side lengths 2ℓ of the square in plan view are 30.48 m. The height h_E of the shell at the mid-points of the four edges is 11.43 m above the plane spanned by the supports. The height h_C of the mid-point of the groined vault is 9.60 m above this plane. The shell has a uniform thickness t of 0.1016 m.

For dead load and snow load it is sufficient to consider only one octant of the hypar groined vault. Fig. 8(b) shows the mapping of the FE grids (coarse grid: FE grid 1, medium grid: FE grid 2, fine grid: FE grid 3) of the considered shell octant on the x - y plane. The medium and the fine grid are successive consistent refinements of the groin region where large stress gradients are expected. Fig. 9 shows the discretizations of the whole hypar groined vault (FE grid 2).

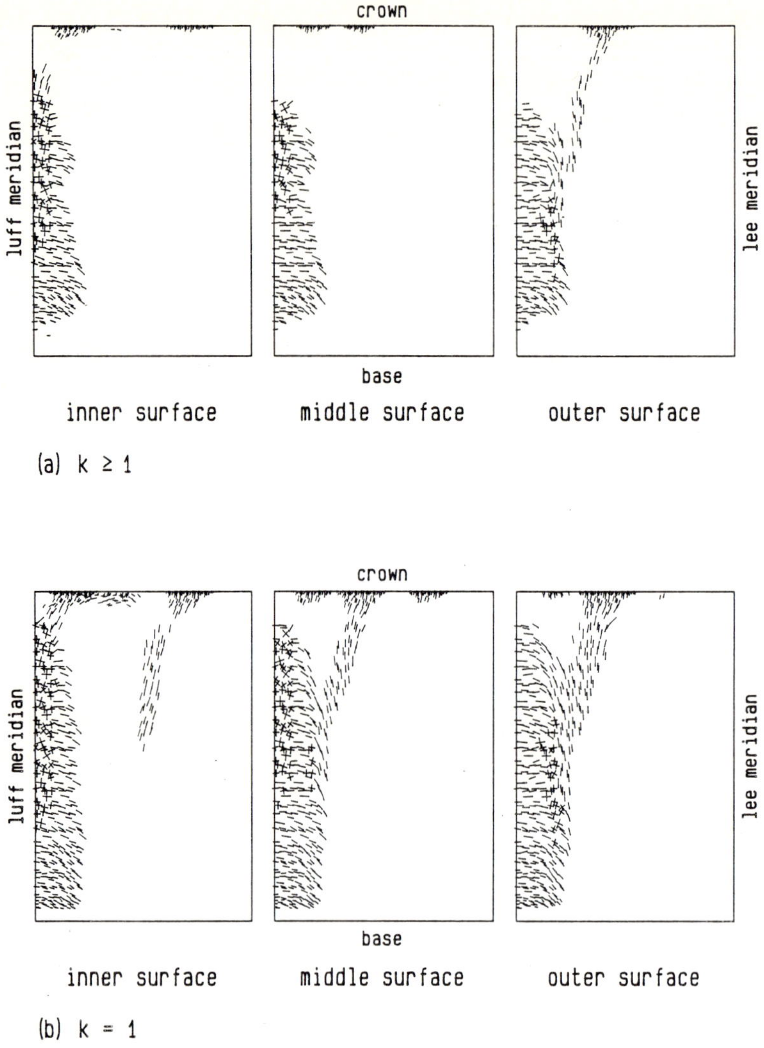

Fig.7. "Smeared cracks" in three different concrete layers, at $\chi = 2{,}35$ [12]

The middle surface of the hypar according to Fig. 8(b) is given, in explicit form on

$$z = \frac{x^2}{a} - \frac{y^2}{b} \tag{16}$$

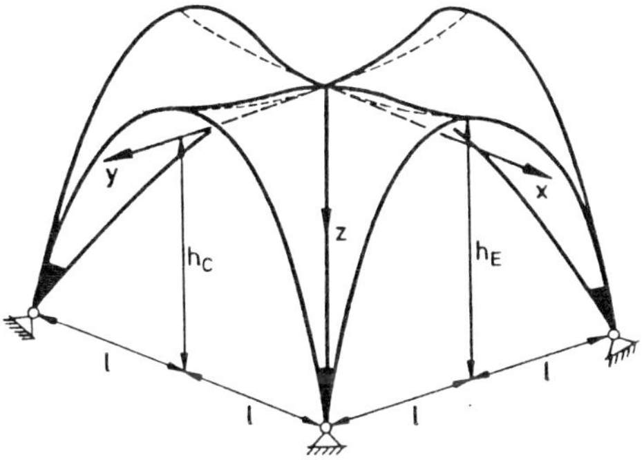

(a) Hypar groined vault

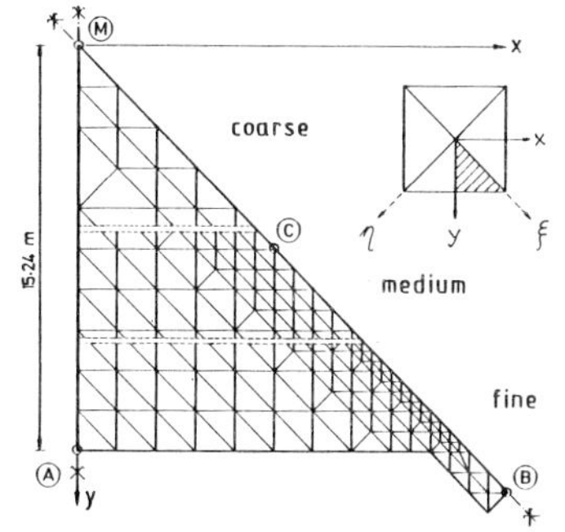

(b) Mapping of FE grids on x-y plane

Fig.8. Hypar groined vault [21],[22]

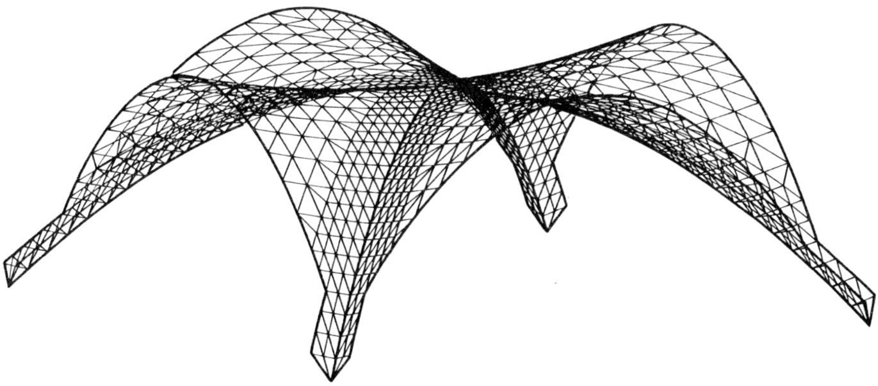

Fig.9. Discretization of hypar groined vault (FE grid 2) [22]

where $a = \ell^2/h_E = 20.23$m and $b = \ell^2/(h_E - h_C) = 126.92$ m. The previously mentioned computer program is restricted to orthogonal parameter lines [13]. Parameter lines $\alpha = x = $ const., $\beta = y = $ const., representing a logical choice for a parametric description of the middle suface of the hypar, however, are nonorthogonal one to another. Employing the concept of discrete orthogonalization of parameter lines at the integration points of the finite elements [23], the available computer program can be used without any modifications. Details concerning the selected parameter form for description of the middle suface of the hypar as well as the material data are given in Ref. [22].

Ultimate load analysis of the hyperbolic groined vault, reported in more detail in Ref. [22], is based on the load combination $g + \chi\bar{s}$ where $g = 2.929$ kN/m² is the dead load per unit area of the middle surface of the shell, $\bar{s} = 1.464$ kN/m² is the reference snow load per unit area of the horizontal projection of the middle surface of the shell and χ is a magnifying factor for $\bar{s}$. The value of χ for which collapse of the shell is signalled by the analysis is denoted as χ_{cr}.

Fig. 10(a) shows diagrams of the vertical displacement of the crown of the shell (point M), w_M, versus χ, based on the new and the conventional strength criterion, respectively, for FE grid 3. This figure illustrates the brittleness of the shell. In comparison with the hypar groined vault, the Voitsberg III cooling tower is a relatively ductile shell (see Fig. 6).

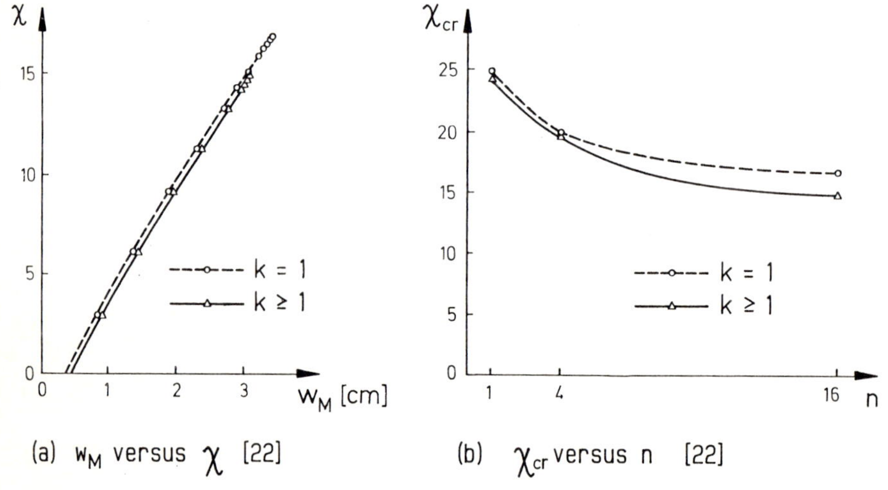

Fig.10. Load-displacement diagram and convergence study for the ultimate load

Fig. 10(b) shows the dependence of χ_{cr} on the FE grids for the new as well as for the conventional strength criterion. n = 4 and 16, respectively, is the number of finite elements into which a typical element of FE grid 1 (n = 1), coinciding at one of its edges with the groin, must be subdivided in order to obtain a typical element of FE grid 2 and 3, respectively, coinciding at one of its edges with the groin [22]. The reason for this "local relation" of n is the restriction of the mesh refinement to the vicinity of the groin arches.

Fig. 10(b) indicates that convergence of χ_{cr} is relatively slow. The figure also shows that the difference of corresponding results for χ_{cr} obtained for $k \geqq 1$ and $k = 1$, respectively, increases with increasing n. These are consequences of the brittleness of the hypar groined vault and of the large stress gradients in the groin region. Because of this situation the hypar groined vault represents a tough test for the applicability of strength criteria for concrete. However, the concern for the objectivity of such criteria, although justified in principle [8],[9],[10], appears to be exaggerated.

By contrast with the results for the cooling tower, for the hypar groined vault, $\chi_{cr}(k = 1) \geqq \chi_{cr}(k \geq 1)$. In the context of this result it is emphasized that the collapse of a structure is a global phenomenon. Therefore, from $\sigma^*_{iu} \geqq \sigma_{iu}$ it cannot be concluded that necessarily $\chi_{cr}(k \geq 1) \geqq$ $\geqq \chi_{cr}(k = 1)$.

Fig. 11 shows plots of "smeared cracks" in two different concrete layers, at $\chi = 14.0$, for $k \geq 1$. The plots refer to FE grid 3. The number of "cracked integration points" is 1532. Altogether there are 12705 integration points. For k = 1, the total number of "cracked integration points" is 2503. The cracks are restricted to the vicinity of the groin arches. Needless to say, for such a concentration of cracks a "smeared technique" for modelling the cracked concrete is less well suited than for crack distributions over relatively large subdomains of structures as is the case for the Voitsberg III cooling tower (Fig. 7).

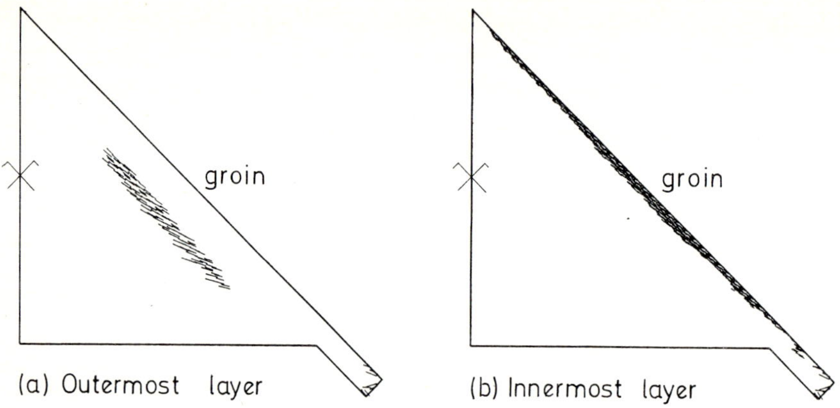

Fig.11. "Smeared cracks" in two different concrete layers, at $\chi = 14.0$ for $k \geqq 1$ (FE grid 3) [22]

Conclusions

Ultimate load analyses of a natural draught hyperbolic cooling tower, representing a relatively ductile shell, and of a hypar groined vault, serving as an example for a brittle shell, have shown that the influence of the nonlocal character of the proposed strength criterion on the global response of the two shells is relatively small. For the conventional as well as for the new strength criterion convergence of results for the ultimate load of the hypar groined vault is relatively slow. Moreover, the difference of corresponding results for the ultimate load obtained by the two criteria was found to increase as the coarse FE grid is refined. These shortcomings follow from the brittleness of the shell and from the large stress gradients in the groin region. They show that ultimate load analysis of the hypar groined vault is a tough test for the applicability of strength criteria for concrete. Nevertheless, it appears that the concern for the objectivity of such criteria is exaggerated.

Acknowledgements

The writers are indebted to Prof. M. Zyczkowski for helpful comments on the presented nonlocal strength criterion. Financial support of the Jubiläumsfonds der Österreichischen Nationalbank (Jubilee Foundation of the Austrian Nationalbank) is gratefully acknowledged.

References

1. Campus, F.: Plastification de l'acier doux en flexion plane simple. Bull. de la Classe des Sciences, Série 5, 49 (1963).

2. Campus, F.: Plastification de l'acier doux en flexion composée. Bull. de la Classe des Sciences, Série 5, 49 (1963).

3. Campus, F.: La plastification de l'acier doux en flexion simple et composée et ses effets sur flambage par compression des pièces droites elastoplastiques. Bull. de la Classe de Sciences, Série 5, 49 (1963).

4. König, J.A.; Olszak, W.: The yield criterion in the general case of nonhomogeneous stress and deformation fields. Topics in Applied Continuum Mechanics (Zeman, J.L.; Ziegler, F.: Eds.) Wien: Springer 1974.

5. Hellmann, H.G.: Beziehungen zwischen Zug- und Druckfestigkeit des Betons. beton 19 (1969) 68 - 70.

6. Mayer, H.: Die Berechnung der Durchbiegung von Stahlbetonbauteilen. Deutscher Ausschuß für Stahlbeton, Vol. 194, Berlin: Wilhelm Ernst 1967.

7. Kupfer, H.: Das Verhalten des Betons unter mehrachsiger Kurzzeitbelastung unter besonderer Berücksichtigung der zweiachsigen Beanspruchung. Deutscher Ausschuß für Stahlbeton, Vol. 229, Berlin: Wilhelm Ernst 1973.

8. Bažant, Z.P.; Cedolin, L.: Blunt crack band propagation in finite element analysis. Journal of the Engineering Mechanics Division, ASCE, 105 (1979) 297 - 315.

9. Bažant, Z.P.; Cedolin, L.: Fracture mechanics of reinforced concrete. Journal of the Engineering Mechanics Division, ASCE, 106 (1980) 1287 - 1306.

10. Bažant, Z.P.; Cedolin, L.: Finite element modelling of crack band propagation. Journal of Structural Engineering, ASCE, 109 (1983) 69 - 72.

11. Bažant, Z.P.: Size effect in blunt fracture: concrete, rock, metal. Journal of Engineering Mechanics, ASCE, 110 (1984) 518 - 535.

12. Mang, H.A.; Eberhardsteiner, J.: Collapse analysis of thin RC shells on the basis of a new fracture criterion. Proceedings of the US-Japan Seminar on Finite Element Analysis of Reinforced Concrete Structures, Japan Concrete Institute, Tokyo, 1985.

13. Floegl, H.: Traglastermittlung dünner Stahlbetonschalen mittels der Methode der Finiten Elemente unter Berücksichtigung wirklichkeitsnahen Werkstoffverhaltens sowie geometrischer Nichtlinerität, Dissertation, Technische Universität Wien, Wien, 1981.

14. Koiter, W.T.: General equations of elastic stability for thin shells. Proceedings of the Symposium on the Theory of Shells to Honor Lloyd Hamilton Donnell. University of Houston, Houston, 1967.

15. Liu, T.C.A.; Nilson, A.H.; Slate, F.O.: Biaxial stress-strain relations for concrete. Journal of the Structural Division, ASCE, 98 (1972) 1025 - 1034.

16. Thomas, G.R.; Gallagher, R.H.: A triangular thin shell finite element: linear analysis. NASA Contractor Report 2482, Washington D.C., 1975.

17. Mang, H.A.; Trappel, F.: Physically linear buckling analysis of reinforced concrete cooling towers: design necessity or academic exercise? Proceedings of the 2nd International Symposium on Natural Draught Cooling Towers. Springer, Berlin, 1984.

18. Mehl, M.: Über das Tragverhalten von Naturzugkühltürmen aus Stahlbeton, Dissertation, Technische Universität Wien, Wien, 1982.

19. Krätzig, W.P.; Peters, H.L.; Zerna, W.: Naturzugkühltürme aus Stahlbeton - Derzeitiger Stand und Entwicklungsmöglichkeiten. Beton- und Stahlbetonbau, 73 (1978) 37 - 42.

20. Dodds, R.H.; Darwin, D.; Leibengood, L.O.: Stress controlled smeared cracking in RC beams. Journal of Structural Engineering, ASCE, 110 (1984) 1959 - 1976.

21. Schaper, G.: Comparative linear and nonlinear analysis of hyperbolic paraboloid groined vaults. Report No. US-SESM 80 - 12, Dept. of Civil Engineering, University of California, Berkeley, California, 1980.

22. Mang, H.A.; Eberhardsteiner, J.; Walter, H.: Development and application of a gradient-dependent fracture criterion for finite-element analysis of reinforced-concrete surface structures. Proceedings of the Europe-US Symposium on Finite Element Methods for Nonlinear Problems. Trondheim, 1985, in press.

23. Eberhardsteiner, J.; Mang, H.: Discrete orthogonalization of parameter lines on curved surfaces for finite element analysis of thin shells. International Journal for Numerical Methods in Engineering, 21 (1985) 837 - 851.

Global Methods for Reinforced Concrete Slabs

A. COMBESCURE - A. HOFFMANN - Ph. JAMET - M. LEPAREUX - A. MILLARD

Commissariat à l'Energie Atomique - Centre d'Etudes Nucléaires de Saclay
DEMT/SMTS/LAMS - 91191 GIF SUR YVETTE CEDEX (France)

Summary

This paper develops the global method strategy to compute elastoplastic thin shells or beams.
It is shown how this methodology can be applied to the case of reinforced concrete structures.
Two cases of applications are presented : one static, the other dynamic.
The numerical results are compared to experimental data.

1. THE GLOBAL METHOD

1.1 - Principle of 'global models'

From a purely intuitive standpoint, it seems natural to define stresses and 'strains' in the usual manner. However, it is also important to consider a formal viewpoint [1]. The basis of this formalism is the notion of generalized stress and generalized 'strain' introduced by Prager [2] which is necessary for the proper introduction of minimum principle. [3]. Briefly, generalized stresses and generalized strains should make it possible to write the expression of the virtual work of internal forces (strain energy virtual variation). In more formal terms, it is necessary to write that the space **E** of 'generalized strains' and the space **S** of 'generalized strains' may be placed in duality by a bilinear $\langle e.s \rangle$ form with real value, which can be called virtual work. The plasticity equations can be expressed directly with generalized stresses and strains : 'yield surface, plastic flow laws and hardening properties'. If work hardening is assumed to be null, we fall back on the method of limit analysis. It is nevertherless important to note that it is theoretically possible to account for work hardening [1] with the notions of generalized stresses and strains.

Inelastic Behaviour of Plates and Shells
IUTAM Symposium Rio de Janeiro 1985
Editors: L. Bevilacqua, R. Feijóo and R. Valid

1.2 - General formulation

A set of points in the structure (the normal of a shell, the cross section of a bar, or of pipe) is defined by a set of (x_i) coordinates. This set is a generalized point. At this point are defined a 'generalized stress' vector with (s_i) components and the dual 'generalized strain' with (e_i) components.

A variation of δe_i in strain e_i correspond to a virtual work :

$$\delta_W = s_i \ \delta e_i$$

The 'yield surface' is defined by an equation containing the generalized stresses s_i and the variables of state μ_k representing the history of generalized point :

$$F(s_i, \mu_k) = 0$$

Plastic or visco-plastic flow is obtained by assuming a law of normality :

$$de_i^P = \frac{\delta F}{\delta s_i} \ d\lambda$$

The computation is then continued by conventional incremental methods.

1.3 - Potential capabilities of 'global models'

Global models have to be governed by qualification procedures. Apart from the saving in computation cost, they offer the advantage of directly accounting for certain heterogeneities in the structure. For example, the tensile (and bending) strength curves may be taken directly from plate samples. Hence they make it possible to take account directly of the mechanical properties of the components used.

Moreover, the potential possibilities of global models are far broader in scope than may be imagined. The use of improved descriptions of work hardening increases the scope considerably. In particular, multilayer models make it possible to increase their effectiveness during cycle changes. They can be used to represent composite shells for instance reinforced concrete. These models have been used in the CASTEM Finite Element System [4], [5], [6], for shells and piping. Consequently, we shall now examine the formulations in these two cases.

1.4 - Applications to shells

Here the 'generalized point' is a normal to the mean surface.

The components of the generalized stresses [7], [8] are (see figure 2) :

- membrane stresses N_1 N_2 N_3 (or N_{11} N_{22} N_{12}),
- bending moments M_1 M_2 M_3 (or M_{11} M_{22} M_{12}),

corresponding to the following components of 'generalized strains'

. membrane strains e_1 e_2 e_3 (or e_{11} e_{22} e_{12})

. curvature variations χ_1 χ_2 χ_3 (or χ_{11} χ_{22} χ_{12}).

In routine practice, the yield surface employed is a generalization of the Von Mises criterion (with, if necessary, several sub-materials). It is expressed as a function of the second order invariants of the deviators of the generalized stresses.

There are three invariants of this type :

$$\begin{cases} N^2 = N_1^2 + N_2^2 - N_1 N_2 + 3\ N_3^2 \\ M^2 = M_1^2 + M_2^2 - M_1 M_2 + 3\ M_3^2 \\ MN \cos\psi = N_1 M_1 + M_2 N_2 - 0{,}5\ (M_1 N_2 + M_2 N_1) + 3\ M_3 N_3 \end{cases}$$

thanks to which the yield surface is expressed as :

$$F(M,\ N,\ \cos\psi,\ \mu_1,\ \ldots) = 0$$

The hypothesis of normality of plastic flow may be written :

$$\begin{cases} \dfrac{de_1}{d\lambda} = \dfrac{\partial F}{\partial N}\ \dfrac{N_1 - 0{,}5\ N_2}{N} + \dfrac{\partial F}{\partial \cos\psi}\ \dfrac{M_1 - 0{,}5\ M_2}{M} \\ \dfrac{de_2}{d\lambda} = \dfrac{\partial F}{\partial N}\ \dfrac{N_2 - 0{,}5\ N_2}{N} + \dfrac{1}{N}\ \dfrac{\partial F}{\partial \cos\psi}\ \dfrac{M_2 - 0{,}5\ M_1}{M} \\ \dfrac{de_3}{d\lambda} = \dfrac{\partial F}{\partial N}\ \dfrac{3N_3}{N} + \dfrac{1}{N}\ \dfrac{\partial F}{\partial \cos\psi}\ \dfrac{3M_3}{M} \\ \dfrac{d\chi_1}{d\lambda} = \dfrac{\partial F}{\partial M}\ \dfrac{M_1 - 0{,}5\ M_2}{M} + \dfrac{1}{M}\ \dfrac{\partial F}{\partial \cos\psi}\ \dfrac{N_1 - 0{,}5\ N_2}{N} \\ \dfrac{d\chi_2}{d\partial} = \dfrac{\partial F}{\partial M}\ \dfrac{M_2 - 0{,}5\ M_1}{M} + \dfrac{1}{M}\ \dfrac{\partial F}{\partial \cos\psi}\ \dfrac{N_2 - 0{,}5\ N_1}{N} \\ \dfrac{d\chi_3}{d\lambda} = \dfrac{\partial F}{\partial M}\ \dfrac{3M_3}{M} + \dfrac{\partial F}{\partial \cos\psi}\ \dfrac{3N_3}{N} \end{cases}$$

In practice, one can often ignore the effect of $\cos\psi$, making it possible to replace the derivative of F by two isotopic work hardening parameters :

$$de^* = \frac{2}{\sqrt{3}} \sqrt{(de_1)^2 + (de_2)^2 + de_1\ de_2 + \frac{1}{4}\ de_3{}^2}$$

$$d\chi^* = \frac{2}{\sqrt{3}} \sqrt{(d\chi_1)^2 + (d\chi_2)^2 + d\chi_1\, d\chi_2 + \frac{1}{4}\, d\chi_3^{\,2}}$$

which leads to the classic expressions :

$$\begin{cases} \dfrac{de_1}{N_1 - \frac{1}{2} N_2} = \dfrac{de_2}{N_1 - \frac{1}{2} N_2} = \dfrac{de_3}{3N_3} = \dfrac{de^*}{N} \\[2ex] \dfrac{d\chi_1}{M_1 - \frac{1}{2} M_2} = \dfrac{d\chi_2}{M_2 - \frac{1}{2} M_1} = \dfrac{d\chi_3}{3M_3} = \dfrac{d\chi^*}{M} \end{cases}$$

and then to the usual computations.
Models of this type are employed in the BILBO, INCA and PLEXUS modules [9], [10], [11] which deal with shells of any shape, or simply axisymetric shells.

1.4 - Application to beam and piping system analysis

Here the generalized point is the cross-section, and its coordinates are simply those of the neutral fiber of the beam or pipe. Consequently, it is possible to use a beam type method [12], [13], [14] in plasticity and viscoplasticity.
The generalized stresses are the overall tension, bending, torsion and pressure in case of pipes, and it is more practicable to select as the components for a tube :

$$\sigma_n = \frac{N}{\pi D t} \qquad \sigma_p = \frac{PD}{2t} \qquad \sigma_b = \frac{Mb}{\pi D^2 \frac{t}{4}} \qquad \sigma_t = \frac{Mt}{\pi D^2 \frac{t}{4}}$$

which corresponds to the following generalized strain components :

$$\varepsilon_n = \frac{\Delta l}{l} \qquad \varepsilon_p = \frac{\Delta D}{D} \qquad \varepsilon_b = \frac{D}{4}\chi \qquad \varepsilon_t = \frac{D}{4}\psi$$

such that the strain energy per unit volume is given by :

$$\delta_W = \sigma_n\, \delta\varepsilon_n + \sigma_p\, \delta\varepsilon_p + \sigma_b\, \delta\varepsilon_b + \sigma_t\, \delta\varepsilon_t$$

where the notations employed are :

- **D** mean diameter,
- **t** thickness,
- **N** tensile force,
- **p** internal pressure,
- M_b bending moment,
- M_t torsion moment,
- $\Delta l/l$ axial elongation,
- χ curvature variation,
- ψ twist variation.

As a rule, tension, bending and torsion are the most important. The useful components are thus reduced to three σ_u, σ_b and σ_t (with ε_b and ε_t).

1.6 - Construction of the 'global' stress-strain curve SAMSON preprocessor

The main problem occuring in this approach is to derive the stress-strain curve in global variables. Non linear homogeneisation has to be carried out. The idea to achieve this, is simple. We present it on beams.
We discretise a current section of the beam so that all the characteristics of the beam can be taken into account (geometry, different materials e-g. steels, concrete ...). Applying St Venant's principle for each normal effort N we can derive a moment curvature curve (see figure 3). We then have a set of stress strain curves associated with each normal effort N (Figure 4). This set of curves is used to compute the equivalent beam. The same procedure can be generalized to shells in the same way.

2. APPLICATIONS

The method is used to compute the static and dynamic limit load of a reinforced concrete slab. The slab is a square plate and the reinforcement is defined on Figure 5. The structure is horizontal and is loaded at its center by a vertical load.

2.1 - Determination of global stress-strain curve

The SAMSON preprocessor has been used to precompute the 'global' stress-strain curve taking into account the concrete and the steels.
Figure 6 gives the stress-strain curve for the concrete. (The limit loads in traction is 4.7 MPa and in compression, 28.4 MPa).
Figure 7 give a typical stress-strain curve for the steels. The Young's modulus and conventional yield stress are $\sigma_{0.2}$ % for the two types of steels :

	E	$\sigma_{0.2}$
Ø 6	170 000 MPa	370 MPa
Ø 4	200 000 MPa	600 MPa

A typical global stress-strain curve for zero, normal stress is given on Figure 8.
The curve is then simplified to be able to go through a standard F E computation.

The equivalent Young's modulus is kept.
The yield stress is chosen as the minimum stress after the peak of stress and a new shape is constructed corresponding to the following drawing :

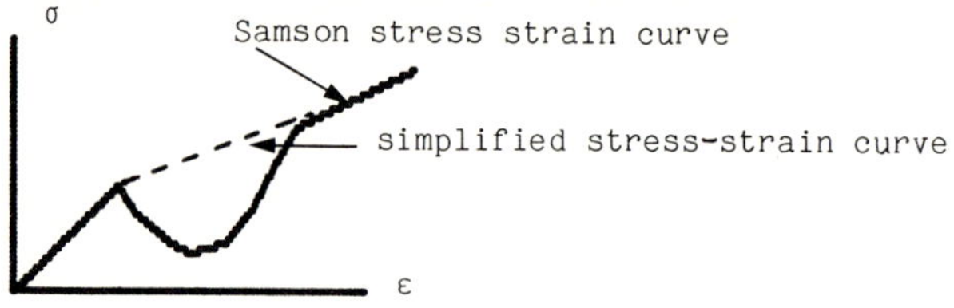

2.2 - Static analysis with BILBO

The computation is done with BILBO code in large displacements and elastoplastic regime. The mesh is given on Figure 9 the load deflection curve is compared with experimental results on Figure 10 and the crack pattern is obtained in Figure 11. The ruin mode and global rigidity is very well represented by this approach.

2.3. - Dynamic analysis with PLEXUS code

The same analysis with the same mesh has been performed with the same characteristics but with an impulsive load.
The explicit dynamic code PLEXUS is used for the analysis unilateral contact are put on nodes 8 and 16. An additional mass of 269.2 Kg is added to the two central elements. The initial speed of the central part was 6.1 m/sec. Figure 12 gives the deformed shape at the time when the maximum displacement occurs. The maximum computed displacement is 22 mm, which compares well with the experimental value of 20 mm.

3. CONCLUSION

The global model is a good model to obtain the limit loads of reinforced concrete structure either in dynamic or in static. It allows to make reasonable computations which would be hardly possible if one would compute these with a mesh representing all the steels and concrete in three dimensions.

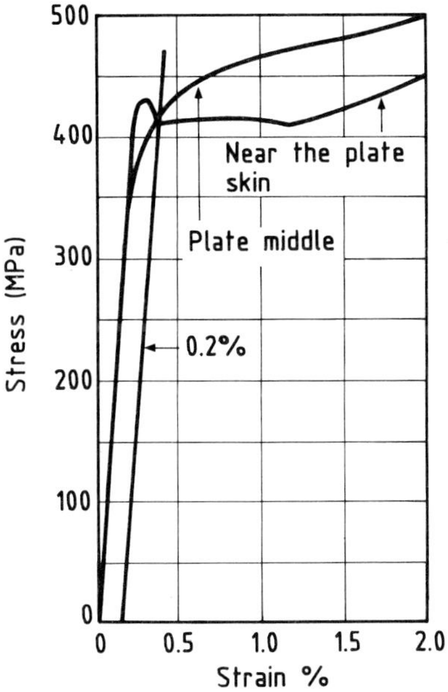

500
400
300
200
100
0
Stress (MPa)
0
0.5
1.0
1.5
2.0
Strain %
Near the plate skin
Plate middle
0.2%

Fig. 1

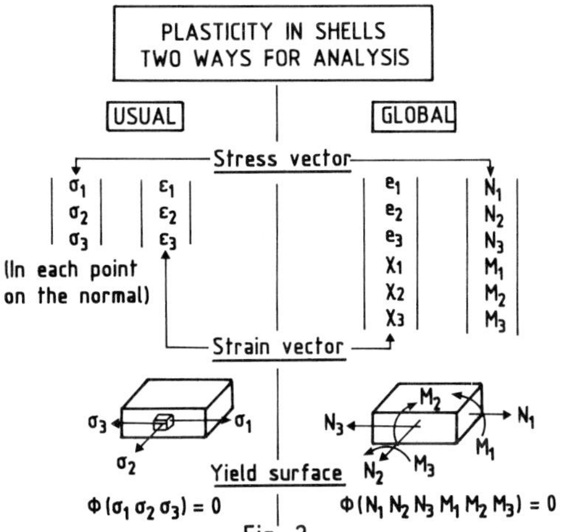

PLASTICITY IN SHELLS
TWO WAYS FOR ANALYSIS
USUAL
GLOBAL
Stress vector
σ1 σ2 σ3
ε1 ε2 ε3
(In each point on the normal)
e1 e2 e3 X1 X2 X3
N1 N2 N3 M1 M2 M3
Strain vector
Yield surface
Φ(σ1 σ2 σ3) = 0
Φ(N1 N2 N3 M1 M2 M3) = 0

Fig. 2

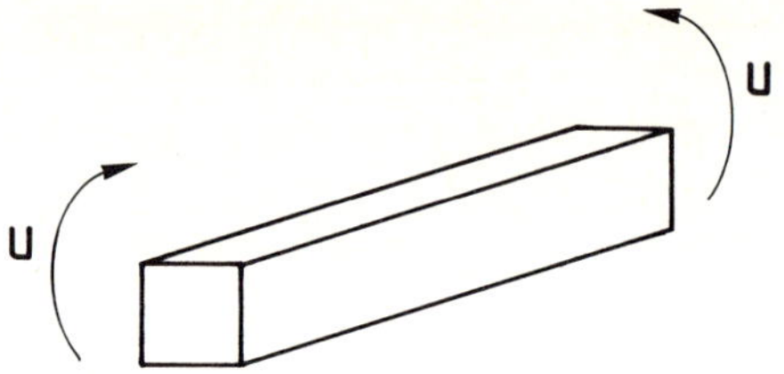

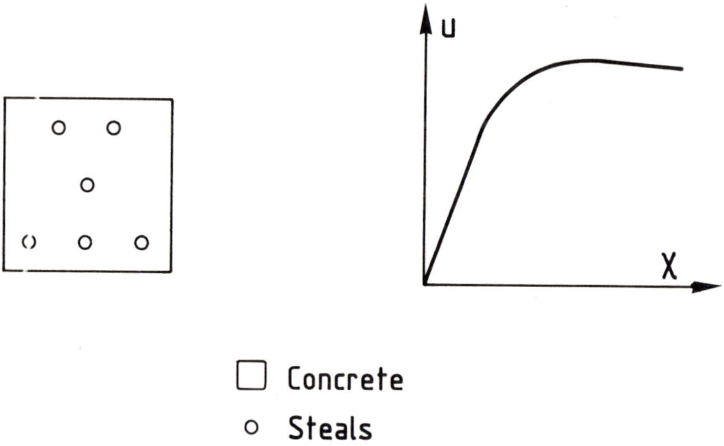

Fig. 3 - TYPICAL CROSS SECTION

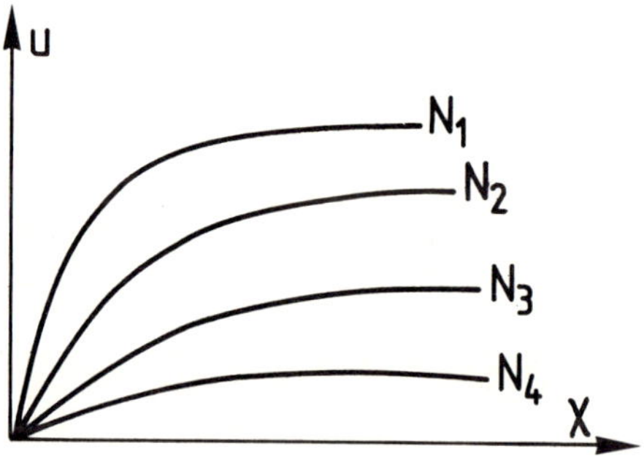

Fig. 4

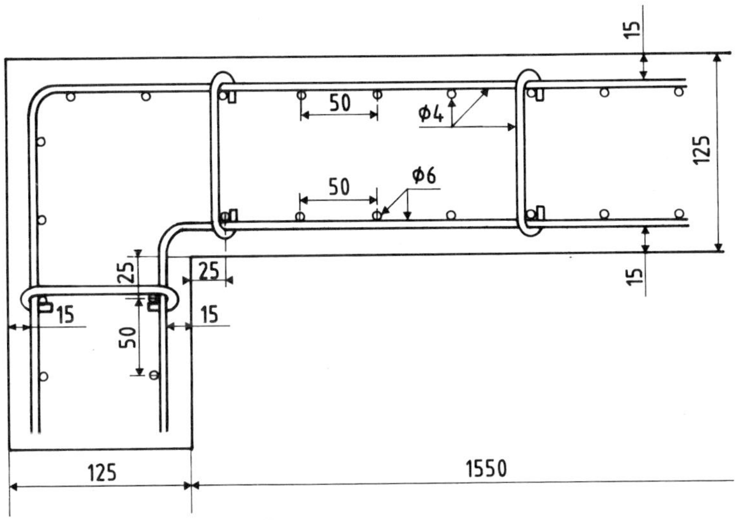

Fig. 5 - REINFORCED CONCRETE SLAB

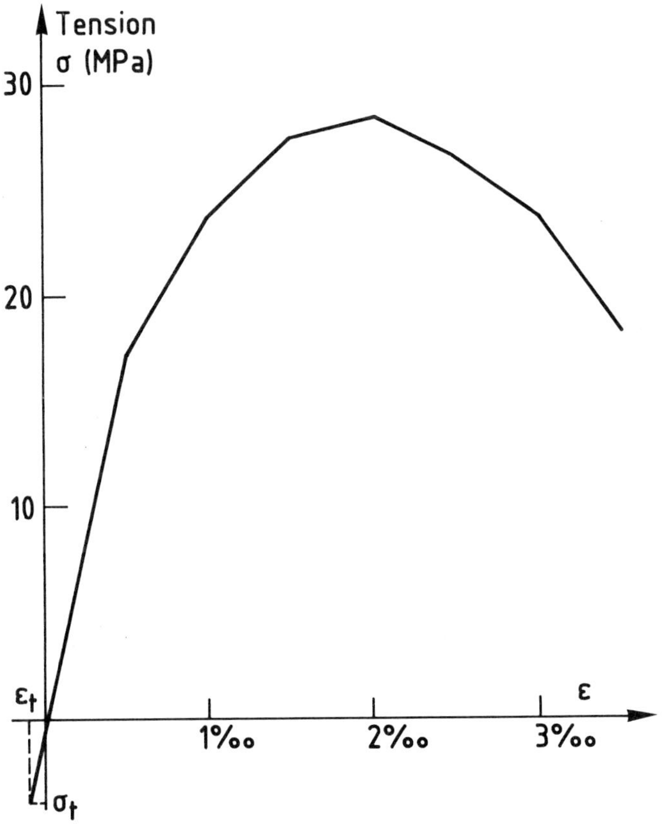

Fig. 6 - CONCRETE STRESS-STRAIN CURVE TENSION

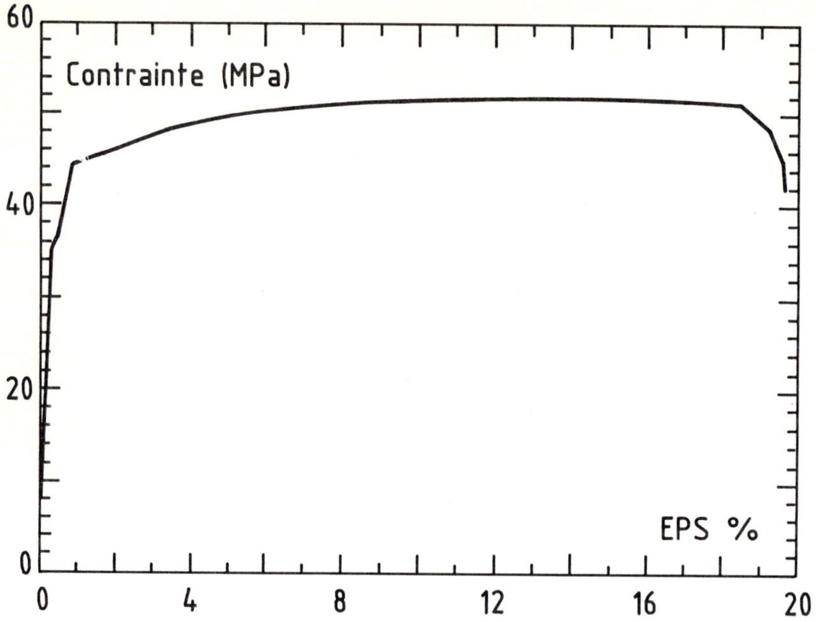

Fig. 7 - STEEL STRESS-STRAIN CURVE

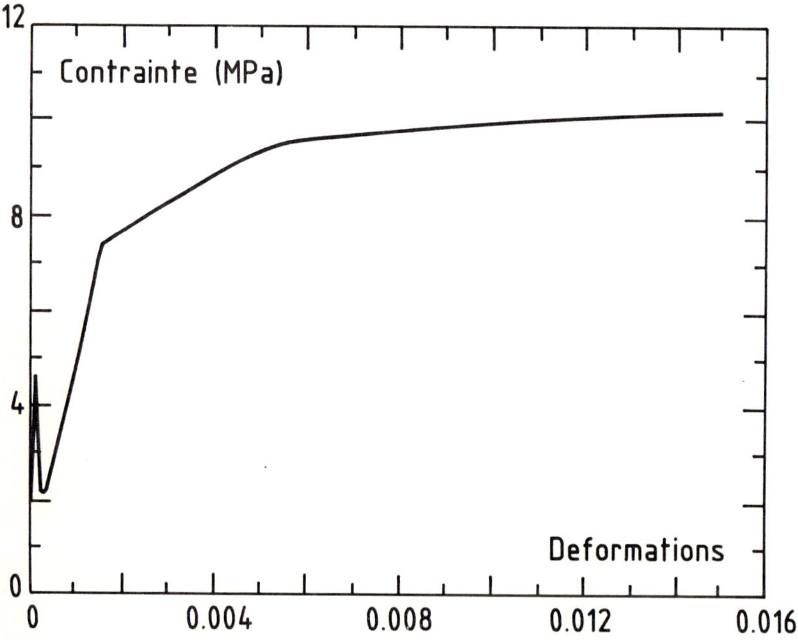

Fig. 8 - HOMOGENEISED STRESS-STRAIN CURVE

Fig. 9 - MESH FOR THE ANALYSIS

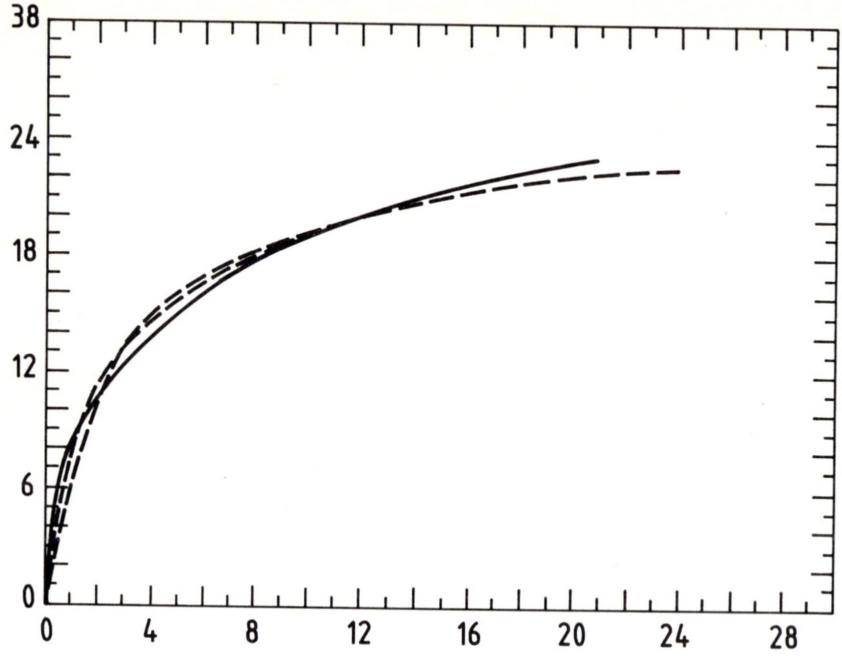

Fig. 10 - COMPARISON EXPERIMENTAL COMPUTED RESULTS STATIC ANALYSIS

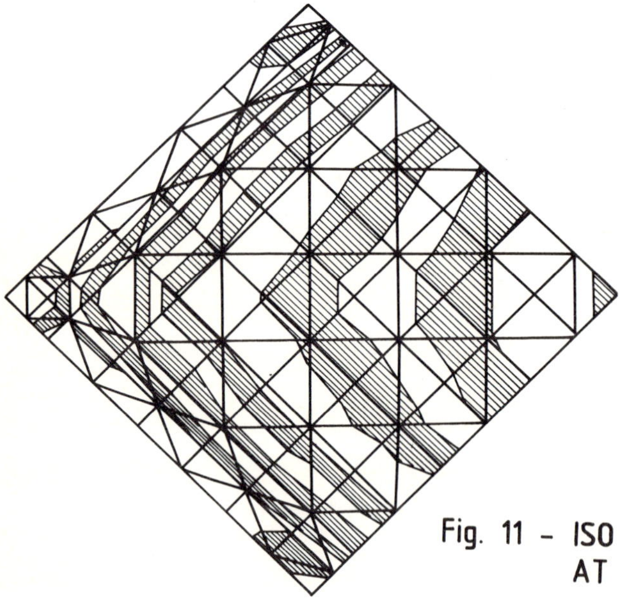

Fig. 11 - ISO VON MISES CURVE AT MAXIMUM LOAD

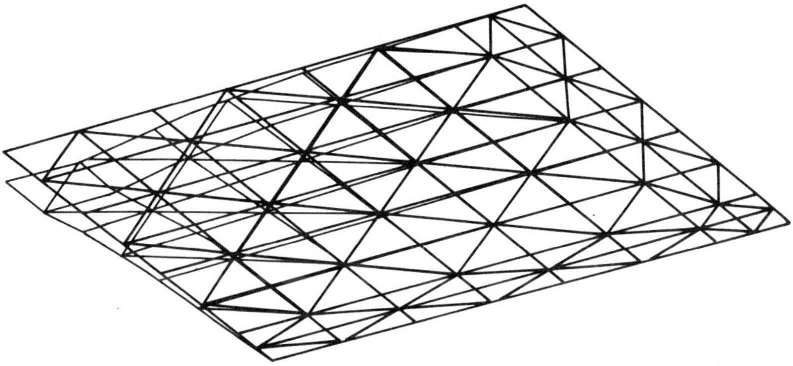

t = 2ms

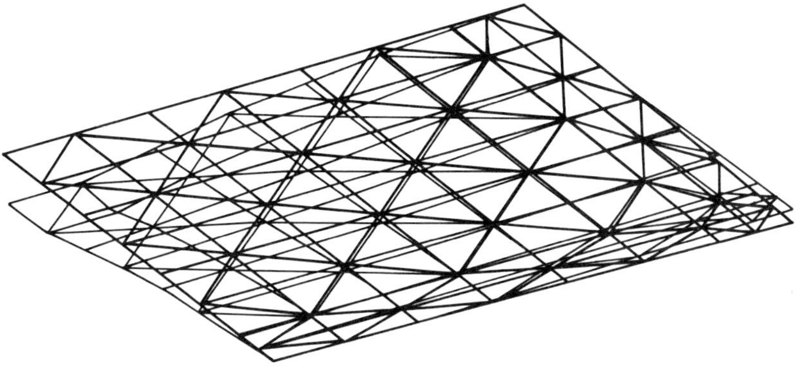

t = 4ms

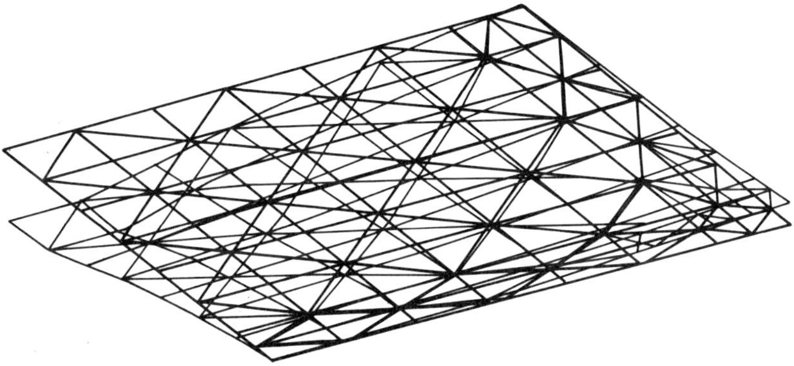

t = 9.7ms

Fig. 12 - DEFORMED SHAPE

References

1. Nayroles, B. : 'Structure algébrique des théories classiques' in 'Plasticité et viscoplasticité' edited by R. Radenkovic and J. Salencon, Edisciencе, Mc-Graw-Hill, Paris 1974.

2. Prager, W. : 'The general thoery of limit design'. Proc. 8th. Int. Congr. Appl. Mach., (Istanbul 1952), 1956.

3. Hodge, P.G. Jr. : 'Numerical applications of minimum principles in plasticity in 'Engineering Plasticity' edited by J. Heymann and F.A. Leckie, The University Press, Cambridge, 1968.

4. Jeanpierre, F. et al : 'CEASEMT System of Finite Element Computer Programs' Use for inelastic analysis in liquid metal cooled reactor components - IAEA-IWGFR Specialist's Meeting on High Temperature Design Technology, Champin, Pennsylvania, April 1976.

5. Jeanpierre, F. et al : 'Système CEASEMT. Ensemble de programmes de calcul de structure à usage industriel' Note CEA-N-1938, Saclay 1976.

6. Hoffmann, A. et al : 'Aperçus théoriques sur les programmes. Eléments. Dynamiques. Non-linéarités géométriques. Flambage. Plasticité' Note CEA-N-1934, Saclay 1976.

7. Hoffmann, A. et al : 'Analyse des coques de forme quelconque dans le domaine plastique par la méthode des éléments finis. Modèles, comparaison avec l'expérience' SMIRT 3, Berlin 1973.

8. Vrillon, B. et al : 'Comparison between experimental and computer analysis of the behaviour under pressure of a 90° bend with an elliptical section' Pressure Vessel Technology, Part III, Discussions, 2nd Int. Conf. on Press. Vessel Techn., San Antonio, October 1973.

9. User's manual of the code BILBO

10. User's manual of the code INCA

11. User's manual of the code PLEXUS

12. Roche, R.L. : 'Modèle simple pour le calcul plastique d'une tuyauterie' - Note CEA-N-1872, Saclay, September 1975

13. Roche, R.L. and Hoffmann, A. : 'Inelatic Piping Flexibility Analysis. A beam type method' Paper II-3, Specialists'Meeting on High Temperature Design Technology of LMFBRs, Champion, Pennsylvannia, April 1976 (IWGFR-IAEA, Vienna)

14. Roche, R.L., Hoffmann, A. and Vrillon, B. : 'Inelastic Piping Systems, a simplified numerical method'. 3rd Int. Conf. on Pressure Vessel Technology, Tokyo, April 1977 (ASME edit.)

On Theories of Elasto-Plastic Shells in Mixed Tensor Formulation

W. WUNDERLICH and H. SPRINGER

Institut für Konstruktiven Ingenieurbau
Ruhr-University Bochum, F.R. Germany

Summary

In extension of a formulation of a theory for thin elastic shells using mixed-variant tensorial components an incremental theory for elasto-plastic shells is derived. In the basic variational principle the flow rule is included, and the corresponding plastic parameter leads to additional plastic resultants. With the mixed components e. g. the elastic part of the constitutive equations can be shifted to the reference surface in closed form. For the plastic part some coefficients are evaluated numerically. In addition to the global variational form the corresponding local equations (equilibrium, strain-displacement-relation, flow-rule) of a two-dimensional elasto-plastic shell theory are given.

1. Introduction

Theories for thin elastic shells have been given by many authors [1, 2, 3]. Although consistent formulations have to satisfy certain invariance requirements, different consistent shell theories seem to be possible due to slightly different definitions of resultants and curvature measures in the reduction from the continuum to the shell surface. Most of the tensorial formulated shell theories use contravariant stress and covariant strain components. It was shown in [4] that the definition of variables through mixed-variant tensor components has advantages in the formulation for linear elastic shells. This holds especially for the reduction of the constitutive equations to the shell surface. The description of generalized Hooke's law with mixed components does not contain metric coefficients explicitly which would lead to series expansions and lengthy expressions in the transformation to the surface.

Inelastic Behaviour of Plates and Shells
IUTAM Symposium Rio de Janeiro 1985
Editors: L. Bevilacqua, R. Feijóo and R. Valid

In this paper the formulation for linear shells is extended into the elasto-plastic range assuming small displacements and small strains and a flow theory of plasticity. Thus, the description has to be given in incremental form. It is now also advantageous in elasto-plastic material laws to employ mixed-variant tensor components, e. g. invariants are most naturally formed in this way. Also, in the reduction from the continuum to the shell surface no metric coefficients are explicitely involved, and it seems to be a straightforward path to derive elasto-plastic shell theories using mixed-type components.

Another problem in the derivation of two-dimensional theories of elasto-plasticity is to express flow rule and yield conditions also in quantities referred to the shell surface (e. g. in forces and moments). This holds especially for stress quantities as plastic zones influence the distribution of the stresses across the thickness essentially. It is for this reason that inelastic shell theories which use constant and linear stress resultants are applicable in certain cases only; e. g. for proportional loading [5, 6, 7]. Theories with higher-order moments were proposed [10] but seldom used in numerical calculations. Instead, in many finite-element-solutions the thickness is subdivided into a number (say 5 to 9) of layers or integration points to trace the yield limits and the changes due to loading and unloading. This approach is used widely, e. g. for the determination of ultimate loads or plastic buckling loads of shells of revolution in [9]. Nevertheless, the use of appropriate global shell quantities could lead to a more efficient solution.

For the derivation of such a shell theory we introduce an additional unknown quantity that represents the distribution of the plastic strain over the thickness which may be expressed by the normality rule of plasticity by a scalar parameter. In addition, we employ the distribution of the total strain across the thickness as unknowns. For the usual theory of plasticity (under small strains) holds the assumption that this distribution is linear. Therefore, it is enough to take global parameters for the constant and linear total strains.

As basis for the derivation a generalized variational principle is used which contains as corresponding local equations not only equilibrium and kinematics but also the flow rule. Some principles in which by independent variation of the plastic strains (or the plastic multiplier) the flow rule can be obtained were given in [8]. In connection with the mixed-type tensor components the generalized variational form is first transformed to the surface by shifting. Then, assumptions regarding the distribution of the increments of the total strains, the displacements and the plastic strains across the thickness are made to render a two-dimensional theory of elasto-plastic shells. On the basis of similar variational forms and hardening rules other variants of elastic-plastic shell theories might be derived. These are not given here in detail.

2. Mixed variant components, shell space and shell surface

For the description of the shell curvilinear coordinates ϑ^α of the surface and the coordinate ϑ normal to it are used. With the base vectors $\underline{g}_i$ of an arbitrary point of the threedimensional shell space the position vector can be expressed by

$$\underline{R} = \underline{r}(\vartheta^\alpha) + \vartheta\ \underline{a}_3(\vartheta^\alpha) \quad . \tag{2.1}$$

It is referred to a point of the two-dimensional shell surface with the base vectors $\underline{a}_\alpha$, $\underline{a}_3$. In the relationship between quantities of the two reference states shell shifters μ^α_β are used, which contain also the curvatures of the shell, for further details see [4].

The use of mixed variant tensorial components is one essential feature of the theory proposed in this paper. Their definition in the shell space is given for the stresses by

$$\tau^{i}_{\cdot k}\, \underline{g}_i\, \underline{g}^k = \tau_k^{\cdot i}\, \underline{g}^k\, \underline{g}_i \quad . \tag{2.2}$$

The relations

$$\tau^{i}_{\cdot k} = \tau^{in} g_{nk} \qquad ; \qquad \tau_{k}^{\cdot i} = \tau^{ni} g_{nk} \tag{2.3}$$

show that only for symmetric contravariant stress components the marking of the corresponding mixed components is immaterial. However, the quantities τ^{i}_{k} are not symmetric themselves, they are also called pseudo-symmetric. In case that the same three-dimensional quantities are referred to the surface coordinates, they are marked by a bar:

$$\bar{\tau}^{i}_{\cdot k} \underline{a}_i \, \underline{a}^k = \bar{\tau}_{i}^{\cdot k} \underline{a}_k \, \underline{a}^i = \bar{\tau}^{ik} \underline{a}_i \, \underline{a}_k \quad . \tag{2.4}$$

Reduction to a two-dimensional theory is accomplished by assumptions across the thickness, related quantities have Greek indices.

It should also be emphasized that incremental quantities remain unmarked but variables of the previous (fundamental) state carry a superscript, e. g. $\overset{o}{n}{}^{\alpha}_{\beta}$ are the normal forces of a shell theory calculated to a certain load level, and these are also given quantities in the next incremental step, in which e. g. n^{α}_{β} are unknowns. Mixed definition may be visualized for two-dimensional quantities [4]. The components $\alpha = \beta$ are normal to the coordinate lines, and their values coincide with those of their physical quantities.

3.Basic equations of an elasto-plastic continuum

As basis for the derivation a generalized variational principle for an elasto-plastic body in static equilibrium is employed. The principle contains the equilibrium conditions, the kinematics and the flow rule as governing local equations.

The dependence of strains in the plastic region on the loading history will be described correctly only by the flow theory of plasticity. Assuming infinitesimal small increments, the governing equations may be linearized. In the framework of a geometrically linear theory this influences the stress-strain relations only, and the well known linear equations of equilibrium, linear strain-displacement relations and related boundary conditions may be adapted, taking the variables as increments:

Equations of equilibrium $\tau_j^i|_i + \bar{k}_j = 0 \quad ,$ (3.1)

Strain-displacement relations $\gamma_j^i = \frac{1}{2}(u^i|_j + u_j|^i) \quad ,$ (3.2)

Boundary conditions $\tau_j^i n_i - \bar{p}_j = 0 \quad \text{on} \quad S_p \, ,$ (3.3)

$$u^j - \bar{u}^j = 0 \quad \text{on} \quad S_u \, .$$

As the most common description of various possibilities the Prandtl-Reuss equations as an associated flow rule on the basis of the v. Mises yield condition are employed here to describe the elasto-plastic stress-strain relations. The expressions needed in the further derivation are briefly summarized:

Assumption $\gamma_k^i = \overset{(E)}{\gamma}{}_k^i + \overset{(P)}{\gamma}{}_k^i$ (3.4)

Yield condition $f = \frac{1}{2}(\mathring{\tau}_k^i \mathring{\tau}_i^k - \frac{1}{3}\mathring{\tau}_i^i \mathring{\tau}_k^k) - \frac{1}{3}\mathring{\sigma}_F^2$ (3.5)

$$= \frac{1}{2} \quad \mathring{T}_k^i \quad \mathring{T}_i^k \qquad - \frac{1}{3}\mathring{\sigma}_F^2 = 0$$

Normality rule $\overset{P}{\gamma}{}_k^i = \frac{\partial f}{\partial \mathring{\tau}_i^k}\lambda$ (3.6)

Flow rule $df = \frac{\partial f}{\partial \mathring{\tau}_k^i}\tau_k^i + \frac{\partial f}{\partial \overset{P}{\mathring{\gamma}}{}_k^i}\overset{P}{\gamma}{}_k^i$

$$= \frac{\partial f}{\partial \mathring{\tau}_j^i}\tau_j^i + \lambda \underbrace{\frac{\partial f}{\partial \mathring{\tau}_j^i}\frac{\partial f}{\partial \mathring{\tau}_i^j}\frac{2}{3}\frac{EE_t}{E-E_t}}_{h} = 0 \quad . \tag{3.7}$$

For isotropic hardening only the yield stress σ_F changes, in case that kinematic hardening shall be described the translation tensor α_j^i of the center of the yield surface has to be accounted for in calculating the deviatoric stress tensor $\mathring{T}^i_j$.

Deviatoric stresses for kinematic hardening $\quad \mathring{T}^i_j = (\mathring{\tau}^i_j - \mathring{\alpha}^i_j) - \frac{1}{3}\delta^i_j(\mathring{\tau}^m_m - \mathring{\alpha}^m_m) \quad ,$ (3.8)

Pragers hardening rule $\quad \alpha^i_j = \frac{2}{3}\frac{EE_t}{E-E_t}\overset{p}{\gamma}{}^i_j \quad .$ (3.9)

Thus, the direction and magnitude of plastic strains will be influenced by the choice of the hardening rule but not the general procedure.

Remarking that

$$E^{ik}_{jl}\frac{\partial f}{\partial \mathring{\tau}^k_l} = 2G\,\mathring{T}^i_j \tag{3.10}$$

the linear relation between stress increments and strain increments can be written as

$$\begin{aligned}\tau^i_j &= \hat{\tau}^i_j - \overset{p}{\tau}{}^i_j \\ &= 2G\,(\gamma^i_j + \frac{\nu}{1-2\nu}\,\delta^i_j\;\gamma^r_r) - \frac{\mathring{T}^i_j(2G)^2\mathring{T}^k_l}{H}\,\gamma^l_k \quad ,\end{aligned} \tag{3.11}$$

$$\gamma^i_j = \frac{1}{2G}\,(\tau^i_j - \frac{\nu}{1+\nu}\,\delta^i_j\;\tau^r_r) + \frac{\mathring{T}^i_j\,\mathring{T}^k_l}{h}\,\tau^l_k \quad . \tag{3.12}$$

It is preferred to calculate the plastic strains in strain space rather than in stress space because there is always a unique relation between stresses and plastic strains. For an elastic-ideal plastic material the quantity h vanishes and the determination of plastic strains in stress space (3.12) is not possible. Thus, plastic strains must be obtained by the last term of (3.11) to

$$\overset{p}{\gamma}{}^i_j = 2G\,\frac{\mathring{T}^i_j\,\mathring{T}^k_l}{H}\,\gamma^l_k \quad , \tag{3.13}$$

$$\text{with} \quad H = \mathring{T}^m_n\,\mathring{T}^n_m(2G + \frac{2}{3}\frac{EE_t}{E-E_t}) \quad . \tag{3.14}$$

The inverse relation of (3.13) is useful for deriving the global form of the elasto-plastic material law and reads

$$\gamma_k^l = \frac{H}{2G} \, \frac{1}{\overset{\circ}{\tau}{}_j^i \, \overset{\circ}{\tau}{}_i^j} \, \overset{p}{\gamma}{}_k^l \, . \tag{3.15}$$

To obtain a consistent approximation for all basic equations, the principle of virtual work can be generalized by adding the variational forms of the kinematic and constitutive equations. This leads to functionals of Hellinger-Reissner or Hu-Washizu type dependent on the choice of independent variables [10].

The generalized principle reads

$$\begin{aligned} \delta J = \int_V \{ & \tau_k^i \, \delta u^k|_i - \bar{k}_i \, \delta u^i \\ & + \delta\tau_k^i (u^k|_i - \gamma_i^k) + \delta\gamma_k^i (\tau_i^k - E_{im}^{kl} \, \overset{E}{\gamma}{}_l^m) \} \, dV \end{aligned} \tag{3.16}$$

$$\int_{S_p} \bar{p}_i \, \delta u^i - \int_{S_u} \{ p_i \, \delta u^i + \delta p_i (u^i - \bar{u}^i) \} \, dS = 0 .$$

Since the shell equations shall be approximated by assumptions for displacements and strains only, the stresses can be eliminated introducing the stress-strain relation as subsidiary condition and replacing all stresses by total and plastic strains using (3.11):

$$\begin{aligned} \delta J = \int_V \{ & (E_{jl}^{ik} \gamma_k^l - 2G \, \overset{p}{\gamma}{}_j^i) \, \delta u^j|_i - \bar{k}_i \, \delta u^i \\ & + \delta\gamma_k^l \, E_{jl}^{ik} (u^j|_i - \gamma_i^j) - \delta\overset{p}{\gamma}{}_k^l \, E_{jl}^{ik} (u^j|_i - \gamma_i^j) \} \, dV \\ & - \textit{boundary terms} = 0 . \end{aligned} \tag{3.17}$$

Inserting (3.15) and substituting γ_j^i by $\overset{p}{\gamma}{}_j^i$ in the last volume term, the principle gets its final form. The independent variables are the displacements, the total and the plastic strains. The corresponding matrix form exhibits the symmetric structure of the theory:

or

$$\delta J = \int_V \begin{bmatrix} \delta \underline{u}^T \\ \delta \underline{\gamma}^T \\ \delta \lambda \end{bmatrix} \begin{bmatrix} & {}_u\underline{D}^T \underline{E} & -{}_u\underline{D}^T \underline{E}\, \partial \underline{f} \\ \underline{E}\, \underline{D}_u & -\underline{E} & \\ -\partial \underline{f}^T \underline{E}\, \underline{D}_u & & \partial \underline{f}^T \underline{E}\, \partial \underline{f} + \partial \underline{f}^T \underline{P}\, \partial \underline{f} \end{bmatrix} \begin{bmatrix} \underline{u} \\ \underline{\gamma} \\ \lambda \end{bmatrix} dV - \begin{matrix}\text{boundary} \\ \text{terms}\end{matrix} = 0 \,, \tag{3.18}$$

with $(\delta \underline{u}\, {}_u\underline{D})^T = \delta u^i|_k$; $\underline{P} = \frac{2}{3} \frac{E E_t}{E - E_t}$; $\partial \underline{f} = \frac{\partial f}{\partial \mathring{\tau}^i_k}$.

Through the normality rule (3.6) the plastic multiplier λ is used instead of the plastic strains.

A slightly simpler form can be obtained if the elasticity tensor $\underline{E}$ is combined with the strains to introduce fictitious stresses $\underline{\hat{\tau}} = \underline{E}\, \underline{\gamma}$.

Since the elastic material law is a subsidiary condition, the content of the principle doesn't change due to the introduction of fictitious stresses but the energy expressions in the related generalized functional get a simpler form:

$$\delta J = \int_V \begin{bmatrix} \delta \underline{u}^T \\ \delta \underline{\hat{\tau}} \\ \delta \underline{\lambda} \end{bmatrix} \begin{bmatrix} & {}_u\underline{D}^T & -{}_u\underline{D}^T 2G\, \partial \underline{\mathring{f}} \\ \underline{D}_u & -\underline{E}^{-1} & \\ -\partial \underline{\mathring{f}}^T 2G\, \underline{D}_u & & H \end{bmatrix} \begin{bmatrix} \underline{u} \\ \underline{\hat{\tau}} \\ \underline{\lambda} \end{bmatrix} dV - \begin{matrix}\text{boundary} \\ \text{terms}\end{matrix} = 0 \,, \tag{3.19}$$

or

$$\delta J = \int_V \delta \Big[u^j|_i \underbrace{(\hat{\tau}^i_j - 2G\, \partial \mathring{f}^i_j\, \lambda)}_{\tau^i_j} - \underbrace{(\tfrac{1}{2} F^{ik}_{jl}\, \hat{\tau}^l_k\, \hat{\tau}^j_i - \tfrac{1}{2} H \lambda \lambda)}_{W^*(\tau)} \Big] dV$$

$$- \int_{S_u} p_j (u^j - \bar{u}^j)\, dS - \int_{S_p} u^j\, \bar{p}_j\, dS - \int_V u^j\, \bar{k}_j\, dV \,. \tag{3.20}$$

The functional (3.20) may be looked at as a generalized form of the Hellinger-Reissner stationary theorem. In the absence of plastic strains, the functional takes the form, which was already used in the derivation of the linear version of the mixed tensor shell theory [4]. It should be noted, that the part of the functional which gives the influence of the plastic strain has to be applied only within the plastic regions of the shell which are determined by the initial yield condition.

The transformation of the three-dimensional functional (3.20) related to general curvilinear coordinates with base vectors $\underline{g}_i$ to the curvilinear coordinates of the shell reference surface with base vectors $\underline{a}_\alpha$, $\underline{a}_3$ is straightforward and follows the procedure outlined in [4]. The reduction is quite simple, because products of the shifter and its inverse vanish for common tensor components. Only the transformations of the volume integral and the expression for the internal work in (3.20) leave some additional coefficients resulting from the shifter. Combination of the additional terms in the expression for the internal work with the strains leads to symmetrical stresses and a special definition for the strains, which can be given in a convenient closed form.

Thus, by the transformation of the internal work expression conjugate stress and strain measures will be defined characterizing the resulting stress and strain resultants and the structure of the shell equations:

$$\begin{aligned}
\int_S\int_\vartheta \Big\{ & (\hat{\bar{\tau}}^\alpha_\beta - \overset{P}{\bar{\tau}}{}^\alpha_\beta)\,\overbrace{\mu\,(\mu^{-1})^\varrho_\alpha\,[\bar{u}^\beta|_\varrho - b^\beta_\varrho\,\bar{u}^3]}^{\bar{\gamma}^\beta_\alpha} \\
&+ (\hat{\bar{\tau}}^\alpha_3 - \overset{P}{\bar{\tau}}{}^\alpha_3)\,\overbrace{\mu\,(\mu^{-1})^\beta_\alpha\,[\bar{u}^3{}_{,\varrho} - b_{\varrho\lambda}\bar{u}^\lambda]}^{\bar{\gamma}^3_\alpha} \\
&+ (\hat{\bar{\tau}}^3_\beta - \overset{P}{\bar{\tau}}{}^3_\beta)\,\overbrace{\mu\,[\bar{u}^\beta{}_{,3}]}^{\bar{\gamma}^\alpha_3} \\
&+ (\hat{\bar{\tau}}^3_3 - \overset{P}{\bar{\tau}}{}^3_3)\,\overbrace{\mu\,[\bar{u}^3{}_{,3}]}^{\bar{\gamma}^3_3} \Big\}\, d\vartheta\; dS
\end{aligned} \qquad (3.21)$$

$$\bar{\tau}^i_k \longleftrightarrow \bar{\gamma}^i_k$$

The barred quantities are related to the surface coordinates, but still are three-dimensional. The physical meaning of the plastic stresses is more clearly exhibited by

$$\overset{P}{\bar{\tau}}{}^{i}_{j} = 2G\, \overset{o}{\bar{T}}{}^{i}_{j}\, \lambda \quad . \tag{3.22}$$

Using mixed tensor components the constitutive equations and the expression for the specific internal energy do not contain the metric coefficients explicitly. This is true for the plastic part, too. Thus, the transformation of the specific internal energy expressions is straightforward, without any approximations, as was mentioned earlier.

The complete generalized functional (3.20), after shifting, has the form

$$\begin{aligned}
\delta J = \int_S \int_\vartheta \delta \{ & -[\widehat{W}^*(\hat{\bar{\tau}}) - \tfrac{1}{2} \bar{H}\, \bar{\lambda}\, \bar{\lambda}\,]\, \mu \\
& + (\hat{\bar{\tau}}{}^{\alpha}_{\beta} - \overset{P}{\bar{\tau}}{}^{\alpha}_{\beta})\, \bar{\gamma}^{\beta}_{\alpha} + (\hat{\bar{\tau}}{}^{\alpha}_{3} - \overset{P}{\bar{\tau}}{}^{\alpha}_{3})\, \bar{\gamma}^{3}_{\alpha} + (\hat{\bar{\tau}}{}^{3}_{\alpha} - \overset{P}{\bar{\tau}}{}^{3}_{\alpha})\, \bar{\gamma}^{\alpha}_{3} \\
& + (\bar{\tau}^{3}_{3} - \overset{P}{\bar{\tau}}{}^{3}_{3})\, \bar{\gamma}^{3}_{3} - [\, \bar{\bar{k}}_{\alpha}\, \bar{u}^{\alpha} + \bar{\bar{k}}_{3}\, \bar{u}^{3}] \}\, d\vartheta\, dS \\
& - \int_{S_p} [(\bar{\tau}^{3}_{\alpha}\, \bar{u} + \bar{\bar{\tau}}{}^{3}_{3}\, \bar{u}^{3})\, \mu \; dS \\
& - \int_{S_p} \int_\vartheta [\, \bar{\bar{\tau}}{}^{\nu}_{\beta} (\mu^{-1})^{\alpha}_{\nu}\, \bar{n}_{\alpha}\, \bar{u}^{\beta} + \bar{\bar{\tau}}{}^{\nu}_{3} (\mu^{-1})^{\alpha}_{\nu}\, \bar{n}_{\alpha}\, \bar{u}^{3}]\; \mu\, d\vartheta\, dS \\
& - \int_{S_u} \int_\vartheta [(\bar{\tau}^{\nu}_{\beta} - \overset{P}{\bar{\tau}}{}^{\nu}_{\beta})(\mu^{-1})^{\alpha}_{\nu}\, \bar{n}_{\alpha} (\bar{u}^{\beta} - \bar{\bar{u}}^{\beta}) + \bar{\tau}^{\nu}_{3} (\mu^{-1})^{\alpha}_{\nu}\, \bar{n}_{\alpha} (\bar{u}^{3} - \bar{\bar{u}}^{3})]\, \mu\, d\vartheta\, dS \} = 0 \, ,
\end{aligned} \tag{3.23}$$

with

$$\begin{aligned}
W^*(\bar{\tau}) &= \widehat{W}^*(\hat{\bar{\tau}}) - \overset{P}{W}{}^*(\bar{\lambda}) \\
&= \frac{1}{2E} [\, (1+\nu)\, \hat{\bar{\tau}}{}^{\alpha}_{\beta}\, \hat{\bar{\tau}}{}^{\beta}_{\alpha} - \nu\, \hat{\bar{\tau}}{}^{\varrho}_{\varrho}\, \hat{\bar{\tau}}{}^{\lambda}_{\lambda} + 2(1+\nu)\, \hat{\bar{\tau}}{}^{\alpha}_{3}\, \hat{\bar{\tau}}{}^{3}_{\alpha} + \hat{\bar{\tau}}{}^{3}_{3} (\hat{\bar{\tau}}{}^{3}_{3} - 2\nu\, \hat{\bar{\tau}}{}^{\varrho}_{\varrho})] \\
&\qquad - \frac{1}{2}\, \overset{o}{\bar{T}}{}^{i}_{j}\, \overset{o}{\bar{T}}{}^{j}_{i} \left(\frac{2}{3}\, \frac{EE_t}{E-E_t} + 2G\right) \bar{\lambda}\, \bar{\lambda}
\end{aligned}$$

where the prescribed values are marked by double bars and $\bar{n}_{\alpha}$ are the components of the outward normal vector at a shell boundary. As was already mentioned above, the incremental functional is quite similar to the linear one, the only difference appears through the splitting of the stresses into a fictitious part and a plastic part. The variation of the additional unknown $\bar{\lambda}$ gives the flow rule (3.7) as additional Euler equation of the variational principle, thus the plastic strains can be approximated independently together with the total strains and displacements.

4. Reduction of the variational principle to the shell surface

For thin shells the assumptions of a plane stress state and a linear distribution of tangential total strains remain valid for the usual theory of plasticity provided that the strains are small. These assumptions are met by introducing a constant and linear distribution of fictitious stress increments (equivalent to total strains) across the shell thickness:

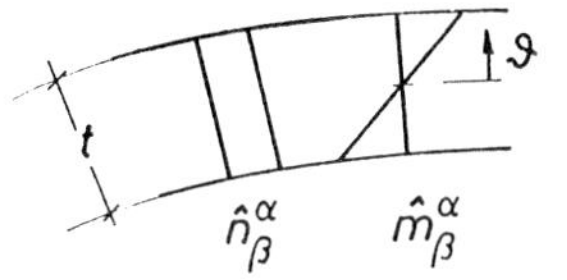

$$\hat{\bar{\tau}}^\alpha_\beta = \frac{1}{t}\,\hat{n}^\alpha_\beta + \frac{6}{t^2}\,\hat{m}^\alpha_\beta\,\frac{\vartheta}{t/2} \quad . \tag{4.1}$$

Fig.1.: Distribution of fictitious stress increments across the shell thickness

Inserting the stress assumptions into the elastic part of the material law (3.11) gives exactly a linear function for the tangential total strain increments

$$\bar{\gamma}^\alpha_\beta = \alpha^\alpha_\beta + \vartheta\,\beta^\alpha_\beta \quad . \tag{4.2}$$

Admissible functions for the displacements are

$$\bar{u}^\alpha = v^\alpha + \vartheta\,w^\alpha \quad , \qquad \bar{u}^3 = v^3 \quad . \tag{4.3}$$

Inserting these functions into the strain-displacement relation, see (3.21), gives a quadratic function for the strains, which on the other hand should have a linear distribution. The quadratic term is not compatible with the strain assumption, but the error will be minimized due to the variation procedure, together with additional errors resulting from approximations for plastic strains. The functional (3.23) relates total strain increments to the difference of fictitious elastic and plastic stress increments. Thus, it seems most natural to define fictitious stress resultants according to total strain resultants and in addition the plastic resultants.
During the loading history the stress distribution across the thickness may become highly nonlinear although the total strains remain linear. This results from the dependence of the plastic strain increments not only of the total strain increments but also of the previous stress state, see (3.13). For this reason the distribution of plastic strain increments necessarily must become nonlinear. Probably the simplest approximation which produces an adequate description of the elasto-plastic stress distribution must subdivide the thickness into an inner elastic layer and one or two outer plastic zones, as was proposed e. g. by Wempner [10].

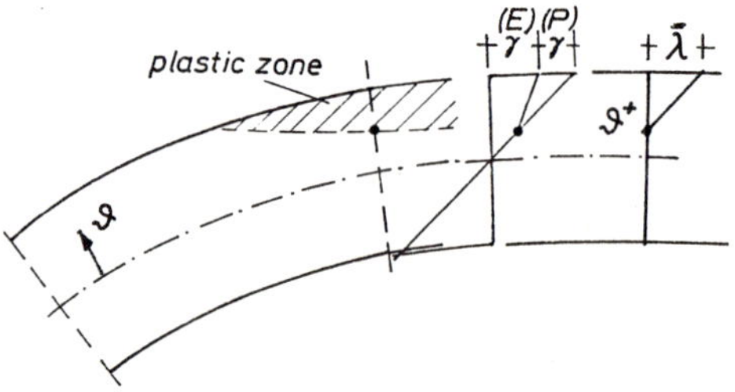

$$\bar{\lambda}(\vartheta) = \lambda^{+}(\vartheta - \vartheta^{+}) > 0 \qquad with \quad \bar{\lambda}(\vartheta \le \vartheta^{+}) = 0. \tag{4.4}$$

Fig.2.: Linear distribution of plastic multiplier λ across the plastic zone

Satisfactory results could then be achieved assuming a linear distribution for the increments of the plastic multiplier $\bar{\lambda}$ independent in each plastic zone. They are obtained in an optimal weighted sense as unknowns of the variational principle.

Fig. 2 shows a possible distribution of increments of total and plastic strains for an upper plastic zone. An assumption for a second lower zone is quite analogous regarding that $\bar{\lambda}$ must always be positive according to the loading condition. For simplicity only one zone will be considered in the further derivation, keeping in mind, that all resulting terms for one zone could be substituted by the sum of all plastic zones. The special case of nearly pure membrane stresses in the shell is included letting ϑ^{+} tend towards $-\infty$.

At this stage we must presuppose that the fundamental state, e. g. the stresses across the thickness and the position of the interface between elastic and plastic zone, is known and remains constant during an infinitesimal small loading increment. Deviations of this state during finite increments must be accounted for by usual iteration or subincrementation techniques. The stress distribution across the thickness may be arbitrary, e. g. stored at some discrete integration points, and it is presupposed that the yield condition holds for the stresses in the plastic zone.

Plastic strains in thin shells usually are determined for a plane stress assumption. This coincides with the choice of a linear tangential strain distribution. Thus for a complete plane stress state the shear stresses $\bar{\tau}_3^{\alpha}$ or $\bar{\tau}_{\alpha}^{3}$ and the lateral stresses $\bar{\tau}_3^{3}$ can be neglected. This would lead to a shell theory of Kirchhoff-Love type. It is also possible to retain the shear stresses, regarding them pure elastic, by introducing a quadratic polynomial for $\bar{\tau}_3^{\alpha}$.

Introduction of the assumptions into the functional (3.23) and performing the integration yields the two-dimensional form of the variational principle for the shell surface (e. g. including average shear deformations):

$$\delta J = \int_S \delta\{-[\hat{W}^*(\hat{n},\hat{m},q) - \overset{P}{W}{}^*(\lambda^+)] + [(\hat{n}^\alpha_\beta - \overset{P}{n}{}^\alpha_\beta)(v^\beta|_\alpha - b^\beta_\alpha\, v^3)]$$
$$+ [(\hat{m}^\alpha_\beta - \overset{P}{m}{}^\alpha_\beta)(w^\beta|_\alpha + (b^\lambda_\alpha - 2H_G\delta^\lambda_\alpha)(v^\beta|_\lambda - b^\beta_\lambda\, v^3))]$$
$$+ [q^3_\alpha\, w^\alpha + q^\alpha_3(v^3_{,\alpha} + b_{\alpha\lambda} v^\lambda)] - [\bar{\bar{p}}_\alpha\, v^\alpha + \bar{\bar{p}}_3\, v^3]\}\, dS$$
$$- \int_S (\bar{\bar{p}}_\alpha\, v^\alpha + p_3\, v^3)\, dS - \int_{S_p} n_\alpha(\bar{\bar{q}}{}^\alpha_\beta\, v^\beta + \bar{\bar{q}}{}^\alpha_3\, v^3 + \bar{\bar{\underset{\sim}{m}}}{}^\alpha_\beta\, w^\beta)\, dS \tag{4.5}$$
$$- \int_{S_u} u_\alpha [(\hat{q}^\alpha_\beta - \overset{P}{q}{}^\alpha_\beta)(v^\beta - \bar{\bar{v}}{}^\beta) + q^\alpha_3(v^3 - \bar{\bar{v}}{}^3) + (\hat{\underset{\sim}{m}}{}^\alpha_\beta - \overset{P}{\underset{\sim}{m}}{}^\alpha_\beta)(w^\beta - \bar{\bar{w}}{}^\beta)]\, dS = 0$$

with

$$W^*(\hat{n},\hat{m},\lambda^+,q) = \frac{1}{2}\{\frac{1}{D}(\hat{n}^\alpha_\beta\ \hat{n}^\beta_\alpha(1+\nu) - \nu\,\hat{n}^\varrho_\varrho\ \hat{n}^\lambda_\lambda)$$
$$+ \frac{1}{B}(\hat{m}^\alpha_\beta\ \hat{m}^\beta_\alpha(1+\nu) - \nu\,\hat{m}^\varrho_\varrho\ \hat{m}^\lambda_\lambda) \tag{4.6}$$
$$- 2H_G\frac{1}{D}(\hat{m}^\alpha_\beta\ \hat{n}^\beta_\alpha(1+\nu) - \nu\hat{m}^\varrho_\varrho\ \hat{n}^\lambda_\lambda)\,2$$
$$+ \frac{1}{D}\,2(1+\nu)\frac{6}{5}\, q^\alpha\ q_\alpha - H\,\lambda^+\lambda^+\}\ .$$

with

$$2H_G = b^\varrho_\varrho$$
$$D = E\cdot t$$
$$B = Dt^2/12$$

The quantities to be prescribed at the boundaries are given by

$$q^\alpha_\beta = n^\alpha_\beta + (b^\alpha_\lambda - 2H_G\delta^\alpha_\lambda)\, m^\lambda_\beta\ ,$$
$$\underset{\sim}{m}{}^\alpha_\beta = m^\alpha_\beta + \frac{B}{D}(b^\alpha_\lambda - 2H_G\delta^\alpha_\lambda)\, n^\lambda_\beta \tag{4.7}$$

where the expressions are to be taken with the appropriate superskript required in (4.5).

The arbitrary distribution of the stresses of the fundamental state requires a special treatment of the coefficients of the plastic multipliers λ^+. These terms, denoted as plastic-stress resultants $\overset{P}{n}{}^\alpha_\beta$, $\overset{P}{m}{}^\alpha_\beta$ and hardening parameter H are determined by integrating the deviatoric stresses multiplied by the form function of λ^+ and the shell shifter μ not across the whole thickness but across the plastic zone. Thus a numerical integration procedure is indicated to prevent cumbersome expressions, e. g. resulting from series expansions of the deviatoric stresses and the use of higher order moments:

$$
\begin{aligned}
\overset{p}{n}{}^{\alpha}_{\beta} &= 2G\lambda^{+}\,\overset{\circ}{N}{}^{\alpha}_{\beta} \quad , & \overset{\circ}{N}{}^{\alpha}_{\beta} &= \int_{\vartheta^{+}}^{t/2} \overset{\circ}{t}{}^{\alpha}_{\beta}(\vartheta-\vartheta^{+}) \quad d\vartheta \;\hat{=}\; \frac{\partial f}{\partial \overset{\circ}{n}{}^{\alpha}_{\beta}} \;, \\
\overset{p}{m}{}^{\alpha}_{\beta} &= 2G\lambda^{+}\,\overset{\circ}{M}{}^{\alpha}_{\beta} \quad , & \overset{\circ}{M}{}^{\alpha}_{\beta} &= \int_{\vartheta^{+}}^{t/2} \overset{\circ}{t}{}^{\alpha}_{\beta}(\vartheta-\vartheta^{+})\,\vartheta\, d\vartheta \;\hat{=}\; \frac{\partial f}{\partial \overset{\circ}{m}{}^{\alpha}_{\beta}} \;, \\
H &= \left(\frac{2}{3}\,\frac{EE_t}{E-E_t} + 2G\right) 2\overset{\circ}{J}_2 \;, & 2\overset{\circ}{J}_2 &= \int_{\vartheta^{+}}^{t/2} \overset{\circ}{t}{}^{\alpha}_{\beta}\, \overset{\circ}{t}{}^{\beta}_{\alpha}(\vartheta-\vartheta^{+})^2 \mu \; d\vartheta \quad .
\end{aligned}
\tag{4.8}
$$

The established variational form (4.5) under the given assumptions is still in closed form with no additional approximations and can be taken as the starting point of numerical discretization techniques. There are some quadratic terms neglected resulting from the mentioned incompatibility of displacement and strain assumptions, the admissibility is discussed extensionally in [4].

5. Basic equations of an elasto-plastic shell theory

Application of Gauß´ divergence theorem to the variational principle (4.5) renders the corrresponding local equations.

The elastic constitutive equations are given as subsidiary conditions of the principle. They follow from the complementary specific energy

$$
\begin{aligned}
\frac{\partial \hat{W}^{*}}{\partial \hat{n}^{\alpha}_{\beta}} &= \frac{1}{D}\left(\hat{n}^{\beta}_{\alpha}(1+\nu) - \nu\,\delta^{\beta}_{\alpha}\,\hat{n}^{\varrho}_{\varrho}\right) &&= \alpha^{\beta}_{\alpha} \quad , \\
\frac{\partial \hat{W}^{*}}{\partial \hat{m}^{\alpha}_{\beta}} &= \frac{1}{B}\left(\hat{m}^{\beta}_{\alpha}(1+\nu) - \nu\delta^{\beta}_{\alpha}\,\hat{m}^{\varrho}_{\varrho}\right) - 2H_{\varrho}\alpha^{\beta}_{\alpha} &&= \beta^{\beta}_{\alpha} \quad .
\end{aligned}
\tag{5.1}
$$

The equilibrium conditions and the kinematic relations follow by variation of displacements and fictitious stress resultants:

Equilibrium (5.2)

$$(\underbrace{\hat{n}^{\alpha}_{\beta} - \lambda\, 2G \mathring{N}^{\alpha}_{\beta}}_{\overset{P}{n}{}^{\alpha}_{\beta}})|_{\alpha} + \left[(b^{\lambda}_{\alpha} - 2H\delta^{\lambda}_{\alpha})(\underbrace{\hat{m}^{\alpha}_{\beta} - \lambda\, 2G \mathring{M}^{\alpha}_{\beta}}_{\overset{P}{m}{}^{\alpha}_{\beta}})\right]|_{\lambda} - b^{\alpha}_{\beta} q_{\alpha} + \bar{\bar{p}}_{\beta} = 0 ,$$

$$b^{\beta}_{\alpha}(\hat{n}^{\alpha}_{\beta} - \lambda\, 2G \mathring{N}^{\alpha}_{\beta}) - K(\hat{m}^{\varrho}_{\varrho} - \lambda\, 2G \mathring{M}^{\varrho}_{\varrho}) + q^{\alpha}|_{\alpha} + \bar{\bar{p}}_{3} = 0 ,$$

$$(\hat{m}^{\alpha}_{\beta} - \lambda\, 2G \mathring{M}^{\alpha}_{\beta})|_{\alpha} - q_{\beta} = 0 .$$

Kinematics (5.3)

$$\alpha^{\alpha}_{\beta} = \tfrac{1}{2}(v_{\alpha}|^{\beta} + v^{\beta}|_{\alpha}) - b^{\beta}_{\alpha} v^{3} ,$$

$$\beta^{\alpha}_{\beta} = -v_{3}|^{\beta}_{\alpha} - b^{\beta}_{\alpha}|_{\lambda} v^{\lambda} - b^{\lambda}_{\alpha} v_{\lambda}|^{\beta} - b^{\beta}_{\lambda} v^{\lambda}|_{\alpha} + b^{\lambda}_{\alpha}\varphi^{\beta}_{\lambda} + b^{\beta}_{\lambda}\varphi^{\lambda}_{\alpha} - 2H\varphi^{\beta}_{\alpha} + K\delta^{\beta}_{\alpha} v^{3}$$

$$\text{with } \varphi^{\beta}_{\alpha} = \tfrac{1}{2}(v_{\alpha}|^{\beta} + v^{\beta}|_{\alpha}) .$$

The equations (5.1)-(5.3) are exactly the same as in the linear elastic case, except that the elastic stress resultants are replaced by the difference of fictitious total and plastic resultants. The plastic resultants are dependent on the plastic multiplier λ^{+} determined by the global form of the flow rule:

Flow rule $$\mathring{N}^{\beta}_{\alpha} 2G \alpha^{\alpha}_{\beta} + \mathring{M}^{\beta}_{\alpha} 2G \beta^{\alpha}_{\beta} - 2\mathring{J}_{2}\left(2G + \frac{2E\,E_t}{3(E-E_t)}\right)\lambda^{+} = 0 . \qquad (5.4)$$

Thus, we have achieved a complete analogy calculating plastic stresses at discrete material points or plastic stress resultants for a special point of the shell reference surface with arbitrary stress distribution. The deviatoric resultants $\mathring{N}^{\alpha}_{\beta}$, $\mathring{M}^{\alpha}_{\beta}$, $\mathring{J}_{2}$ may be regarded as a numerical approximation of a v. Mises yield condition for stress resultants

$$\frac{\partial f}{\partial \mathring{n}^{\alpha}_{\beta}} \hat{n}^{\alpha}_{\beta} + \frac{\partial f}{\partial \mathring{m}^{\alpha}_{\beta}} \hat{m}^{\alpha}_{\beta} - H\lambda^{+} = 0 . \qquad (5.5)$$

An analytical form for such a global yield condition could not be given until now. Many authors tried to develop some semi-empirical yield conditions which are valid at least only for special cases, e. g. monotonic, proportional loading [5, 6, 7]. Under general loading conditions these yield conditions fail totally, and some semi-numerical techniques must be applied.

6. Concluding remarks

An elasto-plastic shell theory is proposed which permits the calculation of plastic stress and strain resultants directly from total strain resultants for the shell reference surface. Therefore linear distributions for total strains and displacements across the shell thickness and linear distributions of plastic multipliers λ in the plastic zones were assumed.

The parameters of the generalized yield condition are to be determined numerically. In the respective numerical integration only the current deviatoric stresses across the thickness are involved. This fact can be used advantageous in iterative solution algorithms like the initial stress method. If one assumes that the current state does not change during a load increment, thus ignoring the incremental expansion of the plastic zone, the evaluation of the numerical integration of the deviatoric quantities $\mathring{N}^{\alpha}_{\beta}$, $\mathring{M}^{\alpha}_{\beta}$, $\mathring{J}_2$ is required only once per increment and not in each iteration step. In contrast, having no global yield condition and employing a layered model, the complete calculation of plastic strains in each integration point across the thickness and the numerical integration of the strains must be performed in each iteration step. Thus, there is a potential reserve in saving computer time when applying an iterative solution technique. Further research could be spend to include the expansion of the plastic zone during a finite increment, e. g. by extrapolation of the generalized yield condition.

In the case of monotonic loading, the determination of plastic zones is quite simple. Inserting the stress assumption (4.1) into the initial yield condition (3.5) gives a quadratic equation for the yield limit

$$\begin{bmatrix} \vartheta^+ \\ \vartheta^- \end{bmatrix} = \frac{-P_{NM} \pm [P_{NM}^2 - 4P_M(P_N - 2\sigma_F^2)]^{1/2}}{2P_M}$$

$$P_N = 3\mathring{N}^{\alpha}_{\beta}\,\mathring{N}^{\beta}_{\alpha} - \mathring{N}^{\rho}_{\rho}\,\mathring{N}^{\lambda}_{\lambda}$$

$$P_M = 3\mathring{M}^{\alpha}_{\beta}\,\mathring{M}^{\beta}_{\alpha} - \mathring{M}^{\rho}_{\rho}\,\mathring{M}^{\lambda}_{\lambda} \qquad \mathring{N}^{\alpha}_{\beta} = \frac{1}{t}\,\hat{n}^{\alpha}_{\beta} \tag{6.1}$$

$$P_{NM} = 6\mathring{N}^{\alpha}_{\beta}\,\mathring{M}^{\beta}_{\alpha} - 2\mathring{N}^{\rho}_{\rho}\,\mathring{M}^{\lambda}_{\lambda} \qquad \mathring{M}^{\alpha}_{\beta} = \frac{12}{t^3}\,\hat{m}^{\alpha}_{\beta}\ .$$

If $\vartheta^+ > t/2$ or $\vartheta^- < -t/2$ then there is still no plastic zone. Another special case is remarkable, e. g. if $\vartheta^+ < -t/2$ then the cross section is fully plastified and if ϑ^+ tends towards infinity, there are no bending strains but pure extensional plastic strains.Some further developement is needed to determine the plastic zones under arbitrary cycling loading conditions. This is possible in general by tracing the yield limits and yield stresses etc. across the shell thickness, but this takes additional computational effort.

7. References

1. Koiter, W.T.: A consistent first approximation in the general theory of thin elastic shells. Proc. Symposium on the Theory of Thin Elastic Shells, Delft 1959.

2. Naghdi, P.M.: Foundations of elastic shell theory. (Progress in Solid Mechanics, Vol. IV). Amsterdam 1963, North Holland Publishing Company.

3. Basar, Y.; Krätzig, W.B.: Mechanik der Flächentragwerke. Braunschweig 1985, Vieweg & Sohn Verlag

4. Wunderlich, W.: On a Consistent Shell Theory in Mixed Tensor Formulation, III. IUTAM Symposium, Tbilisi. In: "Theory of Shells", North-Holl. Pub. Co. (1980).

5. Ilyushin, A.A.: Plasticite. Eyrolles, Paris 1956 (French translation from Russian, Moskwa 1948).

6. Bieniek, M.P.; Funaro, J.R.: Elastic-Plastic Behavior of Plates and Shells. Techn. Rep. DNA 3954T, Weidlinger Associates, New York 1976.

7. Sawczuk, A.: On plastic analysis of shells. In: "Theory of Shells", North-Holl. Pub. Co. (1980).

8. Wunderlich, W.: Incremental Formulation of the Generalized Variational Approach in Structural Mechanics, Proc. Intern. Conf. on Variational Methods in Engineering, Southampton 1972.

9. Wunderlich, W.; Rensch, H.J.; Obrecht, H.: Analysis of Elastic-Plastic Buckling and Imperfection-Sensitivity of Shells of Revolution, In: "Buckling of Shells", Springer-Verlag 1982.

10. Wempner, G.: Discrete Approximations of elastic-plastic bodies by variational methods. Proc. Intern. Conf. on Variational Methods in Engineering, Southampton 1972.

Limit Analysis of Shells of Revolution using Two Finite Element Approaches

N.ZOUAIN PEREIRA*, R.A. FEIJÓO**, E. TAROCO**, L. BEVILACQUA*

* Catholic University of Rio de Janeiro (Brazil)
**Laboratório Nacional de Computação Científica (Brazil)

Summary

Two numerical methods are presented for the limit analysis of shells of revolution with arbitrary shape and using the sandwich Tresca model. The statical method is developed by constructing an approximately equilibrated interpolation of stresses. Equilibrium equations are exactly as derived from the correct geometry of the shell. Plastic admissibility is partially enforced choosing representative points. The kinematical method includes FEM interpolations of velocity and plastic multiplier rates to compute linear expressions for internal and external dissipations. Both interpolations are constrained to fulfil plastic kinematical admissibility in a set of freely selected points. The numerical methods presented are demonstrated in applications related to circular plates, tubes, cones, spherical caps and a cylindrical nozzle.

1. Introduction

The aim of this work is to present two numerical methods suited for limit analysis of shells of revolution with arbitrary shape and axisymmetric loadings. Even for these geometry and loads, and material isotropy, a non-axisymmetric collapse may occur but this possibility, as for any geometrical instability phenomenon, is not considered here.

The numerical methods are developed by transforming the first and second plastic collapse theorems [1], stated for variable static and kinematical fields, into discrete versions where variables are finite dimensional vectors collecting interpolation parameters of FEM approximations of the corresponding fields. In this way two independent numerical methods are generated to solve the same critical load problem.

Finite element interpolation for stresses and velocities are

Inelastic Behaviour of Plates and Shells
IUTAM Symposium Rio de Janeiro 1985
Editors: L. Bevilacqua, R. Feijóo and R. Valid

constructed without any approximation of the geometry of meridian, and consequently using true kinematical and equilibrium equations. The meridian shape must then be prescribed in explicit form, also in the computer codes. A different approach has been adopted for instance by Nguyen Dang Hung et al [2] to treat the same aspect of the problem.

In the static method the stress interpolation is selected to satisfy two of the three equilibrium equations. The remaining one is imposed in a discrete set of points by enforcing a system of linear constraints on variable coefficients of interpolation.

Even when the considered stress fields are related to a finite number of interpolation coefficients, the static theorem still imposes an infinite number of plastic admissibility constraints. A discrete static formulation is only achieved replacing these constraints by a finite set of representative restrictions. We use for this purpose the collocation method and let the number and coordinates of the representative points be chosen independently of interpolation nodes and equilibrium points.

This statical discrete formulation of limit analysis becomes a linear programming problem if the yield function is piecewise-linear. If this is not the case, it is necessary to choose either to proceed with non-linear programming techniques or to linearize the yield condition and solve a linear problem. Recalling that there exists a set of linear equilibrium constraints in the formulation, the linearization of remaining constraints seems preferable. This is achieved with the sandwich Tresca model of the shell [3].

For the kinematical formulation, the problem of obtaining a discrete linear version is completely different. As a first step, the space of possible solution velocity fields is replaced by a finite dimensional one by means of FEM interpolation. Also a linearization of plastic limit is necessary but not sufficient to reach a linear programming formulation.

For a piecewise-linear yield function the internal dissipation is linearly expressed in terms of plastic multiplier rates. For this reason an independent approximation is constructed for plastic multipliers rates, and admissibility constraints are induced, enforcing that the deformation rates related to plastic multipliers are equal to the deformation rates derived from approximate velocity used in external power computations.

2. Model of the Shell

The kinematical and equilibrium equations of a general shell of revolution under axisymmetric loading are settled in this section using cylindrical (e_r, e_θ, e_z) and intrinsic (e_ϕ, e_θ, e_w) reference frames shown in Figure 1.

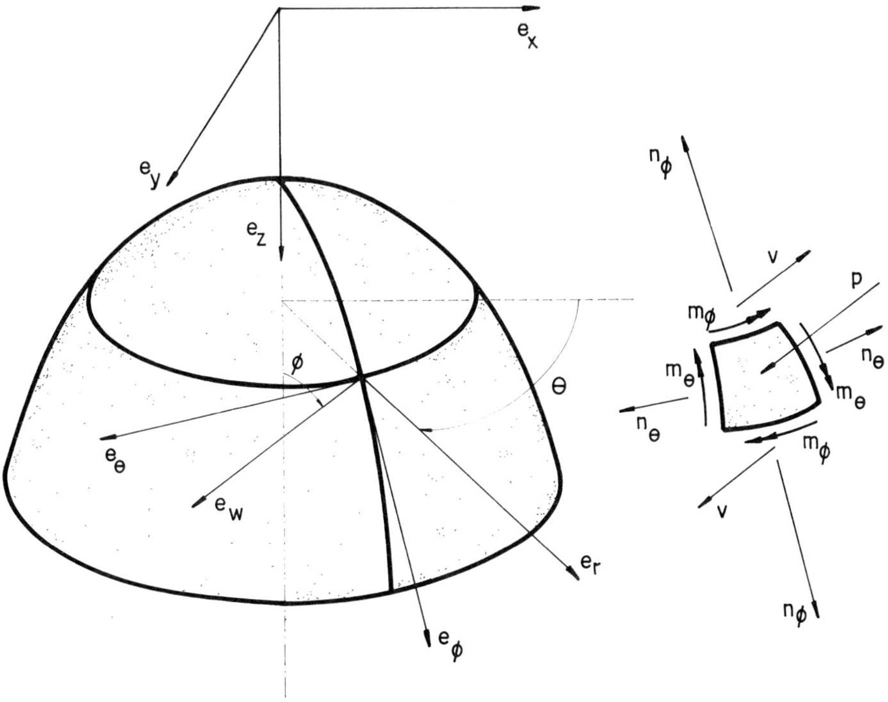

Fig. 1. Shell of revolution

In order to use dimensionless variables the following reference values are adopted

L : characteristic length

N_o: yielding membrane force in pure traction

M_o: yielding moment in pure bending

The dimensionless variables are defined as follows.

Coordinates (divided by L)

r : cylindrical radial coordinate

z : axial coordinate

s : curvilinear coordinate along the meridian

Displacements (divided by L)

$$u = u_\phi e_\phi + w e_w = u_r e_r + u_z e_z \tag{1}$$

Generalized strain

$$q = [\varepsilon_\phi \quad k_\phi \quad \varepsilon_\theta \quad k_\theta]^T \tag{2}$$

$\varepsilon_\phi, \varepsilon_\theta$: membrane deformations in meridian and parallel directions

k_ϕ, k_θ: curvature deformations in meridian and parallel directions (divided by N_o/M_o)

Generalized stress

$$Q = [n_\phi \quad m_\phi \quad n_\theta \quad m_\theta]^T \tag{3}$$

n_ϕ, n_θ: membrane forces in meridian and parallel directions (per unit length of parallel and divided by N_o)

m_ϕ, m_θ: bending moments in meridian and parallel directions (per unit length of parallel and divided by M_o)

v : shear force in meridional direction (per unit length of parallel and divided by N_o)

External loads

p : normal pressure (divided by N_o/L)

f_r, f_z: radial and axial components of a load concentrated along a parallel (per unit length of parallel and divided by N_o)

Dissipation power

d_{int}, d_{ext}: internal and external dissipation (divided by $2\pi L^2 N_o$)

To write the equations for dimensionless variables we denote by ( )' derivatives with respect to s, by r_1 the curvature of the meridian, and we use $\beta = M_o/LN_o$ as dimensionless thickness of the shell.

The kinematical equations defining the deformation operator assumed are

$$\varepsilon_\phi = u'_\phi - \frac{w}{r_1} \qquad k_\phi = \beta\gamma' \tag{4,5}$$

$$\varepsilon_\theta = \frac{1}{r}(u_\phi \cos\phi - w \sin\phi) \qquad k_\theta = \frac{\beta \cos\phi}{r}\gamma \tag{6,7}$$

where $\gamma = -u_\phi/r_1 - w'$ is the rotation of a meridian element.

The corresponding differential equilibrium equations are

$$(rn_\phi)' - \cos\phi\, n_\theta - \frac{r}{r_1} v = 0 \tag{8}$$

$$\sin\phi\, n_\theta + \frac{r}{r_1} n_\phi + (rv)' + rp = 0 \tag{9}$$

$$\beta[(rm\ )' - \cos\phi\, m_\theta] - rv = 0 \tag{10}$$

For the assumed loads the external dissipation is computed as follows

$$d_{ext} = \int_0^{\bar{s}} p\dot{w}r\, ds + \Sigma(f_r\dot{u}_r + f_z\dot{u}_z)r \tag{11}$$

with the summation extended to all points of the meridian where annular loads are applied.

The internal dissipation for a particular deformation rate $\dot{q}$

can be evaluated using the corresponding stress value Q (associated with $\dot{q}$, considered purely plastic, by the constitutive relation) and computing

$$d_{int} = \int_0^{\bar{s}} (n_\phi \dot{\varepsilon}_\phi + m_\phi \dot{k}_\phi + n_\theta \dot{\varepsilon}_\theta + m_\theta \dot{k}_\theta) r ds + \Sigma (n_\phi \Delta\dot{u}_\phi + \beta m_\phi \Delta\dot{\gamma}) r \qquad (12)$$

where summation extends to all points undergoing expansion (and) or rotational hinge deformation of intensities $\Delta\dot{u}_\phi$ and $\Delta\dot{\gamma}$.

A sandwich shell model approximates the homogeneous shell if we set $N_o = h\sigma_Y = 2t\sigma'_Y$ and $M_o = h^2\sigma_Y/4 = th'\sigma'_Y$ where h and σ_Y are the thickness and yielding stress for the homogeneous shell, and h', t and σ'_Y are the thickness, layers thickness and layers yielding stress for the sandwich shell.

Both internal (-) and external (+) membranes of the sandwich shell are in plane stress state, therefore the Tresca yield condition states

$$\max(|\sigma_1|, |\sigma_2|, |\sigma_1 - \sigma_2|) < \sigma'_Y \qquad (13)$$

where the principal stresses σ_1 and σ_2 are either σ_ϕ or σ_θ due to the symmetry of geometry and loads. These stresses are related to generalized ones by

$$\begin{aligned} \sigma_\phi^+ &= \sigma'_Y(n_\phi - m_\phi) \qquad & \sigma_\phi^- &= \sigma'_Y(n_\phi + m_\phi) \\ \sigma_\theta^+ &= \sigma'_Y(n_\theta - m_\theta) \qquad & \sigma_\theta^- &= \sigma'_Y(n_\theta + m_\theta) \end{aligned} \qquad (14)$$

Replacement of these equations in Tresca condition for each skin of sandwich shell leads to the following set of 12 inequalities [3,7]

$$\Phi(Q) = N^T Q - R \leqq 0 \qquad (15)$$

where N is a constant matrix with columns been the normals to each of the twelve linear plastic modes, and R a constant vector containing the corresponding distances of each mode to the origin (that is, one for the present case). Therefore the yielding limit of the model is piecewiselinear.

According to the associated flow rule, the plastic strain rate is a linear combination of the gradients of all modes, that is

$$\dot{q}^p = N\lambda \tag{16}$$

where λ is a vector of twelve non-negative components, each one of them being zero for non-active yielding modes, that is to say

$$\Phi \cdot \lambda = 0 \qquad \lambda \geqq 0 \qquad \Phi \leqq 0 \tag{17}$$

The above equality allows us to compute the specific internal dissipation corresponding to a pure plastic strain rate. Indeed $\Phi \cdot \lambda = N^T Q \cdot \lambda - R \cdot \lambda = Q \cdot \dot{q}^p - R \cdot \lambda$ so that

$$Q \cdot \dot{q}^p = R \cdot \lambda \tag{18}$$

whenever Q, $\dot{q}^p$ and λ are related by the constitutive equations.

3. A Statical Method for Limit Analysis of Shells of Revolution

We develop in this section a numerical method to solve the limit load problem by interpolation of the statical variables and application of the first theorem of plastic collapse.

The base functions for the stress representation are constructed by the FEM and trying to satisfy equilibrium constraints independently of coefficient values. This is only partially achieved. The plastic admissibility of interpolated stresses is then imposed approximately by means of linear constraints on interpolation parameters.

The shape of the meridian curve is not approximated as in [2] for instance, but assumed to be given exactly by functions $r(s)$ and $\phi(s)$. In this way we formulate exact equilibrium equations and propose stress interpolation functions that satisfy identically two of these equations and approximately the remaining one.

The meridian is discretized in n_e finite elements by selecting $n_n=n_e+1$ nodes. In each element i of length $2\ell^i$ it is defined a normalized dimensionless variable $\eta \in [-1,1]$ such that

$$(\cdot)' = \frac{1}{\ell^i}\frac{d\cdot}{d\eta} \tag{19}$$

The stress interpolation is developed in what follows, denoting with $h_n(\eta)$ (n=1,2,3,4) the cubic Hermite polynomials.

3.1. Stress interpolation

i) Interpolation functions $\hat{m}_\phi$ and $\hat{m}_\theta$ for bending efforts are chosen first taking into account continuity requirements. A cubic approximation is used for the product rm_ϕ, and inter-element continuity for m_ϕ is ensured adopting nodal values $r^i m_\phi^i$ as interpolation parameters. It is assumed a piecewise constant interpolation for m_θ because no continuity requirement exists for the variation of m_θ in the meridional direction. Consequently

$$r\hat{m}_\phi = h_1 r^i m_\phi^i + h_2 r^{i+1} m_\phi^{i+1} + \ell^i h_3 (rm_\phi)'^{\,i+} + \ell^i h_4 (rm_\phi)'^{\,(i+1)-} \tag{20}$$

$$\hat{m}_\theta = m_\theta^i \tag{21}$$

Super-index i+ denotes the value of the corresponding function computed as defined in the element i+1 following node i, and super-index i- means that it is computed in the element i before node i. This notation is necessary in the case of discontinuous functions such as $(rm_\phi)'$.

ii) To satisfy identically the third equilibrium equation the shear force v is interpolated by the function $\hat{v}$ such that

$$r\hat{v} = \beta[(r\hat{m}_\phi)' - \cos\phi\ \hat{m}_\theta] \tag{22}$$

Using this equation the expression of m_ϕ given in Eq. 20 transforms into

$$\hat{m}_\phi = \frac{h_1 r^i}{r} m_\phi^i + \frac{h_2 r^{i+1}}{r} m_\phi^{i+1} +$$

$$+ \frac{\ell^i}{r} (h_3 \cos \phi^{i+} + h_4 \cos \phi^{(i+1)-}) m_\theta^i + \frac{\ell^i h_3 r^i}{\beta r} v^{i+} +$$

$$+ \frac{\ell^i h_4 r^{i+1}}{\beta r} v^{(i+1)-} \qquad (23)$$

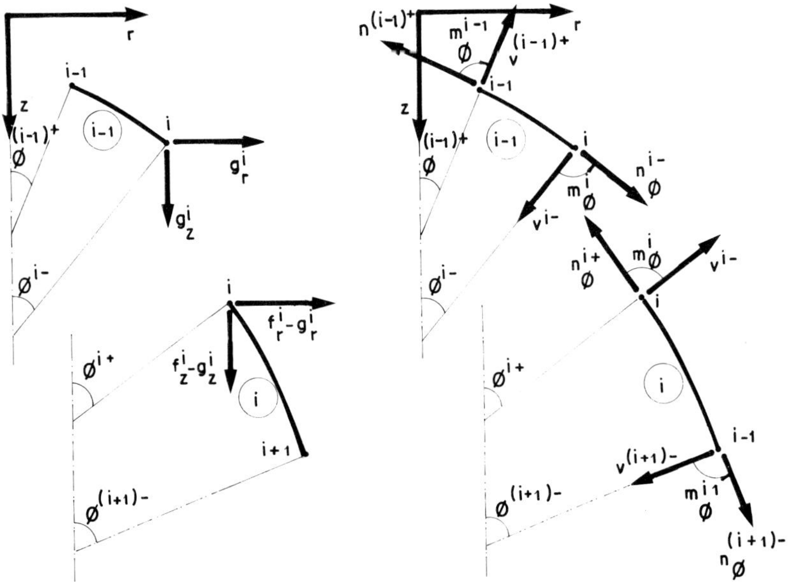

Fig. 2. Loads and stresses in local and global coordinates, for two adjacent elements

iii) A linear approximation is adopted for the function rn_ϕ. Membrane force n_ϕ may be discontinuous in nodes where there is a jump in meridian slope or an annular force is applied. Therefore

$$\hat{n}_\phi = \frac{r^i(1-\eta)}{2r} n_\phi^{i+} + \frac{r^{i+1}(1+\eta)}{2r} n_\phi^{(i+1)-} \qquad (24)$$

iv.1) The first equilibrium equation is ensured in every point along the element if the approximate n_θ is chosen as

$$\hat{n}_\theta = \frac{1}{\cos\phi}\,[(r\hat{n}_\phi)' - \frac{r\hat{v}}{r_1}] \qquad (25)$$

This procedure is not possible when the element has the shape of a cylinder because cos ϕ is zero in this case. We call (P) the interpolation developed in this item (iv.1), suitable for plates and all shell geometries others than cylinder. We develop in the next item (iv.2) an interpolation of type (C), appropriate for cylinder and other shell elements excepting annular plates.

Replacing the expressions of $\hat{n}_\phi$ and $\hat{v}$ previously obtained, Eqs. 24,22 and 23, into the last equation it follows that the interpolation of type (P) for n_θ is

$$\hat{n}_\theta = -\frac{\beta h_1' r^i}{r_1\cos\phi}\,m_\phi^i - \frac{\beta h_2' r^{i+1}}{r_1\cos\phi}\,m_\phi^{i+1} -$$

$$-\frac{\beta(\ell^i h_3'\cos\phi^{i+} + \ell^i h_4'\cos\phi^{(i+1)-} - \cos\phi)}{r_1\cos\phi}\,m_\theta^i -$$

$$-\frac{\ell^i h_3' r^i}{r_1\cos\phi}\,v^{i+} - \frac{\ell^i h_4' r^{i+1}}{r_1\cos\phi}\,v^{(i+1)-} -$$

$$-\frac{r^i}{2\ell^i\cos\phi}\,n_\phi^{i+} + \frac{r^{i+1}}{2\ell^i\cos\phi}\,n_\phi^{(i+1)-} \qquad (26)$$

iv.2) The second equilibrium equation becomes identically satisfied if we adopt

$$\hat{n}_\theta = -\frac{1}{\sin\phi}\,[\frac{r\hat{n}_\phi}{r_1} + (r\hat{v})' + rp] \qquad (27)$$

Only piecewise constant pressure is allowed in what follows

$$p = \alpha\bar{p}^i \qquad (28)$$

where α is the amplifying factor for all external loads, which is the principal variable of the problem. The interpolation of type (C) for n_θ is then obtained replacing Eqs. 24, 22, 23 and 28 in Eq. 27.

The four interpolation expressions for n_ϕ, m_ϕ, n_θ and m_θ constructed so far, Eqs. 24, 23, 26 and 21, define a matrix field $Y_Q^i(\eta)$ with 4 rows and 7 columns, and a vector field $\bar{Q}^i(\eta)$ with 4 components and equilibrated with surface load p. These matrices $Y_Q^i(\eta)$ and $\bar{Q}^i(\eta)$ describe the interpolation operator. The above mentioned four equations are then written in the form

$$\hat{Q} = Y_Q^i(\eta) a^i + \alpha \, \bar{Q}^i(\eta) \tag{29}$$

where a^i is the vector of interpolation parameters of element i, in local coordinates, that is

$$a^i = [n_\phi^{i+} \quad v^{i+} \quad n_\phi^{(i+1)-} \quad v^{(i+1)-} \quad m_\phi^i \quad m_\phi^{i+1} \quad m_\theta^i]^T \tag{30}$$

and

$$\bar{Q}^i(\eta) = [0 \quad 0 \quad -\frac{r\bar{p}^i}{\sin\phi} \quad 0]^T \tag{31}$$

for (P)-interpolation, or all the components of $\bar{Q}^i(\eta)$ are zero when (C)-interpolation is used.

3.2. Equilibrium constraint in an element

In the interpolation scheme (P) previously described, the first and third equilibrium equations have been imposed identically. Since the whole basic approximation functions have been defined so far, we can only hope to satisfy the remaining second equation by restraining the interpolation coefficient values. The collocation method is adopted for this purpose. Consequently, the functions $\hat{n}_\theta$, $\hat{n}_\phi$, $\hat{v}$ and p written in Eqs. 26, 24, 22 and 28 are replaced in the second equilibrium relation, Eq. 9 to give

$$\beta r^{i}(h_1''\cos\phi-\frac{h_1'\sin\phi}{r_1})m_{\phi}^{i}+\beta r^{i+1}(h_2''\cos\phi-\frac{h_2'\sin\phi}{r_1})m_{\phi}^{i+1}+$$

$$+\beta\ell^{i}[(h_3''\cos\phi-\frac{h_3'\sin\phi}{r})\cos\phi^{i+}+(h_4''\cos\phi-$$

$$-\frac{h_4'\sin\phi}{r_1})\cos\phi^{(i+1)-}]m_{\theta}^{i}+\ell^{i}r^{i}(h_3''\cos\phi-\frac{h_3'\sin\phi}{r_1})v^{i+}+$$

$$+\ell^{i}r^{i+1}(h_4''\cos\phi-\frac{h_4'\sin\phi}{r_1})v^{(i+1)-}+$$

$$+\frac{r^{i}}{2}[\frac{(1-\eta)\cos\phi}{r_1}-\frac{\sin\phi}{\ell^{i}}]n_{\phi}^{i+}+\frac{r^{i+1}}{2}[\frac{(1+\eta)\cos\phi}{r_1}+$$

$$+\frac{\sin\phi}{\ell^{i}}]n_{\phi}^{(i+1)-}+r\cos\phi\,\bar{p}^{i}\alpha=0 \qquad (32)$$

For the interpolation alternative (C) we proceed in a similar way. In this case the remaining equilibrium equation is the first one. The resulting equilibrium, to be imposed by adjusting interpolation parameters, coincides with Eq. 32 corresponding to type (P) interpolation.

The last equation, that expresses the only remaining equilibrium condition, can be written in matrix form as

$$Y_E^{i}(\eta)a^{i}+y_E^{i}(\eta)\alpha=0 \qquad i=1,\ldots,n_e \qquad (33)$$

A set of n_{eq} points is prescribed in each element in order to impose this condition, at least for these points. Since this equation is not an identity between polynomial functions, except for some particular shell geometries, it is not possible in general to guarantee this condition identically, no matter how many points are used to enforce its local validity.

3.3. Equilibrium constraints in global coordinates

We define a vector S of all global interpolation parameters

g_r^i, g_z^i, m_ϕ^i and m_θ^i, a vector $\bar{F}$ of prescribed nodal forces (to be multiplied by the load factor α) and a vector F_R of nodal reactions (to be considered as additional variables). Then we can set

$$a^i = T^i L^i S + T_F^i(\alpha \bar{L}^i \bar{F} + L_R^i F_R) \tag{34}$$

where L^i, $\bar{L}^i$ and L_R^i are boolean matrices and T^i and T_F^i are rotation matrices.

This equation is replaced in Eq. 32 and the resulting equilibrium condition is enforced in n_{cq} points per element, that is

$$Y_E^i(\eta_j)T^i L^i S + Y_E^i(\eta_j)T_F^i L_R^i F_R + (Y_E^i(\eta_j)T_F^i \bar{L}^i \bar{F} + Y_E^i(\eta_j)) = 0$$

$$i=1,\ldots,n_e \qquad j=1,\ldots,n_{eq} \tag{35}$$

This is a set of $n_{eq} \times n_e$ linear constraints for parameters S, and F_R that can be cast in matrix notation as

$$B^T S + HF_R - \alpha \mathbb{F} = 0 \tag{36}$$

Variables g_r^i, g_z^i and m_ϕ^i, i.e. the components of vector S, must obey some additional constraints corresponding to the particular conditions existing at both ends of the meridian.

3.4. Plastic admissibility of stresses

The condition that the approximate stress field has plastically admissible values at any point gives rise to an infinite number of linear constraints on interpolation coefficients S, load factor α and reactions F_R. In the discrete version of the problem a selected set of n_p points is used to represent this condition approximately.

The interpolation expression for generalized stress field Q(s), Eq. 29, is introduced in the plasticity inequality Eq. 15 and also the element parameters a^i replaced by global ones to obtain

$$N^T Y_Q^i(\eta_j) T^i L^i S + N^T Y_Q^i(\eta_j) T_F^i L_R^i F_R + (N^T Y_Q^i(\eta_j) T_F^i \bar{L}^i \bar{F} + N^T \bar{Q}^i(\eta_j))\alpha \leqq R$$

$$i=1,\ldots,n_e \qquad j=1,\ldots,n_p \tag{37}$$

This system of $12\times n_e \times n_p$ inequalities is referred to in what follows as

$$\mathbb{N}^T S + MF_R + \alpha\bar{Q} \leqq \mathbb{R} \tag{38}$$

3.5. Limit load computation in the statical approach

The static theorem of plastic collapse is applied to formulate the following linear programming problem

$$\tilde{\alpha} = \max_{\alpha, S, F_R} \alpha \tag{39}$$

under the constraints

$$\mathbb{N}^T S + MF_R + \alpha\bar{Q} \leqq \mathbb{R} \qquad \text{(plastic admissibility)} \tag{40}$$

$$\mathbb{B}^T S + HF_R - \alpha\bar{\mathbb{F}} = 0 \qquad \text{(equilibrium)} \tag{41}$$

$$C_1 S + C_2 F_R + \alpha C_3 \bar{F} = 0 \qquad \text{(boundary conditions)} \tag{42}$$

The number of variables of this problem, $4\times n_e + n_r + 4$ (where n_e is the number of elements and n_r the number of reaction components), changes to $8\times n_e + 2\times n_r + 7$ effective variables when this problem is transformed to the standard form involving positive variables. The number of constraints is essentially $12\times n_e \times n_p$ inequalities plus $n_e \times n_{eq}$ equalities (where n_p is the number of critical points for plastic admissibility) excepting boundary condition restraints.

4. A Kinematical Method for the Limit Analysis of Shells of Revolution

Finite element method is used next to build interpolation fields for displacements. Deformation operations are then applied to these functions to give interpolation expressions for generalized strains. This procedure ensures satisfaction of kinematical admissibility. However, it is not possible to use this strain field to compute internal dissipation as a linear combination of interpolation parameters, as needed to transform the problem into a linear programming one. To accomplish this an independent interpolation is used for plastic factor λ. Consequently, a kinematical compatibility condition appears when it is imposed that purely plastic strains $N\lambda$ used to compute internal dissipation, equals identically the strain field associated with the velocity field used to compute external dissipation.

Once internal and external dissipation are represented as linear combinations of interpolation coefficients, the kinematical theorem is applied to formulate the limit analysis as the problem of finding the coefficient values minimizing internal dissipation for unit external power under the additional kinematical admissibility constraint previously explained.

The kinematical admissibility will be exactly fulfilled only in a discrete number of points so that no guarantee is obtained that the approximate collapse load results are greater than or equal to the exact ones.

The deformation equations are dependent on meridian geometry, which is assumed to be exactly specified as in the static numerical method. Thus the functions $r(s)$ and $\phi(s)$ will appear in the interpolation operator for generalized strains.

4.1. Displacement interpolation

It is assumed a linear interpolation for the tangential

displacement u_ϕ and a quadratic one for the normal displacement w.

Inter-element discontinuities in u_ϕ and w are only allowed to accomplish with nodal jumps in the meridian slope. This means that expansion hinges are not represented in the approximation field set.

The values of the rotation of meridian γ, derived from u_ϕ and w approximations by the kinematical equation, may be different in a particular node when computed in each of the two elements pertaining to this node. Therefore, rotationally non-expansive hinges are included in interpolation fields.

According to previous assumptions

$$\hat{u} = Y_u(\eta)a^i \tag{43}$$

where a^i is the vector of element interpolation coefficients in local coordinates

$$a^i = [u_\phi^{i+} \quad w^{i+} \quad u_\phi^{(i+1)-} \quad w^{(i+1)-} \quad w^{ci}]^T \tag{44}$$

and $Y_u(\eta)$ is the interpolation matrix

$$Y_u(\eta) = \begin{bmatrix} \frac{1-\eta}{2} & 0 & \frac{1+\eta}{2} & 0 & 0 \\ 0 & h_1 & 0 & h_3 & h_2 \end{bmatrix} \tag{45}$$

where the normalized variable $\eta \in [-1,1]$ replaces s in the element i of length $2\ell^i$, and $h_n(\eta)$ (n=1,2,3) denote quadratic Lagrange polynomials.

Variables u_r^i, u_z^i and w^{ci} are collected in a vector U such that

$$a^i = T^i L^i U \tag{46}$$

where T^i is a rotation matrix and L^i a boolean matrix.

Supporting conditions of the shell impose nullity of some of the components of U. Also, variables u_r^1 and u_r^n must be constrained to be zero when the shell has closed ends.

4.2. Strain interpolation

The adopted displacement field Eq. 43 is introduced in the kinematical relation $\gamma=-u_\phi/r_1-w'$ to give

$$\hat{\gamma} = Y_\gamma^i(\eta)\,a^i \tag{47}$$

where

$$Y_\gamma^i(\eta) = \left[-\frac{1-\eta}{2r_1} \quad -h_1' \quad -\frac{1+\eta}{2r_1} \quad -h_3' \quad -h_2'\right] \tag{48}$$

The generalized strain vector q is now approximated applying deformation operations, Eqs. 4-7, to the interpolated u of Eq. 43. Thus

$$\hat{q} = Y_q^i(\eta)\,a^i \tag{49}$$

where the interpolation operator $Y_q^i(\eta)$ is shown in Table 1.

$q \backslash a^i$	u_ϕ^{i+}	w^{i+}	$u_\phi^{(i+1)-}$	$w^{(i+1)-}$	w^{ci}
ε_ϕ	$-\frac{1}{2\ell^i}$	$-\frac{h_1}{r_1}$	$\frac{1}{2\ell^i}$	$-\frac{h_3}{r_1}$	$-\frac{h_2}{r_1}$
k_ϕ	$-\frac{1}{2\ell^i r_1} + {}$ $+ \frac{1-\eta}{2}\left(\frac{1}{r_1}\right)'$	$-\beta h_1''$	$\frac{1}{2\ell^i r_1} + {}$ $+ \frac{1+\eta}{2}\left(\frac{1}{r_1}\right)'$	$-\beta h_3''$	$-\beta h_2''$
ε_θ	$\frac{(1-\eta)\cos\phi}{2r}$	$-\frac{h_1 \operatorname{sen}\phi}{r}$	$\frac{(1+\eta)\cos\phi}{2r}$	$-\frac{h_3 \operatorname{sen}\phi}{r}$	$-\frac{h_2 \operatorname{sen}\phi}{r}$
k_θ	$\frac{\beta(\eta-1)\cos\phi}{2r\,r_1}$	$-\frac{\beta h_1' \cos\phi}{r}$	$-\frac{\beta(1+\eta)\cos\phi}{2r\,r_1}$	$-\frac{\beta h_3' \cos\phi}{r}$	$-\frac{\beta h_2' \cos\phi}{r}$

Table 1. Strain interpolation matrix $Y_q^i(\eta)$

4.3. Interpolation of plastic multipliers

The yielding factors

$$\lambda = [\lambda_1 \ \lambda_2 \ \dots \ \lambda_{11} \ \lambda_{12}]^T \tag{50}$$

are field variables of the problem. They are interpolated quadratically because they are related with strains by the plastic kinematical admissibility. We recall that strain interpolation resulted almost quadratic and the plastic kinematical admissibility is given by the constant matrix N. Therefore

$$\hat{\lambda}_j = h_1 \lambda_j^{i+} + h_2 \lambda_j^{ci} + h_3 \lambda_j^{(i+1)-} \qquad j=1,\dots,12 \tag{51}$$

where $h_n(\eta)$ are quadratic Lagrange polynomials.

The previous equations take the form

$$\hat{\lambda} = Y_\lambda(\eta) \Lambda^i \tag{52}$$

when we define the element vector of interpolation parameters for plastic multipliers

$$\Lambda^i = [\lambda_1^{i+} \ \lambda_1^{ci} \ \lambda_1^{(i+1)-} \ \dots \ \lambda_{12}^{i+} \ \lambda_{12}^{ci} \ \lambda_{12}^{(i+1)-}]^T \tag{53}$$

and the following interpolation matrix, of size 12×36,

$$Y_\lambda(\eta) = \begin{bmatrix} h_1 & h_2 & h_3 & & & & & & \\ & & & h_1 & h_2 & h_3 & & & \\ & & & & & & \ddots & & \\ & & & & & & h_1 & h_2 & h_3 \end{bmatrix} \tag{54}$$

There is also a global vector of parameters Λ, collecting all λ_j^{i+}, λ_j^{ci} and λ_j^{i-}, such that

$$\Lambda^i = L_\lambda^i \Lambda \tag{55}$$

The approximated plastic multiplier fields $\hat{\lambda}(s)$ should be positive or zero in any point along the meridian. Instead of this strong condition it will only be imposed that the interpolation parameters are non-negative, so implying that $\hat{\lambda}(s)$ is also non-negative in three points on each element.

As we mentioned before the independent interpolation of plastic factor λ is needed in order to obtain a linear expression for internal dissipation because this cannot be reached by direct use of approximated strain rates.

We have developed so far a representation of $\lambda(s)$ associated with distributed strain rate, then we must also develop a representation for λ parameters related to the non-expansive hinges admitted in the approximate strain rate fields. In fact, it is easy to compute the meridian slope jump $(\Delta\dot{\gamma})^i$ in any node by using displacement fields in the elements pertaining to this node, however the dissipation expression $M_o|(\Delta\dot{\gamma})^i|$ is not linear due to the absolute value involved.

According to the previous considerations we define nodal variables λ_+^i and λ_-^i respectively related to positive hinges (outward concavity) and negative hinges. These variables are defined by the following properties:

i) they are non-negative and complementaries to $(\Delta\dot{\gamma})^i$, that is

$$\begin{aligned} &\lambda_+^i \geqq 0 \quad \text{and} \quad \lambda_-^i = 0 \quad \text{if} \quad (\Delta\dot{\gamma})^i \geqq 0 \\ &\lambda_+^i = 0 \quad \text{and} \quad \lambda_-^i \geqq 0 \quad \text{if} \quad (\Delta\dot{\gamma})^i \leqq 0 \end{aligned} \tag{56}$$

ii) their difference is the hinge angle

$$(\Delta\dot{\gamma})^i = \lambda_+^i - \lambda_-^i \tag{57}$$

iii) their sum is the concentrated dissipation

$$|(\Delta\dot{\gamma})^i| = \lambda_+^i + \lambda_-^i \tag{58}$$

4.4. Internal dissipation

The power of internal efforts is written as a linear combination of interpolation parameters for λ when Eqs. 18 and 58 are used in the expression of internal dissipation Eq. 12. This procedure leads to

$$d_{int} = \sum_{i=1}^{n_e} \left[\int_{s^i}^{s^{i+1}} R^T Y_\lambda(\eta) r ds \right] \Lambda^i + \sum_{i=1}^{n_n} (\lambda_+^i + \lambda_-^i) r^i \qquad (59)$$

$$d_{int} = \sum_{i=1}^{n_e} J^i \Lambda^i + \sum_{i=1}^{n_n} (\lambda_+^i + \lambda_-^i) r^i \qquad (60)$$

where the vector of 36×1 components

$$J^i = [J_1^i \quad J_2^i \quad J_3^i \overset{12}{\cdots} j_1^i \quad j_2^i \quad j_3^i] \qquad (61)$$

contains the following integrals of Lagrange polynomials

$$j_n^i = \ell^i \int_{-1}^{1} r h_n d\eta \qquad n=1,2,3 \qquad (62)$$

Hence the internal dissipation is the known linear form

$$d_{int} = \mathbf{R} \cdot \Lambda + \mathbf{R}_R \cdot (\lambda_+ + \lambda_-) \qquad (63)$$

in the variables λ_j^i, λ_+^i and λ_-^i collected in global vectors Λ, λ_+ and λ_- respectively.

4.5. External Power

The power of reference external loads, i.e. pressure $\bar{p}(s)$ and the nodal forces $\bar{F}$, is

$$d_{ext} = \sum_{i=1}^{n_e} \int_{s^i}^{s^{i+1}} \bar{p}\dot{w} r \, ds + \sum_{i=1}^{n_n} (\bar{f}_r^i \dot{u}_r^i + \bar{f}_z^i \dot{u}_z^i) r^i \qquad (64)$$

Assuming constant pressure along each element and replacing the velocities by the interpolation given before, Eqs. 43 and

46 we get

$$d_{ext} = \sum_{i=1}^{n_e} \bar{p}^i J_w^i T^i L^i \dot{U} + \sum_{i=1}^{n_n} r^i (\bar{f}_r^i \dot{u}_r^i + \bar{f}_z^i \dot{u}_z^i) \qquad (65)$$

where

$$J_w^i = [0 \quad j_1^i \quad 0 \quad j_3^i \quad j_2^i] \qquad (66)$$

External dissipation is then of the form

$$d_{ext} = \bar{\mathbb{F}} \cdot \dot{U} \qquad (67)$$

4.6. Plastic kinematical admissibility

i) along the interior of an element

The product $R \cdot \lambda$ correctly represents the internal specific dissipation corresponding to a certain strain rate $\dot{q}$, as has been assumed in the previous calculations, only if

$$\dot{q} = N\lambda \qquad (68)$$

Both fields $\dot{q}$ and λ have been independently approximated, then we can only force this equality to be true in a set of selected points, establishing in this way a system of linear constraints for interpolation parameters. Replacing Eqs. 49, 46,52 and 55 in the previous equation it follows that

$$Y_q^i(\eta_k) T^i L^i \dot{U} - N Y_\lambda(\eta_k) L^i \Lambda = 0 \quad i=1,\ldots,n_e \quad k=1,\ldots,n_p \qquad (69)$$

or in matrix notation

$$B\dot{U} - \mathbb{N}\Lambda = 0 \qquad (70)$$

ii) nodal

The rotation rate $\dot{\gamma}$ of an infinitesimal element of meridian has been interpolated independently in adjacent finite elements. So there is a nodal gap $(\Delta\dot{\gamma})^i$ representing the non-expansive hinge deformation rate intensity. This strain rate must be related to coefficients λ_+^i and λ_-^i by Eq. 57 so that internal dissipation is correctly computed by the latter variables, that is

$$(\Delta\dot{\gamma})^i = \lambda_+^i - \lambda_-^i$$

The interpolations of $\dot{\gamma}$ in the elements pertaining to node i are substituted in the above equation to give

$$(Y_\gamma^i(-1)T^iL^i - Y_\gamma^{i-1}(1)T^{i-1}L^{i-1})\dot{U} - \lambda_+^i + \lambda_-^i = 0 \qquad i=2,\ldots,n_e \qquad (71)$$

For the first and last nodes of the meridian this constraint adopts a slightly different form depending on clamping conditions and on whether the shell ends are open or closed.

Equation 71 together with the two equalities for shell ends are then written as

$$G\dot{U} - \lambda_+ + \lambda_- = 0 \qquad (72)$$

4.7. Limit load computation in the kinematical approach

The critical load factor $\tilde{\alpha}$ for prescribed pressure $\bar{p}$ and concentrated (annular) loads $\bar{F}$ is now approximated by solving the linear programming problem

$$\tilde{\alpha} = \min_{\dot{U},\Lambda,\lambda_+,\lambda_-} \mathbb{R}\cdot\Lambda + \mathbb{R}_R\cdot(\lambda_+ + \lambda_-) \qquad (73)$$

under the constraints

$\bar{\mathbb{F}}\cdot\dot{U}=1$	(unit external power)	(74)
$B\dot{U}-\mathbb{N}\Lambda=0$	(internal plastic admissibility)	(75)
$G\dot{U}-\lambda_+ + \lambda_- = 0$	(nodal plastic admissibility)	(76)
$C\dot{U}=0$	(boundary conditions)	(77)
$\Lambda \geqq 0$	(non-negative plastic factors)	(78)
$\lambda_+ \geqq 0 \quad \lambda_- \geqq 0$	(non-negative hinge plastic factors)	(79)

The total number of variables in this problem, $41n_e+5$

($3\times n_e$ in $\dot{U}$, $36\times n_e$ in Λ, n_n in λ_+ or λ_-), becomes $44n_e+8$ in standard formulation of linear programming (n_e is the number of finite elements).

Excluding non-negativity constraints, and boundary conditions the total number of constraints of the problem is $4\times n_e\times n_p+n_e+2$ (n_p is the number of points chosen in each element to enforce admissibility).

This linear programming problem does not coincide with the dual problem of the static formulation but it is of the same form and physical significance. We note that, for instance, matrix B^T used in the statical approach is not the transpose of matrix B deduced in this section, although we have used the same symbol for simplicity.

5. Applications

Circular plates, tubes, spherical caps and cones have been used to compare numerical approximations of the loading capacity with analytical (exact) values, and in some cases with approximations published by other authors [2,4,5].

A nozzle made of portions of cylinder and spherical shell is also treated to demonstrate the advantages derived from the introduction of an exact shape (with singular points) in the equations of the methods developed in this work.

Rules for the appropriate choice of points for plastic admissibility constraints must be obeyed for the particular cases of cylindrical and plate finite elements. These rules are demonstrated for cylinders in the following section. For plate elements $n_{eq}=2$ imposes equilibrium identically but no guarantee of static or kinematic admissibility is achieved.

Method	limit load	diff. %	reactions f_z	reactions m_ϕ	elem. n_e	number of points plast. n_p	number of points equil. n_{eq}	effective variables	constraints $\leq$	constraints $=$
E	.11259	0.	-.0563	-1.00	-	-	-	-	-	-
S	.11952	6.2	-.0598	-1.29	2*	2	2	25	48	7
S	.11084	-1.6	-.0554	-1.15	2*	3	2	25	72	7
S	.11924	5.9	-.0596	-1.32	3	2	2	33	72	9
S	.10964	-2.6	-.0548	-1.16	3	3	2	33	108	9
S	.11784	4.7	-.0589	-1.20	5	2	2	49	120	13
S	.11054	-1.8	-.0553	-1.05	5**	2	2	49	120	13
K	.12000	6.6	-	-	1	2	-	50	-	14
K	.11245	-0.1	-	-	2*	2	-	94	-	23

Table 2. Limit uniform load (divided by N_O/L) for a clamped circular plate of thickness β=0.01. Reference length L is the external radius.
Methods: exact E, statical S and kinematical K.
Nodal coordinates are uniformly distributed exceptfor
* (0.,.7,1.)
**(0.,.2,.4,.7,.95,1.)
Constraints are enforced in Gauss points.

5.1. Cylinders

When the present statical approach is applied in a shell containing a cylindrical finite element, the equilibrium constraint Eq. 32 reduces to

$$n_\phi^{i+} = n_\phi^{(i+1)-} \tag{80}$$

Therefore in these elements we must use a single point (n_{eq}=1) to enforce this identity because otherwise redundant constraints are introduced.

In a cylinder the stress interpolation becomes linear (continuous) for n_ϕ, cubic (continuous with continuous derivative) for m_ϕ, linear (discontinuous) for n_θ, and piece-wise constant for m_θ. Therefore it is impossible to ensure plastic admissibility of stresses as an identity despite of the choice of the set of n_p points.

The kinematical method for the case of cylinders generates an identically admissible pair of fields $(\dot{u},\lambda)$ by using $n_p=3$ as can be demonstrated from the examination of the particular interpolation functions for a straight meridian.

However the approximate collapse load cannot be guaranteed to be a true upper bound because the non-negative condition for $\hat{\lambda}(s)$ is only assured for three points per element.

A tube under different conditions of loads and supports generates several examples suitable to check the two present methods and their corresponding computer programs. For instance, plastic collapse can be induced by simple axial tension, by uniform pressure, or both loads combined. These examples have been treated numerically and trivially verified.

Tube half-length	Exact limit load	Statical method			Kinematical method	
		limit load	number of elem.	n_p	limit load	number of elem.
4.0000	1.7320	1.713	5	2	1.755	9
3.4772	1.7320	1.723	5	2	1.759	7
2.4273	1.5411	1.544	5	2	1.568	7
1.0472	1.2247	1.2247	3	2	1.229	5
0.7227	0.9354	0.9350	1	2	0.938	3
0.1000	0.14142	0.1412	1	2	0.1412	2

Table 3. Limit annular load (per unit length and divided by N_o) of tubes of different lengths. Constraints are enforced in Gauss points; as required $n_{eq}=1$ for statical method and $n_p=3$ for kinematical method.

A tube with a ring load at middle section has also been considered to check the methods in a collapse situation under combined longitudinal bending and circunferential tension.

Results for several tube lengths are shown in Table 3 and compared with analytical solutions [6,7]. Tube length for that ring load case drastically changes plastic collapse characteristics[7,8].

5.2. Spherical cap

A built in spherical cap under uniform pressure $\bar{p}=1$ is analysed for several values of maximum angle $\bar{\phi}$ and dimensionless thickness $\beta=h/4L$, with L and $\bar{\phi}$ shown in Figure 5, and h the thickness of the homogeneous shell approximated by the equivalent sandwich shell.

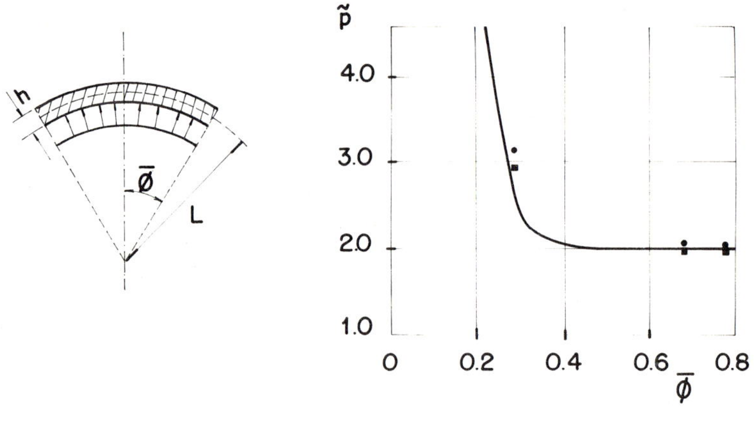

Fig. 3. Limit pressure for a spherical cap of thickness $\beta=0.02$
—— Exact, from Lee and Onat [5]
▪ Statical method . Kinematical method

This case has been solved by Lee and Onat [5] by direct numerical integration of the problem equations so that these results will be called exact and compared with present approximations.

Figure 3 and Table 4 show corresponding results. For the case of the deep shell example, with semiangle $38^{\circ}50'$, the exact solution includes two hinges of the expansion-rotation type, and also an infinite value for a λ component, as pointed out in reference [5]. This may explain the discrepancy of kinematical results for this angle if we recall that these kinds of plastic hinges have been excluded from the set of interpolation functions. The approximate stress fields compare well with exact ones which are essentially analogous to plate collapse stresses.

Method	max. angle $\bar{\phi}$	limit load $\tilde{\alpha}$	number of elem.	number of eff. var.	number of constraints $\leq$	$=$
Exact [5]	$16^o8'$	2.78	-	-	-	-
Static	$16^o8'$	2.98	2	27	48	7
Static	$16^o8'$	2.95	5	51	120	13
Mean S-K	$16^o8'$	3.10	-	-	-	-
Kin.	$16^o8'$	3.24	5	226	-	53
Kin.	$16^o8'$	3.48	2	94	-	26
Exact [5]	$38^o50'$	2.00	-	-	-	-
Static	$38^o50'$	1.81	1	19	24	5
Static	$38^o50'$	1.98	4	41	96	11
Mean S-K	$38^o50'$	2.02	-	-	-	-
Kin.	$38^o50'$	2.07	4	182	-	44

Table 4. Limit pressure for built in spherical caps of thickness $\beta=0.02$ and different angles $\bar{\phi}$. For all cases $n_p=2$ and $n_{eq}=2$. Nodal coordinates are
- 2 elements: $\phi^i=0^o$, 12°, $16^o8'$
- 4 elements: $\phi^i=0^o$, 9° , 30° , $38^o50'$
- 5 elements: $\phi^i=0^o$, 4°, 8°, 12°, 15°, $16^o8'$

5.3. Cone

A built in cone of base radius L, height 2L and thickness $\beta=0.02$ is analysed with respect to plastic collapse produced by internal pressure ($\bar{p}=-1$). The results obtained are compared in Table 5 with values published by Dang Hung et al [2]. These authors used a sandwich Mises model of shell so that our results should be lower than those values.

Authors	shell model	method	limit pressure	diff. %	number of elem. n_e
present	ST	S(P)	1.2402	0.1	5
present	ST	S(C)	1.2402	0.1	5
present	ST	S	1.2204	-1.5	8
present	ST	S	1.2124	-2.1	10
present	ST	mean S-K	1.2385	0.	-
present	ST	K	1.2646	2.1	7
present	ST	K	1.2964	4.7	5
Dang Hung [2]	SM	S	1.2924	0.3	4
Dang Hung [2]	SM	S	1.2540	-3.2	8
Dang Hung [2]	SM	S	1.2406	-4.3	10
Dang Hung [2]	SM	mean S-K	1.2961	0.	-
Dang Hung [2]	SM	K	1.3516	4.3	8
Dang Hung [2]	SM	K	1.3574	4.7	6
Dang Hung [2]	SM	K	1.3694	5.7	4

Table 5. Limit pressure of a clamped cone of base radius L, height 2L and thickness $\beta=0.02$
Shell model: sandwich Tresca ST and sandwich Mises SM
Methods: statical S and kinematical K

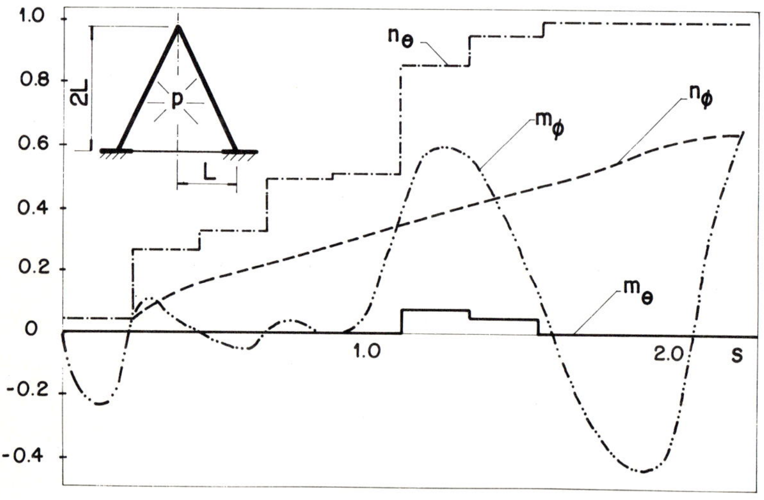

Fig. 4. Collapse stresses in the cone

5.4. Cylindrical nozzle

As a final application we present values for the collapse pressure of a nozzle consisting of a cylinder of radius r_c and a spherical portion of radius r_e. Both cylinder and sphere have the same dimensionless thickness $\beta=h/4L$. Here L is taken to be 1.in. External loads include the axial traction $f_z^1=-0.5r_cp$ corresponding to the internal pressure p.

The meridian curve of the considered shell has then a singular point, where discontinuities and hinges may be expected, and a spherical part that will be exactly represented in the methods presented in this paper.

Table 6 contains limit load approximations obtained by the two present methods and also results due to Biron et al [4], the latter ones corresponding to a sandwich model of the Mises type.

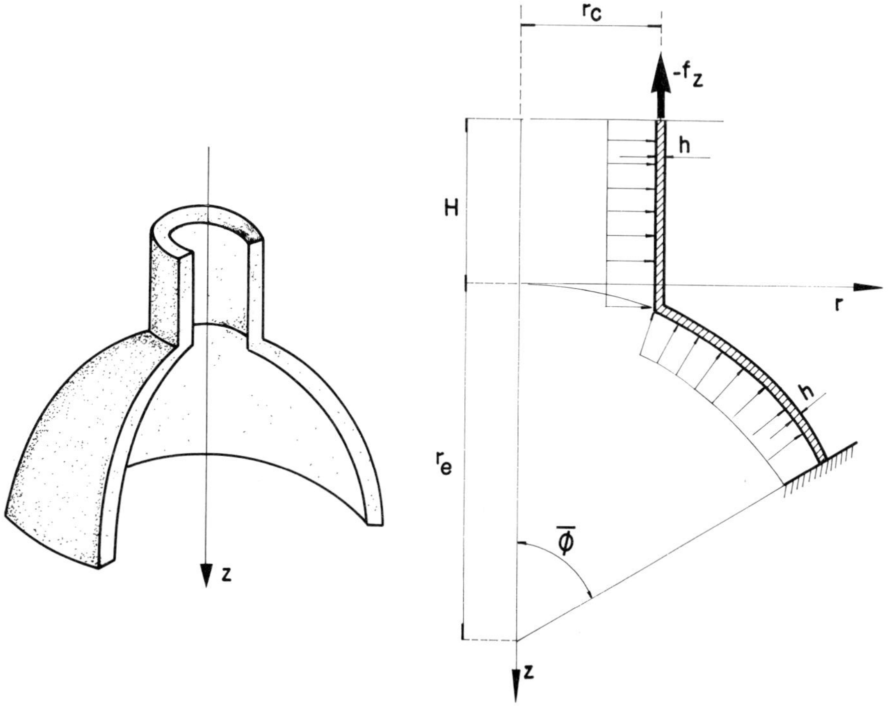

Fig. 5. Nozzle model. $r_e=10.$; $r_c=4.$; $H=5.$ (reference value L=1.in); $\beta=h/4L=0.025$; $\overline{p}=-1.$; $\overline{f}_z=-2.$

Method	limit pressure	diff. %	number of elem.	nodal coordinates cylinder s^i	sphere ϕ^i (degrees)
S	.055024	-13.4	4	0.; 3.; 5.83	23.6; 30; 45
S	.055030	-13.4	5	0.; 3.; 5.83	23.6; 30; 36; 45
S	.056838	-10.5	4	0.; 4.5; 5.83	23.6; 33; 45
S	.055712	-12.4	8	0.; 1.5; 3.; 4.5; 5.83	23.6; 28; 33; 38; 45
S	.056441	-11.2	8	0.; 1.5; 3.; 4.5; 5.83	23.6; 26; 30; 36; 45
mean	.063563	0.	–	–	–
K	.070685	11.2	8	0.; 1.5; 3.; 4.5; 5.83	23.6; 27; 30; 36; 45
K	.073648	15.9	4	0.; 1.5; 5.83	23.6; 33; 45
K	.075401	18.6	5	0.; 3.; 5.83	23.6; 30; 36; 45
S[4]	.06	- 7.8			
mean	.065	0.			
K[4]	.07	7.8			

Table 6. Limit pressure (referred to N_0/L, with L=1.in) for the nozzle of Figure 5.
Methods: statical S and kinematical K
All constraints are imposed in two points placed in Gauss coordinates

6. Conclusions

Present statical and kinematical numerical methods are deduced from interpolation functions where exact geometry is introduced and moderately low degrees of polynomials are assumed. Furthermore, compatibility with respect to equilibrium and deformation differential operations, and continuity conditions, are taken into account to select these approximation functions.

An important characteristic of the present numerical methods is that equilibrium and kinematical equations used are exact, although partially enforced by collocation method.

Applications show that the assumed interpolation fields are appropriate in simple examples where several basic combinations of collapse strains and stresses appear.

Only non-expansive hinges are admitted in the set of interpolation fields for velocities. To introduce also nodal discontinuities for tangential velocity leads to an increase in the number of parameters for concentrated plastic factors.

We note that the great number of inequality constraints of the statical scheme, and of interpolation parameters for λ in the kinematical one, are mainly due to the great number of plastic modes in the shell model. The use of a Mises condition reduces the number of plastic modes but introduces non-linear constraints. We have preferred to linearize this condition because a great number of linear constraints are needed to impose equilibrium or kinematical admissibility and this would complicate the solution of a non-linear formulation. Nevertheless, it is possible to deal with large linear programming problems when appropriate techniques are used.

References

1. Cohn, M.Z.; Maier, G. (eds); "Engineering plasticity by mathematical programming", Proc. NATO Adv. Inst., Canada,1977.

2. Dang Hung, N.; Trapletti, M.; Ransart, D.; "Bornes quasi-inferieures et bornes superieures de la pression de ruine des coques de revolution par la m.e.f. et par la programmation non-lineaire", Int.J. Non-linear Mech., vol. 13 (1978) 79-102.

3. Hodge, P.G.; "Limit analysis of rotationally symmetric plates and shells", Prentice-Hall, 1963.

4. Biron, A.; Charleux, G.; "Limit analysis of axisymmetric pressure vessels intersections of arbitrary shape", Int.J. Mechanical Science, vol. 14 (1972) 25-41.

5. Lee, L.C.; Onat, T.; "Analysis of plastic spherical shell", Engineering Plasticity Conference, Eds. J. Heyman and F.A. Leckie, Cambridge, 1968.

6. Sawczuk, A.; Hodge, P.G.; "Comparison of yield conditions for circular cylindrical shells", J. of the Franklin Institute, 269 (1960) 362-374.

7. Zouain Pereira, N.; "Análise limite de cascas via otimização", Ph.D. thesis, Federal Univ. of Rio de Janeiro, 1982.

8. Demir, H.H.; "Cylindrical shells under ring loads", J. of the Struct.Div. ASCE, ST3 (1965) 71-98.

Inelastic Behaviour of Shells under Concentrated Loads

S. LUKASIEWICZ

Department of Mechanical Engineering
University of Calgary
Calgary, Alberta, Canada T2N 1N4

Summary

The paper is devoted to the description of the present state of knowledge in the theory of inelastic shells under concentrated and local loads. Problems of current interest, and in the broad sense, the analytical and numerical methods that can be brought to bear, are discussed. Other problems involving stress concentration in shells, such as crack propagation and concentration of stresses caused by irregularities of the shape, or holes and cutouts, are not the subject of the paper.

The action of concentrated loads on shells has drawn the attention of many researchers for the last 20 years. The problems concerning stress and displacement distribution in linear, elastic thin shells under the static action of concentrated loads applied at one point of shell surface, or subjected to locally distributed loads, were examined at first and were the subjects of those early investigations. Many papers from the western countries as well as from the U.S.S.R. were published in those years.

Mostly, the linear theory of shells was the starting point of these works. This area seems to be sufficiently explored at present. However, as far as we consider the similar problems in inelastic, non-linear shells, we recognize that many problems are not yet solved or are not solved in a satisfactory, general manner.

The rapid expansion of numerical methods and techniques in recent years has enabled the solution of many difficult problems. However, these solutions usually concern particular situations and are often difficult to generalize.

The state of knowledge concerning the following problems of inelastic shells is presented in the paper.

1) The geometrical non-linearities, large deformations of elastic shells under local loads and complex loads when the local, concentrated load is accompanied by a distributed general load.
2) Contact problems in shells involving the geometrical and physical non-linearities
3) Problems related to concentrated loads acting on shells made of inelastic materials
4) Elasto-plastic behaviour of shells under concentrated loads associated with large defomrations of shells, collapse and

Inelastic Behaviour of Plates and Shells
IUTAM Symposium Rio de Janeiro 1985
Editors: L. Bevilacqua, R. Feijóo and R. Valid

limit local loads.
5) Simplified solutions of dynamical problems.
6) Optimum design of shells loaded by concentrated forces.
The paper covers both the results of experimental investigations in each of the above-mentioned areas, as well as the description of the analytical and numerical methods used.

1. Introduction

The paper reviews problems concerning the action of concentrated forces on shells. It is based on works published in the last 20 years. The paper discusses nonlinear solution whose nonlinearity has geometrical (large deflections) or material character. The action of concentrated forces on shells has aroused interest of researchers for more than 20 years. Numerous papers were published based on the linear theory of shells. Their discussion can be found in the publications of the present author [1,2]. This area seems to be well explored. However, as far as inelastic, nonlinear shells are concerned, many problems are not yet solved or are not solved in a satisfactory general manner. The area of geometrical type nonlinearities, and elastic post-buckling behaviour of shells has been best explored. There are also several solutions concerning the problems of load carrying capacity of shells under concentrated forces. The situation looks worse in the case of problems concerning material and geometrical nonlinearities, elasto-plastic and visco-elasto-plastic problems. There are very few solutions available and most of them obtained using the finite element method. Also solutions concerning local dynamic loads acting on shells with large displacements made of nonlinear materials are scarce. The nonlinear, three-dimensional solution for the region surrounding the point of application of the concentrated force on shell is nonexistent too. The experimental investigations in this field are very limited.

Rapid expansion of numerical methods and techniques, as for example finite element method, in recent years, has enabled the solution of many difficult problems. However, these solutions usually concern particular situations and are often difficult to generalize and verify. Therefore more analytical approaches would be valuable. The following topics concerning the action of concentrated forces are discussed in the paper:

- The geometrical nonlinearities and large deflections of elastic shells under local loads and complex loads,
- Problems related to elasto-plastic behaviour of shells under concentrated loads associated with large deflections, collapse and limit loads,
- Contact problems of shells involving the geometrical and physical nonlinearities,
- Optimum design of shells loaded by concentrated forces.

2. Geometrical Nonlinearities, Stability and Large Deflections

2.1 Spherical Shell Under Concentrated Normal Force

The nonlinear behaviour of spherical shells under a central concentrated force was the subject of the early papers concerning the action of concentrated loads on shells. In this case the problem is axially symmetrical which makes theoretical analysis much easier. The experiments prove that already for relatively small deflection of the order of several thicknesses, the force-deflection relation becomes nonlinear and is not compatible with that calculated using Linear Theory of Shells. When the concentrated force is applied to the spherical elastic shell a "circular dimple", i.e. a region of reversed curvature, appears in the shell and spreads outwards as the load increases. This problem was first solved by Biezeno (1935) [3]. He obtained a nonlinear relation between load and deflection. Several researchers later treated the same case using nonlinear shallow shell equations and energy methods and obtained similar results, (see Chien and Hu [4], and D.G. Ashwell [5]). Ashwell solved the nonlinear problem of large displacement of shells by means of the combination of two linear solutions. R. Archer [6] reduced the problem of the spherical shell loaded by a concentrated force to the solution of three algebraic equations by use of the nonlinear Reissner equations and the method of finite differences. A method proposed by A.V. Pogorelov [7], resembling that of Ashwell, is interesting and worth wider discussion. Pogorelov noticed that when the shell deflects elastically, it takes the form of its isometric transformation. That makes possible to predict to a certain degree the shape of the deflected shell by looking for it among the isometric transformations [17]. In the case of the spherical shell the simplest isometric transformation corresponds to mirror-like reflection (Fig. 1).

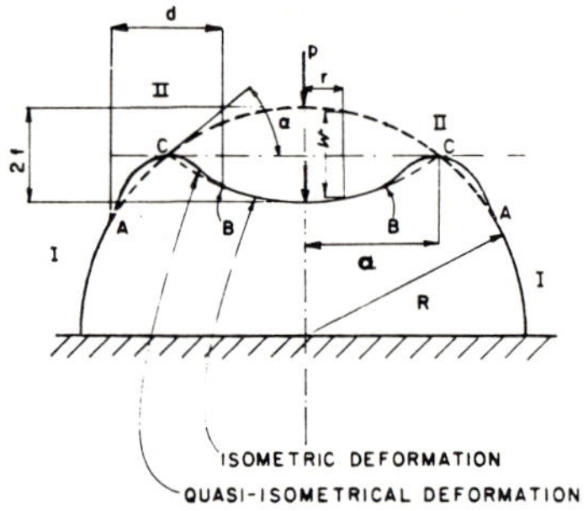

Fig. 1 Regions in a deformed spherical shell.

The problem was solved variationally with the assumption that the displacement u and w in the ridge area are of local character and vanish at a certain distance from the ridge. Finally, a very simple relation for the energy caused by the change of the shape of the shell was obtained

$$U = 2\pi c\ E(2f)^{3/2}\ h^{5/2}\ \frac{1}{R} \tag{1}$$

where 2f is total deflection at the point of application of the load. c - constant value equal to c ≅ 0.19, E, h and R are Young modulus, thickness and radius of the curvature of the shell respectively. Knowing relation (1), only one step more was required to obtain the load-deflection relation i.e. the calculation of the work of the external load. Equating the variation of the functional with respect to f, to zero, results in the following simple formula for the deflection of the shell.

$$w = 2f = \frac{R^2}{9\pi^2 c^2 E^2 h^5} P^2 \quad \text{or} \quad P = \frac{3\pi c E h^{5/2}}{R} \sqrt{2f} \tag{2}$$

This relation is presented graphically in Fig. 2. The results of experiments by F.A. Penning [9] are also shown there. These experiments were performed with a number of shallow spherical shells with clamped edges made of aluminum and loaded at the apex through a small circular contact area. The geometry of the shell was defined by a parameter $\mu = a/\ell$, a being the radius of the clamped edge and ℓ a characteristic length of the shell.

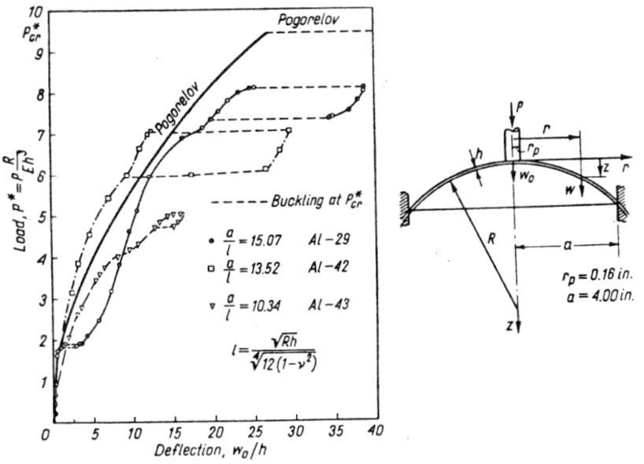

Fig. 2 Comparison of results by Pogorelov's theory with results by F.A. Penning's experiments

This revealed that the behaviour of the spherical cup depends on the radius of the clamped edge. Different patterns were observed during the testing. The thicker shells suffered plastic yielding in the vicinity of the load and did not buckle. Thus permanent deflections remained at the apex. The thinner walls buckled and did not show any evidence of permanent deflections after having been loaded to their highest values of the load. Deflection calculated from linear theory showed good agreement for the initial shape of 4 load-deflection curves. The deflection given by Pogorelov's formula (2) corresponds quite well to the curve resulting from experiments for small values of the load. The formula, Eq. 3, was obtained by Pogorelov using many simplifications. One of the most important seems to be the assumption that the strain energy can be calculated for the unit length of the ridge which is considered as very narrow not taking into account the real dimensions of that area. Recent, more precise calculations performed by the present author confirm the result (Eq. 3) obtained by Pogorelov. The membrane energy in the ridge can be obtained solving the set of nonlinear shells equations. Then the energy of the whole system can be obtained in terms of the deflection and the stress function.

Pogorelov's result can be checked calculating the coefficient c from the equation.

$$c = \frac{U}{2\pi E h^{5/2} w^{3/2} R^{-1}} \tag{3}$$

where U is the total strain energy associated with the dimple on the shell surface, h - thickness of the shell, R - radius of the shell, E - Young modulus, w - deflection at the point of application of the load. Finally, it has been obtained that indeed c is almost a constant value, for

$R/h = 1200$ and for $P^* = P/\frac{Eh^3}{R} = 2$, $c = 0.174$, for $P^* = g$, $c = 0.184$.

It has also been found that the value of c is independent of the value of h/R. Pogorelov's value for the constant c is $c = 0.19$ which is a sufficiently close value. The maximal bending stress in the ridge area is defined by Polorelov

$$\sigma_{max} = \frac{Eh}{2} \max (w'') = 0.9E \frac{(wh)^{1/2}}{R} \tag{4}$$

This value of constant coefficient $c' = 0.9$ was compared to the more precise solution and the following was obtained: $c' = 0.56 \div 0.61$ from the equation $c' = \frac{\sigma_m R}{E\sqrt{wH}}$ where σ_m is the maximum stress obtained using the above described more precise method.

The experiments proved that the deflection patterns are symmetrical only when the force is smaller than a certain critical value. After surpassing this value the shape of the deflection becomes non-symmetrical and its magnitude grows rapidly. The deformed area takes the form similar to the triangle, quadrangle, etc. The critical force corresponding to that phenomenon can be found analysing the stability of the shape of the ridge appearing on the shell surface. Pogorelov solved this problem presenting this shape in the form of the equation

$$\frac{r}{R} = \rho_o (1 + p \cos k\,\theta) \qquad \rho_o << 1 \tag{5}$$

where ρ_o is the mean nondimensional position radius of the ridge, k - number of waves on the circumference, θ - angle measured in the plane of the ridge, p - a certain small arbitrary parameter $p < 1$.

The following critical value for $k = 3$ was obtained.

$$P_{cr} = 3\pi \frac{Eh^3}{R} , \qquad \rho_o = \frac{1}{c} \sqrt{\frac{h}{R}} \tag{6}$$

In Fig. 2 this force is given by the horizontal straight line. Fig. 3, [12], presents the load as a function of the parameter λ for a shallow shell with the built-in edge. We observe that for small value of λ only symmetrical mode is possible. For larger values of λ the nonsymmetrical deformation patterns are possible. No rapid jumps of the load were observed when the shell changed its configuration.

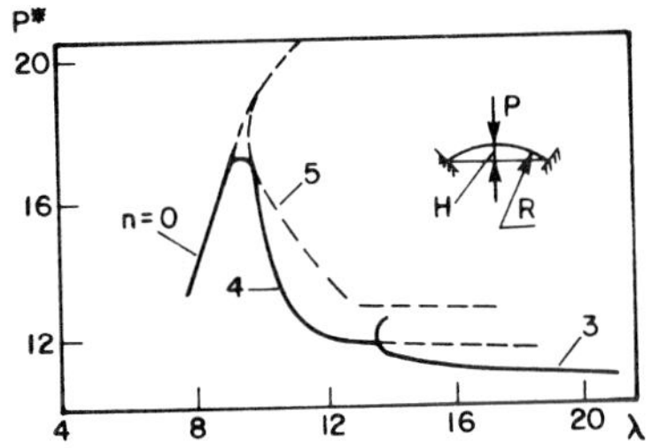

Fig. 3 Buckling mode as a function of λ

2.2 Concentrated force and interal pressure

The combined effect of uniform pressure and central concentrated load was theoretically discussed and experimentally measured by Evan-Iwanowski, Cheng and Loo [14]. The typical results obtained by them are presented in Fig. 4 where the correlation between the critical buckling pressure q_{cr} and the concentrated force P_{cr} acting on the shell can be observed. It is seen that for small values of external pressure the shell behaviour is similar to that for a concentrated force alone, but that as the pressure increases the behaviour rapidly changes to that for uniform pressure alone. The above problem can also be solved using Eq. 3 for the strain

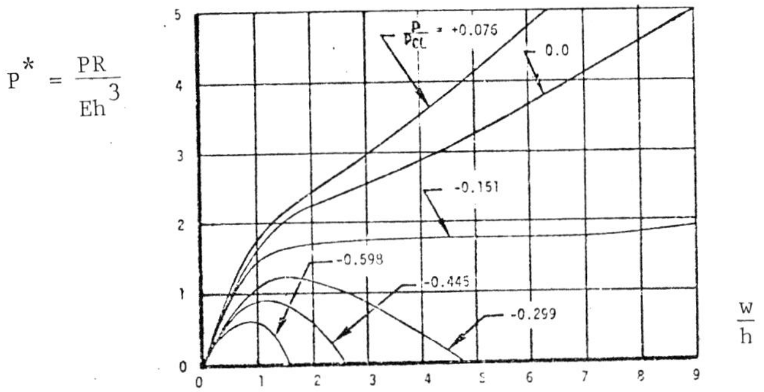

Fig. 4 Typical load deflection curves for a clamped spherical shell subjected to a central concentrated load combined with uniform pressure, Loo and Evan-Iwanowski (1964)

energy. The following relation between critical buckling pressure and the critical buckling force was obtained [15],

$$\frac{P}{P_c} \cdot \frac{q}{q_c} = \frac{3}{8} c^2 \sqrt{3(1-\nu^2)} = \text{const.} \tag{7}$$

where $P_c = \frac{3\pi E h^3}{R}$ $\qquad q_c = \frac{2E}{\sqrt{3(1-\nu^2)}} \frac{h^2}{R^2}$

and P, q are the concentrated force and external pressure reacting on the shell, respectively. In the above solution the radius of the base of the shell does not have the effect on the value of the critical load. It is obvious that this result can be applied only if a is relatively large, several times larger than the characteristic length, ℓ. The results from the experiments support this conclusion. The relation, (Eq. 7), (a straight line) compared in Fig. 5 with the results by Ta Cheng Loo and

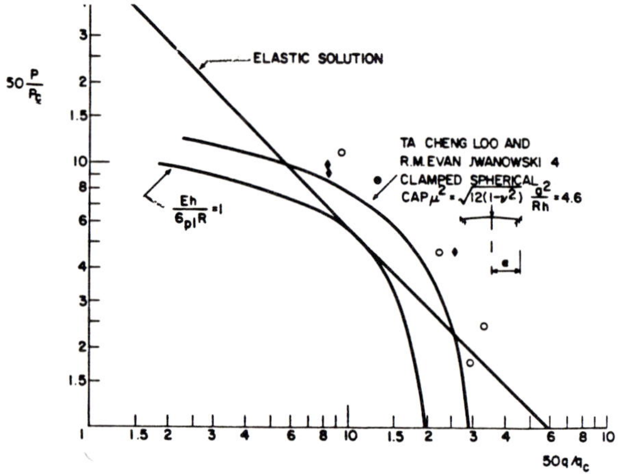

Fig. 5 Correlation between critical load and critical pressure

Evan Iwanowski [14]. The curved lines were obtained by the assumption that the first plastic deformation limits the movement of the ridge on the shell surface. The analysis of the spherical cup subjected to a local axisymmetric pressure load was performed by Fitch and Budiansky [8], (1970). The ring loaded shell was the subject of the paper by Akkas and Bould [17], (1971). The effect of the geometrical parameter

$$\lambda = \sqrt{\frac{H}{h}}\, 2[3(1-\nu^2)]^{1/4} \tag{8}$$

was examined using method of Fitch and Budiansky. Here H is the height of the cup. The parameter λ is related to the previously used parameter μ.

For small $H = a^2/2R$, a - radius of the cup base, $\mu = \lambda$. The buckling of truncated or complete spherical shell was discussed by Onoda [25], (1973). The maximum stress for a locally loaded spherical shell was calculated by Kao and Perrone [21], (1973) using nonlinear shallow shell equations. The results show the effect of the value of the radius of the small area on which the load is distributed r_o. The application of the nonlinear theory shows that the linear theory gives higher stresses, (about 1.5 times higher than those resulting from the linear theory for $w/h = 0.6$ and $r_o/\sqrt{Rh} = 0.5$). The solution was obtained using nonlinear relaxation method. The nonlinear differential equations were replaced by a set of two nonlinear algebraic finite-difference equations. Mescoll [26] (1966) solved the problem of the spherical shell under the concentrated load assuming small deformations but finite rotations using Reissner equations and the method of finite differences. He obtained positions of equilibrium in the postbuckling state corresponding to positive and negative value of the load, see Fig. 6. Large deflections of deep spherical shells under

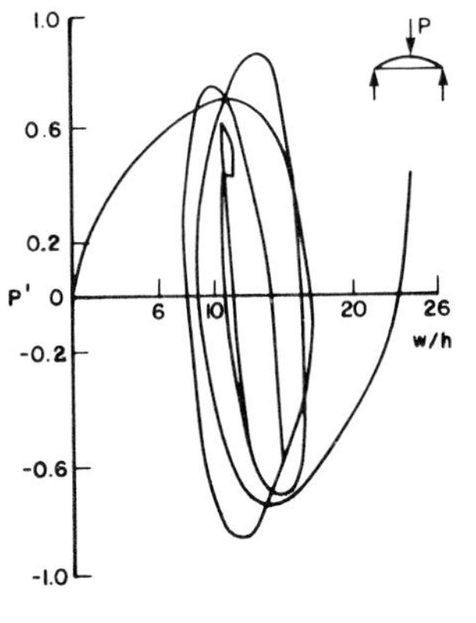

Fig. 6 Load deflection curve

concentrated force were discussed by Ranjan and Steel [22], (1977). The shell filled with fluid was the subject of the paper by Taber [23], (1982), as was the compression of the spherical shells and membranes filled with fluid by a rigid cylindrical intender [24], (1983). The solutions were obtained using bending shell theory as well as a membrane model where only stretching was included. The experiments were conducted on rubber shells (raquetballs). Isometric mode of displacements could be observed. A spherical sandwich shell under concentrated force and a centrally distributed pressure was the object of the paper by Akkas and Bauld [27]. The critical buckling loads for clamped shallow spherical shell under centrally distributed pressure and concentrated load at the apex were obtained using perturbation technique proposed by Koiter [28]. The post-buckling behaviour was investigated using series representations for the Airy stress function and normal deflections.

2.4 Dynamic application of the load

A simplified solution to the dynamic problem can be also obtained using geometrical method. In the case when the load is applied dynamically and the displacements are the functions of time, we have to include to the functional T, the kinetic energy of the shell. This can be done easily if we notice that this energy is almost entirely contained in the cases of internal value I and the ridge zone II.

If we assume that the displacements in the deformed area are purely isometrical then

$$w \cong 2f\left(1 - \frac{r^2}{a^2}\right) \cong 2\left(f - \frac{r^2}{2R}\right) \tag{9}$$

then $a^2 = 2Rf$ and $\dot{w} = 2\dot{f}$ where $\dot{w} = dw/dt$.

The stationary value of the Hamiltonian gives the following equation

$$\frac{3\pi c\ Eh^{5/2}}{R}(2f)^{\frac{1}{2}} + 2\pi R\left(\dot{f}^2 + 2f\ddot{f}\right) + 2M\ddot{f} = P \tag{10}$$

where M is the mass of the body applied at the apex. This equation can be easily solved using numerical integration routine as for example Runge-Kutta routine. The above described solution is very simple. However, it should be considered only as an approximation which is true only if the assumed shape of the deflection function is correct. If the inertia forces of the shell have a large effect, i.e. for very fast excitation, this result cannot describe the real behaviour of the shell.

2.5 Shell of positive double curvature

The nonlinear deformation of a shell of double Gaussian curvature can be considered in a manner similar to that for a sphere. An elliptical dimple which has the form of the original surface, but negative reversed curvature is observed during the experiments with the shell loaded by a concentrated normal force, Fig. 7. Also for this case Pogorelov obtained a simple result calculating the total energy as the effect of the dimple, adding the bending energy of the isometrically deformed area and the energy stored in the area along the ridge. The total strain energy in the shell is given by a similar simple solution. Calculating the work of the external force $W = P \cdot 2f$ and equating the variation of the fundamental $T = U - W$ to zero, he found the relation between the deflection of the shell and the load

$$P = \frac{3\pi c}{2} Eh^{5/2} \left(\frac{1}{R_1} + \frac{1}{R_2}\right) \sqrt{2f} \; . \tag{11}$$

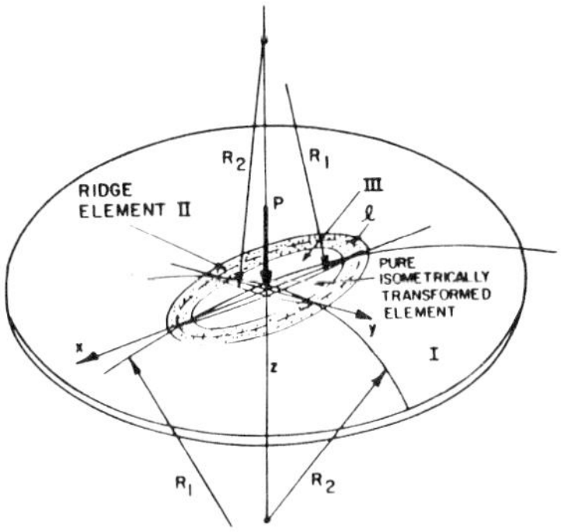

Fig. 7 Isometrical deformations of the shell of positive Gaussian curvature, decomposition into elements

The maximum stresses appear near the ridge and are the effect of bending in the direction perpendicular to the ridge. The stresses can be calculated approximately from

$$\sigma_o = \frac{Eh}{2} \max \left(\frac{\partial^2 w}{\partial n^2}\right) = c' \; E\left(\frac{2fh}{R_1 R_2}\right)^{\frac{1}{2}} \tag{12}$$

For a concentrated force acting through the rigid insert, the similar formula of the critical buckling force [7] was obtained.

2.6 Shells of Revolution

Geometrically nonlinear deformations of multilayered shells of revolution under the action of local loads were discussed in the paper by Grigorenko and Timotin [29], (1982). The shells were closed in the circumferential direction and characterized by a small shear stiffness. Transverse stress was included. The expansion in trigometric series was used and the problem was reduced to boundary value problem for a system of nonlinear first-order ordinary differential equations. The cylindrical and hemispherical shells were discussed as the examples.

2.7 Ponding of the Shell

The problem of stability of shells of positive Gaussian curvature subjected to a concentrated load, internal or external pressure and an accumulating ponding fluid in the depression caused by the concentrated load was discussed by Lukasiewicz and Glockner [38]. Critical values of the load were calculated using the above described geometrical method. The simple relations for the critical loads as well as the postbuckling behaviour of the shell were obtained. The load curve-deflector was defined by the equation

$$P = \frac{3}{2} \pi c \, Eh^{5/2} \left(\frac{1}{R_1} + \frac{1}{R_2}\right) \sqrt{w} + \pi q \sqrt{R_1 R_2} \, w - 0.175 \, \pi \, \gamma \sqrt{R_1 R_2} \, w^2 \qquad (13)$$

where w - deflection of the point of application of the load. The maximum value of $P_{max} = P_{cr}$ (a critical load) can be obtained from the condition

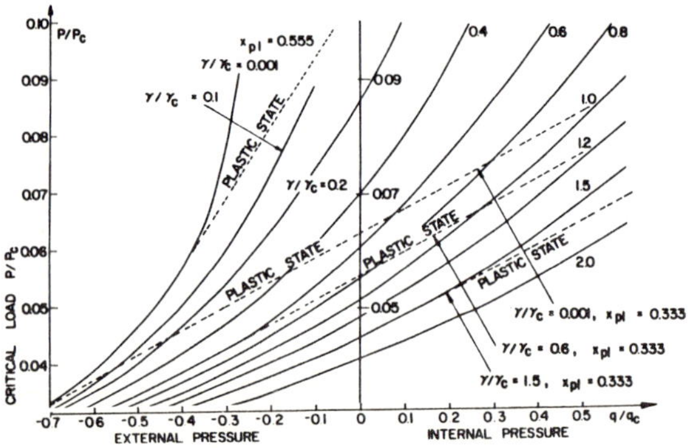

Fig. 8 Critical load as a function on internal and external pressure for different fluid densities

$\frac{\partial P}{\partial w} = 0$. The results are presented in Fig. 8. The dotted lines in this figure present the values of the force taking into account the first plastification of the ridge area. Assuming that the movement of the ridge over the surface of the shell is limited by the plastic deformations in the ridge, we obtain from (4)

$$\sqrt{\frac{w}{h}} + x + \frac{\sigma p\ell}{c'Eh}\sqrt{R_1R_2} = x_{p\ell} \tag{14}$$

Introducing the value into Eq. 13 we find the corresponding critical force which is presented in the Fig. 18 by dotted straight lines. The assumption that the ridge does not move after the first plastic deformation appears [7], is only approximate. The more exact analysis of this case discussed in the ϕ 3.1 shows that this zone changes its position, however much slower than in the elastic state.

2.8 Cylindrical Shell Under Normal Point Load

The literature concerning the cylindrical shells under concentrated loads is extremely rich. However, if we consider the nonlinear large deformations the number of papers is limited. Experiments performed with the cylindrical shells under a concentrated load proved the load deflection curves become nonlinear, if deflection is of the order of several thicknesses of the shell. Then the deformation pattern corresponding to the small deflection linear solution (ellipse elongated in the direction of the generator) changes rapidly into a new configuration which takes the form resembling the isometric transformation of the cylindrical surface. This form consists of one diamond segment of Yoshimura pattern observed during the experiments with cylindrical shells under compressive axial load accompanied by inextensionally deformed large cylindrical panels. The solutions to the nonlinear problems of cylindrical shells were usually based on the nonlinear shell equations. For example, in the paper by Almroth and Brogan [31] a cylindrical panel under concentrated transverse load was discussed and the linear buckling load was obtained. There are also numerous solutions using Finite Element Method, FEM., see paragraph (2.11). The method which can be applied in the case of analysis of very large deflections, based on the geometrical approach was discussed by Lukasiewicz and Szyszkowski [17], [32]. The main idea of that method will be presented here. We know that the cylindrical surface is a surface of zero Gauss curvature. Therefore basing on the Gauss theorem, we conclude that the ideally isometrically transformed cylindrical surface can be built only from the surfaces and planes of zero Gauss curvature. Looking on the deformed cylinder under the action of concentrated load (see Fig.

9) we can distinguish the areas of different shapes as conical surfaces, planes and cylindrical surfaces. The ideal isometric transformation of the cylindrical surface is not possible without the discontinuities of the surface which in real shell is prevented by its bending rigidity. Therefore the shell deforms only to a nearly-isometric form; there appear the areas where the deformation is not possible without the change of the Gaussian curvature. That leads to the change of the metric tensors which causes the membrane stresses in that region. That kind of region is called "apex" in Fig. 10 where we observe a strong double-curvature of the surface. The above observations are the basis for the so called "Geometrical Method" [40], GM, which can be considered a particular version of

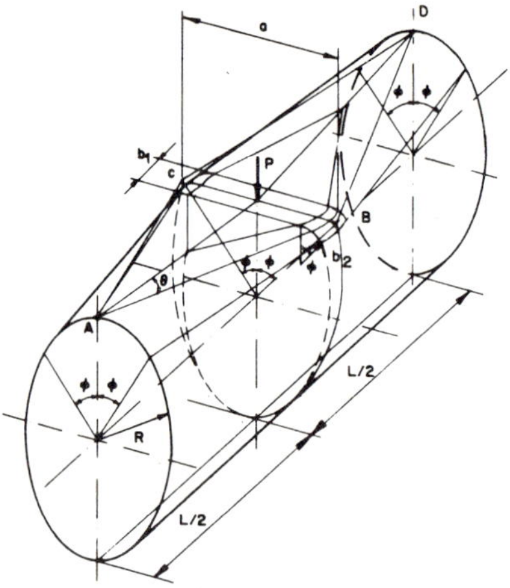

Fig. 9 Deformed shape of the cylindrical shell.

FEM. The approach is based on Lagrangian variational principle and the assumption that during large deformation the shell structure deforms in a nearly-isometrical manner. The structure is subdivided into elements in which the strain energy can be calculated as a function of a small number of geometrical parameters. This energy is calculated for the geometrical boundary conditions taking into account the fact that the membrane stresses produced by the deformation of the surface can exist only in the non-isometrically transformed areas. Thus the "library" of different elements can be established and the corresponding strain energy determined. The method is quite general. The dynamic as well as elasto-plastic

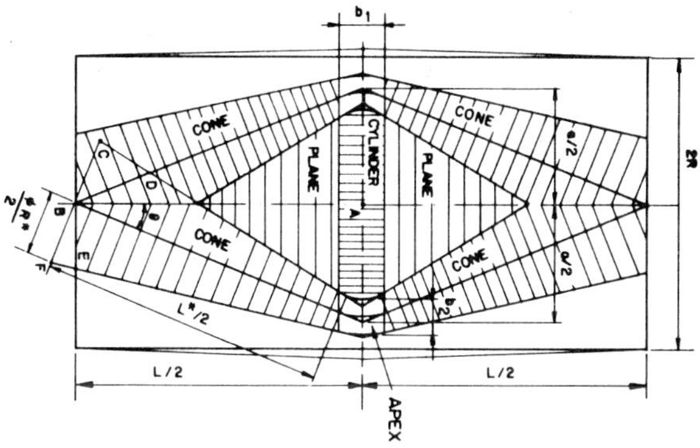

Fig. 10 Subdivision of the cylindrical shell into elements

problems can be solved if we include the kinetic energy and energy of plastic deformations. The kinetic energy can be obtained easily if the deformations of the structure are known. The most important feature of the method is that there are no limitations as to the magnitude of the displacements. The details for the cylindrical shell loaded by a concentrated force can be found in the paper by Lukasiewicz [40]. The load-deflection curve, corresponding ratio $\lambda = b_1/b_2$ and $b = b_1$, obtained numerically minimizing the total potential energy are presented in Fig. 11. In this figure we find a load-deflection curve obtained experimentally for a cylindrical shell supported by rings at the ends and loaded by a concentrated force at the centre. The shell was made of mylar for which the coefficient $\sigma y/E$ is high. We observe a very good agreement for the very large deflection $\frac{w}{R} = 0.1 \div 0.5$. The results predicted by GM for the initial small deflection show some discrepancies. However, this is caused by the fact that the non-isometrical deflections corresponding to the local effect of the concentrated load in the linear stage are not taken into account in the assumed deformation pattern. It can be easily improved adding this deflection pattern and introducing more parameters defining that effect. The experiments with cylindrical shells of metal and fiberglass reinforced plastic have been performed by Rabotnov, [34], (1982). The shells were fixed at the ends and loaded at the central cross-section by a piston with rubber gaskets. The load-deflection curves, as well as the deflections in circumferential direction and longitudinal direction were measured. The measured maximum displacements were in the range of $w/h < 1.0$. The load-deflection curves slightly deviated from the straight lines predicted by the linear theory. The deformation of cylindrical shell local loads were measured using holographic interferometry by Matsumoto, Iwota and Nagata, [35]. Behaviour of the shell with concentrated masses under dynamic external pressure was experimentally measured in [36]. Stresses in a pressure vessel flanges near the threaded holes for studs were obtained in [37]. Stresses and displacements were measured in cylindrical pressure vessels loaded by loads applied to the nozzles, [39].

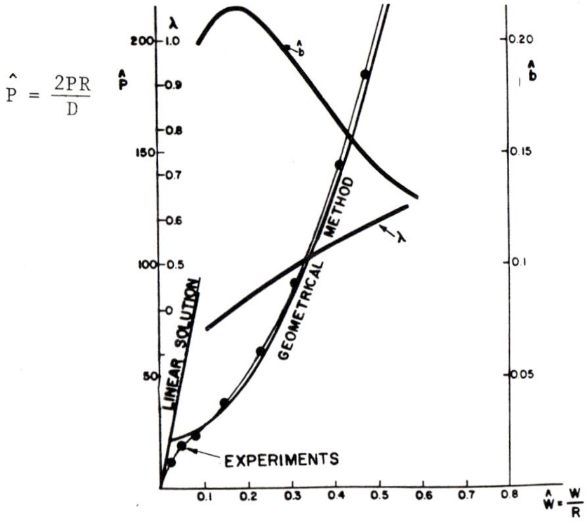

Fig. 11 Load deflection diagram for the cylindrical shell under concentrated static load.

2.9 Shallow Shell

Large deflection of shallow paraboloid shells were the subject of the paper by Neng-Ming Wong, Hilario Oh and Kuo-Kuang Chen [42] who obtained the solution for the geometrically nonlinear shell using Ritz method. The results were presented in the form of diagrams for a variety of shells of clamped edges. The satisfactory convergence of the computations was observed only for shells of relatively small rise to thickness ratio (less than 15).

2.11 Method of Finite Elements

The Finite Element Method, FEM, was applied to the analysis of the stability and postbuckling behaviour of shells under concentrated loads. The literature devoted to the FEM is extremely rich. It is difficult to discuss in detail all aspects of that approach. However, we should mention here the papers by G. Horrigmoe and P. Bergana [43], A.B. Sabira and A.C. Lock [44], (1973), M.A. Crisfield [45] and T.Y. Chang and K. Swamiphakdi [46], (1982) where the results concerning concentrated forces are presented. In the last paper, the finite element formulation was based on the 9-mode shell element, and on the updated Lagrangian description of the deformed state. The effect of deformation dependent loads was treated in an approximate manner by changing the right-hand side of the equilibrium equations in reference to the updated geometry. The post-buckling analysis was conducted by New-Raphson procedure with a constant-displacement length method when the limit load of the structure was reached. The results for the case of thin cylindrical shell segment hinged supported along the longitudinal edges and free along the curved edges and subjected to the concentrated force applied at the top were obtained. The other example of the application of FEM is given in the paper the Brendal and Ramm [47] (1980). The linear and nonlinear stability of the cylindrical panel under a concentrated transverse load is discussed in this paper which describes the application of the curved isoparametric shell element to large displacement analysis. The Lagrangian formulation has been adopted using the incremental - iterative solution procedure. The linear stability analysis performed for the initial position has been repeated at several advanced equilibrium positions on the nonlinear prebuckling path. Finally, a current estimate of the failure load was given. The load-deflection curves were presented in the diagram. The longitudinal edges of the examined panel were free and the curved boundaries were either clamped or simply supported i.e. in the latter case, the edges were not restrained from motion in the axial direction. The results show a large effect of the boundary conditions on the behaviour in the post buckling state.

3. Elasto-plastic behaviour of Shells

Performing the experiments with shells made of the plastic materials under local loads, we observe that at a certain value of the applied load the shell undergoes permanent deformation. The first zone which becomes plastic is the close vicinity of the point of application of the load. The dimensions of the area on which the load is distributed are the most important factors deciding the value the load producing first plastification. When the deflection of the shell further increases, the shell deforms according to its isometric transformation. Then the local, strongly bent area with plastic strains appears at the distant points on the shell surface. The next step of the mechanism of the deformation consists in the motion of these zones and creation of new plastic regions accompanied by the unloading zone in the previously plastified areas. This mechanism can be easily investigated on the example of the spherical shell.

3.1 Large elasto-plastic deflections of the spherical shell

The similar approach to that described in the previous paragraph can be used to examine the behaviour of the shell. We observe that if the deflection increases, the curvature change in the ridge area also increases which is the reason for the plastic deformation in that place. The other zone where the stress can reach the plastic limit is, of course, the vicinity of the point of application of the load.

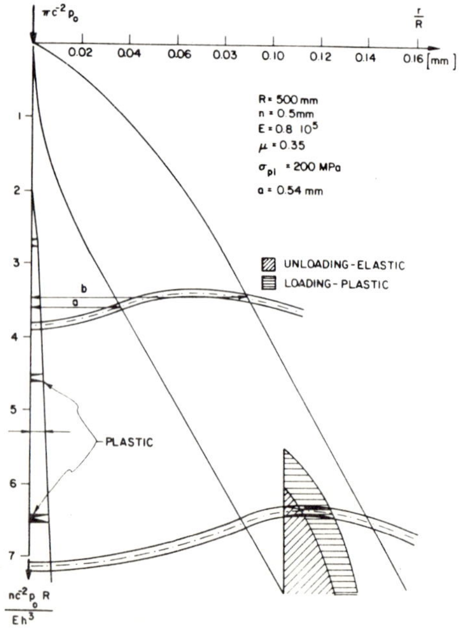

Fig. 12 Elasto-plastic behaviour of the spherical shell.

The experiments with the spherical shells show that the bending of the surface plays the most important role also during the elasto-plastic deflections. The mechanism of the deformations is presented in Fig. 12. We observe that the ridge area in which the largest deformation appears moves in outwards direction on the shell surface. The region which was before under the largest bending flattens and finally is being bent into the reverse direction when it becomes a part of the central isometrically transferred area. The GM used before in the solution of the elastic case can also be easily applied here. The description of the method can be found in the paper by Lukasiewicz and Opalinski [51].

The stationary values of the total potential energy of the system were obtained using optimization routine. The optimization was performed with respect to geometrical parameters defining the shape of the deformed shell. The results of calculations are presented in Fig. 12, [51] for the shell with the following parameters: radius of the sphere, R = 500 mm, thickness h = 0.5 mm, Young modulus $0.8 \cdot 10^5$ MPa, Poisson ratio $\nu = 0.35$, yield stress σ_p = 200 MPa, radius of the loaded area c_o = 0.54 mm, which corresponds to R/h = 1000, $\sigma p/E = 0.0025$. We observe that the first plastic zone appears near the point of application of the load for $P^* \cong 2.0$. Then the plastic zone spreads more and more as the load increases but this zone is always very localized, Fig. 13. At the load value $P^* = 5.5$ the plastic zone at the ridge can be observed. With the increase of the load this zone spreads outwards. As the ridge area moves, there remains the unloaded area with residual strains.

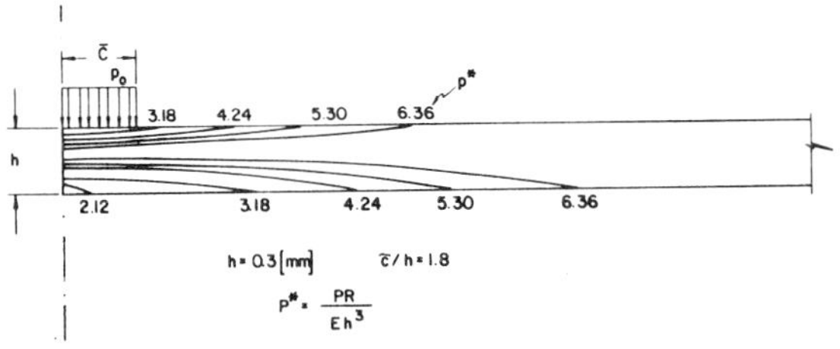

Fig. 13 The plastic zones for different values of the load.

Leckie [49] (1967) studied the snap-through behaviour of a spherical shell with a uniform thickness loaded at its apex with a concentrated load applied through a rigid boss and obtained a corresponding limit load. The study was based on so-called two-moment-limited-interaction criterion which allows the moment resultants and force resultants to assume their respective ultimate values separately. This is not justified physically. Recently Podovan and Tovichakchaikul [50], (1983) presented results for the elasto-plastic postbuckling collapse of a spherical cap under central point load obtained using FEM. However, the results obtained by them seem to be wrong or very inaccurate. The nondimensional deflection w/h calculated for the non-dimensional force $P^* = PR/(Eh^3) = 12.3$ should be about $w/h = 30$, (using Pogorelov's relation for elastic deflection). They obtained only $w/h = 2.0$, the progress of the elastoplastic zones presented does not seem to be reasonable.

3.2 Elastoplastic behaviour of the double curved shell

A simplified solution for an elastoplastic case can be easily obtained using assumption that the energy stored in the shell can be calculated for the unit length of the ridge and basing on the local configuration of this area. The results obtained in this way by Lukasiewicz and Opalinski [51] are given in Figs. 14 a,b where the plastic and unloaded zones are presented. This simplified analysis for the double curved shell is based on

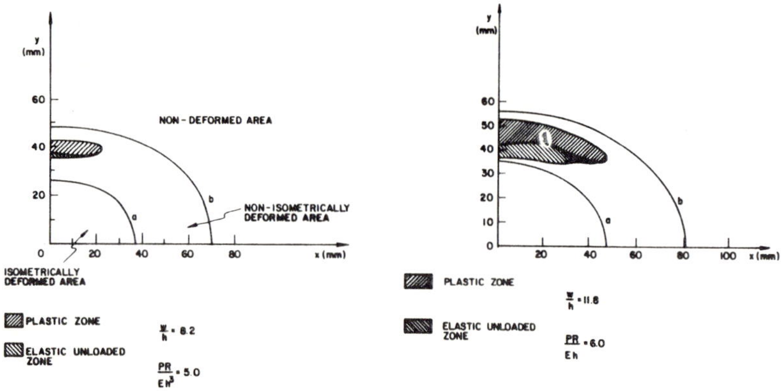

Fig. 14 Elasto-plastic zones in the shell of positive Gaussian curvature.

the observations and results obtained for the sphere where the boundaries of the zones were obtained as a function of the applied load and the local values of the curvature of the ridge. Calculating the total energy contained in the elastic and plastic zones it was found that the load-deflection relations is in the case of not very much developed plastic state $zp < h/4$ almost the same as for the elastic deformations (Eq. 2.) Fig. 14 a and b present the results obtained for two values of the applied loads. We observe that the ridge area moves in the outward direction. The plasticized zones move forward leaving behind the unloaded zones with residual stresses.

3.4 Load carrying ability of thin shells loaded by concentrated forces

The normal force

The maximum limit load which can be carried by a shell can be obtained examining plastic failure mechanism. If we define the kinematically admissible field of deflection, we can obtain analytical expressions relating the limiting values of the concentrated loads with the shell geometric parameters for the corresponding plastic failure mechanism. In the first paper devoted to this problem by Drucker and Shield [53], [1959], the symmetrically loaded shells of revolutions were discussed. Kulikov and Khomyakov [54], (1976) applied that method to the solution of spherical and cylindrical shell under the action of a concentrated normal load Q and the internal pressure p. The material of the shell was assumed rigid-plastic. The problem of crushing of a tube between rigid plastic plates was discussed in [55]. The modes of collapse of rigid-plastic circular cylindrical shell not supported at the edge subjected to applied opposed point loads were analysed by S.R. Reid [56]. He considered the mode of collapse of the cylinder to be composed of a number of inextensional regions separated by plastic hinges. The work dissipated in the region of plastic hinges was calculated as the work of membrane forces and bending moments on the corresponding deformation rates. Three modes of deformation were examined (Fig. 14). For the regions IV and V the inextensibility conditions were assumed. The collapse loads for each plastic mode were obtained using interaction as (Fig. 14). The analogy with the rigid-plastic beam on elastic foundation was discussed. Author claims that reasonably well defined limit loads were obtained only for relatively short tubes when the ring-like deformation mode was applied. The lack of success was probably caused by the assumed collapse modes which do not satisfy conditions of inextensibility in the central cross-section. Rather the mode of deformations presented in Fig. 9 should be applied.

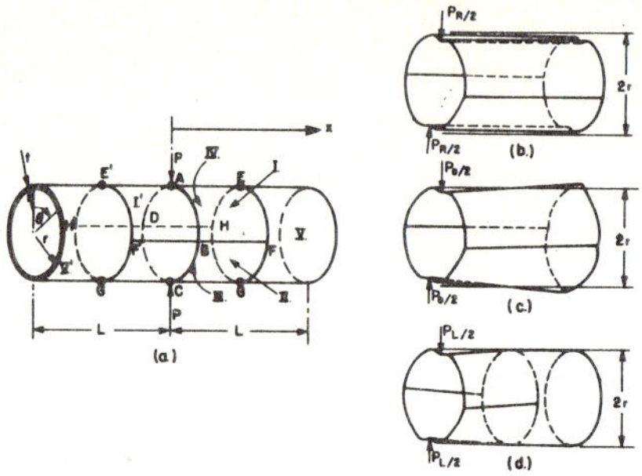

Fig. 15 The modes of collapse of rigid-plastic cylindrical shell

3.4.3 Shallow Shells

The paper by Dekhtyar and Tyutyunnik [57] (1974) contains the calculations for the limit loads for shallow shells, the middle surface of which is formed by a quadratic parabola of revolution and hyperbolic paraboloid with a negative Gaussian curvature. All the upper bounds of limiting loads obtained for shells of positive and negative Gaussian curvature are in agreement with the result for plates and shells. The results are shown in Fig. 16 where the hatched portion corresponds to the region of the parameters $\varepsilon = h/R$ and $\gamma = f/R$ for which the exhaustion of load carrying capacity is of local nature. In the next paper by Dekhtyar and Kotova [58] (1980) the problem of a force concentrated at an arbitrary point of the shell was studied. To solve the problem, the normal displacement w was sought in the form of two functions usually used to describe the Gaussian probability distribution. Finally the limit loads were obtained for elliptical and hyperbolic paraboloids.

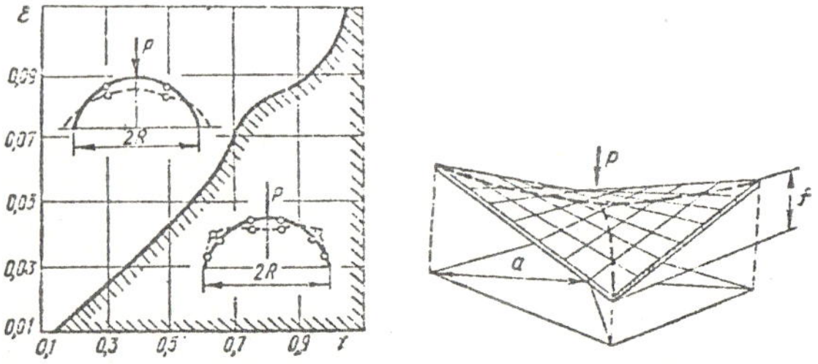

Fig. 16 Collapse modes of the shallow shell

Experimental investigation of the load carrying capacity of reinforced concrete ribbed shallow shells at large deflections under combined action of concentrated and distributed loads was the subject of the article [89] by Shugaev and Luydkovsky. It was shown that the local failure takes place in the zone of concentrated load application. The obtained results were compared with the results of calculations. The results of experiments with the shallow reinforced concrete shell under concentrated loads were presented in [79].

4. Optimum design

The optimum design of shells under concentrated loads is usually performed in order to eliminate concentration of stress which is the result of the application of the local load. This can be achieved in different ways. The shell can easily be strengthened at the loaded place by welding strips, ribs or other strengthening elements. Optimum shell can also be achieved by a proper forming of its thickness and middle surface. The objective functions which are used are minimum weight, minimum volume, surface, minimum or maximum stiffness, maximum stability, frequencies of vibrations, etc. First papers in the field of shell optimisation used the most simple form of the condition of uniform strength $\sigma_1 = \sigma_2 \pm k = \text{const.}$ This condition together with the equations of equilibrium defines uniquely the shape of the shell, i.e. thickness and the shape of the middle surface. Assuming that the thickness of the shell is constant, its shape can be defined using membrane equations of equilibrium and the condition of minimum volume

$$\delta V = h\delta \int_S \sqrt{g}\, ds = 0$$

which gives the equation of minimum surface of the shell. In the case of shallow shells this equation can be simplified to

$$\Delta z = p/h\sigma_o$$

where p(x,y) is the load distribution, and z coordinate of the middle surface. For shells of revolution of constant thickness loaded by a concentrated force and constant pressure simple solution can be found, [2].If the more general condition is used as for example

$$\sigma_r^2 = \sigma_1^2 - \sigma_1\sigma_2 - \sigma_2^2 = k^2 = \text{constant}$$

it is possible to use the objective function for further optimisation, to achieve, for example, a minimum volume of the shell. Lukasiewicz [59], (1969) obtained in this way the shapes of the symmetrical shells of variable thickness loaded by a concentrated force and external or internal

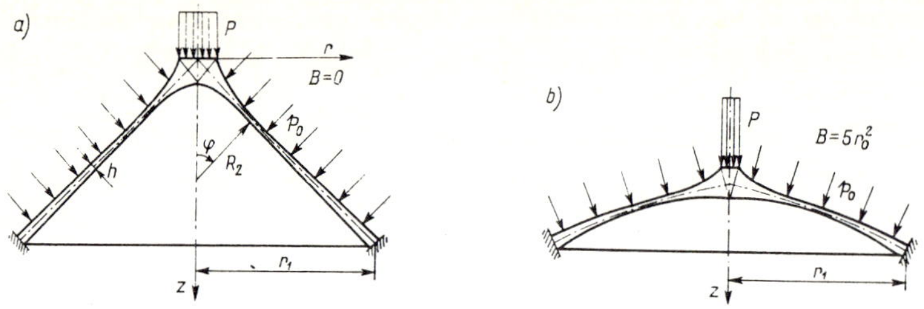

Fig. 17 Shapes of shells of minimum weight subjected to the single force P and the pressure p_o

pressure, Fig. (15). The shape of the shell in the vicinity of the place of application of the load takes the form resembling the conical shell with the thickness decreasing with the distance from that point. The paper by Agafonov is devoted to the design of a structurally orthotropic cylindrical shell [66], (1979). Grigorenko [67] considered shells of revolution of variable stiffness (1973). Thevendran [68], (1982) considered the maximum weight design of a spherical shell under a concentrated load at the apex. The finite element was used to discretize the design parameters. The nonlinear minimization problem is solved using optimisation routine. In this paper, the bending state of stress was not eliminated. Maksimenko [69] examined the problems of determining the minimum weight of discretely reinforced cylindrical shells under the local loads applied to curvilinear contour. The number of ribs in both directions and their stiffness was different with the regular and irregular arrangement. The problem was solved by the method of finite differences using the shallow shells equations. As the example a cylindrical shell reinforced with eight stringers and rings was discussed. The shell was loaded with four longitudinal symmetrical compressible forces. It was indicated that the sections of the stringers located far from the load application had almost no effect on the stress state of the shell. A similar problem was also discussed by Drugach, Polyakov and Maksimenko [70]. The stress distribution in the area where the cylindrical shell is supported were calculated in the paper by Modestova, Simakin, and Samoilenko [71]. This area was reinforced by increasing by steps the thickness of the shell. The results were compared with experiments. Optimum design of shallow shells and plates taking into account the physical nonlinearities was the subject of the paper by Bocharev and Krysko [70] (1981). The thickness of the shell was obtained by a variational Vlasov-Kantorovich method. Two partial differential equations were replaced by a system of two ordinary differential equations with variable coefficients which were solved using finite differences. The process of the design was performed by successive iterations. The optimum design of circular shallow shells with respect to stability was presented in an interesting paper by Plaut and Johnson [73] (1984). The objective function was the maximum of the critical buckling load under a uniformly distributed load or a concentrated load at the center. The form of the middle surface was varied for clamped and simply supported boundary conditions. The set of nonlinear differential shallow shell equations were solved for primary equilibrium state under the applied load. Then the small vibrations about this state were examined. The critical load was found when the vibration frequency decreased to zero. The optimality conditions derived by the calculus of variations was used to improve the shell form by successive iterations. The form of the shell and the deflected shape at a critical concentrated load for clamped edge and λ = 8.5 are presented in Fig. 18.

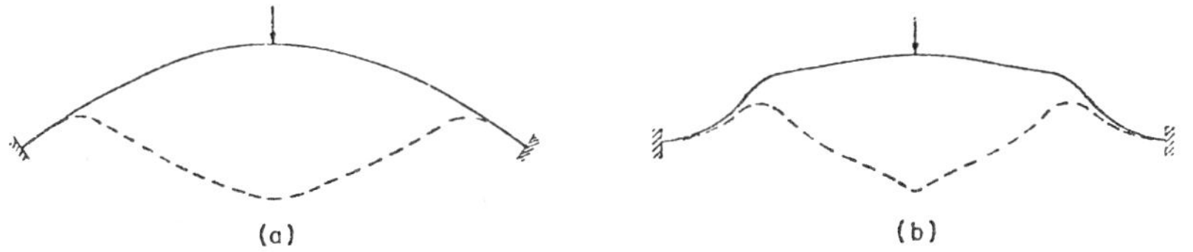

Fig. 18 Shell form and deflected shape at critical concentrated load for clamped edge, λ = 8.5: (a) spherical form; (b) optimal form

The shape of the shell corresponding to the given stress was determined by Alpe, Bozzo, Corsauega, and Del Grosso, [74]. The shells rectangular in plane on point-like supports were considered. The solution was obtained using finite differences method.

The problem of the optimum design of the attachment introducing the load into the spherical shell was discussed by Lukasiewicz [76,77] (1974). The shape of the attachment was obtained by solving the equations of continuity of strains allowing the line of the contact between the shell and attachment for the assumed distribution of constant stresses. The stress distribution was chosen in such a way that the shell was uniformly loaded along the line of contact with the attachment. Similar problem was discussed by Popov and Uzhva [78] (1981). The concentrated normal force was introduced into the shell by means of a designed rib of variable cross section. The method of nonlinear programming was applied. Optimisation of dynamic effects in shells of revolution under axisymmetric dynamic force was discussed by Burak and Domanski [75]. The example solved was a circular cylindrical shell under dynamic application of the symmetrically applied load. The distribution of that load was optimised in order to minimise the kinetic energy of the shell.

5. Contact Problems

A very important domain, but not very well explored, are the contact and impact problems. The load is always applied to the shell by means of another body which can have just an arbitrary shape and properties. The solution of the contact problem requires the determination of the form and rise of the contact region, usually small in comparison to the dimensions of the shell, and distribution of the contact forces. During impact the contact forces are transmitted in time depending on contact zones. Shell contact problems have not been yet extensively studied as compared to solid body contact. Because of large deflections and variable boundary of the contact zone these problems are essentially nonlinear. Updike and Kalnins [80], (1972) studied contact between an elastic spherical shell and a rigid plate. A comparison of the results of linear and large deflection shell theory was given in this paper as well as the important question of the effect of transverse shear deformations was investigated. A similar problem was studied by Kitching, Houlston and Johnson [81], (1975) and a good agreement between experiment and large deflection shell theory was found. Tielking and Schapery [82], (1981) studied a static contact of a toroidal shell (a pneumatic tire) with a rigid plane. The linear and nonlinear solution was obtained using discrete Fourier transform. The contact pressure between two co-axial cylindrical shells under axisymmetrical loads was investigated by Paczelt and Herpai [83], (1977). The problem was discretized and reduced to quadratic programing problem. Variational formulation of a contact problem for linearly elastic and physically nonlinear shallow shells was presented by Lvov [84]. The author reduced the problem to the minimizing a Lagrange functional in a set of allowable displacements. No example was given. The same author studied contact problems of shell creep [85]. The methods of the solution of contact problems in the theory of plates and shells were presented by Mossakovskii Gudramovich and Makeev [87], (1978) and by Grigoluk and Tolkachev [86], (1980). Impact-contact problems of thin elastic shells taking into account geometrical nonlinearities within the contact region were the subject of the paper by Stein and Wriggers [88], (1982). The paper used FEM formulation. The example contained the solution for a spherical shell impacting a rigid plane. The comparison of the results obtained from linear and nonlinear approach was given. Several papers were devoted to the analysis of stresses in the saddle supported cylindrical vessels.

6. References

1. Lukasiewicz, S.A., Introduction of concentrated loads in plates and shells, Progress in Aerospace Science, Vol. 17, No. 2, pp. 109-146 (1976).
2. Lukasiewicz, S.A., Local Loads in Plates and Shells, Noordhoff International Publishing, Leyden (1979).
3. Biezeno, C.B., Über die Bestimmung der Durchschlagkraft einer schwachgekrummten kreisformigen Platte, Z. Angew, Math Mech. 15.10 (1935); C.B. Biezeno and R. Grammel, Engineering Dynamics, 484, London 1965; 2 Blackie, London, 484, 1956.
4. Chien Wei-Zang and Hu Hai-Chang, On the snapping of a thin spherical cap, 9th Int. Congr. Appl. Mech., 6, University of Brussels, 309, 1957.
5. Ashwell, D.G., On large deflection of a spherical shell with an inward point load, Proc. of IUTAM Symp. Theory Thin Elastic Shells Delft, 1959, North Holland Publish., Amsterdam 43-63, 1960.
6. Archer, R.R., On the numerical solutions of the nonlinear equations for shells of revolution, J. of Math. Phys. 41.3 (1962) 165-178.
7. Pogorelov, A.G., Geometrical method in nonlinear theory of elastic shells, Izd. Nauka, Moscow 1967 (in Russian).
8. Thurston, G.A. and Penning, F.A., Effect of initial imperfections on the stability of shallow spherical shells, AFOSR, Sci. Report No. 64, 1627, August 1964.
9. Penning, F.A. and Thurston, G.A., The stability of shallow spherical shells under concentrated load, NASA CR-265, July 1965.
10. Penning, F.A., Experimental buckling modes of clamped shallow shells under concentrated load, J. Appl. Mech. 33.2 (1966) 297-304.
11. Evan-Iwanowski, R.M., Cheng, H.S. and Loo, T-C., Deformations and Stability of Spherical Shells Under the Action of Concentrated Loads", SURI Report No. 834-2, September, 1961.
12. Budiansky, B., Hutchinson, J.W., A survey of some buckling problems, AIAA Journal 1966, Vol. 9, pp. 1505-1510.
13. Bushnell, D., Bifurcation phenomena in spherical shells under concentrated and ring loads, AIAA J. 5.11 (1967) 2034-2040.
14. Evan-Iwanowski, R.M., Cheng, H.S. and Loo, T.C., Experimental investigation of deformation and stability of spherical shells subjected to concentrated loads at the apex, Proc. of the Fourth U.S. Nat. Congr. Appl. Mech. ASME (1962) 563-575.
15. Voss, W.C., Peabody, D., Staley, H.R. and Dietz, A.G.H., Thin shallow domes loaded excentrically, Prc. ASCE 73 (1947) 1173-1195.

16. Loo, Ta-Cheng, Evan-Ivanowski, R.M., Interaction of Critical Pressure and Critical Concentrated Loads Acting on Shallow Spherical Shells. Transactions of the ASME Journal of Applied Mechanics, September 1966, pp. 612-615.
17. Lukasiewicz, S.A., Szyszkowski, W., W. Geometryczne metody nieliniowej teorii powlok, Proceedings of the symposium, Konstrukcje Powlokowe, teoria i Zastosowania PAN, Krakow, 25-27 April, (1974).
18. Fitch, J.R., Budiansky, B., Buckling and post buckling behaviour of spherical caps under axisymmetric load, AIAA J. 8 (1970), 686-693.
19. Akkas, N., Bauld, N.R. Jr., Buckling and post buckling behaviour of clamped shallow spherical shells under axisymmetric ring loads. J. Appl. Mech. 38 (1971), pp. 996-1002.
20. Teregulov, I.G. and Mogilevskii, L.I., The influence of the geometrical nonlinearity in asymptotic formulae for circular plates and shallow spherical dome loaded by a concentrated force, Tr. seminara po teorii obolochek. KFT AN SSSR, Kazan.3 (1973) in Russian.
21. Kao, R., Perrone, N., Stresses in spherical shells due to local loadings, Journal of Ship Research, March (1973) 19-22.
22. Ranjan, G.V., and Steele, C.R., Large deflection of deep spherical shells under concentrated load, Proceed. AIAA/ASME 18th Structures, Structural Dynamics, Materials Conference, San Diego, Mar. 21-23 (1977), 265-278.
23. Taber, L.A., Large deflections of fluid-filled spherical shell under a pointload, ASME, Journal of Applied Mechanics,vol 49, (1982), 121-128.
24. Taber, L.A. Compression of Fluid-Filled spherical shells by rigid intenders, Journal of Applied Mechanics, Vol 50/717, (1983).
25. Onoda, J., An analysis of the buckling of truncated or Complete spherical shells under axial compressive loads. Trans. Japan. Soc. Aero. Space Sci. Vol 26, No. 73.
26. Mescoll, J., Numerical solutions of nonlinear equations for shells of revolution. AIAA Jour. (1966), vol 4, No. 11, pp. 2041-2063.
27. Akkas, N., Bauld, Nrjr, Buckling and post-buckling behaviour of shallow spherical sandwich shells under axissymmetrical loads, Developments in Mechanics, Vol. 6. Proceedings of the 12th Midwestern Mechanic Conference.
28. Koiter, W.T., Elastic Ability and Post-buckling behaviour. Proc. Symp. Nonlinear Problems., Ed. R.E. Conger p. 257., University of Wisconsin Press (1963).
29. Grigorenko, Ya. M., Timotin A.M., Numerical solution of non-axissymetrical problems in the nonlinear theory of laminated shells of revolution, Prikladnaya Mekhanika. Vol. 18, No. 5, pp. 43-48, May (1982).
30. Goldenveizer, A.L., Geometrical theory of stability of Shells, Izv. AN. SSSR Mekhanika Tverdego Tela, Vol. 18, No. 1, pp. 143-154, (1984).
31. Almroth, B. and Brogan, F.A., Bifurcation buckling as an approximation of the collapse load for general shells. AIAA J., 10, 463-467 (1972).
32. Lukasiewicz, S.A., On the postbuckling behaviour of the Cylindrical Shell. Report No. 468, (1973), Laboratory of Engineering Mechanics, Delft, Netherlands.
33. Lukasiewicz, S.A., Szyszkowski, W., Geometrical analysis of large elastic deflections of axially compressed cylindrical and conical shells. Int. J. Non-linear Mechanics Vol. 14, pp. 273-284, (1979).

34. Rabotnov, N.A., Effect of a local load on a cylindrical shell of fiberglass-reinforced plastic. Prikl. Mekhanika, Vol 18. No 6, pp. 63-68, (1982).
35. Matsumoto, T., Jwata, K. and Nagoto, R., Measurement of deformation in a cylindrical shell by holographic interferometry. Applied Optics Vol. 13, No. 5, (1974).
36. Andreev, L.V., Antsiferov, A.V., Dubovik O.M., Experimental Study of the behaviour of cylindrical Shells with concentrated masses under a dynamic external pressure. Izv. AN. SSSR, Mekhanika Tverdogo Tela, Vol. 18, No. 3., pp. 187-191 (1983).
37. Malova, D.G. and Ventskovskii, Effect of design features of pressure vessel flange on its state of stress. Khimicheskoe i Neftyanoe Mashinostroenie, No. 4, pp. 5-7, (1983).
38. Lukasiewicz, S.A. and Glockner, P.G., Collapse by ponding of shells, Int. J. Solids Structures, Vol. 19, No. 3, pp. 251-261, (1983).
39. Whipple, R.A., Hagstrom, J., Dykstra, H., Experimental Investigation of cylindrical shell stresses due to penetration loads where R/T=1264. J. Pressure Vessel Tech, Vol. 105 (1983), pp. 201-208.
40. Lukasiewicz, S.A., Geometrical elements method for the solution of nonlinear shell problems. Report No. 268, of the Department of Mechanical Engineering, The University of Calgary, (1983).
41. Krysko, V.A. and Kolomoets, A.A. Stability of Local Loading of a cylindrical shell, Prikladnaya Mekhanika, Vol 17, No. 10, pp. 86-92, (1981).
42. Neug-Ming Wong, Hilario Oh, and Kno-Kuiug Cheu, Large deflection of Shallow Paraboloid Shells, Journal of the Engineering Mechanics Division, EM4, 1973.
43. Horrigmoe, G. and Bergau, P., Nonlinear analysis of free form shells by flat finite elements, Comput. Meths. Appl. Meths. Appl. Mech. Eng. 16 (1978), pp. 11-35.
44. Sabir, A.B. and Lock, A.C., The application of finite elements to the large deformation geometrically nonlinear behaviour of shells, in Brebbia, C.A. and Tottenham, H., Eds; Variational Methods in Engineering (Southampton University Press, London (1973).
45. Crisfield, M.A., A fast incremental/iterative solution procedure that handles "snap-through". Comput. and Structures 13, (1981), pp. 55-62.
46. Chang, T.Y. and Sawamiphakdi, Large deflections and post-buckling analysis of shell-structures. Computer Methods in Applied Mechanics and Engineering 32 (1982), pp. 311-326.
47. Brendel, Bl, and Ramm, E., Linear and nonlinear Stability analysis of cylindrical shells, Computers and Structures, Vol 12, pp. 549 - 558, (1980).
48. Leckie, F.A., Plastic instability of a spherical shell, Proc. IUTAM Symposium on Theory of Thin Shells, Copenhagen (1967).
49. Leckie, F.A. The plastic analysis of a spherical shell subjected to a radial load applied through a rigid boss. Brown Univ. Rep. NSFGP-1115/23 (1965).
50. Padovan, J., Tovichakchaikul, S., On the Solution of elastic-plastic and dynamic postbuckling collapse of general structure. Computers and Structures, Vol. 16, No 1-4, pp. 199 - 205, (1988) or Advances and Trends in Structural and Solid Mechanics Proceedings of the Symposium, Washington, D.C., (1982) Pergamon Press.
51. Lukasiewicz, S.A., Opalinski, W., An elasto-plastic behaviour of the double-curved shell under concentrated load. Report No. 328 of the Department of Mechanical Engineering, The University of Calgary, (1985).

52. Waszczyszyn, Z., Radwanska, M., Pabisek, E., Application of the initial value method to analysis of elasto-plastic plates and shells of revolution, Computers and Structures, Vol. 16, No. 6., pp. 761-771, (1983).
53. Drucker, D.C., Shield, R.T., Limit analysis of symmetrically loaded thin shells of revolution J. Appl. Mech. 1, 16, pp. 61-68 (1959).
54. Kulikov, Yu., Khomyokov, A.M., Load carrying ability of thin shells loaded by local normal force, Izvestiya VUZ Aviatsionnaya Tekhnika, Vol. 19, No. 2., pp. 131-136, (1976).
55. De Runtz, J.A., and Hodge, P.G., Crushing a tube between rigid plates, J. Appl. Mech. 30, 391-395, (1963).
56. Reid, S.R., Influence of Geometrical Parameters on the mode of collapse of a pinched rigid-plastic cylindrical shell, Int. J. of Solids, Structures, Vol. 16, pp. 1027-1043 (1978).
57. Dekhtyar, A.S. and Tyutyunnik, A.M., Maximum concentrated load on a shell, Priklodnaya Mekhanika, Vol. 11, No. 10, pp. 35-73 (1975).
58. Dekhtyar, A.S. and Kotova, L.B., Limiting point load applied to an arbitrary point of a shell, Prikladnaya Mekhanika, Vol. 19, No. 1, pp. 47-51, (1983).
59. Lukasiewicz, S., On the optimum design of shells loaded by concentrated forces, in Theory of Thin Shells, Proc. IUTAM Symp. Copenhagen 1967, Springer Verlag, (1969).
60. Read, W., Equilibrium shapes for pressurised domes, J. Appl. Mech. Trans. ASME, ser. B, II (1963).
61. Shield, R.T., On the optimum design of shells, J. Appl. Mech. Trans, ASME (1969), 316-322.
62. Prager, W., and Shield, R., A general theory of optimum plastic design, J. Appl. Mech. Trans, ASME, E34 (1967) 184-186.
63. Freiberger, W., Minimum weight design of cylindrical shells, J. Appl. Mech. Trans. ASME 23, 14 (1956).
64. Save, M.A. Shield, R.T., Mimum weight design of sandwich shells subjected to fixed and moving loads, Proc. Eleventh Int. Congr. Appl. Mech., Munich 1964.
65. Życzkowski, M., Optimisation of shell structures, in Proceedings of the Symposium on Shell Structures, Theory and Applications, Kraków (1974).
66. Agafonov, A.V., Designing a structurally orthotropic cylindrical role for concentrated loads, Raschet Prostranstv. Konstrukt. No. 18, 94-104, (1979).
67. Grigorenko, Ya. Isotropic and unisotropic layered shells of revolution of variable stiffness, [in Russian], Nautova Duko, Kiev, (1973).
68. Thevendran, V., Minimum weight design of Spherical Shell under a concentrated load of the apex, Inter. J. for Numerical Methods in Engineering, Vol. 18, 1091-1111 (1982).
69. Maksimenko, V.P, Equal-stress cylindrical shell under the influence of locally applied longitudinal forces, Prikl. Mekh. 15, No. 12, 59-64 (1979).
70. Drugach, M.I., Polyakov, P.S., and V.P. Maksimenko, Study of the strength of cylindrical shells loaded with longitudinal forces. Prikl. Mekh., 4, No. 4, 25-36 (1968).
71. Modestova, R.W., Simakin, A.M., and Samoilenko, E.N., Problemy Prochnosti, 5, Nos. 1-6 (1973).
72. Bocharev, V.V., and Krysko, V.A., Optimum design of plates and shells taking into account the physical nonlinearities, Prikladnaya Mekhanika, Vol. 18, No. 7, pp. 52-57, (1982).

73. Plaut, R.H., and Johnson, L.W., Optimal forms of Shallow shells in the circular boundary, Part 2. Maximum buckling load, Journal of Applied Mechanics, Vol. 51, Sept. (1984).
74. Alpa, G., Bozzo, E., Corsauszo, A., and Del Grosso, A. Shape determination for shell structure on pointlike supports. Bull Int. Assoc. Shell. Spot. Struct. V19-2 No. 67 Aug (1978), pp. 3-9.
75. Burak, Ya. I, Domanski, P.P., Optimisation of dynamic effects in shells of revolution under axisymmetric force load, Prikladnaya Mekhanika, Vol. 18, No 2, pp. 7-16, (1982).
76. Lukasiewicz, S., and Borajkiewicz, W., Optiumum design of an element introducing the load into the spherical shell, Int. J. of Solids and Structures, (1974).
77. Lukasiewicz, S., An introduction of a concentrated tangential force into a shell, Proceeding of IUTAM Symposium, Optimisation in Structural Design, Warsaw, Springer Verlag, 1975.
78. Popov, B.G., and Uzhva, V,V, Vybor ratsionalnykh parametrov podkrepleniya tsylindricheskoi obolochki v zone sosredotochnego vozdeistvia, Izvestia Vysshikh Uchebnykh Zavedenii, Mashinostroenie, Nos. 4-6 (1981).
79. Paduart, A. and Halleux, P., Flexunel behaviour of a micro-concrete lypar shell.
80. Updike, D.P. and Kalnins, A., Contact pressure between an elastic spherical shell and a rigid plate., J. Appl. Mech., 39 (1972), pp. 1110-1114.
81. Kitching, R., Houlston, R., and Johnson, W., A theoretical and experimental study of hemispherical shells subjected to axial loads between flat plates, Inter. J. Mech. Sci. 17 (1975) pp. 693-703.
82. Tielking, J.T. and Schapery, R.A., A method for shell contact analysis,Computer methods in applied mechanics and engineering, 26, (1981) pp. 181-195.
83. Paczelt, J. and Herpai, B., Some remarks on the solution of contact problems of elastic shells. Archiwum Budowy Maszyn XXIV (1977), 2, pp. 187-201.
84. Lvov, G.I., Variational formulation of a contact problem for linearly elastic and physically nonlinear shallow shells, Prikl. Matem. Mekhan, 45 No. 5, (1982), pp. 861-846.
85. Lovov, G.I., Certain contact problems of shell creep, Transactions of the XII All-Union Conference on the theory of shells and plates, Vol 3, Trol. Erevan Univa, (1980) pp. 31-37.
86. Grigoluk, E.I. and Tolkachev, V.M., Contact Problems of the Theory of Plates and Shells, Mashinostroenie, Moscow (1980).
87. Mossakovski, V.L., Gudramovich, V.S., and Makeev, E.M., Contact problems of the theory of shells and rods, Mashinostroenie, Moscow (1978).
88. Stein, E. and Wriggers, P., Calculation of impact-contact problems of thin elastic shells taking into account geometrical nonlinearities within the contact region. Computer Methods in Applied Mechanics and Engineering 34, (1982), pp. 861-880.
89. Shugaev, V.V. and Lyudkovsky, A.M., Carrying capacity of reinforced concrete ribbed shells at large deflections, IASS World Congress on Space Enclosures, Colec. Univ. Montreal, July 1976.

Axially Loaded Metal Tubes as Impact Energy Absorbers

S.R. REID and T.Y. REDDY

Department of Mechanical Engineering,
University of Manchester Institute of Science and Technology,
P.O. Box 88,
Manchester M60 1QD.

Summary

Plastic collapse mechanisms involving axial deformation of metal circular tubes provide efficient means of absorbing kinetic energy following impact. Experimental data concerning a variety of these mechanisms including axial buckling, tube inversion, axial splitting and the crushing of tubular rings are described and discussed. Areas where further theoretical work is required are identified.

1. Introduction

The design of metal shell structures and structural components capable of sustaining predictable loads under conditions of gross plastic deformation is one of the prime means of improving the crashworthiness of vehicles [1,2]. In addition, such components find extensive use in other areas where one needs to mitigate the effect of impact/impulsive loading, for example in the design of pipe-whip restraint systems used in the nuclear industry [3]. Comprehensive reviews of the properties of many of these devices are to be found in [4,5] whilst the recent reviews by Reid [6,7] pay particular attention to the characteristics of metal tubes, both circular and non-circular.

Thin-walled circular tubes ($D/t > 20$, where D is the mean diameter of the tube cross-section and t its wall thickness) compressed axially provide a number of particularly efficient energy absorbing mechanisms. These include axial buckling, inversion and axial splitting, the last two usually requiring the use of a radiused die on to which the tube is compressed. The first two modes of deformation have received considerable attention in the literature and formulae have been produced which enable the operating load and energy absorption capacity to be estimated. However rigorous analyses of these gross deformation fields have yet to be produced. Axial splitting of circular tubes has been identified as a

Inelastic Behaviour of Plates and Shells
IUTAM Symposium Rio de Janeiro 1985
Editors: L. Bevilacqua, R. Feijóo and R. Valid

failure mode, for a tube undergoing external inversion [8]. It has however been investigated and developed recently as a primary energy absorbing mechanism. Experiments will be described in which each of these three primary axial deformation modes have been produced using the same tube stock so providing a direct comparison between them.

A further example will be provided of a device which uses metal tubes of circular cross-section whose response is strongly influenced by the axial loads generated in it during gross deformation. This device has been termed a tubular ring [9] and consists of four lengths of circular tube welded together at mitred joints to produce a "square torus". The ring is crushed between flat plates in a direction perpendicular to its plane and exhibits significantly greater energy absorbing capacities than other tube configurations loaded laterally. An experimental and theoretical treatment of an axisymmetric toroidal shell undergoing the same type of loading has been produced recently by Sugita [10]. Both of these studies show that the improved performance stems from the axial (or circumferential in the case of the toroidal shell) stresses generated by the mutual constraint between the neighbouring elements.

Attention will be concentrated upon the quasi-static, large deformation response of the various components. However reference will be made to the influence of dynamic effects and some data provided on the major changes produced by impact loading. In the main the paper provides an account of experimental data and observations although, where appropriate, available theoretical models are described and applied and suggestions are made regarding areas where further work is required.

2. Circular Tubes Under Direct Axial Loads

2.1 Experiments and Results

Seamless mild steel tubes, 50.8 mm outside diameter and 1.6 mm wall thickness, were used in the experiments conducted using an Instron 1185 universal testing machine at a cross-head speed of 0.167 mms^{-1}. All the specimens were of 100 mm length. The buckling mode was produced by compressing a tube specimen axially between two flat plates. The inversion and splitting modes were produced by compressing nominally identical specimens on to a lubricated die. The different modes were generated by using hardened dies of different fillet radius.The yield stress, σ_0, (or plane strain flow stress) was measured by performing a lateral compression test [11] on a tube of 100 mm length. This provided a value for σ_0 of 732 N/mm^2.

2.1.1 Axial buckling

Fig. 1 shows the load-compression characteristic of the test and the specimen is also shown in the inset. The axisymmetric (concertina) mode of buckling occurred. A mean load of 80 kN, was obtained with a load fluctuation having a peak-to-peak amplitude of 45 kN. The maximum compression possible was 70 mm after which the specimen produced a compression resistance in excess of 500kN. There was a straight, undeformed, length at the end of the tube.

2.1.2 External inversion

Fig. 2a shows the load-compression curves from the inversion tests. With dies of radii 4 mm and 6 mm, inversion was accomplished successfully. After an initial transient phase during which the leading edge rotates through 270°, the inversion load achieved an essentially constant value. With a 10 mm radius die, the tube flared and cracked. With a 3.2 mm radius die, the tube inverted but the leading edge of the tube pressed against the tube. The consequent increase in resistance to deformation resulted in this tube buckling. Fig. 2b shows the deformed specimens for the dies of fillet radius 10 mm, 6 mm and 4 mm.

2.1.3 Axial splitting

A tube identical to the one which cracked in the inversion test was compressed on to the 10 mm radius die. The tube flared and cracked at five locations around the circumference. The cracks did not appear simultaneously. However, only two cracks propagated and three stopped within 5 mm. It is interesting to note, however, that the two propagating cracks bifurcated successively to produce a total of five cracks. The cracks realigned and were propagating axially again at a compression of approximately 60 mm. The other splitting specimens each had four saw cuts of 3 mm depth introduced symmetrically around their circumference. Cracks started from the tips of these stress raisers and the strips so formed curled up as the tubes were compressed on to the dies. The strips were also flattened in the circumferential direction in the bending process.

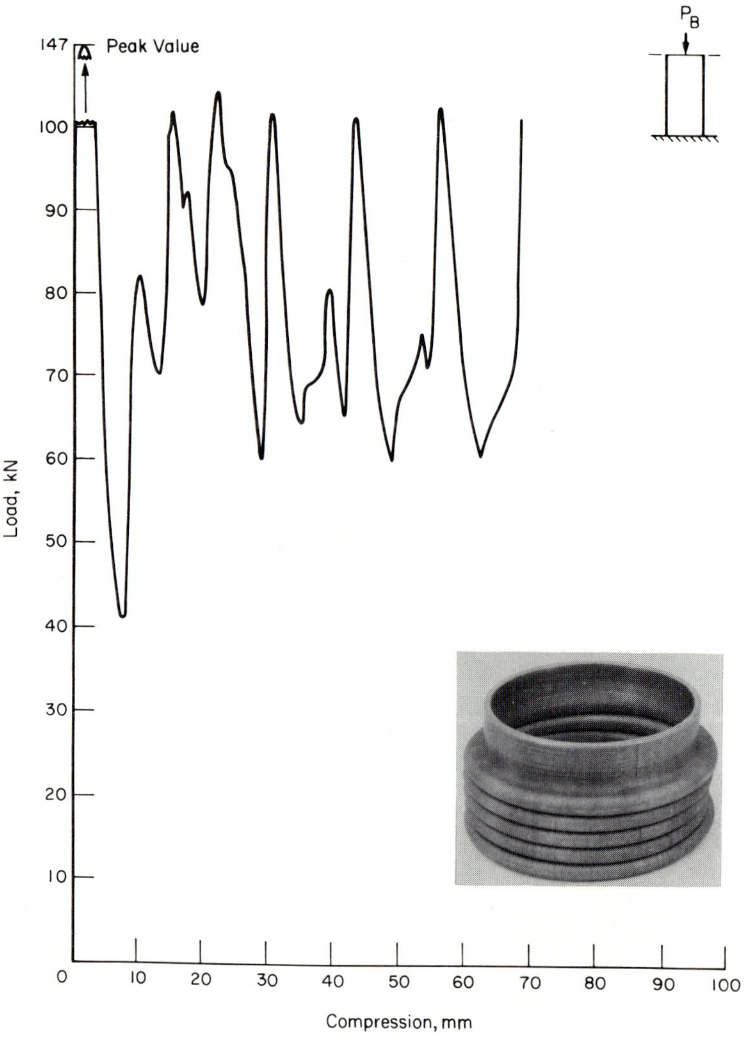

Fig.1. Load-deflection curve for axial buckling test. Inset shows deformed specimen.

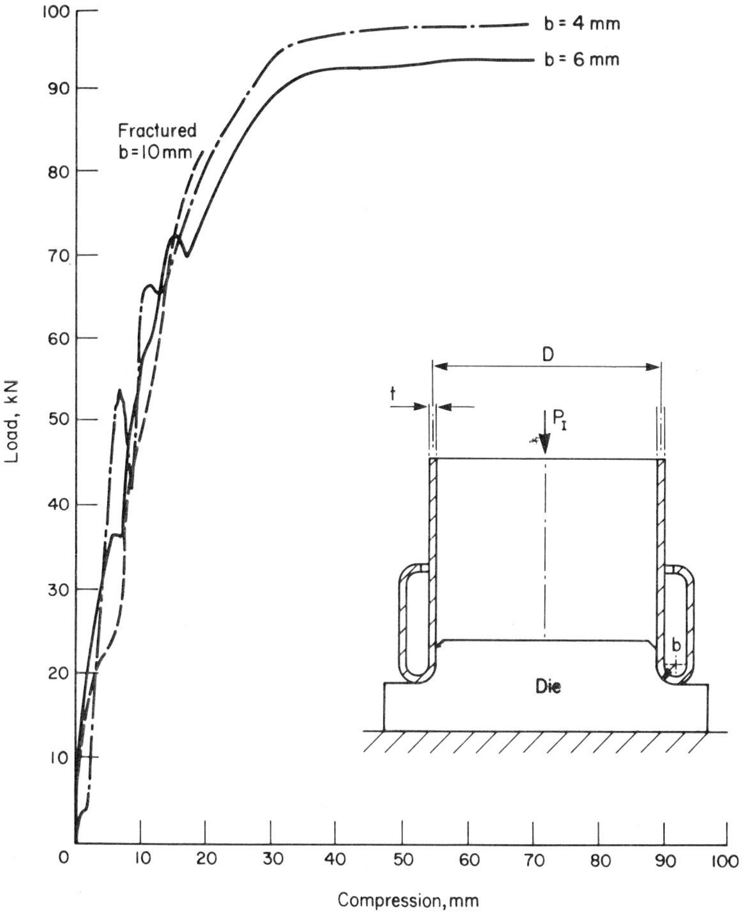

Fig.2(a) Load-deflection curves for external inversion tests.

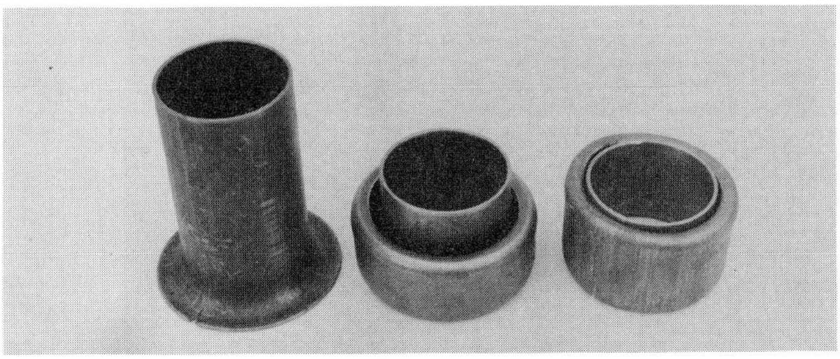

Fig.2(b) Deformed inversion specimens for b=10mm, 6mm, and 4 mm (left to right)

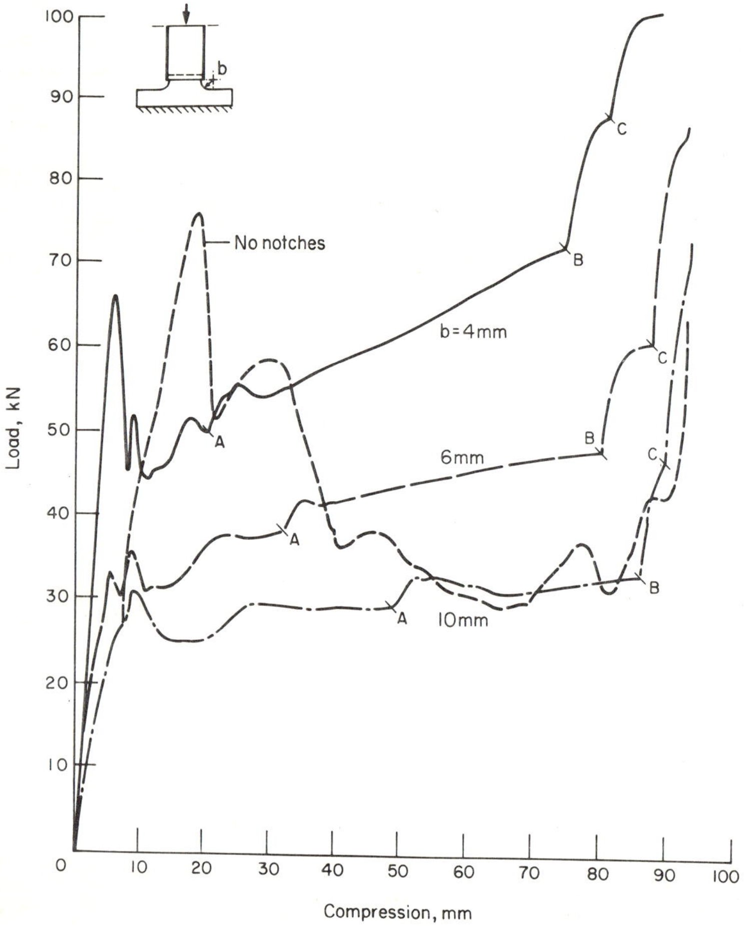

Fig.3. Load-deflection curves for splitting and curling tubes.

The load-compression curves for these tests and the one involving the tube without initial cuts are presented in Fig. 3. The curling strips made contact with the tube when they completed one convolution, the load slightly increased at this point, (A in Fig. 3) and decreased again. However, after this instant, i.e. strips completing one convolution, the load-deflection curves hardened, the slope being larger the smaller the die radius. At points marked B, the strips came into contact with the top plate, first to be straightened at the base and then the curls themselves

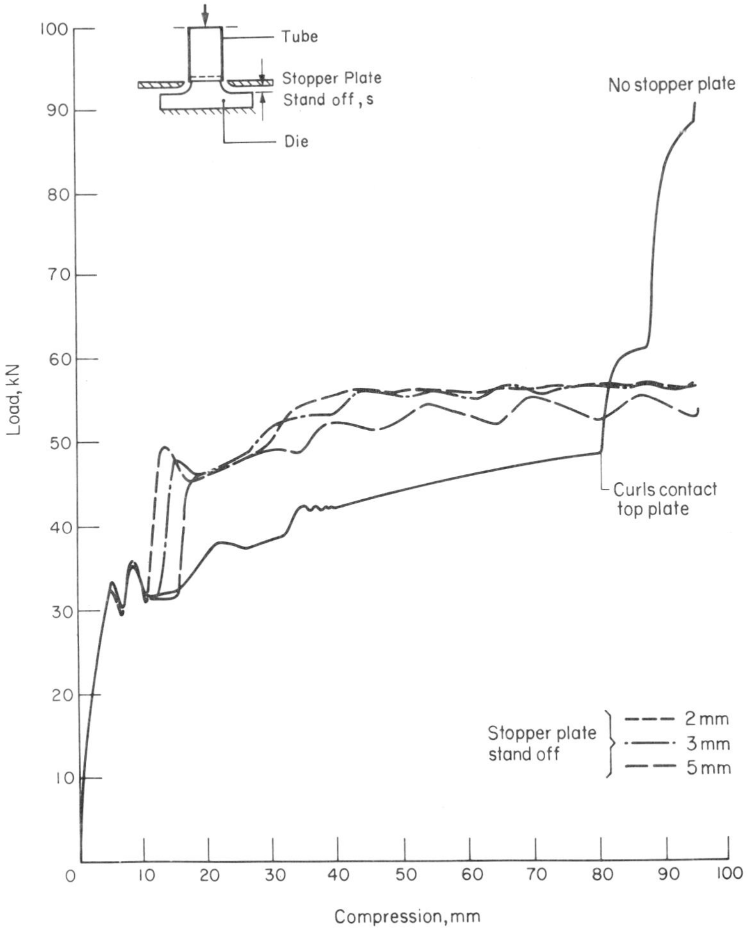

Fig. 4.Effect of curl-stopper plate at various stand-off distances.

were compressed beyond point C. These test results are quite typical of the basic splitting mechanism more details of which can be found in

[12].In terms of performance as an energy absorber, an improved arrangement has been devised. To illustrate this, tests were conducted using the 6 mm die and identical tube specimens but fixing a "curl-stopper" plate at a certain stand-off distance, s, from the surface of the die. Essentially, this plate prevents the strips from curling and forces them to move radially as the compression progresses. Fig. 4 shows the load-deflection curves for different values of s. It may be noted that in the two larger spacing tests the strips developed a wavy pattern at their edges (see Fig. 4(b)). This decayed towards the centre of the strips. The amplitude of these waves (peak-to-peak) was equal to s and the wavelength corresponded to the distance between peaks in the load-compression curves. With a 2 mm stand-off no such waviness was observed. The deformed specimens whose load-compression characteristics are shown in Figs. 3 and 4 can be seen in Fig. 5.

(a)

(b)

Fig.5. Axial splitting specimens.
(a) Free to curl. Left to right b=10 mm,6 mm,4 mm;b=10 mm, no initial cuts
(b) Curl stopper tests, b=6 mm. Left to right, s=2 mm,3 mm,5 mm.

2.1.4 Comparison of mechanisms vis a vis energy absorbing capacity

A summary of the principal features of the test results is given in Table 1. The mean load calculations are accurate to 5%. The similarities and differences in the energy absorbing mechanisms will be discussed below but their global effects are evident in the data presented in Table 1. The two modes which do not involve fracture possess a high mean load reflecting the efficient mechanisms consisting of circumferential stretching and axial bending about circumferential hinges. The stroke or effective crushing length is of the order of 70% of the tube length. For the splitting mechanism the lower mean loads are offset by the increased stroke of 95% resulting in comparable energy absorbing capacities for the smaller die radii. Particular benefits (e.g. an essentially constant steady state force) accrue from the use of the curl-stopper plate which actually provides a device with comparable if not better performance than the other two mechanisms and allows for a degree of tuning in the performance of the component.

Table 1

Summary of the axial tube compression tests

SPECIMEN NUMBER	DEFORMATION MODE		MEAN LOAD (kN)	COMPRESSION (= stroke %)	ENERGY ABSORBED (J/mm)
1	axial buckling		80.0	70	56
2	inversion	b=4mm	85.0	66	56.1
3	inversion	b=6mm	80.0	70	56
4	splitting: curls form	b=4mm	60.0	95	57
5	-ditto-	b=6mm	42.5	95	40.4
6	-ditto-	b=10mm	30.0	95	28.5
7	splitting: curls prevented	s=5mm	50.0	95	47.5
8	-ditto-	s=3mm	52.0	95	49.4
9	-ditto-	s=2mm	52.5	95	49.9

2.2 Discussion of deformation mechanisms and comparison with theoretical models

Both axisymmetric buckling and external inversion involve a combination of and interaction between meridional (axial) bending and circumferential stretching. Additionally for inversion using a die one should make some allowance for the effects of friction. Much has already appeared in the literature on these two mechanisms and so, in the main, attention will be drawn to areas which would benefit from further study.

2.2.1 Axisymmetric buckling

Abramowicz and Jones [13] have provided a comprehensive review and substantial data concerning both axisymmetric and non-axisymmetric buckling modes. They re-worked the rigid-plastic analysis due to Alexander [14], the essence of which is contained in Fig. 6(a). The buckling process is assumed to take place in a section of length 2H, and to consist of a set of three stationary (relative to the material) plastic hinges separating two outward moving portions which undergo circumferential stretching. Common to much of the current work in the analysis of crushing structures, this kinematic field is used to calculate the mean crushing force, P_B, in terms of the parameter H defining the extent of the buckling region. The relevant value of H and consequently the value of P_B are determined in terms of the material properties and the geometry (R and h) of the tube by minimisation, invoking a global minimum work hypothesis, to give

$$P_B/M_o = 20.79\,(2R/t)^{1/2} + 11.90 \qquad (1)$$

and

$$H/R = 1.76\,(t/2R)^{1/2} \qquad (2)$$

where $M_o = \sigma_o t^2/4$, σ_o being the plane strain yield stress.
For the above experiment, equations (1) and (2) give P_B = 59.6 kN and H = 7.8 mm. Experimentally H = 7.4 mm approximately but P_B = 80 kN.

This significant underestimate is fairly typical of the predictions of equation (1) and was ascribed by Abramowicz and Jones to the assumption that the convolutions flatten into discs as shown in Fig. 6(a). Primarily because of the effects of strain hardening the convolutions remain and an approximate analysis gives an effective crush length per convolution, δ_e, given by

$$\delta_e/2H = 0.86 - 0.568\ (t/2R)^{1/2}. \qquad (3)$$

Using this reduced crush length instead of 2H, as assumed in the Alexander analysis, produces a value for P_B of 78.6 kN, much closer to the experimental value.

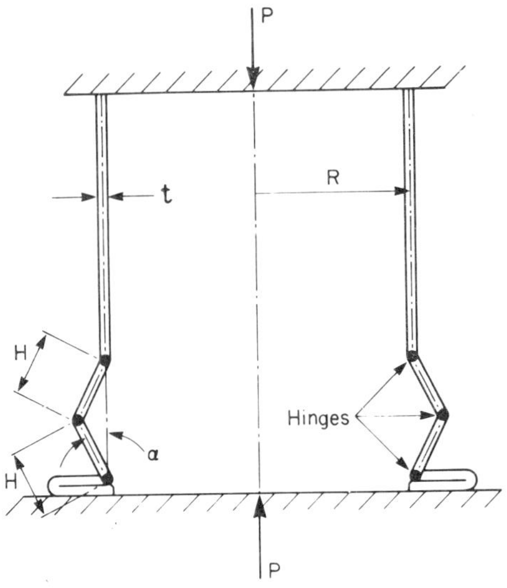

(a)

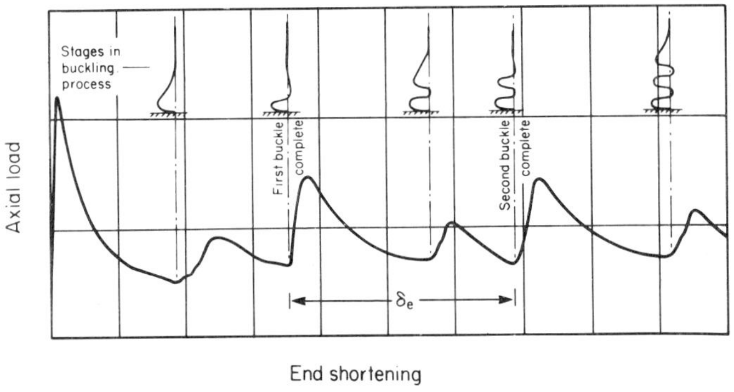

(b)

Fig.6. (a) Axi-symmetric axial buckling collapse mechanism due to Alexander [14]
(b) Relationship between generator shape and shape of load-deflection curve for axially symmetric buckling mode (adapted from [16])

Equation (3) results from an analysis of the geometry of the plastic "hinge" region involving the effects of strain hardening. The analysis, taken from an earlier paper by Abramowicz [15], demonstrates the significant effect that strain hardening can have on the geometry of a structure undergoing gross plastic bending deformation. Similar effects have been analysed in other problems involving regions of intense plastic bending by Reid and Reddy [16]. The important point to note is that the effects of strain hardening show themselves not only in enhanced values for the yield stress (a feature normally accounted for by the use of a flow stress or by estimating a mean strain level appropriate to the process) but also in local geometry changes which can have significant effects on the loads involved and the way in which the load varies during progressive deformation.

The theory presented by Abramowicz and Jones provides a good method for estimating the mean crushing load and the stroke of an axially buckling tube even in the non-axisymmetric range. However, a complete theory for predicting the shape of the load-deflection curve is still wanting. The theory as presented gives no estimate for the amplitude of the oscillation of the load about the mean level let alone the shape of a typical cycle of load. The variation in load shown in Fig. 1 shows a double wave structure within each portion of stroke δ_e. Fig. 6(b), adapted from the interesting article by Allan [17], gives an impression of the variation with stroke of the shape of the generator of an axisymmetric buckling tube. These observations are consistent with those made in the present test. The waves are laid down in a manner which clearly does not have the symmetry or the simplicity of the model suggested by Alexander. Slight inward as well as the dominant outward movement of the shell is noticeable as indeed it was in the cylinder tested

Several factors suggest that an exploration of the analogous beam-foundation problem may provide a useful step towards understanding this complex problem. In this context the papers by Reid [18] and Yun and Kyriakides [19] may prove useful although it is clear that any successful theory would require both large beam deflections and strain hardening to be included in the formulation.

2.2.2 External tube inversion

Several authors [8,20-23] have analysed both external (as depicted in Fig. 2) and internal inversion. An estimate of the steady state inversion load can be made using the principle of virtual work assuming a deformation field (see Fig. 7) in which the tube wall undergoes axial (meridional) bending at A, the point of contact with the die, circumferential stretching in the toroidal region and unbending at B.

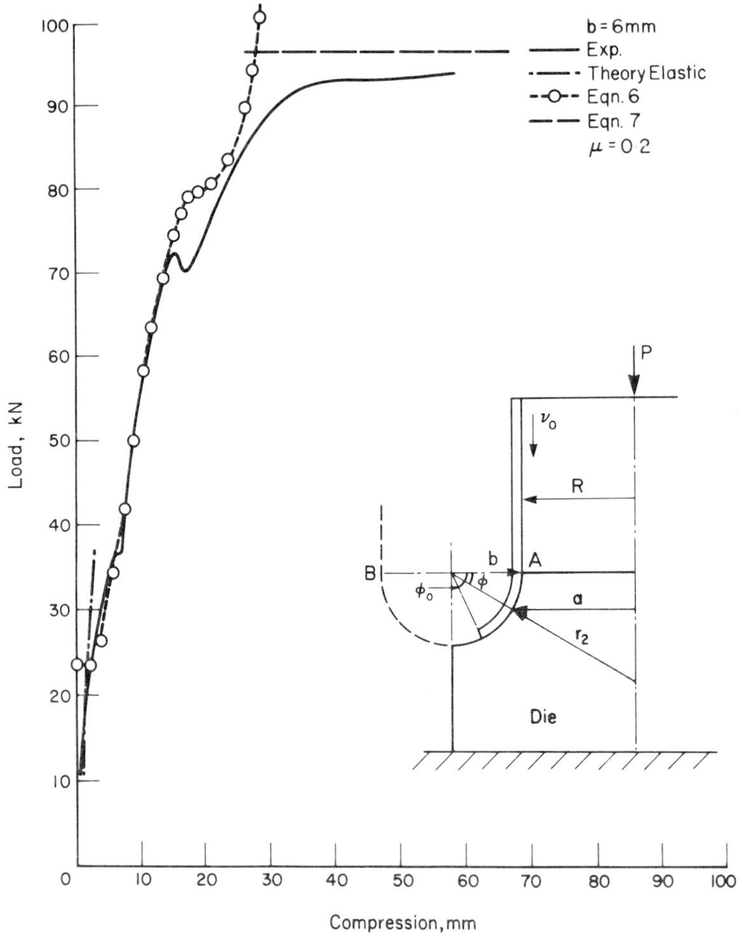

Fig.7. Comparison between theory and experiment for external tube inversion

Allowance can be made for the effect of friction by estimating the contact pressure, using Coulomb friction and treating the tube-die interface as a tangential velocity discontinuity. The applied force P can be estimated by equating its rate of working with the sum of the rates of energy dissipation in the three principal mechanisms, bending, stretching and interface friction. It is assumed that the material is rigid perfectly plastic, that there is no interaction between membrane force resultants and bending moments, that the contribution from circumferential bending is negligible and that the interface pressure is given simply by $p=N_o/r_2$ where $N_o=\sigma_o t$ is the membrane yield force per unit length.

In general the rate equation gives

$$Pv_o = \int_A (N_\theta \dot{\varepsilon}_\theta + N_\Phi \dot{\varepsilon}_\Phi)\, dA + \int_L M_o \dot{\Phi}\, dl + \int_{A^*} \mu p v\, dA^* \tag{4}$$

where N_θ and N_Φ and the circumferential and meridional membrane forces and $\dot{\varepsilon}_\theta$ and $\dot{\varepsilon}_\Phi$ their respective strain rates, A is the surface area of the part-toroidal shell into which the cylinder has been deformed, l is the length of the hinges at A (and B if $\Phi_o > \pi$), A^* is the contact area between the shell and the die and $v=v_oR/a$ is the local tangential velocity of the shell over the die surface.

Using the Tresca yield criterion and associated flow rule with $\sigma_\Phi < p < 0 < \sigma_\theta$ (which implies that $\dot{\varepsilon}_\Phi = -\dot{\varepsilon}_\theta$ and thickness, t, is constant) and $\dot{\varepsilon}_\theta = (v\sin\Phi)/a$, the first term reduces as follows,

$$\int_A (N_\theta \dot{\varepsilon}_\theta + N_\Phi \dot{\varepsilon}_\Phi)\, dA = \int_0^{\Phi_o} \frac{2\pi R N_o v_o b \sin\Phi}{a}\, d\Phi = 2\pi R N_o v_o \ln(a_o/R) \tag{5}$$

since $a = R + b(1-\cos\Phi)$ and putting $a_o = a(\Phi_o)$. Thus we have,

$$P = 2\pi R N_o \left[\ln(a_o/R) + t/4b + \mu b \int_0^{\Phi_o} \frac{\cos\Phi}{a}\, d\Phi \right] \tag{6}$$

for $0<\Phi<\pi/2$. For $\pi/2<\Phi_0<\pi$ the upper limit of the integral is set at $\pi/2$. For $\Phi_0>\pi$ the tube has passed hinge B and equation (6) is replaced by the steady state equation

$$P_I = 2\pi RN_0 \left[\ln(1 + \frac{2b}{R}) + \frac{t}{2b} + \mu b \int_0^{\pi/2} \frac{\cos\Phi}{a} \, d\Phi \right] \qquad (7)$$

As shown in Fig. 7, these equations provide a reasonable representation for the transient and steady state load sustained by specimen 3 in Table 1. A value of 732 N/mm^2 has been used for σ_0 together with $\mu=0.2$. Also included in Fig. 7 is the initial elastic response derived using the classical ring edge load solution for a circular cylinder [24]. The dip in the experimental curve and the change of slope corresponds to the formation of the lip visible at the leading edge of the inverted tube, the reduction in load being associated with circumferential unloading . In this region equation (6) simply shows a change in slope as circumferential stretching gives way to bending at B.

There is an interesting stability problem associated with the transition from forming the toroidal shell to the development of the hinge at B and subsequent steady state inversion. As with the axial buckling problem, whilst the technological theory described above (which itself is a minor variant of those found in the literature) provides estimates of P_I which are suitable for design purposes, it would appear that a full understanding of the stability problem referred to above would only result from more sophisticated analyses than the kinematic methods which dominate the literature. Again one would expect that strain hardening would play a significant role given the fact that πb is of the order of the length of a typical plastic "hinge" in a real shell. The assumption that all the bending is concentrated at A and B may therefore mask a major feature of the stress field which controls these more detailed features of the inversion process. Calladine has discussed this in the context of free-tube inversion elsewhere in these proceedings.

2.2.3 Axial Splitting

The analysis of the mechanism of tube splitting is at a preliminary stage. Reference [12] contains a description of circular tubes (with and without initial slots cut into the leading edge) being split by compression on to dies of various radii. The results shown in Figs. 3 to 5 convey the qualitative features of these tests including the effect of using a curl stopper plate. The latter is a particularly useful modification to the basic splitting and curling mechanism since it removes the hardening characteristic of the load-deflection curve. This results from the increased frictional contributions and additional plastic deformation which occurs when the curls are forced to reduce in radius as they coil following point A in Fig. 3.

If no slots are cut into the leading edge of the tubes the experimental evidence [12] is that a characteristic number of fractures is produced in the steady state. The five fractures produced in the specimen compressed on to the 10 mm radius die provides a typical example of this, the steady state being achieved following a crack initiation stage and a series of bifurcations. The linking of the number of fractures to the material properties requires explanation but it is relatively unimportant from the point of view of energy absorption and indeed it would be preferable to remove the load fluctuations associated with this transient phase by pre-triggering a certain number of cracks. One practically important feature of the characteristic number of cracks for a given tube is that it provides an upper limit on the number of fractures that can be maintained in the tube. Attempts to exceed this number usually result in a number of the cracks not propagating. A simple minimum energy argument has established that a bound can be placed upon the number of fractures initiated [25].

Since five cracks were produced in the un-triggered tube, four slots were cut into the remaining specimens. This creates a situation similar to that examined by Stronge et al [26] who performed splitting and curling tests on square tubes of aluminium and mild steel, the deformation being achieved using contoured dies. In earlier work (referred to in [26]) the mechanisms had been achieved by compressing tubes with sawcuts at each corner on to a flat plate. Stronge et al [26] developed a rate equation for the latter situation which contained contributions to the energy dissipation rate from plastic bending, fracture and friction. In this the contact force between the tube walls and the plane was relatively easy to define and locate. An interesting discussion was provided of the role of strain hardening in the deformation of the curled plates although these effects were ignored in deducing quantitative information from the model.

It has been noted that in the tests performed that meridional bending, circumferential flattening and crack propagation occurred. Incorporating these into a rate equation for splitting and curling leads to

$$P_s v = 2\pi N_0 R v\,[t/4b + t/4R] + nG_c t v + \dot{W}_f \qquad (8)$$

where b is the die radius as before, n is the number of fractures and G_c is the fracture toughness (approximately 100kJ/m^2 for mild steel) of the material. $\dot{W}_f$ is the rate of dissipation of energy by friction and this clearly depends upon the nature of the contact between the strips and the die. Using the model of ref. [26], $\dot{W}_f=\mu P_s v$. If the resultant reaction force on each strip is S and this is inclined at an angle α to the horizontal, then $\dot{W}_f=n\mu Sv/(1+\mu^2)^{½}=\mu P_s v/\sin\alpha(1+\mu^2)^{½}$. Thus equation (8) gives

$$P_s = \frac{2\pi N_0 R[\frac{t}{4b} + \frac{t}{4R}] + nG_c t}{1-\mu/\sin\alpha(1+\mu^2)^{½}} \qquad (9)$$

The case assumed by Stronge et al corresponds to $\alpha=\pi/2+\Phi$ where $\tan\Phi=\mu$ and the denominator reduces to $1-\mu$. The estimate is clearly sensitive to both μ and α. In [12] it is suggested that $\alpha=\pi/4$ is a reasonable first estimate representing the direction of the resultant force. For a different batch of mild steel tubes (σ_0=800N/mm^2) splitting into twelve strips, μ=0.2 provided a good estimate for P_s. In the present tests, Fig. 8 shows that a higher value of approximately 0.45 is required to match the experimental data. This probably reflects the more severe loading conditions at the edges of the four strips compared with the case when twelve strips are produced.

The curl prevention tests involve both extra bending plastic work (t/4b is replaced by t/2b in equation (9)) as the strips are unbent and, more problematical, extra frictional dissipation. The latter presumably leads to a modification of the die contact force distribution which requires further study. The addition of the extra bending term into equation (9) results in an estimate of 60 kN for the three curl prevention tests.To summarise, the analysis of curl splitting is still at an elementary stage.

Consideration of the bending moment distribution in the curls, as outlined in [26], leads to the conclusion that strain hardening again plays an important role in controlling the deformation field in the bending regions.

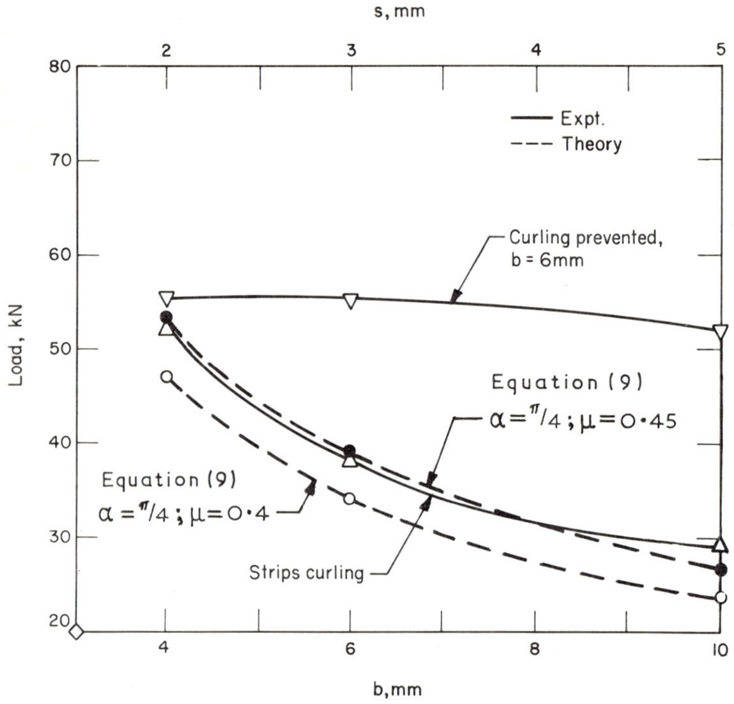

Fig.8. Operating loads for tube splitting and curling compared with predictions of equation (9) and operating loads for curl-stopper tests using a 6mm die radius.

2.3 Effects of loading rate

The initial yield stress of mild steel increases significantly with strain rate and this is reflected in the operating loads of energy absorbing devices whose behaviour is dominated by plastic deformation. The operating loads, P_B and P_I, axial buckling and inversion respectively are each increased under dynamic loading conditions. Abramowicz and Jones [13] suggest a dynamic enhancement factor, m, given by

$$m = 1 + \left[\frac{0.25V}{6844R\{0.86-0.568(t/2R)^{1/2}\}}\right]^{1/3.91} \qquad (10)$$

for axi-symmetric axial buckling, where V is the impact speed. This makes allowance for the fact that the strain rate enhancement reduces for large strains. A similar factor applied to the inversion tube data produces results consistent with the experimental data.

Stronge et al [26] estimated a factor of 2.5 to 3 for mild steel square tubes. This was partly ascribed to strain rate enhancement of the yield stress and partly to increases in the value of μ. The results of Reddy and Reid [12] indicate that where a larger number of fractures are produced (say eight to twelve) the operating force may not differ significantly from the quasi-static value. It should be noted that in [12] dynamic forces were measured directly whereas in [26] they were estimated from energy considerations. The result in [12] may not be too surprising since the increase in flow stress resulting from increasing the strain rate is often accompanied by a reduction in the strain to fracture which would be reflected by lower dynamic G_c values. It is worth pointing out that Thornton and Dharan [27] report that the flow stress of an aluminium magnesium alloy foam reduces under high rate loading. They state that "notch sensitivity increases with increase in strain rate, and so the strength of the foam would be decreased." Where fracture is a significant mechanisms in the deformation field, caution should be exercised in predicting the influence of increased strain rate.

3. Tubular Rings

3.1 Experimental data

Fig. 9(a) shows a typical tubular ring. It is fabricated from four lengths of mild steel tube (D=50.8mm, t=1.6mm) welded together at mitred joints. The behaviour of single components of this type and multilayered systems has been described by Reid et al [9]. Fig. 10(a) shows the quasi-static load per unit length versus deflection curves for a series of single elements of different side lengths, W, compared with a single length of tube (curve 1) under lateral compression between flat plates. The interaction between the tubes in the ring (which stems from the mutual constraint at the mitred joints which prevents the warping shown in Fig. 9(b)) clearly produces a dramatic increase in the energy absorbing capacity of the tubes. Fig. 10(b) and (c) shows that the initial collapse loads and the energy absorbing capacities per unit length arrange themselves in a hierarchy of efficiency with an optimum corresponding to a side length of approximately 114 mm (central hole of side 13 mm).

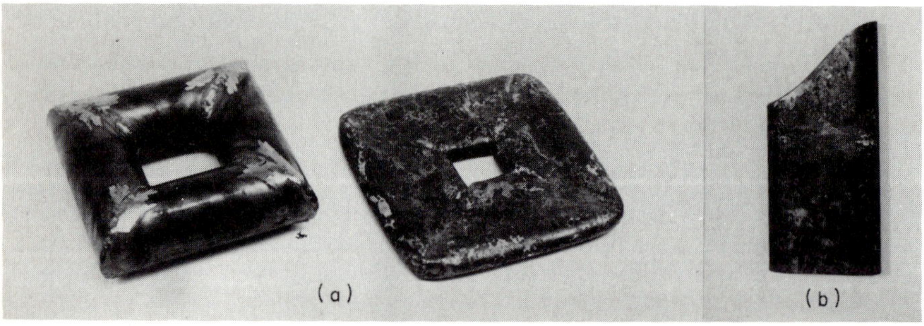

Fig. 9.(a) Tubular ring (w=152.4mm) before and after lateral compression
(b) Warping of edge of tube cut at 45° to axis following lateral compression.

3.2 Deformation mechanisms

As described in [9], as the tubular ring is compressed laterally, the warping constraint generates axial stresses and in-plane moments. These result in several features in the history of deformation. Fig. 11 shows the state of a specimen of side W=127 mm (curve 5 in Fig. 10(a)) at

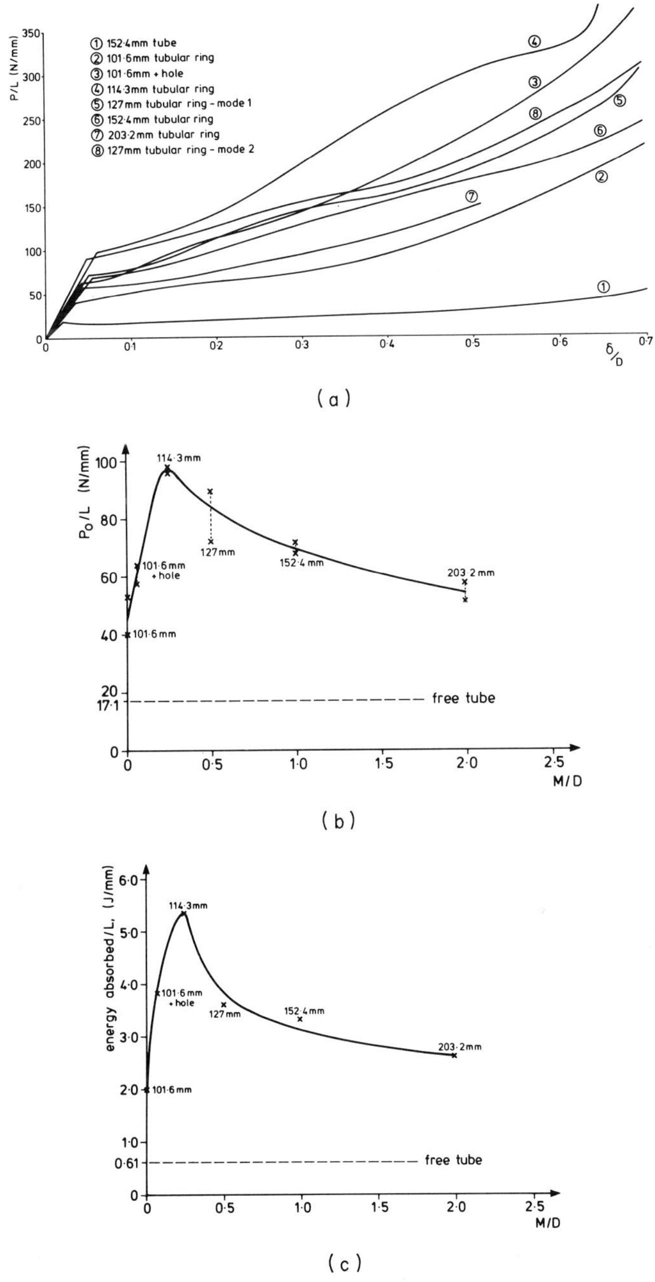

Fig.10.(a) Load per unit length versus deflection for set of tubular rings
(b) Collapse load per unit length versus M/D where M = size of central hole
(c) Energy absorbing capacity per unit length versus M/D

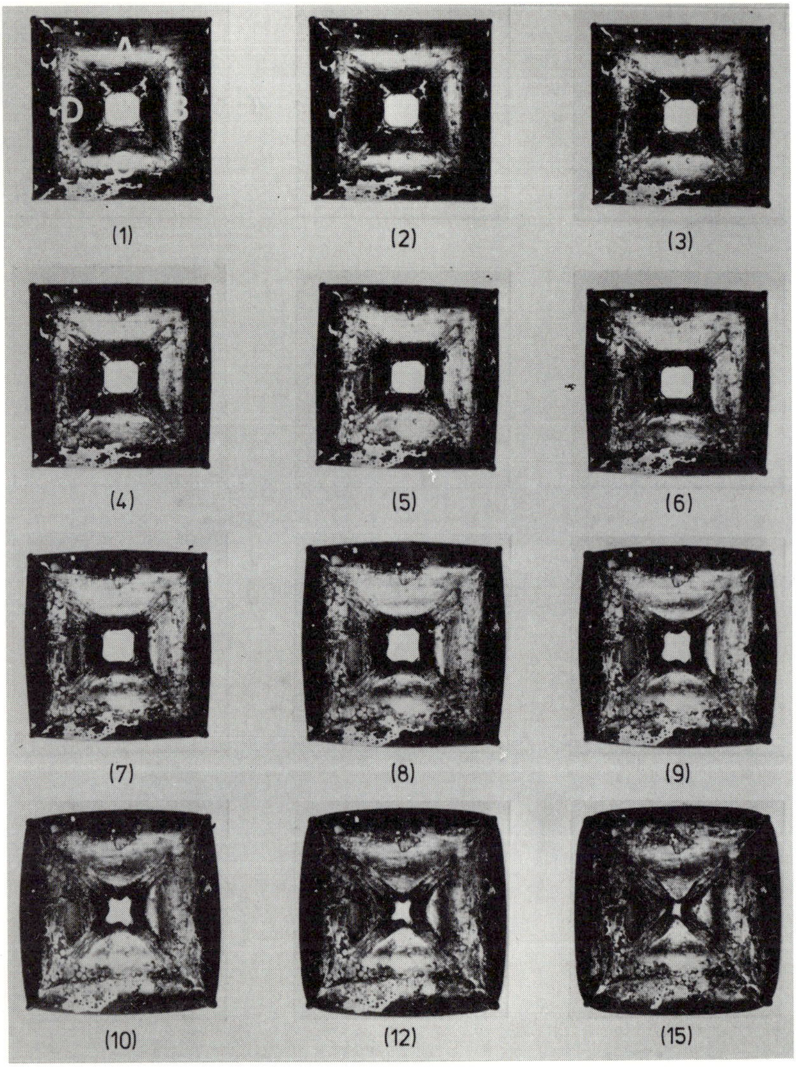

Fig.11. Stages in the deformation of a 127mm tubular ring specimen.

approximately 2.3 mm deflection increments from the undeformed state (1). Stage 2 represents the initial collapse state (deflection, δ=2.8 mm) which occurs at a load of 27.8 kN. As is evident from stages 2 and 3 the lack of deformation on the inner and outer peripheries reveals that the collapse mechanism is localised close to the upper and lower generators which are in contact with the loading platens. This type of collapse mechanism and the consequent enhancement in collapse load are reminiscent of the collapse of braced metal tubes [28]. In this case the bracing results from the pinch effect produced by the in-plane bending moments as described in a different context by Calladine [29].

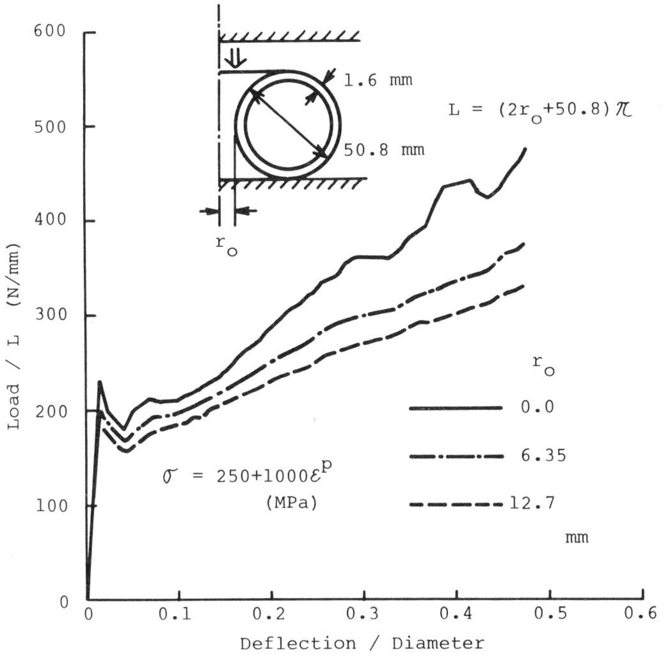

Fig.12. Load-deflection curves forlaterally compressed toroidal shells with small central hole diameters (due to Sugita [30])

The outer generators of the tube go into a tensile state whilst the inner generators are in compression. Eventually the inner generators buckle and a form of internal inversion occurs as is evident in the later stages of the deformation shown in Fig. 11.

3.3 Theoretical models

The increase in the collapse load permit length shown in Fig. 10(b) is consistent with the predictions of simple upper bound calculations similar to those found in [28] apart from the existence of a local maximum. As yet no large deflection analysis for the tubular ring has been formulated.

Sugita [10] has produced an interesting computational technique for the axisymmetric deformation of toroidal shell loaded laterally between flat plates. This is based on an incremental Rayleigh-Ritz method in which the meridional curvature distribution is approximated by a trigonometric series. Large deflections and the effects of strain hardening are included in the analysis which shows good agreement with experiments. In these experiments on toroidal shells there is some evidence of non-axisymmetric (local buckling) behaviour. Recent calculations (see Fig. 12) performed by Sugita [30] aimed at investigating the behaviour of toroidal shells for small values of the central hole radius (r_0 in Fig. 12) do not show the local maximum in the collapse load or indeed any reduction in the energy absorbing capacity per unit length. It seems clear therefore that the observed phenomenon is due to the onset of local buckling at an early stage and hence to non-axisymmetric behaviour. This is currently under investigation as is the general phenomenon of the onset of local buckling in toroidal shells.

4. Conclusions

A full understanding of the performance of metallic impact energy absorbing components depends upon the ability to analyse large deflection plastic deformation mechanisms of shells. In this paper emphasis has been placed upon those mechanisms which result from axial loads applied to circular tubes. The resulting modes of deformation generally involve local, intense regions of bending and stretching. Biaxial strain fields are produced and in many instances strain hardening plays a significant role both in controlling the force levels involved as well as the local geometry in the deforming region. Whilst simple kinematic analyses do provide formulae which are useful for design purposes it is clear that there are many aspects of the deformation mechanisms which require more detailed analysis.

5. Acknowledgements

The authors would like to express their thanks to Mr. R. Smith who performed the detailed tests on the tubular rings and to Dr. Y. Sugita for providing Fig. 12. Thanks are also due to Mrs. M. Mellor and Mr. J. Howe for preparing the paper for publication and to the Departments of Transport and Energy for support by means of a research contract administered by the Transport and Road Research Laboratory.

References

1. Johnson, W. and Mamalis, A.G., Crashworthiness of Vehicles, Mechanical Engineering Publications, I.Mech.E., London 1978.

2. Jones, N. and Wierzbicki, T. (Eds.), Structural Crashworthiness, Butterworths, 1983.

3. Reid, S.R., Johnson, W. and Reddy, T.Y., Pipe whip restraint systems, Chartered Mechanical Engineer 55-60, 1980.

4. Ezra, A.A. and Fay, R.J., An assessment of energy absorbing devices for prospective use in aircraft situations, Dynamic Response of Structures, eds. G. Herrmann and N. Perrone, Pergamon Press, 225-246, 1972.

5. Johnson, W. and Reid, S.R., Metallic energy dissipation systems, Appl.Mech.Rev. 31, 277-288, 1978.

6. Reid, S.R. Laterally compressed metal tubes as impact energy absorbers, Structural Crashworthiness, ed. N. Jones and T. Wierzbicki, Butterworths, 1-43, 1983.

7. Reid, S.R., Metal tubes as impact energy absorbers, Metal Forming and Impact Mechanics, ed. S.R. Reid, Pergamon Press, ch. 14, 1985.

8. Al-Hassani, S.T.S., Johnson, W. and Lowe, W.T., Characteristics of inversion tubes under axial loading, J.Mech.Eng.Sci., 14, 370-381, 1972.

9. Reid, S.R., Austin, C.D. and Smith, R., Tubular rings as impact energy absorbers, Structural Impact and Crashworthiness, Vol.2., ed. J. Morton, Elsevier Applied Science Publishers, 555-563, 1984.

10. Sugita, Y., Non-linear load-deflection relations of toroidal-shells subjected to axisymmetric compression between rigid plates, to appear, Int.J.Mech.Sci., 27, 1985.

11. Reddy, T.Y. and Reid, S.R., On obtaining material properties from the ring compression test, Nuc.Engng.Design, 52, 257-263, 1979.

12. Reddy, T.Y. and Reid, S.R., Axial splitting of circular metal tubes, to appear Int.J.Mech.Sci., 27, 1985.

13. Abramowicz, W. and Jones, N., Dynamic axial crushing of circular tubes, Int.J.Impact Engng., 2, 263-281, 1984.

14. Alexander, J.M., An approximate analysis of the collapse of thin cylindrical shells under axial loading, Quart.J.Mech.Appl.Math., 13, 10-15, 1960.

15. Abramowicz, W., The effective crushing distance in axially compressed thin-walled metal columns, Int.J.Impact Engng., 1, 309-317, 1983.

16. Reid, S.R. and Reddy, T.Y., Effects of strain hardening on the lateral compression of tubes between rigid plates, Int.J.Sol.Struct., 14, 213-225, 1978.

17. Allan, T., Investigation of the behaviour of cylindrical tubes subject to axial compressive forces, J.Mech.Eng.Sci., 10, 182-197, 1968.

18. Reid, S.R., Influence of geometrical parameters on the mode of collapse of a "pinched" rigid-plastic cylindrical shell, Int.J.Sol.Struct., 14, 1027-1043, 1978.

19. Yun, H.D. and Kyriakides, S., Localised buckling of a heavy beam on a contacting surface: a model for beam mode buckling of buried pipelines, ASME Paper 84-PVP-71, 1984.

20. Guist, L.R. and Marble, D.P., Prediction of the inversion load of a circular tube, NASA Tech.Note TN D-3622, 1966.

21. Al-Qureshi, H.A. and DeMorais, G.A., Analysis of multi-inversion of tube ends, ASME Paper 77-DE-35, 1977.

22. Kinkead, A.N., Analysis for inversion load and energy absorption of a circular tube, J.Strain Anal., 18, 177-188, 1983.

23. Wierzbicki, T. Crushing behaviour of plate intersections, Structural Crashworthiness, ed. N. Jones and T. Wierzbicki, Butterworths, 66-95, 1983.

24. Timoshenko, S.P. and Woinowsky-Krieger, S., Theory of Plates and Shells, McGraw-Hill, ch. 15, 1959.

25. Atkins, A.G. Private communication.

26. Stronge, W.J., Yu, T.X. and Johnson, W., Energy dissipation by splitting and curling of tubes, Structural Impact and

Crashworthiness, Vol.2, ed. J. Morton, Elsevier Applied Science Publishers, 576-587, 1984.

27. Thornton, P.H. and Dharan, C.K.H., The dynamics of structural collapse, Mats.Sci.Engng., 18, 97-120, 1975.

28. Reid, S.R. Drew, S.L.K. and Carney, J.F., Energy absorbing capacities of braced metal tubes, Int.J.Mech.Sci., 25, 649-6677, 1983.

29. Calladine, C.R., Plastic buckling of tubes in pure bending, Collapse, ed. J.M.T. Thompson and G.W. Hunt, Cambridge University Press, 111-124, 1983.

30. Sugita, Y., Private communication.

The Deformation of Thin Plates Subjected to Impulsive Loading

G.N. NURICK, H.T. PEARCE and J.B. MARTIN

UNIVERSITY OF CAPE TOWN, Faculty of Engineering, Rondebosch 7700
South Africa.

Summary

Experimental and theoretical studies of thin circular, square and rectangular plates subjected to impulsive loading are described. The range of interest is that where permanent deflections are large compared to the plate thickness. Experiments have been carried out on a ballistic pendulum using sheet explosive to provide the loading: displacement-time histories were measured using a light interference device which is mounted on the pendulum. Experimental results are compared to theoretical predictions based on the assumption of rigid-visco-plastic behaviour. A mode approximation technique was used in the theoretical model: membrane effects were considered predominant, and the mode shape is updated as the geometry of the plate changes during deformation.

INTRODUCTION

The deformation of thin plates subjected to large impulsive loads is a classical problem in plastic structural dynamics. Experimental studies have been reported by Witmer, Balmer, Leech and Pian (1), Florence (2), Jones, Uran and Tekin (3), Wierzbicki and Florence (4), Duffey and Key (5), Jones, Griffin and van Druzen (6), Bednarski (7), Jones and Baeder (8), Ghosh and Weber (9), Bodner and Symonds (10), (11) and Nurick and Martin (12). A number of theoretical studies have also been reported, covering the range of analytical approaches, finite element analyses and approximate methods.

A new series of experimental results is reported in this paper. The experiments cover fully clamped circular, square and rectangular plates, over a range of deformations from 3 to 12 plate thicknesses. Maximum permanent deformations were measured, and detailed contour plots of the final deformed shape were obtained. The deflection-time history at the centre of the

Inelastic Behaviour of Plates and Shells
IUTAM Symposium Rio de Janeiro 1985
Editors: L. Bevilacqua, R. Feijóo and R. Valid

plates was measured using a light interference device.

Theoretical studies were based on the mode approximation method, proposed by Martin and Symonds (13) and extended to the finite deformation of plates by Chon and Symonds (14),(15), Symonds and Wierzbicki (16), Guedes Soares (17) and Perrone and Bhadra (18), (19). The attraction of this approach lies not only in its potential to provide reasonably accurate predictions of the permanent deformations, but also in its potential to isolate the governing mechanisms occurring in the deformation and thus to provide insight into the mechanics of the problem.

The mode approximation presented in this paper is based on the assumption that membrane stresses predominate, and that the behaviour can be modelled as rigid-viscoplastic. It is also assumed that at any instant the shapes of the displacement field and the velocity field are the same: the shape is calculated by means of a modification of the algorithm for the computation of the mode given by Martin (20). Very simple kinematic and dynamic conditions are imposed, and an iterative forward integration scheme is used to advance the solution in time.

The tests were carried out on mild steel plates, and the results of Manjoine (21) were adopted to describe the relation between dynamic yield stress σ and strain rate $\dot{\varepsilon}$, in the form

$$\frac{\sigma}{\sigma_o} = 1 + \left[\frac{\dot{\varepsilon}}{\dot{\varepsilon}_o}\right]^{1/n}$$

where σ_o is the static yield stress and $\dot{\varepsilon}_o$, n are material parameters with the values of $40s^{-1}$ and 5 respectively.

EXPERIMENTAL RESULTS

A total of 111 tests were carried out, consisting of 26 circular, 40 square and 45 rectangular plates. The specimens were cut from two sheets of 1.6mm thick commercially avilable cold rolled structural steel plate. After clamping, the specimens were of equal area, with the circular plates being 100mm diametre, the square plates 89mm x 89mm, and the rectangular plates 113mm x 70 mm .

Typical stress-strain curves from tension tests are shown in Fig.1 . The static yield stress was computed using the results of the tests and substituting into equn. (1) : the average static yield stresses for the two plates were 282 MPa and 296 MPa respectively.

The experimental procedure adopted in the tests has been described by Nurick and Martin (12), and the light interference device has been described by Nurick (22). Metabel sheet explosive was used to provide the impulsive load, and was arranged in two concentric annuli of the shape of the plate, as shown in Fig. 2 . A 16mm thick polystyrene pad was placed between the sheet explosive and the plate, both to protect the plate surface and to provide an approximately uniform impulse on the plate surface.

The test data and the results are given in Tables 1-3 . The displacement-time history at the centre of each plate was recorded over a period of 14ms, and Figs. 3-5 show typical responses for plates of different geometries subjected to similar impulses. The maximum deformation and deformation time were obtained from these traces.

The permanent deformation at the centre of the plate was measured mechanically. Independently, profiles of the final deformed shape were measured by the use of a Reflex Metrograph (Nurick and Adams (23)). This permitted contour plots of the final displacements to be plotted, and typical results are shown in Fig. 6 .

Figs. 7-9 show the measured permanent central deflection to thickness ratio plotted against impulse for each of three plate geometries. The deflection-thickness ratio is replotted in Fig.10 against the dimensionless parameter $\gamma = I^2/\rho M_o A^2$.

MODE APPROXIMATION

Mode approximations have been applied to impulsively loaded thin plates in which bending is ignored by Symonds and Wierzbicki (16) (who compute a mode shape which was fixed throughout the motion), Guedes Soares (17), Duffey (24) and Perrone and Bhadra (18), (19) (all of whom assumed the mode shape). The model adopted in this paper differs primarily in that the mode shape is recomputed at each time step. We assume that the displacement u at a point on the plate is vertical (i.e. transverse to the original plate position), and can be written at a particular instant of time t as

$$u = u_c \, \phi \qquad (2)$$

where u is a function of spatial coordinates defining a point on the original plate surface and of time, u_c is the central displacement of the plate depending only on time, and ϕ is a mode shape which may vary with time. The mode shape is normalised so that its magnitude at the centre of the plate is unity.

We further assume that the velocity and acceleration fields at time t can be written as

$$\dot{u} = \dot{u}_c \, \phi \quad , \quad \ddot{u} = \ddot{u}_c \, \phi \qquad (3)$$

where $\dot{u}_c$, $\ddot{u}_c$ are respectively the velocity and acceleration at the centre of the plate. This assumption implies that equn. (2) can be differentiated at the instant t without taking into account the variation of the mode shape ϕ with time: this has been referred to as an instantaneous or stationary mode assumption.

It is evident from equn. (3) that at the instant t we can write

$$\ddot{u} = - \lambda \dot{u} \quad , \quad \lambda = - \ddot{u}_c / \dot{u}_c \quad , \qquad (4)$$

where λ is a scalar multiplier. We assume that, at any instant t, $u_c(t)$, $\dot{u}_c(t)$ are known, and that λ and ϕ must be determined.

In order to carry out this computation, we discretise the problem by assuming contours along which ϕ is constant, as shown diagrammatically by the full lines in Fig. 11 . At the centre of the plate $\phi_o = 1$, and we label contour levels from the centre as $\phi_1, \phi_2, \ldots \phi_n$. For the circular plate the contours are circular, as is dictated by axial symmetry. For the square plate, symmetry also dictates square contours. For the rectangular plate, the contours have been formed by drawing lines at 45° to the clamped edges from the corners: this is approximately in accord with the contours found experimentally and shown in Fig. 6 . At time t the number of unknowns is thus (n+1), made up of $\lambda, \phi_1, \phi_2, \ldots \phi_n$.

The mode shape ϕ is assumed to vary linearly with distance between the contour levels, so that between contours ϕ_i and ϕ_{i+1} the mode shape is that of a frustum of a cone for the circular plate and the frustum of a pyramid for the square and rectangular plates. We also draw intermediate contours, shown dotted in Fig. 11, which are equidistant between $\phi_1, \phi_2, \ldots \phi_n$ and which are labelled $\bar{\phi}_1, \bar{\phi}_2 \ldots, \bar{\phi}_{n+1}$.

The mass m_i of the plate lying between contours $\bar{\phi}_i$, $\bar{\phi}_{i+1}$ is lumped at contour ϕ_i, with the lumped mass m_o at the centre of the plate being the mass contained within the contour $\bar{\phi}_1$. We can now write (n+1) dynamic equations for the lumped masses; the i-th contour or node is shown diagrammatically in Fig. 12. The force $\lambda m_i \phi_i \dot{u}_c$ is the inertia force arising from the deceleration of the node, and it must be equilibrated by forces $\bar{F}_i$, $\bar{F}_{i+1}$. The forces $\bar{F}_i$, $\bar{F}_{i+1}$ are taken as the vertical components of the membrane stresses integrated along the contours $\bar{\phi}_i$, $\bar{\phi}_{i+1}$ respectively.

Along the contour $\bar{\phi}_{i+1}$, the membrane stresses are assumed to be perpendicular to the contour. The strain component perpendicular to this contour is

$$\varepsilon_{i+1} = \frac{u_c^2(\phi_{i+1} - \phi_i)^2}{2\ell_i^2} \tag{5}$$

where ℓ_i is the shortest horizontal distance between the contour ϕ_i and ϕ_{i+1}. The length ℓ_i is constant for the contour shapes shown in Fig.11. The strain rate is then taken as

$$\dot{\varepsilon}_{i+1} = \frac{\dot{u}_c\, u_c(\phi_{i+1} - \phi_i)^2}{\ell_i^{\,2}} \tag{6}$$

with the assumption the ϕ_{i+1}, ϕ_i are fixed at time t in the differentiation of equn.(5). The stress component in the plate normal to the contour line $\bar{\phi}_{i+1}$ is then computed from equn.(1), giving

$$\sigma_{i+1} = \sigma_o \left(1 + \left[\frac{\dot{\varepsilon}_{i+1}}{\dot{\varepsilon}_o} \right]^{1/n} \right) \tag{7}$$

The in-plane force per unit length is $h\sigma_{i+1}$, where h is the thickness of the plate. The inclination of the force to the horizontal is θ_{i+1}, where

$$\tan\theta_{i+1} = \frac{u_c(\phi_{i+1} - \phi_i)}{\ell_i} \tag{8}$$

It then follows that

$$\bar{F}_{i+1} = \int_{L_{i+1}} h\,\sigma_{i+1} \sin\theta_{i+1}\, dL \quad , \tag{9}$$

where L_{i+1} is the length of the contour $\bar{\phi}_{i+1}$.

The dynamic equations then take the form

$$\lambda m_i \phi_i \dot{u}_c = -\bar{F}_i + \bar{F}_{i+1} \;, \qquad i = 1,2,\ldots,n \; . \tag{10}$$

At the centre of the plate we have an additional equation,

$$\lambda m_o \dot{u}_c = + \bar{F}_1 \tag{11}$$

At time t we regard u_c, $\dot{u}_c$ as known, and equns.(10) and (11) provide (n+1) equations for the multiplier λ and the modal coefficients $\phi_1, \phi_2, \ldots, \phi_n$. The equations are highly non-linear, and are solved by a Newton-Raphson technique.

The motion is integrated forward in time using an implicit constant average acceleration algorithm. We write

$$\dot{u}_c(t+\Delta t) = \dot{u}_c(t) + \frac{\Delta t}{2}\left[\ddot{u}_c(t) + \ddot{u}_c(t+\Delta t)\right] \tag{11a}$$

$$u_c(t+\Delta t) = u_c(t) + \frac{\Delta t}{2}\left[\dot{u}_c(t) + \dot{u}_c(t+\Delta t)\right] , \tag{11b}$$

and $$\ddot{u}_c(t+\Delta t) = - \lambda \dot{u}_c(t+\Delta t) . \tag{12}$$

A second, outer, iteration loop is used to solve these equations. At the beginning of a time step we set $\dot{u}_c(t+\Delta t) = \dot{u}_c(t)$ and take the value of λ from the last iteration of the previous time step: we use equns.(11a) and (11b) to obtain corrected values of $u_c(t+\Delta t)$, $\dot{u}_c(t+\Delta t)$, and then recompute λ from the Newton-Raphson solution of equns.(10) and (11). This iterative procedure continues until a prespecified tolerance is met, and we then continue to the next time step.

The initial conditions for the problem as a whole are $u_c(0)=0$, $\dot{u}_c(0)=\dot{u}_{co}$. The initial central velocity is computed from the initial total impulse I, which is assumed to impart a uniform velocity to the plate $\dot{u}_o$, given by

$$\dot{u}_o = \frac{I}{\sum_{i=0}^{n} m_i} . \tag{13}$$

The initial mode velocity, $\dot{u}_{co}$, is then obtained from a generalised momentum balance:

$$\dot{u}_{co} = \dot{u}_o \left[\left(m_o + \sum_{i=1}^{n} m_i \hat{\phi}_i \right) \Big/ \left(m_o + \sum_{i=1}^{n} m_i \hat{\phi}_i^2 \right) \right], \tag{14}$$

where $\hat{\phi}_i$ is the initial mode shape. The initial mode shape cannot in fact be computed, since equns.(10) and (11) are singular for $u_c=0$. For this reason, $\hat{\phi}_i$ was taken to vary linearly, implying a conical or pyramidal initial mode shape. Further, for the first time step t_1, the stiffness of the plate was assumed to be zero, and we put

$$\dot{u}_c(t_1) = \dot{u}_c(0) = \dot{u}_{co}, \quad u_c(t_1) = \dot{u}_{co} t_1 \ . \tag{15}$$

The iterative solution then began on the second time step. The forward integration terminated at time t_f when $\dot{u}_c(t_f) = 0$. If in a particular time step the predicted velocity at the end of the step was negative, this step was taken as the final step and the $\dot{u}_c$ was assumed to vary linearly with time over this final step.

Solutions were found for a range of impulses for each of the plate geometries, using n=4, an initial time step of 1 ms and subsequent time steps of 15 μs. The final time was approximately 100 μs, so that in all 6-8 time steps were used. The predicted results are shown on Figs. 7-9. Convergence of the algorithm was rapid, and an increase in the number of nodes or the number of time steps did not significantly change the solution.

Computationally, the algorithm is very efficient: each computation runs for about 10 seconds on the Sperry 1100/80 mainframe computer on which the calculations were carried out.

Predicted permanent central deflections computed by this method have been plotted on Figs. 7, 8 and 9, together with the predictions, for the circular plate, of Symonds and Wierzbicki (16), Guedes Soares (17), Perrone and Bhadra (19) and Duffey (24).

DISCUSSION

The central permanent deflection predicted for the circular plate, shown in Fig.7, compares favourably with the experimental results, particularly at the higher deflection thickness ratios where the membrane effect is predominant. By comparison, the predictions of Symonds and Wierzbicki (16), who computed a mode shape which remained fixed through the motion, tend to underestimate the deflection at higher-deflection thickness ratios. The predictions of Guedes Soares (17), Perrone and Bhadra (19) and Duffey (24) are less accurate, showing the advantage of including in the mode approximation at algorithm for computing the mode shape.

The predicted deflections for the square plates, shown in Fig. 8, also compare favourably with the experimental data. No other large deflection membrane mode approximation results are known to the authors for this case. The final mode shapes predicted by the model for the circular and square plates also show good agreement with the experimental results, as can be seen in Fig.13.

Predictions for the rectangular plates show a slightly different trend from the experimental results, although the correlation is reasonably good. Further attention will be given to this case, where the consequences of the choice of the shape of the contours (Fig.11) has not been fully assessed.

The dimensionless plot shown in Fig.10 shows quite clearly that for plates of equal area subjected to equal total impulses the midpoint deflection of the circular plate exceeds that of the square plate, which in turn exceeds that of the rectangular plate.

Acknowledgements

The authors are indebted to R Beverton for his assistance in preparing and detonating the explosive charges, to H Tomlinson for preparing the plate specimens, to L P Adams and A M Tregidga for their assistance and guidance in the use of the Reflex Metrograph, to AECI (Pty) Ltd for the donation of the explosive material and to the Council for Scientific and Industrial Research for financial support.

References

1. Witmer, E.A.; Balmer, H.A.; Leech, J.W.; Pian, T.H. Large dynamic deformation of beams, rings plates and shells. AIAA J 1 (1963) 1848-1857.

2. Florence, A.L.: Circular plates under a uniformly distributed impulse. Int J Solids Structures 2 (1966) 37-47.

3. Jones, N.; Uran, T.O.; Tekin, S.A.: The dynamic plastic behaviour of fully clamped rectangular plates. Int J Solids Structures 6 (1970) 1499-1512.

4. Wierzbicki, T.; Florence, A.L.: A theoretical and experimental investigation of impulsively loaded clamped circular viscoplastic plates. Int J Solids Structures 6, (1970) 555-568.

5. Duffey, T.A.; Key, S.W.: Experimental-theoretical correlations of impulsively loaded clamped circular plates. Research Report. Sandia Laboratories SC-RR-68-210 (1968).

6. Jones, N.; Griffin, R.N.; Van Duzer, R.E.: An experimental study into the dynamic plastic behaviour of wide beams and rectangular plates. Int J Mech Sci 13 (1971) 721-735.

7. Bednarski, T.: The dynamic deformation of a circular membrane. Int J Mech Sci 11 (1969) 949-959.

8. Jones, N.; Baeder, R.A.: An experimental study of the dynamic plastic behaviour of rectangular plates. Symposium on Plastic Analysis of Structures, Rumania (1972).

9. Ghosh, S.K.; Weber, H.: Experimental-theoretical correlation of impulsively loaded axisymmetric rigid-plastic membrane. Mech Res Comm 3 (1976) 423-428.

10. Bodner, S.R.; Symonds, P.S.: Experiments on dynamic plastic loading of frames. Int J Solids Structures 15 (1979) 1-13.

11. Bodner, S.R.; Symonds, P.S.: Experiments on viscoplastic response of circular plates to impulsive loading. J Mech Phys Solids 27 (1979) 91-113.

12. Nurick, G.N.; Martin, J.B.: The measurement of the response of clamped circular plates to impulsive loading. Int Phys Conf Ser No 70 (1984) 495-502.

13. Martin, J.B.; Symonds, P.S.: Mode approximations for impulsively loaded rigid plastic structures. Proc ASCE J Eng Mech Div 92 (EM5) (1966) 43-66.

14. Chon, C.T.; Symonds, P.S.: Large dynamic plastic deflection of plates by mode method. Proc ASCE J Eng Mech Div 103 (EM1) (1977) 169-187.

15. Symonds, P.S.; Chon, C.T.: Finite viscoplastic deflection of an impulsively loaded plate by the mode approximation technique. J Mech Phys Solids 27 (1979) 115.

16. Symonds, P.S.; Wierzbicki, T.: Membrane mode solutions for impulsively loaded circular plates. J Appl Mech 46 (1979) 58-64.

17. Guedes Soares C.: A mode solution for the finite deflections of a circular plate loaded impulsively. Rozprawy Inzynierskie Engineering Transactions 29 (1981) 99-114.

18. Perrone, N.; Bhadra, P: A simplified method to account for plastic rate sensitivity with large deformations. J Appl Mech 46 (1979) 811-816.

19. Perrone, N.; Bhadra, P.: Simplified large deflection mode solutions for impulsively loaded, viscoplastic, circular membranes. J Appl Mech 51 (1984) 505-509.

20. Martin, J.B.: The determination of mode shapes for dynamically loaded rigid-plastic structures. Meccanica 16 (1981) 42-45.

21. Manjoine, M.J.: Influence of rate of strain and temperature on yield stress of mild steel. J Appl Mech (1944) 211-218.

22. Nurick, G.N.: A new technique to measure the deflection-time history of a structure subjected to high strain rates. Int J Impact Engng 3 (1985) 17-26.

23. Nurick, G.N.; Adams, L.P.: The use of a reflex metograph to measure engineering structural deformations in three dimensions. Faculty of Engineering, University of Cape Town Internal Report (1985).

24. Duffey, T.A.: The large deflection dynamic response of clamped circular plates subjected to explosive loading. Sandia Laboratories Research Report No SC-RR-67-532 (1967).

Table 1. Test Data For Circular Plates

Test No.	Impulse (Ns)	Measured Mid-Point Deflection (mm)	Deflection-Thickness
1006851	9,0	10,62	6,64
2	9,5	10,90	6,81
3	5,6	6,14	3,84
4	10,0	11,96	7,48
5	9,9	12,08	7,55
1106851	10,8	12,26	7,66
2	10,6	12,20	7,63
3	10,8	12,80	8,00
4	11,5	13,62	8,51
5	11,7	13,22	8,26
6	11,4	14,00	8,75
1306851	12,6	14,90	9,31
2	12,4	14,60	9,13
3	12,9	15,28	9,55
4	12,8	16,08	10,05
5	13,6	16,30	10,19
6	13,4	16,38	10,24
1706851	14,1	17,70	11,06
2	13,8	17,06	10,66
1906851	14,7	18,56	11,60

Table 2 Test Data for Square Plates

Test No.	Impulse (Ns)	Measured Mid-Point Deflection (mm)	Deflection Thickness
1704851	9,5	10,46	6,54
2	11,2	11,58	7,24
3	12,1	12,54	7,84
4	8,7	9,44	5,90
2204851	11,2	12,78	7,99
2	11,4	12,78	7,99
3	11,4	13,10	8,19
4	12,8	13,98	8,74
5	13,4	14,44	9,03
6	13,8	15,66	9,79
2304851	15,3	17,78	11,11
2	16,6	19,14	11,96
3	16,0	17,96	11,23
2404951	15,3	17,40	10,88
2504851	14,0	15,18	9,49
2	12,9	14,10	8,81
3	12,2	12,76	7,98
4	13,9	15,50	9,69
5	10,5	11,14	6,96
2904851	9,6	9,62	6,01
2	9,4	9,84	6,15
3	10,3	11,72	7,33
4	9,3	9,92	6,20
5	11,2	13,10	8,19
6	9,1	10,40	6,50
0605851	9,5	10,62	6,64
2	9,7	10,62	6,64
3	10,4	12,12	7,58
4	11,7	13,00	8,13
5	13,1	14,32	8,95
6	14,1	15,70	9,81
7	14,9	16,94	10,59
0705851	15,3	16,72	10,45
2	14,7	16,36	10,23
3	15,0	17,09	10,68
4	15,3	17,34	10,84
5	16,1	17,86	11,16

Table 3 Test Data for Rectangular Plates

Test No.	Impulse (Ns)	Measured Mid-Point Deflection (mm)	Deflection-Thickness Ratio
0801851	5,2	4,68	2,93
2	9,1	7,58	4,74
0901851	7,1	6,38	3,99
2	5,8	5,30	3,31
1501851	7,4	7,56	4,73
2	6,1	7,26	4,54
1601851	8,8	9,04	5,65
2	9,6	9,86	6,16
1701851	11,6	12,02	7,51
2201851	11,2	11,88	7,43
2	12,2	13,40	8,38
3	12,2	12,68	7,93
2301851	8,8	9,12	5,70
2	9,5	9,98	6,24
3	7,8	9,48	5,93
4	13,1	13,88	8,68
2501851	12,8	13,62	8,51
2	13,6	14,50	9,06
3	13,4	14,60	9,13
4	14,1	14,88	9,30
5	13,7	14,82	9,26
2801851	13,8	14,26	8,91
2	13,6	14,52	9,08
2901851	14,5	14,86	9,29
2	14,5	15,30	9,56
3	14,9	15,84	9,90
4	15,3	16,72	10,45
0102851	15,7	16,72	10,45
2	16,7	17,88	11,16
3	17,4	18,24	11,40
0402851	15,2	16,58	10,36
2	8,3	8,40	5,25
3	8,4	8,82	5,51
4	9,1	9,84	6,15
5	7,8	9,42	5,89
0502851	10,5	10,76	6,73
2	9,2	10,36	6,48
3	11,6	11,98	7,49
4	11,3	12,00	7,50
5	11,6	12,58	7,86
0702851	11,4	12,12	7,58
2	10,8	11,02	6,89
3	-	10,98	6,86
4	16,0	16,60	10,88
5	16,3	17,86	11,16

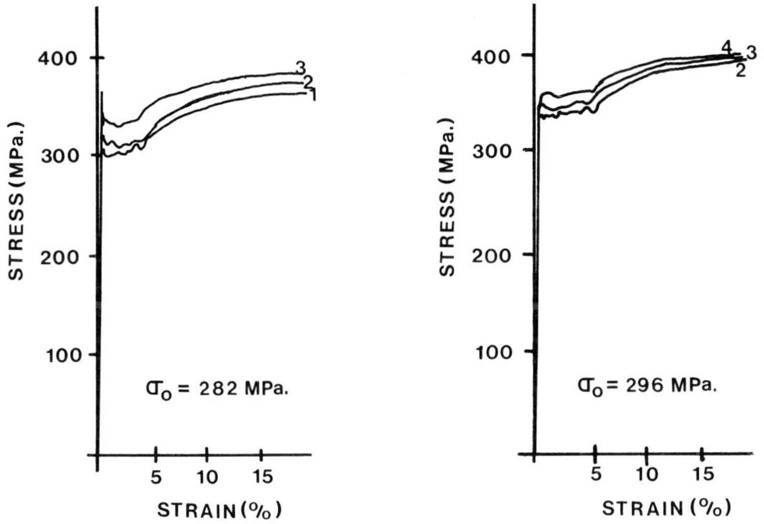

Fig.1. Stress-Strain Tensile Tests.
Strain Rates : 1 - $3{,}33\times10^{-4}$; 2 - $1{,}33\times10^{-3}$;
3 - $6{,}66\times10^{-3}$; 4 - $3{,}33\times10^{-2}$.

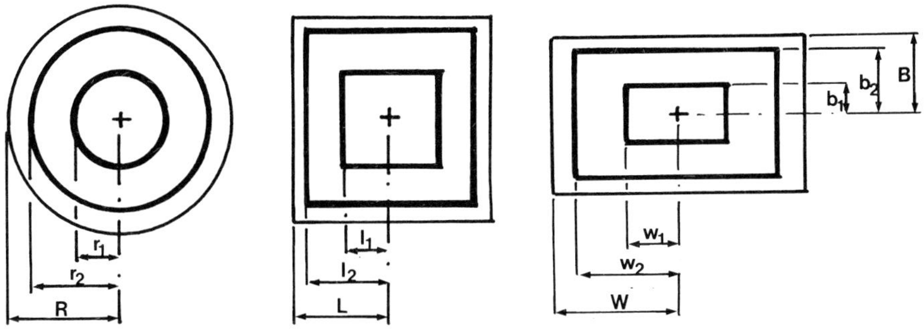

Fig.2. Explosive Layout.
Circular: $r_1=0{,}41R$; $r_2=0{,}82R$
Square: $\ell_1=0{,}49L$; $\ell_2=0{,}84L$
Rectangular: $w_1=0{,}50W$; $w_2=0{,}87W$; $b_1=0{,}50B$; $b_2=0{,}87B$.

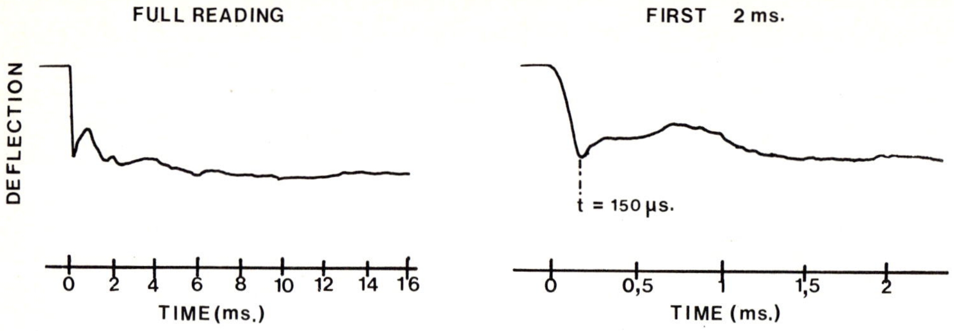

FIGURE 3 CIRCULAR PLATE; TEST NO. 1306854

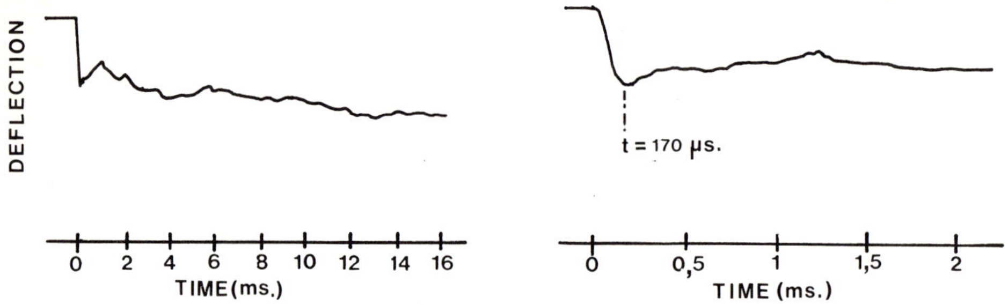

FIGURE 4 SQUARE PLATE; TEST NO. 2204854

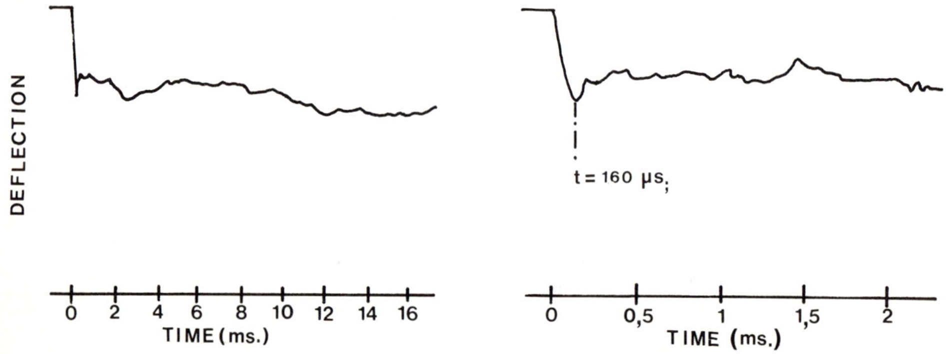

FIGURE 5 RECTANGULAR PLATE; TEST NO. 2101853

Figs.3-5. Typical Deflection-Time Curves.

TEST NO 0705853

TEST NO 0102851

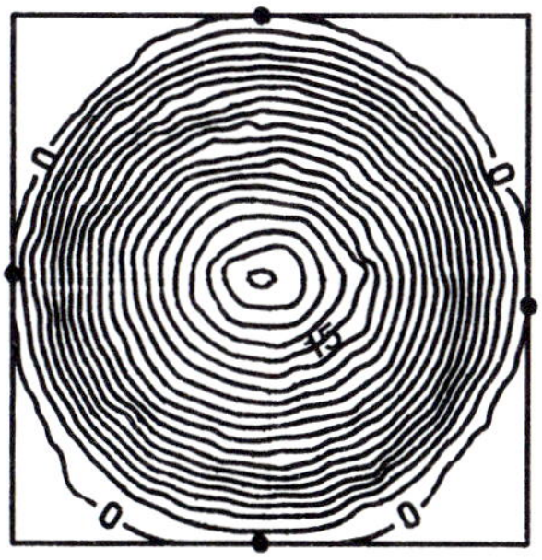

TEST NO 1906854

Fig.6. Contour Plots of Deformed Plates.

Fig.7. Graph of Measured Mid-Point Deflection-Thickness Ratio vs. Impulse for Circular Plates.

(0- Experimental Data;
1- Present Theory;
2- Symonds & Wierzbicki (Ref.16)
3- Perrone & Bhadra (Ref.19)
4- Duffey (Ref.24)
5- Guedes Soares (Ref.17)

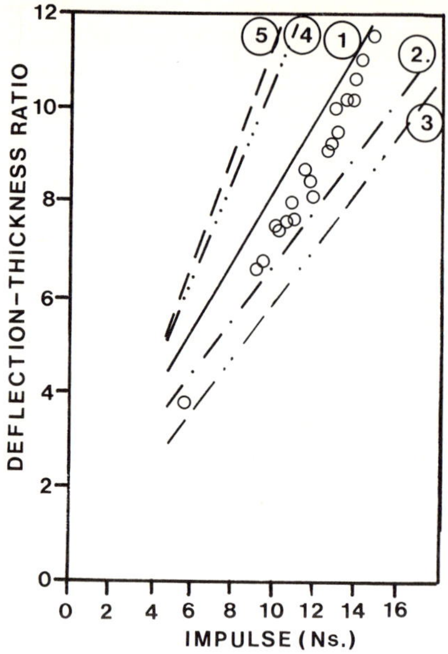

Fig.8. Graph of Measured Mid-Point Deflection-Thickness Ratio vs. Impulse for Square Plates.

(□- Experimental Data;
1- Present Theory;
2- Small Deflection Theory, Jones et al (Ref.6)

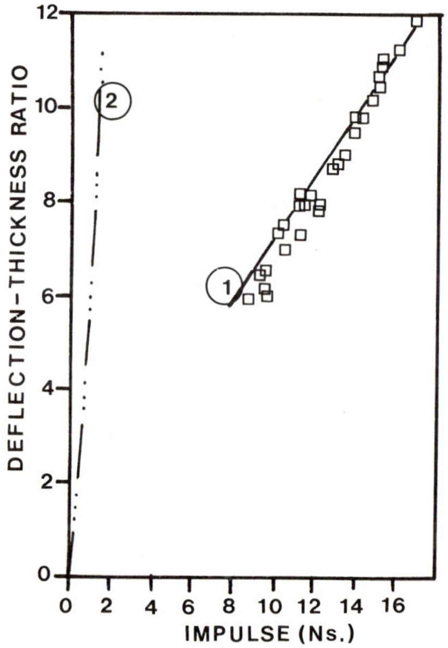

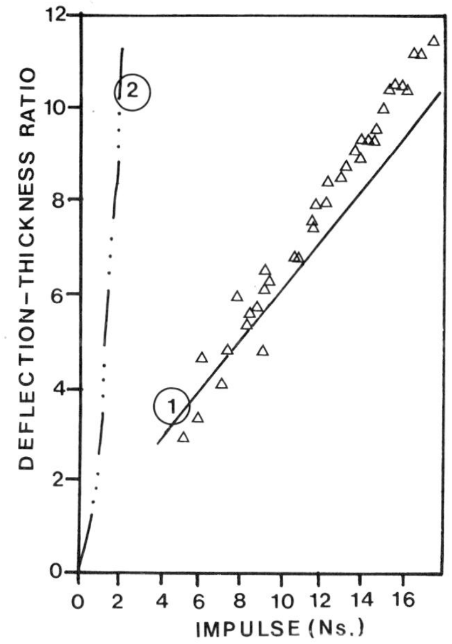

Fig.9. Graph of Measured Mid-Point Deflection Thickness Ratio vs. Impulse for Rectangular Plates.

(Δ- Experimental Data;

1- Present Theory;

2- Small Deflection Theory, Jones et al (Ref.6)

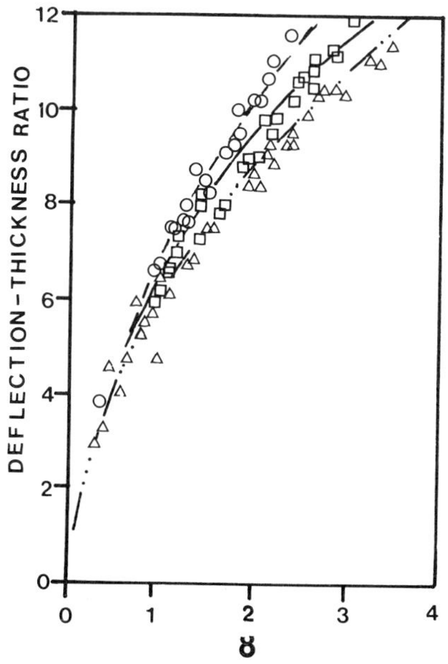

Fig.10. Graph of Measured Mid-Point Thickness Ratio vs. $\gamma = I^2/\rho A^2 M_o$

(O- Circular Plates;

□- Square Plates;

Δ- Rectangular Plates.

Lines represent least squares curve;

--- Circular Plates;

___ Square Plates;

-···- Rectangular Plates).

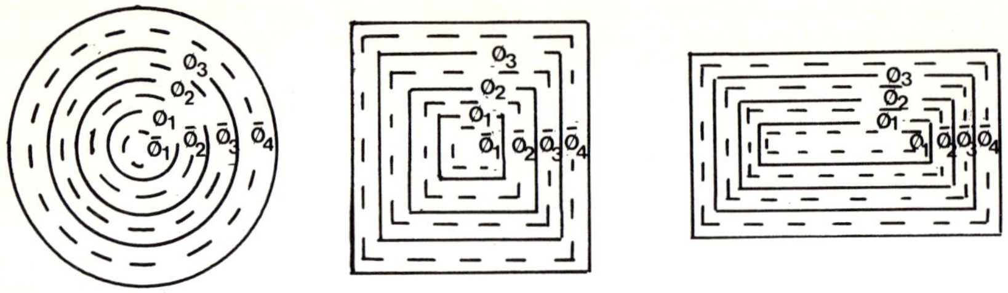

Fig.11. Discretisation of the Plates.

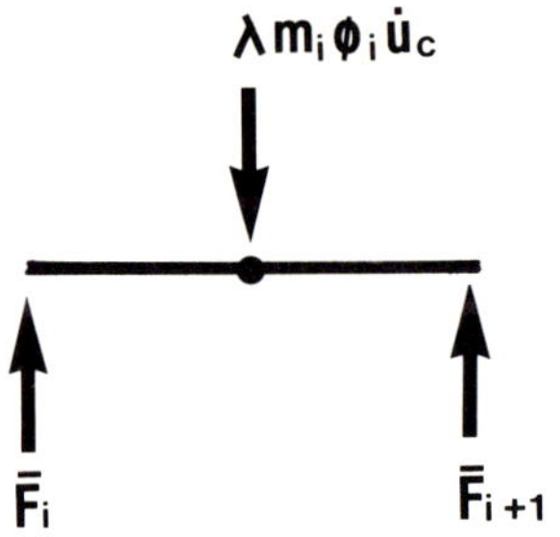

Fig.12. Forces acting on the i-th Node.

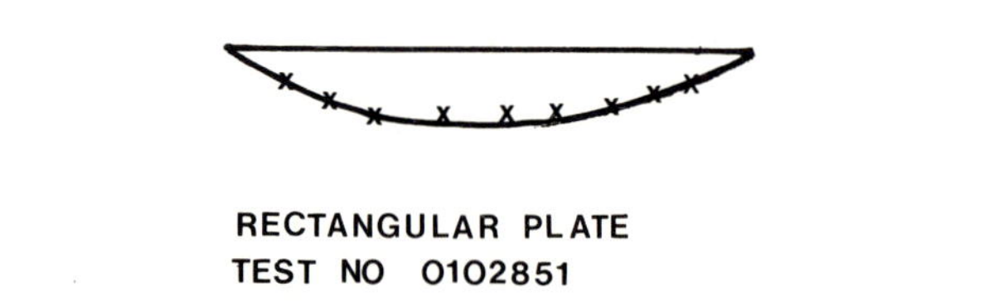

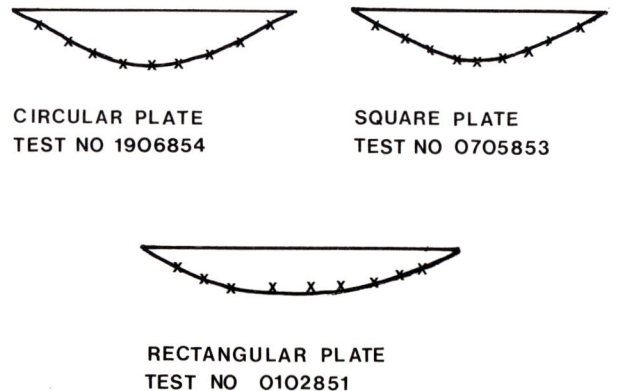

Fig.13. Final Mode Shapes.

——— Experimental Shape. **x** Present theory prediction.

Collapse Behaviour of Circular Aluminium Plates

G. H. LITTLE

Department of Civil Engineering
University of Birmingham
P.O. Box 363, Birmingham B15 2TT, U.K.

Summary

Circular aluminium plates are analysed to collapse under two separate loading conditions, one being uniform lateral pressure, and the other uniform radial in-plane compression. Attention is restricted to simply supported plates with axial symmetry. The method of analysis incorporates classical strain-displacement equations for moderately large deflections, together with the incremental stress-strain relationship based on J_2 flow theory, with isotropic hardening for the in-plane loading cases and with kinematic hardening for the lateral loading cases. For kinematic hardening, a method is proposed for determining the material state relative to the uniaxial stress-strain curve. This curve is characterised for aluminium by the Ramberg-Osgood formula. Some plates are initially perfect and some imperfect, and the main features examined in the results are the effects of the initial imperfection amplitude and of the sharpness of the "knee" of the material stress-strain curve. Some interesting cases of snap-through failure are evident for both types of loading.

INTRODUCTION

The paper is concerned with the numerical analysis of the behaviour of thin circular aluminium plates, simply supported around the edge, when subjected to either a uniform lateral pressure loading, or a uniform in-plane radial edge compression. The plate behaviour is restricted to being axially symmetric in all cases, and the material behaviour is assumed to be characterised by a uniaxial stress-strain curve of the Ramberg-Osgood type. Some steel plates are also analysed, as an extreme case for which the curve becomes two straight lines.

Circular plates are of interest for two reasons. Firstly, they are a widely used structural component, appearing in structures of many different sizes and types. Secondly, they are of special interest to the analyst, since, for the axisymmetric

Inelastic Behaviour of Plates and Shells
IUTAM Symposium Rio de Janeiro 1985
Editors: L. Bevilacqua, R. Feijóo and R. Valid

case, they form probably the simplest case of plane stress combined with non-linear geometrical and material behaviour. Thus, although a circular plate might be regarded as a "simple model" compared to other more complicated plated structures, basic analytical techniques developed, cheaply, for a circular plate are likely to be effective for more complicated structures.

For the above reasons, circular plates have received a great deal of attention from research workers over the past fifty years, both theoretically and experimentally, but most of this has been devoted to the case of lateral pressure loading rather than in-plane compression. Early non-linear analyses were restricted to geometrical non-linearity, the material remaining elastic, and this work is summarised in ref. 1. The advent of the digital computer made it possible to include plastic behaviour of the material, and various results including both geometrical and material non-linearity have appeared since about 1960. A useful summary is given in ref. 2. The theoretical work referred to so far has all assumed either elastic or elasto-plastic material behaviour. A further line of approach is via the assumption of rigid-plastic behaviour in an upper bound calculation of the type described by Calladine [3]. Such an approach gives a simple analysis for sufficiently stocky plates, but does not include elastic strains or membrane effects, which become important in more slender plates.

Despite the large volume of literature on circular plate analysis, the number of available results based on rigorous non-linear theory is surprisingly small. There appears to have been no systematic study made of the effect of initial imperfections and of the shape of the material stress-strain curve on collapse behaviour. These are the two effects that are particularly examined in the present work, both for lateral pressure and for in-plane compression. The method of analysis is basically the same as was used in previous studies of columns [4] and of rectangular plates [5]. The method is described briefly here, but in greater detail in papers currently in preparation. The basic equations include standard strain-displacement relationships for moderately large deflections, together with incremental stress-strain relations based on J_2 flow theory. Although isotropic hardening has been assumed for in-plane compressive loading, this was recognised to be unrealistic for certain cases of lateral pressure loading, and kinematic hardening was used for all the lateral pressure cases analysed. A simple improvement to previous kinematic hardening formulations is proposed. Some of the results show snap-through buckling actions, at constant load for lateral pressure, and at constant radial shortening for in-plane compression.

PLATE DETAILS AND BOUNDARY CONDITIONS

A typical circular plate as considered in the analysis is illustrated in Fig. 1. The plate is of radius R and uniform

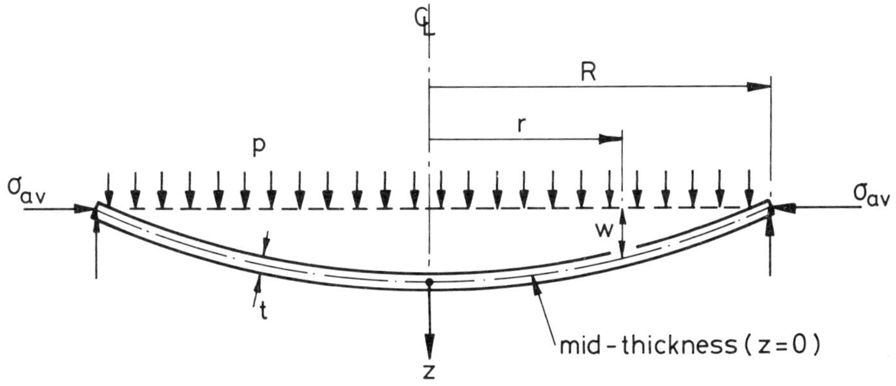

Fig. 1 Diametral section through plate

thickness t, and is simply supported around the edge. The r and θ axes are radial and circumferential, and the z axis is lateral, with z zero at mid-thickness. As usual with thin plate analysis, plane stress is assumed. The assumption of axially symmetric behaviour then has the following consequences:

(a) No quantity varies with θ.

(b) There is no circumferential displacement, the only non-zero displacements being radial (with u at mid-thickness), and lateral (w). On a diametral plane, the displacements u and w are anti-symmetrical and symmetrical, respectively, about $r = 0$.

(c) There is no shearing stress $\tau_{r\theta}$, the only non-zero stresses being the direct stresses σ_r and σ_θ which are the principal stresses. Similarly the shearing strain $\gamma_{r\theta}$ is zero, with only the direct strains ε_r and ε_θ present.

The stress and strain vectors at (r,z) are therefore:

$$\underset{\sim}{\sigma}^T = \{\sigma_r \;\; \sigma_\theta\} \quad \text{and} \quad \underset{\sim}{\varepsilon}^T = \{\varepsilon_r \;\; \varepsilon_\theta\} \tag{1}$$

And the stress resultants per unit length at radius r are:

$$\underset{\sim}{N} = \int_{-t/2}^{+t/2} \underset{\sim}{\sigma}\, dz \quad \text{and} \quad \underset{\sim}{M} = \int_{-t/2}^{+t/2} z\, \underset{\sim}{\sigma}\, dz \tag{2}$$

Any vector contains **two** components: a radial component (subscript r) and a circumferential component (subscript θ).

With regard to the loading, there may be a uniform lateral pressure of magnitude p, and a uniform radial in-plane edge **compressive** stress of magnitude σ_{av}. A lateral central **ring** load is considered in a later section for comparison with previous work.

The mean lateral displacement w_{av} is defined as:

$$w_{av} = \frac{2}{R^2} \int_0^R w\, r\, dr \tag{3}$$

and the radial edge shortening strain e is:

$$e = -u_{r=R}/R \tag{4}$$

The boundary conditions (BCs) may now be stated as, at $r = R$:

$$\left.\begin{aligned} w &= 0 \\ M_r &= 0 \end{aligned}\right\} \tag{5}$$

and

Either $\quad u = -eR \quad$ (6a)

or $\quad N_r = -\sigma_{av} t \quad$ (6b)

Equation (5) gives the BCs corresponding to simple support against lateral deflections. Equation (6a) is the in-plane BC for the prescribed-e case (e.g. a stiff testing machine) and (6b) is the in-plane BC for the prescribed-σ_{av} case (e.g. dead loading).

The plate is assumed to be initially stress-free, and may have an initial geometrical imperfection w_o of amplitude δ_o. The lateral **deflection** is therefore $(w - w_o)$. The material behaviour is initially elastic with Young's modulus E and Poisson's ratio ν, which is taken to be 0.33 throughout. The material behaviour is described with reference to the uniaxial stress-strain curve as shown in Fig. 2(a). This curve will be referred to as the MSSC, which is an abbreviation of

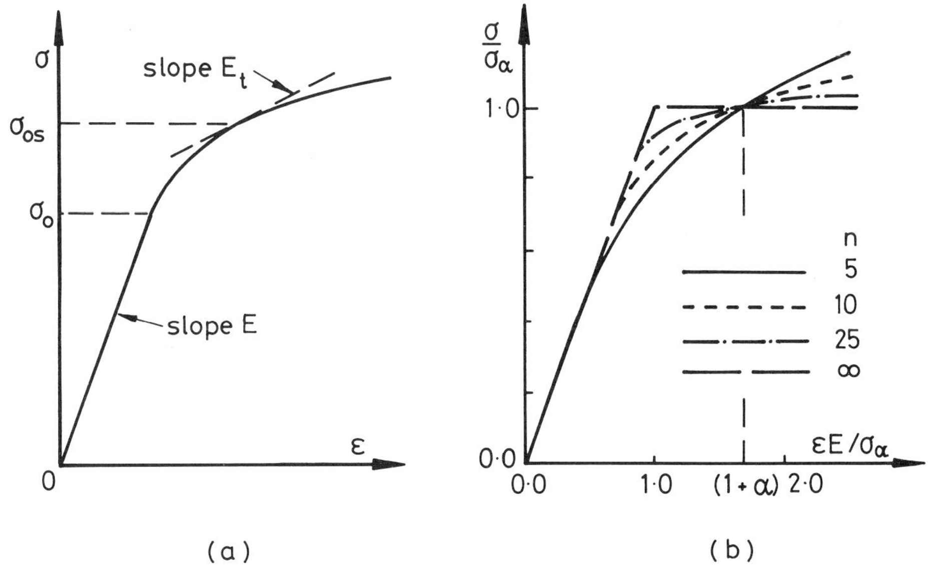

Fig. 2 Material stress-strain curves.
(a) A typical curve.
(b) Ramberg-Osgood curves for various values of n

"material stress-strain curve". Note from Fig. 2(a) that the initial yield stress is σ_o, and the tangent modulus in the plastic range is E_t. The MSSC will be defined by the Ramberg-Osgood formula as follows:

$$(\varepsilon E/\sigma_\alpha) = \sigma/\sigma_\alpha + \alpha(\sigma/\sigma_\alpha)^n \tag{7}$$

where n = factor which determines the degree of sharpness of the "knee" of the MSSC

σ_α = stress at which the plastic strain is equal to α times the elastic strain.

In the present work α is taken as 0.7 throughout, and Fig. 2(b) shows typical curves plotted from equation (7) for various values of n. Note that $n = \infty$ implies an elastic-perfectly-plastic MSSC typical of mild steel, for which $\sigma_\alpha = \sigma_o$. An apparent inconsistency between equation (7) and Fig. 2(a) is that equation (7) does not exhibit an initial elastic range. In fact (7) is applied in the following, slightly modified, form:

$$\sigma \leq \sigma_o \quad ..... \quad \varepsilon E/\sigma_\alpha = \sigma/\sigma_\alpha \tag{8a}$$

$$\sigma > \sigma_o \quad ..... \quad \varepsilon E/\sigma_\alpha = \sigma/\sigma_\alpha + \alpha(\sigma/\sigma_\alpha)^n - \alpha(\sigma_o/\sigma_\alpha)^n \tag{8b}$$

which has the required elastic range. For isotropic hardening, σ_o is chosen so that the final term in equation (8b) is negligible [5]. For kinematic hardening, however, it is essential to have a realistic value of σ_o, which is chosen such that:

$$\alpha(\sigma_o/\sigma_\alpha)^n = 0.02 \tag{9}$$

and this gives σ_o/σ_α values which seem appropriate by inspection of Fig. 2(b).

DETAILS OF THE ANALYSIS

Accurate collapse analysis of structures is necessarily of an incremental nature, and that gives rise to the following two items of notation:

(a) The bar superscript (e.g. $\bar{\sigma}$) denotes the quantity value at the beginning of the current increment - the **initial** value, and

(b) The prefix Δ denotes the change during the current increment. Thus, for example:

current stress σ = initial stress $\bar{\sigma} + \Delta\sigma$

And a third item of notation is as follows:

$$(\ )' = \frac{d}{dr}(\ ) \quad \text{and} \quad (\ )'' = \frac{d^2}{dr^2}(\ )$$

Basic Equations

The **equilibrium equations** are:

$$\begin{aligned} &\text{radial direction} \quad ..... \quad (rN_r)' - N_\theta = 0 \\ &\text{lateral direction} \quad ..... \quad (rM_r)'' + (rN_r w' - M_\theta)' + pr = 0 \end{aligned} \tag{10}$$

The **strain-displacement** equations are:

$$\begin{aligned} \Delta\varepsilon_r &= \Delta u' + \Delta w'(\tfrac{1}{2}\Delta w' + \bar{w}') - z\Delta w'' \\ \Delta\varepsilon_\theta &= \frac{1}{r}\Delta u - \frac{z}{r}\Delta w' \end{aligned} \tag{11}$$

The non-linear equations (11) account for deflections of magnitude similar to t but small compared to R, and incorporate the Love-Kirchhoff hypothesis.

The **stress-strain** relations may be written:

$$\Delta\underset{\sim}{\sigma} = \underset{\sim}{D}\ \Delta\underset{\sim}{\varepsilon} \qquad (12)$$

where, for elastic behaviour, $\underset{\sim}{D} = \underset{\sim}{D}^e$ the ordinary elastic isotropic stress-strain matrix for plane stress. For plastic behaviour $\underset{\sim}{D}$ may be deduced from the following:

$$\Delta\underset{\sim}{\sigma} = \underset{\sim}{D}^e\ [\Delta\underset{\sim}{\varepsilon} - \lambda\underset{\sim}{S}] \qquad (13)$$

$$\underset{\sim}{S} = \underset{\sim}{A}\ \underset{\sim}{\sigma} \qquad (14)$$

$$\underset{\sim}{A} = \frac{1}{3}\begin{bmatrix} 2 & -1 \\ -1 & 2 \end{bmatrix} \qquad (15)$$

$$\lambda = \underset{\sim}{S}^T\underset{\sim}{D}^e\Delta\underset{\sim}{\varepsilon}/(\underset{\sim}{S}^T\underset{\sim}{D}^e\underset{\sim}{S} + \frac{4}{9}\,m\,\sigma_e^2) \qquad (16)$$

$$\sigma_e^2 = \frac{3}{2}\,\underset{\sim}{\sigma}^T\ \underset{\sim}{S} \qquad (17)$$

$$m = E_tE/(E-E_t) \quad \text{and } E_t \text{ is taken from the MSSC at } \sigma = \sigma_{os} \qquad (18)$$

$$\sigma_{os} = \text{greater of } \sigma_o \text{ and } (\sigma_e)_{max} \qquad (19)$$

In the above, $\underset{\sim}{S}$ is the vector of stress deviators, $\lambda\underset{\sim}{S}$ are the plastic strain increments, σ_e is the von Mises effective stress, σ_{os} is the yield function, and $(\sigma_e)_{max}$ is the maximum value attained by σ_e throughout the stress history. The conditions for plastic flow to occur are as follows:

$$\left.\begin{aligned} \sigma_e^2 &= \sigma_{os}^2 \\ \lambda &> 0 \end{aligned}\right\} \qquad (20)$$

Otherwise, λ is set to zero resulting in an elastic response. Equations (12) to (20) summarise J_2 flow theory for plane stress with **isotropic hardening**. In the numerical analysis, the equations are applied using **initial** stress values, thus linearising the incremental relationship of equation (12). Plastic flow automatically causes an increase in σ_e, and correspondingly in σ_{os} (provided $E_t>0$, of course).

Kinematic Hardening

Isotropic hardening means that any increase in effective yield stress (σ_{os}) due to strain hardening is equally effective under reversed loading conditions. In other words, no account is taken of the Bauschinger effect. Therefore, if reversed stressing is a prominent feature of any case being analysed, it is necessary to calculate plastic flow using a **kinematic** hardening theory based on Ziegler´s modification [6] of Prager´s original proposal [7]. Such a theory assumes that, as plastic flow occurs, the yield surface remains of constant size and shape, and does not rotate. Strain hardening causes a shift of the **position** of the yield surface, this being defined by $\underset{\sim}{c}$, the position vector of the centre of the surface. This model gives a "complete" Bauschinger effect, which may not precisely conform to the material behaviour. The yield criterion becomes:

$$\hat{\sigma}_e^2 = \sigma_o^2 \tag{21}$$

where
$$\left.\begin{aligned} \hat{\sigma}_e^2 &= \frac{3}{2}\,\hat{\underset{\sim}{\sigma}}^T \hat{\underset{\sim}{S}} \\ \hat{\underset{\sim}{S}} &= \underset{\sim}{A}\,\hat{\underset{\sim}{\sigma}} \\ \text{and} \quad \hat{\underset{\sim}{\sigma}} &= \underset{\sim}{\sigma} - \underset{\sim}{c} \end{aligned}\right\} \tag{22}$$

During each increment when plastic flow occurs, changes occur both in $\underset{\sim}{\sigma}$ and in $\underset{\sim}{c}$ such that the yield criterion (21) is not violated. Thus:

$$\Delta(\hat{\sigma}_e^2) = \left\{\frac{\partial(\hat{\sigma}_e^2)}{\partial\underset{\sim}{\sigma}}\right\}^T \Delta\underset{\sim}{\sigma} + \left\{\frac{\partial(\hat{\sigma}_e^2)}{\partial\underset{\sim}{c}}\right\}^T \Delta\underset{\sim}{c} = 0$$

$$= 3\,\hat{\underset{\sim}{S}}^T \Delta\underset{\sim}{\sigma} - 3\,\hat{\underset{\sim}{S}}^T \Delta\underset{\sim}{c} = 0 \tag{23}$$

Equations (13) to (18) give kinematic hardening if $\underset{\sim}{S}$ is replaced by $\hat{\underset{\sim}{S}}$ in (13),(14) and (16), $\underset{\sim}{\sigma}$ is replaced by $\hat{\underset{\sim}{\sigma}}$ in (14) and σ_e^2 by σ_o^2 in (16). In determining λ, m is still found from the MSSC at a stress level σ_{os}, but the problem is how to determine σ_{os}. Pifko, Levine and Armen [8] have suggested various ways of doing this, involving some rather involved functions of the stresses. However, the method now described is simple, logical, and does not appear to have been used previously. As in the case of isotropic hardening, plastic flow requires $\lambda > 0$, and this ensures that the term in $\Delta\underset{\sim}{\sigma}$ in equation (23) is positive, i.e. there is an **increase** of $3\,\hat{\underset{\sim}{S}}^T \Delta\underset{\sim}{\sigma}$ in $\hat{\sigma}_e^2$ associated with $\Delta\underset{\sim}{\sigma}$. This represents an increase $(\Delta\hat{\sigma}_e)_\sigma$ in $\hat{\sigma}_e$ given by:

$$(\Delta\hat{\sigma}_e)_\sigma = \left(\frac{3}{2\sigma_o}\right) \hat{\underset{\sim}{S}}^T \Delta\underset{\sim}{\sigma} \tag{24}$$

since $\hat{\sigma}_e = \sigma_o$ at yield. The present proposal is to set σ_{os}

originally to σ_o, and then to increment it each time plastic flow occurs by the value of $(\Delta\hat{\sigma}_e)_\sigma$ given by equation (24), i.e.

$$\Delta\sigma_{os} = (\Delta\hat{\sigma}_e)_\sigma \tag{25}$$

The original setting for $\underset{\sim}{c}$ is $\underset{\sim}{0}$, and $\Delta\underset{\sim}{c}$ is found from equation (23) together with Ziegler's proposal [7] that:

$$\Delta\underset{\sim}{c} = \mu\hat{\underset{\sim}{\sigma}} \tag{26}$$

where μ must be positive, and this is ensured by a positive λ. In fact equations (21)-(26) give:

$$\mu = \Delta\sigma_{os} / \sigma_o \tag{27}$$

The proposed formulation for kinematic hardening satisfies the necessary condition of being correct for uniaxial stressing. The formulation is again applied using initial stress values, giving a linear incremental stress-strain relationship.

Solution by a Rayleigh-Ritz Method

Adopting a Rayleigh-Ritz procedure, displacement functions satisfying the geometric BCs are used to describe the deformed shape of the middle surface (u and w). For any set of values of the displacement function amplitudes, the strain increment distribution can be found via equations (11), and the corresponding stress increments from (12). This enables the value of a particular energy function F (defined below) to be calculated. For any prescribed value of p (or w_{av}) and σ_{av} (or e), the displacement function amplitudes are varied until F has been minimised. This gives the combination of the given displacement functions which best satisfies equilibrium throughout the plate together with the static BCs. If w_{av} and/or e had been prescribed, then p and/or σ_{av} must be calculated, before moving on to the next stage of loading. The energy function F is defined as follows:

$$F = \int_{vol} (\bar{\underset{\sim}{\sigma}} + \tfrac{1}{2}\Delta\underset{\sim}{\sigma})^T \Delta\underset{\sim}{\varepsilon}\, d(vol) - (\pi R^2)p w_{av} - (2\pi R^2 t)\sigma_{av} e \tag{28}$$

If w_{av} is prescribed, the w_{av} term is omitted; if e is prescribed the e term is omitted.

The displacement functions used in the present work are Fourier series as follows:

$$\left.\begin{aligned} u &= -\,er + \sum_{m=1,2,3,\ldots} U_m \sin\frac{m\pi r}{R} \\ w &= \sum_{m=\frac{1}{2},1\frac{1}{2},2\frac{1}{2},\ldots} W_m \cos\frac{m\pi r}{R} \end{aligned}\right\} \qquad (29)$$

and the first four terms of each series for u and w have been used in the numerical analysis. The initial imperfection w_o has been taken as:

$$w_o = \delta_o \cos(\pi r/2R) \qquad (30)$$

The stress-strain state over a radial section is monitored using a mesh of grid points numbering 9 and 5 in the r and z directions respectively. This enables F to be evaluated by numerical integration using Simpson's rule. It is essential to store $\bar{\underset{\sim}{\sigma}}$ and $\bar{\sigma}_{os}$ at each grid point, and, for kinematic hardening $\bar{\underset{\sim}{c}}$ must also be stored, considerably increasing the storage requirements. The minimum of F is found using a quasi-Newton algorithm - a type of procedure that is particularly effective for the present problem [5]. Although the solution procedure is based on the equations given above, those equations are actually used in a normalised form having various computational advantages [5]. For the arrangement described, a typical CPU time per increment on a CDC7600 computer in single precision is only 0.015 secs.

COMPARISON WITH PREVIOUS WORK

At the end of the previous section, details were given of the precision of the numerical model used for the present results. That arrangement was decided on as a result of previous experience [4, 5], and of many comparisons made with previous work, particularly with classical elastic large-deflection results for circular plates such as those included in ref. 1. Space does not permit the details of all these comparisons to be included here. The one comparison which will be given is that of the present results with those of Winter and Levine [9], theirs being both theoretical and experimental for an aluminium plate subjected to a central lateral load from a rod of diameter 9.53 mm = 0.144R. The edge is both simply supported and free-to-pull-in (σ_{av} = 0). The comparison is shown in Fig. 3. Winter and Levine helpfully provide all the necessary experimental data, including the Ramberg-Osgood equation for the MSSC. Their theoretical results were obtained using the PLANS finite element system [8] with kinematic hardening. Because of the similarity of the loading to a central point load, and the fact that centrally loaded plates tend towards a conical shape as deflections become large, an additional displacement function for w was introduced into the present analysis. This consists of a short parabolic section from $r = 0$ to $r = \gamma R$, joining to a straight section which then extends to the edge. The slope is zero at $r = 0$ and continuous throughout. This was used together with the first three terms of the series in equation (29), making four functions in all, as before. It was necessary to increase the number of grid stations in the r direction from 9

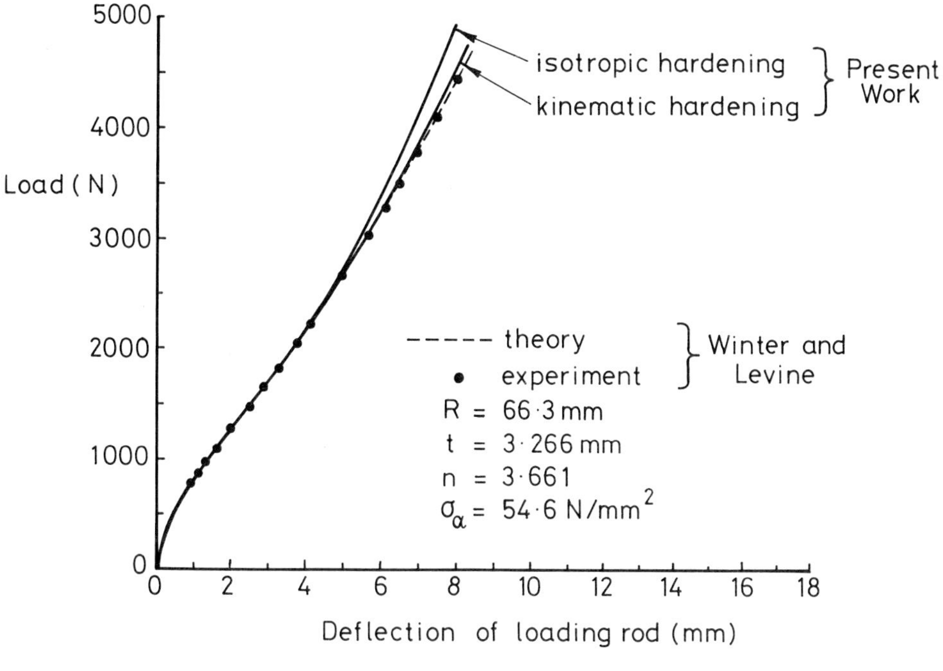

Fig. 3 Comparison of present results with those of Winter and Levine [9]

to 11 in order to integrate accurately over the central region where the parabola is. The value of $\gamma = 0.05$ was found to be suitable. The present results incorporate the loading as a ring load of diameter 9.53mm, requiring a minor adjustment to the energy formulation. The lateral deflection of the ring load was prescribed, with an increment size of 0.02t, and the plate was analysed as initially flat.

Fig. 3 shows the present results for both isotropic and kinematic hardening, together with Winter and Levine's experimental and theoretical results. Clearly, even for this initially flat plate, the difference between the two hardening formulations becomes significant as deflections become large. This is due to reversed plastic stressing in the region between mid-thickness and the loaded surface. In the early stages of loading, this region yields due to **compressive** bending stresses, but as deflections increase **tensile** membrane stresses become dominant. The present kinematic hardening results agree very satisfactorily with both the theory and experiment of Winter and Levine.

BEHAVIOUR OF PLATES UNDER LATERAL PRESSURE

Results

Lateral pressure-displacement curves, calculated using the present method, are given in Fig. 4 for various simply supported plates with in-plane edge movement prevented ($e = 0$). The curves are plotted in normalised form with pR^4/Et^4 as ordinate and w_{max}/t as abscissa, where w_{max} is w at $r = 0$. In each case the plate slenderness β is unity, where:

$$\beta = (\sigma_\alpha / \sigma_{cr})^{\frac{1}{2}} \tag{31}$$

$$\text{and} \quad \sigma_{cr} = \text{elastic buckling stress}$$

$$= 0.4E/(R/t)^2 \text{ when } \nu = 0.33 \tag{32}$$

$\beta = 1.0$ means an intermediate plate slenderness and corresponds to $R/t = 11.83$ for an alloy having $\sigma_\alpha = 200$ N/mm^2 and $E = 70$ kN/mm^2. The following values of the parameters δ_o/t and n have been used:

$$\delta_o/t = -1.0,\ 0.0,\ +1.0 \qquad n = 5,\ 10,\ 25 \tag{33}$$

A negative δ_o/t means that the lateral pressure is applied to the **convex** surface (Fig. 1). For all three n values, kinematic hardening was employed. In addition to the n values given above, for each δ_o/t value a curve is given for a steel plate, together with one based on completely elastic material behaviour. There is therefore a total of 15 curves in Fig. 4. Each curve was computed using the prescribed-w_{av} method to define each load stage, with $w_{av}/t = 0.01$. The curves are plotted as a series of straight lines joining the computed points, and are presented exactly as plotted (i.e. they have not been traced).

Effect of the Initial Imperfection

Plates carry lateral pressure loading through a combination of bending and membrane effects. Membrane stresses help to carry the load in association with the curvature and slope of the plate. Considering firstly the results for initially flat

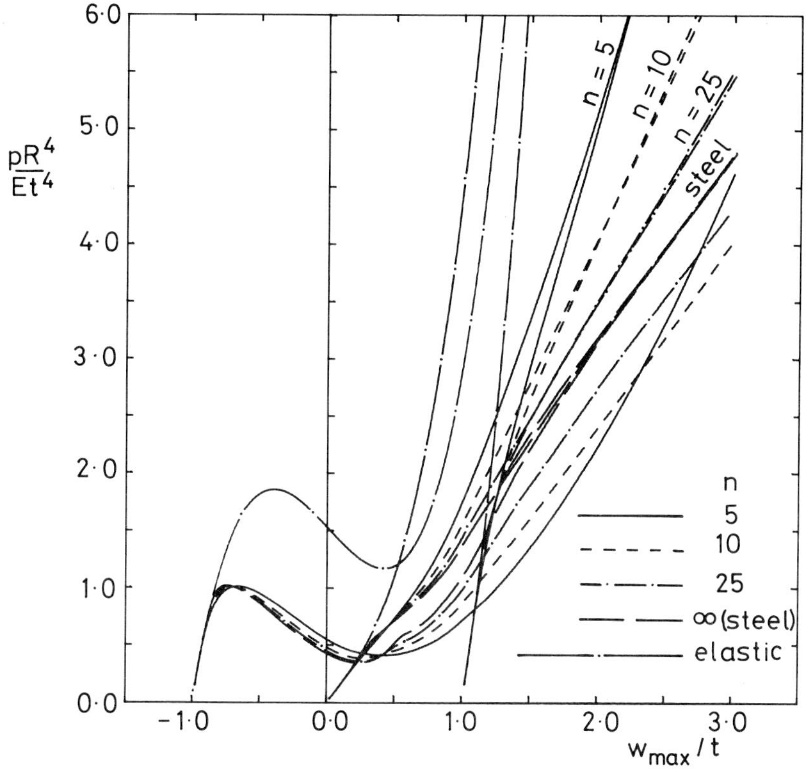

Fig. 4 Normalised lateral pressure-central displacement curves for $\beta = 1.0$. The w_{max}/t value at zero pressure indicates the initial imperfection amplitude δ_o/t in each case.

plates ($\delta_o/t=0$), for small deflections both the membrane stresses and the curvature are negligible, and so virtually all the load is carried through bending of the plate. As deflections increase, however, the membrane action becomes steadily more significant, especially for in-plane edge fixity, and the plate response becomes correspondingly more stiff. Even with plasticity allowed for, the load is predicted to rise continuously with no maximum being reached.

The results for a positive initial imperfection ($\delta_o/t = +1.0$) show behaviour broadly similar to that of an initially flat plate, except that the initial stiffness is considerably larger (by a factor of five). This increased stiffness is because the presence of the imperfection results in an immediate load-carrying action from membrane stresses. Interestingly though, as displacements increase, and plasticity spreads, corresponding curves for $\delta_o/t = 0$ and $\delta_o/t = +1.0$ come together. Clearly at large displacements, and at the same w_{max}/t value, two such plates have similar stress distributions, the effect of the initial imperfection having effectively been "washed out".

We turn now to the results for a **negative** initial imperfection. An immediately striking aspect of the curves in Fig. 4 is that, for $\delta_o/t = -1.0$, a definite stationary maximum load is reached at negative w_{max}, followed by a minimum at positive w_{max}, followed by steadily increasing load as w_{max} becomes larger in the positive direction. This behaviour is a well-known phenomenon found in shallow shells and arches (e.g. ref. 10), which is difficult to observe **completely** in an experiment. For pressure loading the tendency would, of course, always be to snap through from the maximum point, at constant load, to the rising curve at positive w_{max}. The general shape of these negative δ_o curves is explained by reference to membrane effects. Initially membrane compression, together with the negative curvature, carries a significant proportion of the load. As the load increases, the compression increases, but the negative curvature decreases. At the maximum point, the latter effect is about to become dominant, and the load begins to fall. However, as w_{max} increases through the positive range, the membrane compression decreases and eventually changes to tension, the curvature is then increasingly positive, and this combination is again making significant and increasing contributions to the load carried by the plate.

As mentioned above, corresponding curves for $\delta_o/t = 0$ and $+ 1.0$ come together at large w_{max}, but the same does not happen for the corresponding negative δ_o case. More will be said about that in the next section.

It is worth noting that the increase in initial stiffness associated with an initial imperfection is exactly the same for a negative imperfection as for a positive one of the same magnitude.

<u>Effect of the MSSC (n) and of Kinematic Hardening</u>

As might be expected, the effect of the shape of the MSSC increases as displacements increase and the degree of plastic straining increases. Thus at large w_{max}, the curves for $n = 5$ lie considerably higher, and stiffer, than those for $n = \infty$. For small deflections (up to about 0.25t), the effect of n is small.

The cases covered in Fig. 4 were also analysed assuming isotropic hardening, although those results are not presented here. A comparison of them with the kinematic hardening results in Fig. 4 shows relatively small differences (<5 %) in all cases except those with negative δ_o. For the negative δ_o cases, agreement is close until w_{max}/t exceeds about $+ 0.5$, whereupon the isotropic curves rise above the kinematic, and "home in" on the corresponding curves for $\delta_o/t = 0$ and $+ 1.0$. The corresponding kinematic curves, by contrast, lie considerably lower , as is evident from Fig. 4. This is not surprising, as for negative δ_o, a considerable amount of reversed plastic stressing occurs through a large proportion of the plate volume, if the plate is loaded beyond the maximum and minimum points. This is therefore a good example of a case where isotropic hardening is inappropriate, and an unsafe assumption.

The difference between results for isotropic and kinematic

hardening decreases as n increases, and is, of course, zero at $n = \infty$, when there is no hardening. The fact that the comparison between the results of the two hardening theories suggests such reasonable conclusions, gives support to the present method of incorporating kinematic hardening into the analysis.

BEHAVIOUR OF PLATES UNDER IN-PLANE COMPRESSION

Results

Theoretical in-plane load-radial shortening curves (abbreviated to LS curves) are given in Fig. 5 for various simply supported plates under zero lateral loading. The curves are plotted in normalised form with $\sigma_{av}/\sigma_{\alpha}$ as ordinate, and eE/σ_{α} as abscissa. As in the case of Fig. 4, all cases considered have slenderness $\beta = 1.0$, and the same three n values, together with steel, are used. The imperfection amplitudes are different though, and are 0.0, 0.05, 0.15, 0.35, as shown in Fig.5. For all three n values, isotropic hardening was employed. The imperfect cases were also analysed with kinematic hardening, and the difference found to be negligible, as would be expected. All curves, except two (to be mentioned later), were computed using the prescribed-e method to define each load stage, with $(\Delta e)E/\sigma_{\alpha} = 0.01$. The curves are plotted as a series of straight lines joining the computed points, as before.

Behaviour of Initially Flat Plates

By contrast to the case of lateral pressure loading, for the case of in-plane compression the most complex behaviour, and the most difficult to analyse, is that of initially flat plates. The analysis begins by determining the plastic buckling stress, σ^P_{cr}, based on the Shanley concept, using formulae stated by Needleman [11]. For $\beta = 1.0$, the value of σ^P_{cr} for each n value is as follows:

	n = 5	n = 10	n = 25
$\sigma^P_{cr}/\sigma_{\alpha}$ =	0.673	0.746	0.841

The process of incrementing e then begins at a level corresponding to $\sigma_{av}/\sigma_{\alpha} = (\sigma^P_{cr}/\sigma_{\alpha}) - 0.1$, at which the plate will remain flat. In the early stages, at each increment of e, the initial guess for w is a small sinusoidal out-of-flatness, and iterations to the equilibrium state recover the flat form.

Then the plate is in a state of uniform biaxial compression, i.e. $\sigma_r = \sigma_\theta = -\sigma_{av}$ throughout, and the normalised in-plane stiffness of the flat plate is $1/(1 - \nu)$. The uniform biaxial compression does not, of course, affect the load at first yield, which occurs when $\sigma_{av} = \sigma_o$. The plate remains flat until the

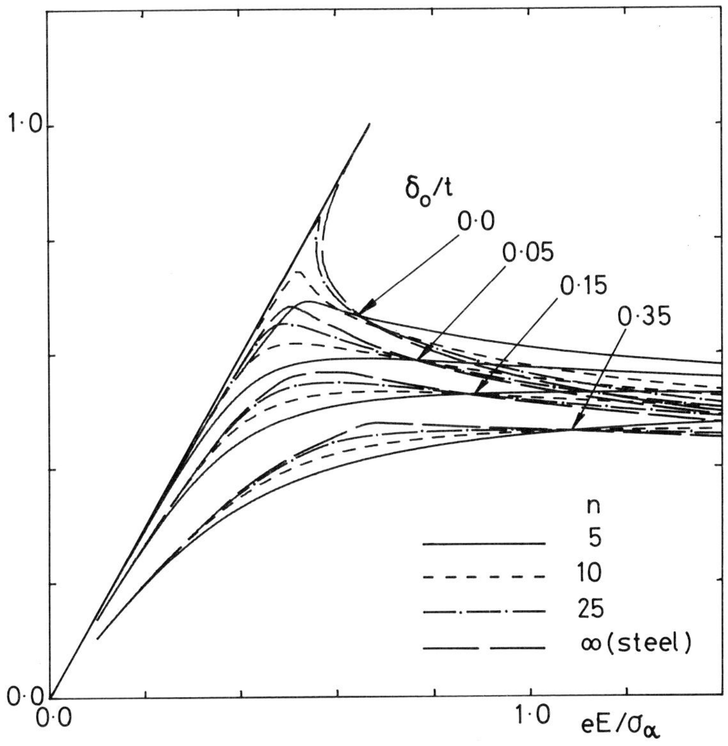

Fig. 5 Normalised in-plane load-radial edge shortening curves for $\beta = 1.0$. Each δ_o/t value indicates a group of four curves, all of that δ_o/t value, but of different n values as given by the key.

prescribed-e value reaches a level corresponding to an edge stress slightly greater than σ^P_{cr}. At this stage the equilibrium state remains slightly unflat, with a very small region of material unloading from yield, the remainder continuing to load. In other words, buckling has occurred at σ^P_{cr}, providing further evidence in support of the Shanley concept. Buckling occurs closely at σ^P_{cr} for all three n values.

As loading proceeds into the post-buckling range, the region of material unloading from yield spreads at a rate which increases as n increases. Indeed at n=25, this rate is so rapid that there is a catastrophic snap-through buckling effect to a lower load even with e prescribed. In the extreme case ($n=\infty$), the steel plate unloads throughout its volume at the instant of buckling, and even if it were held in a perfectly rigid testing machine, σ_{av} would suddenly snap from σ_α to $0.65\,\sigma_\alpha$ at a radial shortening given by $eE/\sigma_\alpha = 1 - \nu = 0.67$. This type of behaviour is a consequence of the plate slenderness β being such that the elastic buckling stress is equal to the yield stress of the steel, and has been described by the author with reference to steel columns in previous work [4], where it is an even larger effect. In order to follow the complete equilibrium path accurately in the analysis, it is necessary to make use of both prescribed-σ_{av} **and** prescribed-e loading types in a manner similar to that described in ref. 4. The present results confirm that similar behaviour is to be expected of aluminium plates with a sufficiently sharp "knee" on the MSSC (i.e. $n \gtrsim 25$ approx.) and proportioned such that $\beta \simeq 1.0$. For a more rounded MSSC, no such catastrophic post-buckling behaviour occurs.

Behaviour of Initially Imperfect Plates

As is evident from Fig. 5, the effect of the imperfection amplitude δ_o/t on the in-plane behaviour of circular plates is basically similar to the effect of imperfection size on columns and rectangular plates [4, 5]. An increase in δ_o/t causes a decrease in both strength and initial stiffness, for all n values. This weakening effect is very marked in Fig. 5 since the plate slenderness ($\beta = 1.0$) is that of greatest imperfection-sensitivity. The imperfection-sensitivity is greatest also at low δ_o/t, so that, for example, even the curves for $\delta_o/t = 0.05$ show maximum loads considerably lower than the curves for initially flat plates, and at high n, i.e. the more rounded the MSSC, the less the effect of δ_o.

The effect of the n value on the LS curves is shown to be

broadly similar to its effect on the MSSCs in Fig. 2. Thus in the initial stages of loading the curves for a high n value lie above those for low n; however, for e larger than a certain value, this situation reverses. It is perhaps strange that this reversal does seem to occur at a point, all four curves for a given δ_o value apparently intersecting at a well-defined point (which varies with δ_o), except for the initially flat plates. The author cannot see an obvious explanation for this, although the same thing was predicted for rectangular plates [5].

For most plates considered, the maximum load is reached in the region of $eE/\sigma_\alpha = 0.6$. However, for plates with large imperfections and low n values, the LS curves are still rising at $eE/\sigma_\alpha = 1.4$, the maximum value considered.

CONCLUSIONS

Numerical results for the collapse analysis of thin, simply supported, circular plates, of slenderness $\beta = 1.0$, gave rise to the following conclusions:

For lateral pressure loading:

1. The main effect of a positive imperfection δ_o is to increase plate stiffness, due to membrane action, in the initial stages of loading.
2. A negative imperfection of magnitude $\delta_o/t = -1.0$ is large enough to result in a stationary maximum load point at negative w_{max}, at which, at constant load, the plate will snap through to the rising load path at positive w_{max}. For such a plate, if the complete equilibrium path is followed, a large amount of reversed stressing occurs, and accurate analysis requires a kinematic, rather than isotropic, hardening theory.
3. The proposed kinematic hardening formulation is simple to apply, has a rational basis, and has so far given reasonable results.
4. The effect of n, the "knee" factor of the MSSC, is greatest at large displacements, when the lower the value of n (the more rounded the knee), the stiffer the plate.

For in-plane compression:

5. The results provide further confirmation that flat plates buckle at the load predicted using the Shanley concept.
6. Initially flat plates of large n show snap-through post-buckling behaviour even if loaded by a rigid testing machine.
7. Imperfection-sensitivity is greatest at low δ_o/t and high n.

REFERENCES

1. Timoshenko, S.P.; Woinowsky-Krieger, S.: Theory of plates and shells, 2nd. edn. New York: McGraw Hill 1959.

2. Turvey, G.J.; Lim, G.T.: Axisymmetric full-range analysis of transverse pressure-loaded circular plates. Int.J.Mech.Sci. **26** (1984) 489-502.

3. Calladine, C.R.; Simple ideas in the large-deflection plastic theory of plates and slabs. Engineering Plasticity (Ed. Heyman, J. and Leckie, F.A.). Cambridge: Cambridge University Press 1968, 93-127.

4. Little, G.H.; Complete collapse analysis of steel columns. Int.J.Mech.Sci. **24** (1982) 279-298.

5. Little, G.H.; Collapse analysis of plates with strain hardening. Int.J.Mech.Sci. **23** (1981) 561-576.

6. Ziegler, H.: A modification of Prager's hardening rule. Quart.App.Math. **17** (1959) 55-65.

7. Prager, W.: A new method of analyzing stresses and strains in work-hardening plastic solids. J.App.Mech. **23** (1956) 493-496.

8. Pifko, A.; Levine, H.S.; Armen, H.: PLANS - A finite element program for nonlinear analysis of structures; Vol.I - Theoretical Manual. NASA CR-2568 1975.

9. Winter, R.; Levine, H.S.; Nonlinear behavior of circular plates with work hardening. Exp.Mech. **18** (1978) 281-291.

10. Gjelsvik, A.; Bodner, S.R.: Nonsymmetrical snap buckling of spherical caps. J.Engng.Mech.Div. A.S.C.E **88** EM5 (1962) 135-165.

11. Needleman, A.: Postbifurcation behavior and imperfection sensitivity of elastic-plastic circular plates. Int.J.Mech.Sci. **17** (1975) 1-13.

Shakedown Analysis of Thin Tube Under Cyclic Loading Cases

O. MAHRENHOLTZ

Arbeitsbereich Meerestechnik II/Strukturmechanik, TU Hamburg-Harburg, FR Germany

K. LEERS

Institut für Mechanik, Universität Hannover, FR Germany

J. A. KÖNIG

Institute of Fundamental Technological Research, Warsaw, Poland

Summary

Tubular elements are oft-used in various structural systems. To evaluate the safety of those structural elements under cyclic loading cases and, in particular, in the presence of variable temperature fields, the shakedown analysis should be employed rather than plastic limit analysis. A shakedown procedure for axisymmetric cylindrical shells is outlined in connection with a discussion of experimental investigations. The specimens were exposed to cyclic load and temperature variations by means of a specially designed set-up with automatic load and temperature variation control and data recording. Then, a proposal is presented for a more refined analysis, accounting for strain-induced geometry changes as well as for material strainhardening. It seems to be able to explain some peculiarities observed in the experiments.

1. Remarks on Structural Safety Assessment

Any engineering structure has to satisfy safety conditions. They were usually based on an elastic brittle material model. However, the majority of metals exhibit non-negligible inelastic deformations without immediate material failure. Therefore, the further development of structural mechanics has tried to account

- for immediate inelastic (plastic) deformations. Plastic yielding results in redistribution of the stress field making it more homogeneous and, therefore, allows for a more economic design. However, in the course of such a redistribution, structural deformations may happen to enlarge significantly.

- for irreversible (viscous) strains which grow up with time even under constant stress. They show up at higher stresses and - especially - at more elevated temperatures, though, by means of more precise measurements, they can be detected even at low temperatures and low stresses. Occurence of them (creep) may lead, after a sufficiently long time,

Inelastic Behaviour of Plates and Shells
IUTAM Symposium Rio de Janeiro 1985
Editors: L. Bevilacqua, R. Feijóo and R. Valid

to excessive structural deformations and, eventually, to failure. On the other hand, viscous deformations may cause a relaxation of higher stressed elements, thus also contributing to the stress redistribution.

Plastic limit analysis allows to determine the maximum load which can be carried by the structure. However, this approach may fail to give a proper answer if the loads acting upon the structure vary in a non-monotonous way within too wide limits. A better approach in this case, especially if there are also temperature variations, is the shakedown analysis which allows to detect a possible accumulation of plastic deformations as well as low-cycle material fatigue.

All the above general remarks were made in a deterministic context. It is obvious that a probabilistic approach would be more relevant. However, the existing theories accounting for the random character of loads, structural dimensions, material strength etc. do not account for peculiarities of the inelastic structural analysis.

2. The Structural Behaviour in the Case of Cyclic Influences

In the presence of inelastic deformations the stress-strain dependence is no more a one-to-one relation and, therefore, response of a given structure to cyclic loading or temperature changes may vary from cycle to cycle. Moreover, presence of a temperature field

- o alters the stress field within a structure because of appearance of additional thermal stresses;
- o influences the material properties such like its viscosity (creep strain rates will be higher), the yield stress (drops down with temperature, usually), or elastic moduli.

Moreover, cyclic changes of temperature contribute to material fatigue analogously like stress variations do.

For a broad spectrum of materials and structures acted upon by cyclic loads and temperature fields resulting in inelastic deformations, the following pattern of structural response can be observed, see Fig. 1 (deformations at the end of each cycle are marked, only):

a) First, we have a transition period. The deformation increments in a single load/temperature cycle are high but they diminish fast with the number of cycles.

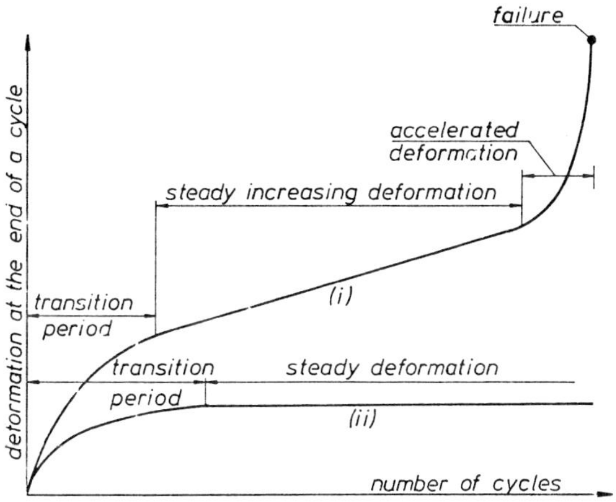

Fig. 1. Structural response to cyclic external actions
(i) Non-vanishing deformation increments
(ii) Stabilisation of deformations

b) Then, the deformation increments in subsequent cycles become nearly constant and the stress state becomes a cyclic function of time, with the same period as of the external actions, curve (i). Some authors call the particular case when the strain increment per cycle vanishes, curve (ii), adaptation or shakedown though this term has been employed in the literature only for the case when inelastic strain variations cease. The process of a linear deformation accumulation (curve (i) in Fig. 1) is called ratcheting or incremental collapse.

c) In the case of non-vanishing deformation increments, after a certain (rather high) number of cycles, the deformation process accelerates. This may take place due to progressive material degradation (appearance of defects, voids, pores etc.) but may be, also, due to geometric effects which begin to be non-negligible at higher deformations.

The above-sketched structural behaviour pattern was observed, e.g. in our experimental investigations, cf. [1, 2, 3], performed on aluminium specimens exposed to cycles of mechanical loads as well as of temperature field, Fig. 2. These investigations were aimed to check if the shakedown theory, confirmed experimentally in the case of mechanical loads, holds also in the presence of simultaneous temperature field variations. If stabilisation

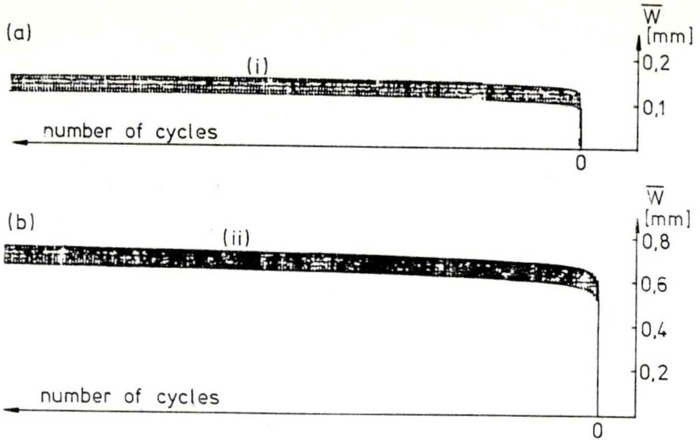

Fig. 2. Experimental data of tests performed on tubular specimens subjected to cycles of mechanical loads (internal pressure, local pressure) and of the internal temperature: (i) stabilisation of deformation
(ii) ratcheting

of the deformations was attained for certain load and temperature amplitudes e.g. like in curve (i) of Fig. 2, this means that the corresponding loads and temperatures belong to the shakedown range.

In further tests, their magnitudes were enlarged to attain a clearly non-zero deformation increment per cycle. In such a way, the border between shakedown and ratcheting ranges could be approximated by bounding it from below and from above. An example of such a shakedown surface is given in Fig. 3. A fairly good agreement of the theory and of the experimental data can be seen.

3. The Classical Shakedown Analysis of Axisymmetric Cylindrical Shells

In the case of axial symmetry, usually, there is no need to employ more sophisticated numerical methods since reasonable approximate solutions can be obtained analytically or semi-analytically. Moreover, one should keep in mind that the shakedown analysis is based on rather rough assumptions and, therefore, precise computations do not necessarily mean a better description of the actual structural response. The assumptions are as follows:

o deformations are small,

o there is no material strainhardening,

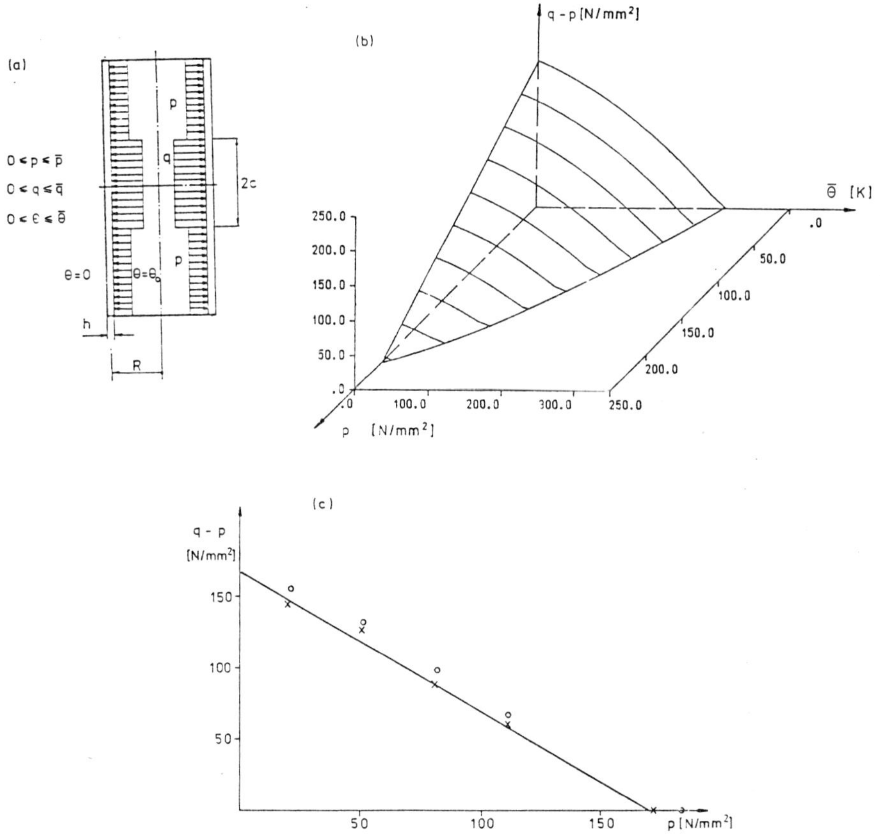

Fig. 3. Shakedown surface for a tubular specimen
(a) scheme of the test
(b) theoretical shakedown surface
(c) a cross-section for $\overline{\Theta}$ = 100° C
—— theory
x experimental points at which shakedown has been observed
o experimental points at which ratcheting has appeard

- o the elastic moduli as well as the thermal expansion coefficient do not vary with temperature,
- o there are no viscous strains i.e. no creep.

To determine the load and temperature variation limits ensuring safety against plastic ratcheting within the framework of the above-listed assumptions, one can employ the formula resulting from the kinematic shakedown theorem of Koiter [4, 5]. The original Koiter's theorem, in the case of thermal effects, reads (the formulation is given for the case of axially symmetric cylindrical shells):

A given structure is safe against inadaption in a mechanism defined by a kinematically admissible field of displacement increments Δw, Δu over a certain time period $[0,T]$ if

- o for any load/temperature cycle, admitted by their limits and resulting in thermoelastic stresses $\sigma_\varphi^E(z,x,t)$, $\sigma_x^E(z,x,t)$, and
- o for any cycle of plastic strain rates $\dot{\bar{\varepsilon}}_\varphi(z,x,t)$, $\dot{\bar{\varepsilon}}_x(z,x,t)$ - independent of the stress cycle - such that

$$\Delta \bar{\varepsilon}_\varphi = \int_0^T \dot{\bar{\varepsilon}}_\varphi \, dt = \frac{\Delta w}{R} \quad , \quad \Delta \bar{\varepsilon}_x = \int_0^T \dot{\bar{\varepsilon}}_x \, dt = \frac{d}{dx}(\Delta u) - z \frac{d^2}{dx^2}(\Delta w) \quad , \tag{3.1}$$

the following inequality holds:

$$\int_0^T \int_L \int_{-h/2}^{h/2} (\sigma_\varphi^E \dot{\bar{\varepsilon}}_\varphi + \sigma_x^E \dot{\bar{\varepsilon}}_x) \, dz \, dx \, dt \leq \int_0^T \int_L \int_{-h/2}^{h/2} D(\dot{\bar{\varepsilon}}_\varphi, \dot{\bar{\varepsilon}}_x, \Theta) \, dz \, dx \, dt \, . \tag{3.2}$$

Here: w and u are radial and axial displacements of the shell middle surface, z is the coordinate along the shell radius measured from that surface, x is the coordinate along the shell axis, t denotes time, R is the mean shell radius, h is its wall thickness, $\Theta(z,x,t)$ describes the temperature field and L symbolises the shell length.

$D(\dot{\bar{\varepsilon}}_\varphi, \dot{\bar{\varepsilon}}_x, \Theta)$ denotes plastic energy dissipation depending on the form of yield condition and on temperature. We assume that the yield stress Y varies linearly with temperature,

$$Y(\Theta) = Y_o(1-e\Theta) \quad . \tag{3.3}$$

Then, the dissipation does this also:

$$D(\dot{\bar{\varepsilon}}_\varphi, \dot{\bar{\varepsilon}}_x, \Theta) = D_o(\dot{\bar{\varepsilon}}_\varphi, \dot{\bar{\varepsilon}}_x) \cdot (1-e\Theta) \quad , \tag{3.4}$$

where e is a material constant.

The formulae (3.1) mean that the plastic strain rates result in kinematically admissible displacement increments whereas (3.2) states that the work done by stress field σ_φ^E, σ_x^E on the above-said strain rates over the time interval $[0,T]$ is not higher than the plastic energy dissipated with these strain rates and integrated over the same time period.

Occurence of alternating plasticity can be analysed in terms, solely, of thermoelastic stress. Therefore, let us transform the formula (3.2) to a form appropriate for incremental collapse, exclusively. In this case, the plastic strain rates are proportional to the total increments:

$$\dot{\bar{\varepsilon}}_\varphi(z,x,t) = \dot{\Lambda}(z,x,t)\,\frac{\Delta w(x)}{R}\;, \qquad \dot{\bar{\varepsilon}}_x(z,x,t) = \dot{\Lambda}(z,x,t)\Big[\frac{d}{dx}\,\Delta u(x) - z\,\frac{d^2}{dx^2}\,\Delta w(x)\Big]\;,$$

$$\text{where} \quad \Lambda(z,x,0) = 0, \quad \dot{\Lambda}(z,x,t) \geqq 0, \quad \Lambda(z,x,T) = 1 \;. \tag{3.5}$$

The result of substituting of (3.5) into (3.2), in view of (3.4), is as below

$$\int_0^T \int_L \int_{-h/2}^{h/2} [\sigma_\varphi^E(z,x,t)\,\Delta\bar{\varepsilon}_\varphi(z,x) + \sigma_x^E(z,x,t)\,\Delta\bar{\varepsilon}_x(z,x) + e\Theta(z,x,t)\,D_o(\Delta\bar{\varepsilon}_\varphi,\Delta\bar{\varepsilon}_x)]\cdot$$

$$\cdot\,\dot{\Lambda}(z,x,t)\;dz\;dx\;dt \leqq \int_L \int_{-h/2}^{h/2} D_o(\Delta\bar{\varepsilon}_\varphi,\Delta\bar{\varepsilon}_x)\;dz\;dx \;. \tag{3.6}$$

The most stringent condition follows if the left-hand side of (3.6) assumes its maximum. This takes place when $\dot{\Lambda}(z,x,t)$ is Dirac impulse function, concentrated for each material point (z,x) at that instant $t_o(z,x)$ at which the expression in brackets assumes its maximum:

$$\dot{\Lambda}(z,x,t) = \delta[t - t_o(z,x)] \tag{3.7}$$

and

$$\sigma_\varphi^E(z,x,t_o)\Delta\bar{\varepsilon}_\varphi(z,x) + \sigma_x^E(z,x,t_o)\Delta\bar{\varepsilon}_x(z,x) + e\Theta(z,x,t_o)D_o(\Delta\bar{\varepsilon}_\varphi,\Delta\bar{\varepsilon}_x) = \max \;. \tag{3.8}$$

In practice a given structure is usually subjected to m independent load and temperature agents and the thermoelastic stresses as well as the temperature fields associated with them are linear functions of parameters $\beta_k(t)$, $k=1, \ldots, m$, characterising intensities of the agents:

$$\sigma_\varphi^E(z,x,t) = \sum_{k=1}^{m} \beta_k(t)\sigma_\varphi^{Ek}(z,x)\;, \qquad \sigma_x^E(z,x,t) = \sum_{k=1}^{m} \beta_k(t)\sigma_x^{Ek}(z,x)\;,$$

$$\Theta(z,x,t) = \sum_{k=1}^{m} \beta_k(t) \cdot \Theta^k(z,x) \quad , \tag{3.9}$$

whereas the limits of variation of these parameters are independent of each other

$$a_k \leqq \beta_k(t) \leqq b_k \quad , \qquad k = 1, \ldots, m \tag{3.10}$$

a_k, b_k being constants.

In such a case the maximum magnitude (3.8) is equal to

$$\sum_{k=1}^{m} \alpha_k(z,x)[\sigma_\varphi^{Ek}(z,x)\Delta\bar{\varepsilon}_\varphi(z,x) + \sigma_x^{Ek}(z,x)\,\Delta\bar{\varepsilon}_x(z,x) + e\,\Theta^k(z,x)\,D_o(\Delta\bar{\varepsilon}_\varphi,\Delta\bar{\varepsilon}_x)] =$$

$$= \sum_{k=1}^{m} \alpha_k(z,x)\cdot J_k(z,x)$$

where

$$\alpha_k(z,x) = \begin{cases} b_k & \text{if} \quad J_k(z,x) > 0 \\ a_k & \text{if} \quad J_k(z,x) < 0 \end{cases} \quad , \tag{3.11}$$

cf. [4].

Thus, when employing the above result to incremental analysis of a particular cylindrical shell, the following stages are to be performed:

a) Thermoelastic stress fields $\sigma_\varphi^{Ek}(z,x)$, $\sigma_x^{Ek}(z,x)$ are to be determined for each one of the external agents which are going to vary. For those remaining constant this is not needed.

b) Some expectable mechanisms of incremental collapse are to be selected. Practically, there is no purpose to look for very complicated ones. They should be of simple form, rather, but should contain some free parameters allowing for a further optimisation of the safety bounds.

c) For an assumed mechanism $\Delta w(z,x)$, $\Delta u(z,x)$, strain increments (3.1), the dissipation $D_o(\Delta\bar{\varepsilon}_\varphi, \Delta\bar{\varepsilon}_x)$ as well as the expressions $J_k(z,x)$ and $\alpha_k(z,x)$ are to be calculated.

d) All the above-calculated terms are to be plugged into the safety criterion resulting from (3.6). It reads,

$$\int_L [\bar{p}_r(x)\ \Delta w(x) + \bar{p}_x(x)\ \Delta u(x)]dx + \sum_{k=1}^{m} \int_L \int_{-h/2}^{h/2} \alpha_k(z,x)\ J_k(z,x)\ dz\ dx \leqq$$

$$\leqq \int_L \int_{-h/2}^{h/2} D_o(\Delta\bar{\varepsilon}_\varphi, \Delta\bar{\varepsilon}_x)\ dz\ dx \quad , \qquad (3.12)$$

where $\bar{p}_r(x)$, $\bar{p}_x(x)$ denote constant loads, the radial and axial ones, respectively. Difficulties may appear with integration over the shell volume as well as for determining the ranges of constant sign of the J_k-expressions.

e) The final result (3.12) is to be optimised with respect to the free parameters of the mechanism of incremental collapse.

f) Steps c), d), e) are to be repeated for other mechanisms as to assess the lowest upper bound to the actual safety condition against incremental collapse.

Let us notice that in the case of constant loads only and no thermal effects, the formula (3.12) becomes identical with that appropriate for the kinematic method of the plastic limit analysis.

To illustrate the procedure, let us consider the case depicted in Fig. 3a but for the constant local pressure $q = \bar{q}$ = const. There are two load/ temperature parameters p and Θ_o varying within the limits

$$0 \leqq p \leqq \bar{p} \quad , \qquad 0 \leqq \Theta_o \leqq \bar{\Theta} \quad . \qquad (3.13)$$

The corresponding unit thermoelastic stresses and temperature fields are

$$\sigma_\varphi^{Ep} = \frac{R}{h} \quad , \qquad \sigma_x^{Ep} = 0 \quad , \qquad \Theta^p = 0 \quad ,$$

$$\sigma_\varphi^{E\Theta} = \sigma_x^{E\Theta} = \frac{E\alpha z}{(1-\nu)h} \quad , \qquad \Theta^\Theta = \frac{1}{2}(1-\frac{2z}{h}) \qquad (3.14)$$

where E is Young modulus, ν is Poisson ratio, α is linear thermal expansion coefficient and, due to $h<<R$, the temperature has been assumed to vary linearly across the shell wall thickness.

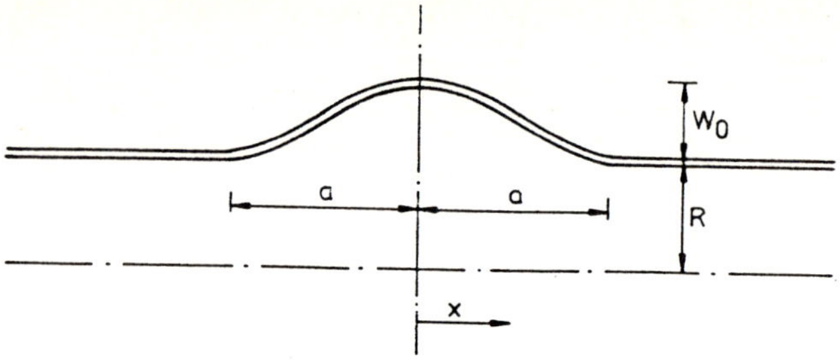

Fig. 4. An incremental collapse mechanism for the shell of Fig.3a

Let us investigate the incremental collapse mechanism given in Fig. 4,

$$\Delta\bar{w}(x) = \frac{w_o}{2}\left(1 + \cos\frac{\pi x}{a}\right) , \qquad (3.15)$$

containing the free parameter a . The associated strain increments are

$$\bar{\varepsilon}_\varphi = \frac{w_o}{2R}\left(1 + \cos\frac{\pi x}{a}\right) , \qquad \bar{\varepsilon}_x = \frac{w_o \pi^2}{2a^2}\cos\frac{\pi x}{a}\cdot z . \qquad (3.16)$$

Let us employ the Huber-Mises yield condition. The corresponding dissipation is (for plane stress)

$$D_o = \frac{2Y}{\sqrt{3}}\sqrt{\bar{\varepsilon}_\varphi^2 + \bar{\varepsilon}_\varphi\bar{\varepsilon}_x + \bar{\varepsilon}_x^2} = \frac{Yw_o}{R\sqrt{3}}\left[\left(1+\cos\frac{\pi x}{a}\right)^2 + \frac{zR}{a^2}\pi^2\cos\frac{\pi x}{a} + \frac{z^2R^2}{a^4}\pi^4\cos^2\frac{\pi x}{a}\right]^{1/2}$$
$$(3.17)$$

where Y is the yield stress at pure tension.

The expressions defined by (3.11) are

$$J_p(z,x) = \sigma_\varphi^{Ep}\bar{\varepsilon}_\varphi + \sigma_x^{Ep}\bar{\varepsilon}_x = \frac{w_o}{2h}\left(1+\cos\frac{\pi x}{a}\right)$$

$$J_\Theta(z,x) = \sigma_\varphi^{E\Theta}\bar{\varepsilon}_\varphi + \sigma_x^{E\Theta}\bar{\varepsilon}_x + e\,\Theta^\Theta D_o =$$

$$= \frac{E\alpha w_o}{2Rh(1-\nu)}\, z\left(1+\cos\frac{\pi x}{a}\right) + \frac{E\alpha w_o}{2a^2h(1-\nu)}\pi^2 z^2\cos\frac{\pi x}{a} + \frac{e}{2}\left(1-\frac{2z}{h}\right)\frac{Yw_o}{R\sqrt{3}}\left[\left(1+\right.\right.$$

$$\left.\left.+\cos\frac{\pi x}{a}\right)^2 + \frac{zR}{a^2}\pi^2\cos\frac{\pi x}{a} + \frac{z^2R^2}{a^4}\pi^4\cos^2\frac{\pi x}{a}\right]^{1/2} . \qquad (3.18)$$

One can see that $J_\rho \geqq 0$ everywhere for $|x| \leqq a$ whereas the regions of positive and negative J_Θ have to be specially determined. When integrating the dissipation (3.17) one had to make a certain approximation but with a very small error. This allowed to arrive at the following closed form result (given here only for $e=0$):

$$\bar{p} + \bar{q}\left(\frac{c}{a} + \frac{1}{\pi}\sin\frac{\pi c}{a}\right) + \frac{E\alpha\bar{\Theta}h}{4R(1-\nu)}\left[1 + \frac{2\pi h}{3a^2}\right] = \frac{2Y_o h}{\sqrt{3}\,R} \qquad (3.19)$$

where c denotes the length of the region where the local pressure is applied, see Fig. 3a.

The final optimisation with respect to the free parameter a had to be performed numerically.

4. A Proposal of Non-Classical Shakedown Analysis

It can be seen in Fig. 3c that the discrepances between the experimental data and the results of theoretical analysis are not very dramatic. Nevertheless, they exist and one should try to explain them. In our opinion, the main reasons of the differences are

- o the influence of strainhardening,
- o the geometry changes.

What concerns the former, there exist shakedown theorems, cf. e.g. [7] accounting for linear strainhardening. In such a formulation (with positive definite hardening matrix) the occurence of incremental collapse is principally excluded. Therefore, they seem to be inapplicable for our purpose.

What concerns the latter effect, nothing like a geometrically non-linear shakedown theory has been worked out. Some cases of geometric effects were analysed by Davies [6] and Maier [7] but their approach would not suffice.

In our opinion, the most promising is the proposal formulated first in [8]. Its idea consists of investigating of the current shakedown load and temperature bounds at every stage of the deformation process. If one neglects the initial transition period (cf. Fig. 1 and 2) then the further structural geometry changes are only due to a kinematically admissible mechanism which can be determined by the procedure outlined in section 3. In such a way the current geometry changes are defined by a parameter characterising the mechanism. Thus, the changes of thermoelastic stresses σ_φ^{Ek}, σ_x^{Ek} associated with various load and temperature agents can be evaluated. Moreover, also the strainhardening influence on the yield stress can be accounted for. Such an approach holds under the condition that the structural geometry change in a single cycle is small. But this requirement is usually fulfilled (cf. Fig. 2).

In the proposed procedure one employs the formula (3.12) at each stage of the deformation process but the elastic stresses as well as the plastic dissipation are to be adjusted according to the geometry changes and to the resulting strainhardening.

Let us notice that the bounds resulting from the above described analysis can be broader or narrower than for the virgin, undeformed structure. The latter case deserves a special attention as geometry changes make the results of the geometrically linear analysis invalid and unsafe.

To illustrate the general considerations let us regard the axisymmetric shell of Fig. 3a if acted upon only by variable repeated internal pressure p and internal temperature Θ_o, varying within the limits (3.13). The only reasonable shell geometry change consists of increasing of its radius from R to $R+\Delta R$, of decreasing of the wall thickness from h to $h-\Delta h$ and of the shell length from L to $L-\Delta L$. Moreover, the plastic incompressibility yields

$$\frac{\Delta R}{R} = \frac{\Delta h}{h} + \frac{\Delta L}{L} \quad . \tag{4.1}$$

Such a deformation pattern would result from the following incremental collapse mechanism:

$$\Delta\bar{w}(z,x) = w_o = \text{const} \quad , \quad \Delta\bar{u}(z,x) = u_o\frac{x}{L} \quad ,$$

thus

$$\bar{\varepsilon}_\varphi = \frac{w_o}{R} \quad , \quad \bar{\varepsilon}_x = \frac{u_o}{L} \quad , \quad \bar{\varepsilon}_r = -\bar{\varepsilon}_\varphi - \bar{\varepsilon}_x = -\frac{w_o}{R} - \frac{u_o}{L} \quad . \tag{4.2}$$

We shall employ the Huber-Mises yield condition but, for the sake of simplicity, neglecting the temperature dependence of the yield stress i.e. assuming $e=0$ in formula (3.3). We account, however, for the increase of the latter with deformations following the isotropic hardening rule,

$$Y = Y_o(1 + \beta \cdot \varkappa) \quad , \tag{4.3}$$

where

$$\varkappa = \frac{2}{\sqrt{3}} \sqrt{\bar{\varepsilon}_\varphi^2 + \bar{\varepsilon}_x^2 + \bar{\varepsilon}_r^2} = \frac{2}{\sqrt{3}} \sqrt{\bar{\varepsilon}_\varphi^2 + \bar{\varepsilon}_\varphi \bar{\varepsilon}_x + \bar{\varepsilon}_x^2} \tag{4.4}$$

is the Odqvist parameter integrated with respect to time and β is the hardening modulus. One can easily calculate that

$$\varkappa = \frac{2}{\sqrt{3}} \frac{w_o}{R} \sqrt{1-\mu+\mu^2} \quad \text{where} \quad \mu = -\frac{u_o}{w_o}\frac{R}{L} \quad . \tag{4.5}$$

Employing the thermoelastic solutions (3.14) we have

$$J_p = \frac{w_o}{R} \quad , \quad J_\Theta = \frac{E\alpha z\, w_o}{Rh(1-\nu)}(1-\mu) \quad , \tag{4.6}$$

thus $J_p > 0$ everywhere and $J_\Theta > 0$ for $z > 0$, provided $\mu < 1$. Moreover, the dissipation is

$$D_o = \frac{2Y}{\sqrt{3}} \sqrt{\bar{\varepsilon}_\varphi^2 + \bar{\varepsilon}_\varphi \bar{\varepsilon}_x + \bar{\varepsilon}_x^2} = \frac{2Y}{\sqrt{3}} \frac{w_o}{R} \sqrt{1-\mu+\mu^2} \quad . \tag{4.7}$$

In view of (4.2), (4.6) and (4.7), the classical shakedown approach (geometrically linear, with no strainhardening) would give us the following condition of safety with respect to incremental collapse

$$\bar{p} + \frac{E\alpha\bar{\Theta}h}{4R(1-\nu)}(1-\mu) = \frac{2\,Y_o h}{R\sqrt{3}} \sqrt{1-\mu+\mu^2} \tag{4.8}$$

which has to be optimised with respect to the parameter μ . This optimisation gives:

$$\mu_{opt} = \frac{1}{2 + \frac{E\alpha h}{4R(1-\nu)} \cdot \frac{\bar{\Theta}}{\bar{p}}} \quad . \tag{4.9}$$

Now, let us repeat this incremental collapse analysis but for a deformed shell of radius $R + \Delta R = R(1+\frac{w_o}{R})$, of wall thickness $h - \Delta h = h[1-\frac{w_o}{R}(1-\mu)]$ and of length $L - \Delta L = L(1-\mu\frac{w_o}{R})$ and with the current yield stress defined by (4.3) and (4.5). This would give

$$\bar{p} + \frac{E\alpha\bar{\Theta}(1-\mu)}{4(1-\nu)} \cdot \frac{h[1-\frac{w_o}{R}(1-\mu)]}{R(1+\frac{w_o}{R})} = \frac{2Y_o\sqrt{1-\mu+\mu^2}}{\sqrt{3}} \left(1+\beta\frac{2}{\sqrt{3}}\frac{w_o}{R}\sqrt{1-\mu+\mu^2}\right) \frac{h[1-\frac{w_o}{R}(1-\mu)]}{R(1+\frac{w_o}{R})} . \tag{4.10}$$

It can be seen that for a constant temperature amplitude $\bar{\Theta}$ and sufficiently high strainhardening modulus β, the internal pressure amplitude $\bar{p}$, in the course of deformation development i.e. for increasing $\Delta R = w_o$, would first grow up then attain a maximum at a certain $\Delta R = \Delta R_{opt}$ and decrease, afterwards.

Existence of such a maximum implies that an incremental collapse process going on at a certain constant amplitude $\bar{p}$ would have a tendency to slow down at lower deflections $\Delta R < \Delta R_{opt}$ and to accelerate for $\Delta R > \Delta R_{opt}$.

The formula (4.10) could be analysed, in the same way, also in the case of other hardening laws e.g. if β were a function of $\varkappa$. One could put here even the actual hardening function taken directly from a material test. Also, at least in principle, one could account for the effect of cyclic hardening but this can be done only in a semi-empirical way since information on the number of cycles is needed in this case.

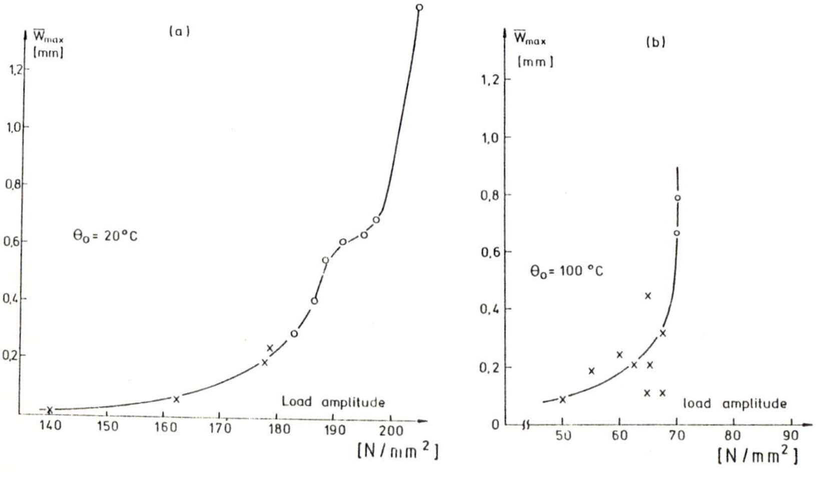

Fig. 5. Maximum radial deflection as function of the load amplitude and temperature. x shakedown, o incremental collapse

(a) tests at room temperature (b) tests at elevated temperature

The above outlined effect can explain some peculiarities observed in our experimental investigations, cf. Fig. 5a. Namely, the $\bar{w}_{max}$-load amplitude curve possesses a turning point in the vicinity of the shakedown load. The general shape of that curve fits well into the theoretical upper bound to shakedown deflections for lower load amplitudes. At the shakedown load the theoretical upper bound loses its physical sense - the maximum theoretical deflection tends to infinity. The actual one grows up, also, and the strainhardening effect begins to play a more important role. Because of its influence the deflection rate per cycle falls down, temporarily and, then, begins to increase again. In this way the turning point on the $\bar{w}_{max}$-load amplitude curve appears. At more elevated temperatures this turning point was not clearly visible or just did not occur, Fig. 5b. This is understandable because the hardening modulus β falls with temperature and, therefore, over a certain critical temperature, the whole effect vanishes.

5. Remarks on Experimental Shakedown Investigations

The theory of shakedown seems to be sufficiently confirmed by experimental investigations for structures subjected to cycles of mechanical loads. However, there are only few reports on such tests made in the presence of the temperature field variations. Our experience from investigations, cf. [1, 2, 3], performed at the Institut für Mechanik (Universität Hannover, Hannover, Federal Republic of Germany) suggest the following directions:

1° Shakedown experiments intended to include thermal field variations require a special set-up, properly designed. It should enable to perform high numbers (at least a few thousand) of load and temperature cycles. Therefore, the stearing of the load and temperature variations, must be possible to be preprogrammed and, then, executed automatically. Also, the data recording should be automatised.

2° The appearance for creep can be seen under higher stress intensities even at room temperature if the deformations are recorded with a sufficient precision. At higher temperatures, this influence becomes more pronounced. Therefore, in the course of a shakedown test, even if plastic strains tend to stabilise, the total permanent deformation increments still do not cease, usually, and some rules enabling to recognise a particular result as shakedown or inadaptation must be adopted.

3° Controlling of the creep deformation may require high-precision deflection measurements. Therefore, we have used inductive gauges to record the radial displacements.

4° Tests to be made in conditions of axial symmetry require a very precise machining of specimens as to avoid any imperfections of their shape. The latter would distort the results by causing a non-symmetric deformation pattern development.

References

1. Mahrenholtz, O.; Leers, K.; König, J.A.: Shakedown of Tubes: a Theoretical Analysis and Experimental Investigations. Metal Forming and Impact Mechanics. Commemorative volume in honour of Prof. W. Johnson. Prof. S. R. Reid editor, Dept. Eng., University of Aberdeen, 1984.

2. Leers, K.; Klie, W.; König, J.A.; Mahrenholtz, O.: Experimental Investigations on Shakedown of Tubes. Plasticity Today, A. Sawczuk & G. Bianchi editors, Elsevier Applied Science Publishers Ltd., Amsterdam 1984, 259-274.

3. Leers, K.: Experimentelle und theoretische Shakedownuntersuchung an Rohren. Ph.D.thesis, University of Hanover, 1985.

4. König, J.A.: On the Incremental Collapse Criterion Accounting for Temperature Dependence of Yield Point Stress. Arch. Mech., 31 (1979) 317-325.

5. Koiter, W.T.: A New General Theorem on Shakedown of Elastic-Plastic Structures. Koninkl. Ned. Ak. Wett. B 59 (1956) 24-34.

6. Davies, J.M.: Collapse and Shakedown Loads of Plane Frames. J. Struct. Div. 93 (1969) 35-50.

7. Maier, G.: A Shakedown Matrix Theory Allowing for Workhardening and Second-Order Geometric Effects. Symp. Foundations of Plasticity, A. Sawczuk editor, Noordhoff Int. Publishers, Leyden 1973, 417-433.

8. König, J.A.: On Stability of the Incremental Collapse Process. Arch. Inz. Lad. 26 (1980) 219-229.

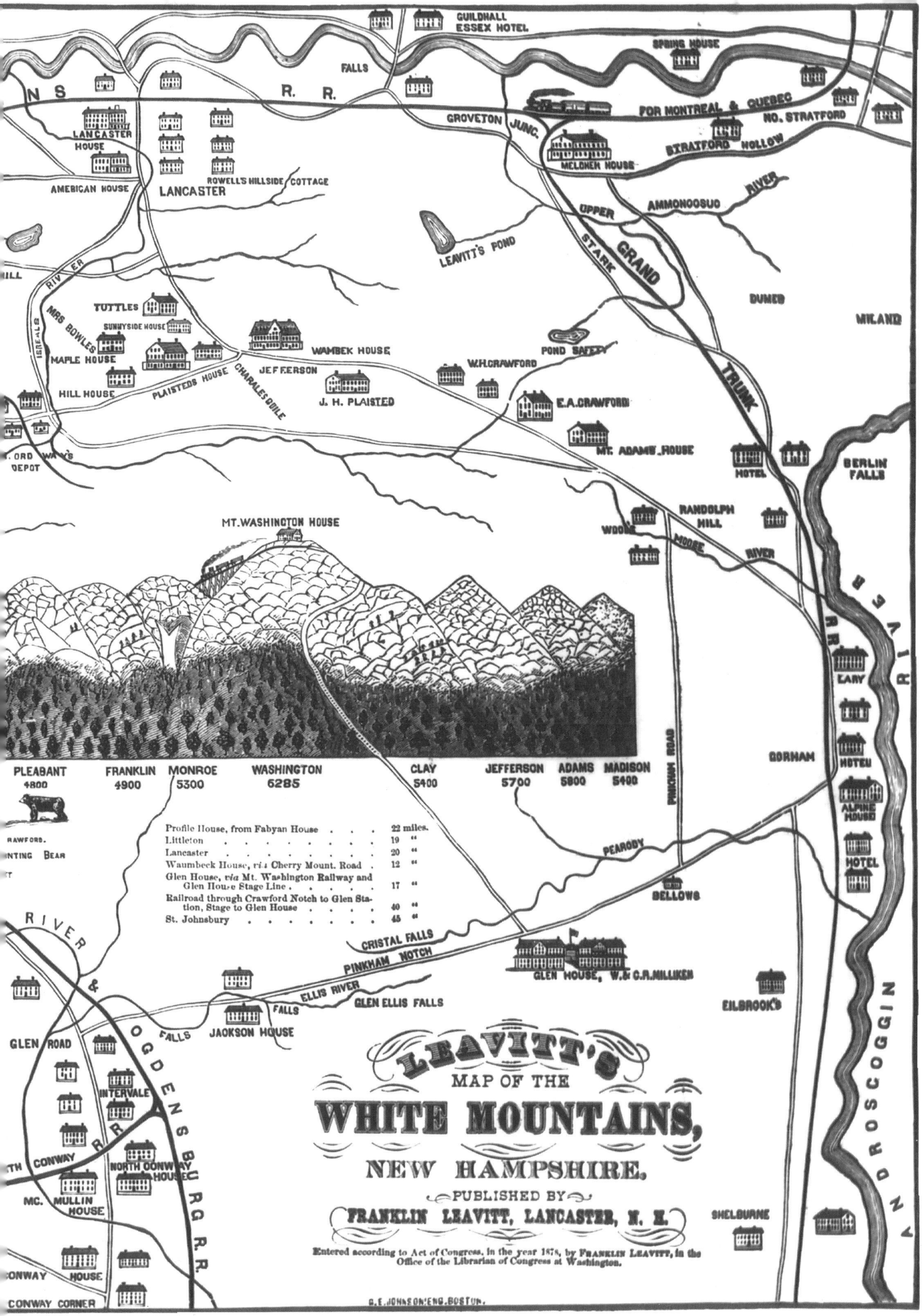
GUILDHALL
ESSEX HOTEL
SPRING HOUSE
FALLS
R. R.
GROVETON JUNC.
FOR MONTREAL & QUEBEC
NO. STRATFORD
STRATFORD HOLLOW
MELCHER HOUSE
LANCASTER HOUSE
AMERICAN HOUSE
LANCASTER
ROWELL'S HILLSIDE COTTAGE
UPPER AMMONOOSUC RIVER
LEAVITT'S POND
STARK
GRAND TRUNK
DUMER
MILAND
TUTTLES
SUNNYSIDE HOUSE
MRS BOWLES
ISREALS RIVER
MAPLE HOUSE
HILL HOUSE
WAMBEK HOUSE
JEFFERSON
PLAISTEDS HOUSE
CHARLES GILE
J. H. PLAISTED
W.H.CRAWFORD
POND SAFETY
E.A.CRAWFORD
MT. ADAMS' HOUSE
HOTEL
BERLIN FALLS
DEPOT
RANDOLPH HILL
MOOSE RIVER
MT.WASHINGTON HOUSE
PLEASANT 4800
FRANKLIN 4900
MONROE 5300
WASHINGTON 6285
CLAY 5400
JEFFERSON 5700
ADAMS 5800
MADISON 5400
PINKHAM ROAD
GORHAM
HOTEL
ALPINE HOUSE
HOTEL
Profile House, from Fabyan House . . . . 22 miles.
Littleton . . . . 19 "
Lancaster . . . . 20 "
Waumbeck House, via Cherry Mount. Road . 12 "
Glen House, via Mt. Washington Railway and Glen House Stage Line . . . . . 17 "
Railroad through Crawford Notch to Glen Station, Stage to Glen House . . . . 40 "
St. Johnsbury . . . . 45 "
PEABODY
BELLOWS
CRISTAL FALLS
PINKHAM NOTCH
GLEN HOUSE, W.& C.R.MILLIKEN
ELLIS RIVER
GLEN ELLIS FALLS
FALLS
JACKSON HOUSE
EILBROOK'S
RIVER
GLEN ROAD
INTERVALE
OGDENSBURG R.R.
NORTH CONWAY HOUSE
MC. MULLIN HOUSE
HOUSE
CONWAY CORNER
SHELBURNE
ANDROSCOGGIN RIVER
LEAVITT'S
MAP OF THE
WHITE MOUNTAINS,
NEW HAMPSHIRE.
PUBLISHED BY
FRANKLIN LEAVITT, LANCASTER, N. H.
Entered according to Act of Congress, in the year 1878, by FRANKLIN LEAVITT, in the Office of the Librarian of Congress at Washington.
G.E. JOHNSON ENG. BOSTON.